Bacterial Toxins

Genetics, Cellular Biology and Practical Applications

Edited by

Thomas Proft

Department of Molecular Medicine & Pathology
School of Medical Sciences
Maurice Wilkins Centre for Molecular Biodiscovery
University of Auckland
Auckland
New Zealand

Caister Academic Press

Caister Academic Press
Norfolk, UK

www.caister.com

British Library Cataloguing-in-Publication Data
A catalogue record for this book is available from the British Library

ISBN: 978-1-908230-28-7

Cover design adapted from Figure 8.1

Printed and bound in Great Britain

Contents

Current books of interest

Pathogenic *Escherichia coli*: Molecular and Cellular Microbiology	2014
Burkholderia: From Genomes to Function	2014
Myxobacteria: Genomics, Cellular and Molecular Biology	2014
Next Generation Sequencing: Current Technologies and Applications	2014
Omics in Soil Science	2014
Mollicutes: Molecular Biology and Pathogenesis	2014
Genome Analysis: Current Procedures and Applications	2014
Bacterial Membranes: Structural and Molecular Biology	2014
Cold-Adapted Microorganisms	2013
Fusarium: Genomics, Molecular and Cellular Biology	2013
Prions: Current Progress in Advanced Research	2013
RNA Editing: Current Research and Future Trends	2013
Real-Time PCR: Advanced Technologies and Applications	2013
Microbial Efflux Pumps: Current Research	2013
Cytomegaloviruses: From Molecular Pathogenesis to Intervention	2013
Oral Microbial Ecology: Current Research and New Perspectives	2013
Bionanotechnology: Biological Self-assembly and its Applications	2013
Real-Time PCR in Food Science: Current Technology and Applications	2012
Bacterial Gene Regulation and Transcriptional Networks	2012
Bioremediation of Mercury: Current Research and Industrial Applications	2012
Neurospora: Genomics and Molecular Biology	2012
Rhabdoviruses	2012
Horizontal Gene Transfer in Microorganisms	2012
Microbial Ecological Theory: Current Perspectives	2012
Two-Component Systems in Bacteria	2012
Foodborne and Waterborne Bacterial Pathogens	2012
Yersinia: Systems Biology and Control	2012
Stress Response in Microbiology	2012
Bacterial Regulatory Networks	2012
Systems Microbiology: Current Topics and Applications	2012
Quantitative Real-time PCR in Applied Microbiology	2012
Bacterial Spores: Current Research and Applications	2012
Small DNA Tumour Viruses	2012
Extremophiles: Microbiology and Biotechnology	2012

Full details at www.caister.com

Contributors

Milena M. Awad
Department of Microbiology
Monash University
Clayton, VIC
Australia

milena.awad@monash.edu

Juliane Bubeck Wardenburg
Departments of Pediatrics and Microbiology
University of Chicago
Chicago, IL
USA

jbubeckw@peds.bsd.uchicago.edu

Benjamin Busch
Max von Pettenkofer-Institut für Hygiene und Medizinische Mikrobiologie
Ludwig-Maximilians-Universität
Munich
Germany

busch@mvp.uni-muenchen.de

Glen P. Carter
Department of Microbiology
Monash University
Clayton, VIC
Australia

glen.carter@monash.edu

Sheng Chen
Department of Applied Biology and Chemical Technology
The Hong Kong Polytechnic University
Hung Hom
Kowloon
Hong Kong SAR

bcschen@polyu.edu.hk

Julie Claudinon
Institute of Biology II and BIOSS Centre for Biological Signalling Studies
Albert-Ludwigs-University Freiburg
Freiburg
Germany

julie.claudinon@bioss.uni-freiburg.de

Wolfgang Fischer
Max von Pettenkofer-Institut für Hygiene und Medizinische Mikrobiologie
Ludwig-Maximilians-Universität
Munich
Germany

fischer@mvp.uni-muenchen.de

J. Ross Fitzgerald
The Roslin Institute and Centre for Infectious Diseases
University of Edinburgh
Edinburgh
UK

ross.fitzgerald@roslin.ed.ac.uk

John D. Fraser
The School of Medical Sciences and the Maurice Wilkins Centre for Molecular Biodiscovery
The University of Auckland
Auckland
New Zealand

jd.fraser@auckland.ac.nz

Ries J. Langley
The School of Medical Sciences and the Maurice Wilkins Centre for Molecular Biodiscovery
The University of Auckland
Auckland
New Zealand

r.langley@auckland.ac.uk

Clifford Lingwood
Research Institute Hospital for Sick Children
Toronto, ON
Canada

cling@sickkids.ca

Nils Lycke
MIVAC – Mucosal Immunobiology & Vaccine Center
Department of Microbiology and Immunology
Institute of Biomedicine
University of Gothenburg
Gothenburg
Sweden

nils.lycke@microbio.gu.se

Dena Lyras
Department of Microbiology
Monash University
Clayton, VIC
Australia

dena.lyras@monash.edu

Johan Mattsson
MIVAC – Mucosal Immunobiology & Vaccine Center
Department of Microbiology and Immunology
Institute of Biomedicine
University of Gothenburg
Gothenburg
Sweden

johan.mattsson@gu.se

José R. Penadés
Instituto de Biomedicina de Valencia (IBV-CSIC)
Valencia
Spain

jpenades@ibv.csic.es

Winfried Römer
Institute of Biology II and BIOSS Centre for Biological Signalling Studies
Albert-Ludwigs-University Freiburg
Freiburg
Germany

winfried.roemer@bioss.uni-freiburg.de

Julian I. Rood
Department of Microbiology
Monash University
Clayton, VIC
Australia

julian.rood@monash.edu

Gustaf E. Rydell
Institut Curie
Centre de Recherche
Traffic, Signaling and Delivery group
CNRS UMR 144
Paris
France

gustaf.rydell@curie.fr

Rodney K. Tweten
Department of Microbiology and Immunology
University of Oklahoma Sciences Center
Oklahoma City, OK
USA

rod-tweten@ouhsc.edu

James Whisstock
Department of Biochemistry and Molecular Biology
Monash University
Melbourne
Australia

james.whisstock@monash.edu

Preface

Since the discovery of diphtheria toxin 125 years ago, remarkable progress has been made in the field of bacterial toxins. Driven by significant technical advances in various disciplines, such as molecular microbiology, complete genome sequencing, protein crystallography and experimental animal models, our knowledge on microbial toxins has increased dramatically over recent decades. More recent advances include the discovery of metalloprotease ADAM10 as the receptor for *Staphylococcus aureus* α-haemolysin, novel insights into the immune evasion mechanisms of the staphylococcal superantigen-like toxins, a better understanding of the diverse role of the *Helicobacter* CagA protein, and the role of the *Clostridium difficile* toxins in disease. This progress not only provides important insights into virulence mechanisms and how certain bacteria cause disease, but also equips scientist with a 'toolkit' to decipher metabolic pathways and intracellular transport mechanisms in eukaryotic cells. Important to mention are also technical and clinical applications of bacterial toxins, e.g. the use of B-subunits of cholera toxin and *E.coli* heatlabile toxin as vaccine adjuvants, and the application of botulinum toxin for various cosmetic and medical procedures.

The aim of this book is to highlight recent achievements in toxin research. The nine chapters are written by a panel of 20 international experts from Australia, France, Canada, Germany, New Zealand, Hong Kong, Spain, Sweden, United Kingdom, and the United States of America, who describe the latest insights from this rapidly expanding field of research.

Thomas Proft, PhD

Receptor-related Risk Factors for Verotoxin Pathogenesis

1

Clifford A. Lingwood

Abstract

The family of *E. coli*-derived verotoxins (Shiga toxins) has been extensively studied over the last 25 years because of its primary role in the infectious aetiology of haemolytic–uraemic syndrome (HUS). This acute paediatric renal disease, defined by the triad of thrombocytopenia, glomerular endothelial damage and haemolytic anaemia is mediated by toxin B subunit pentamer binding to its receptor glycosphingolipid, globotriaosyl ceramide (Gb_3) within lipid rafts in the glomerular endothelial cell plasma membrane and the subsequent endothelial pathology. HUS is preceded by a self-limiting haemorrhagic colitis prodrome which develops into a systemic pathology in ~10% of patients, after a several days of apparent recovery. Symptoms remain fatal in ~10% of HUS cases and no specific effective therapy has yet been devised.

Understanding of the molecular basis and risk factors for HUS following gastrointestinal infection with VT producing *E. coli*, although incomplete and still a matter of controversy, has suffered a severe setback with the most recent highly virulent O104:H4 outbreak which showed more frequent incidence of HUS and death and preferentially targeted the adult female as opposed to the paediatric/elderly population. Since the toxin involved, VT2, was the same as involved in other more typical HUS cases, new concepts to explain this variation in VT-induced pathology are required.

Verotoxins and their receptors

The members of the Verotoxin family of *E. coli* AB5 subunit toxins (also termed Shiga toxins, due to virtual identity with Shiga toxin from *Shigella dysenteriae*), comprise a single 37 kDa A-subunit responsible for inhibition of protein synthesis by depurination of a specific adenine base in the 28S RNA of the 60S ribosomal subunit (Endo *et al.*, 1988), non-covalently linked to a pentameric array of smaller 7.5 kDa B subunits responsible for receptor binding (Lingwood, 1996b), and hence tissue and cell targeting *in vivo* and *in vitro*. Of the described members of this family, only three are associated with human disease, VT1, VT2 and VT2c (Friedrich *et al.*, 2002). VT2 is ~60% homologous to VT1 but is not neutralized by antibodies against VT1. VT2c is 95% homologous to VT2. VT2 is primarily associated with human disease but *E. coli* producing VT1 and VT2 are common. Interestingly, verotoxin-producing *E. coli* (VTEC) producing more than one type of VT2 have yet to be described. 0157:H7 is the *E. coli* serotype most commonly associated with production of these toxins. However, many other serotypes are of significant clinical importance and the most recent outbreak of disease inducing infections in Germany show that enteroaggregative *E. coli* strains can harbour this toxin also (Frank *et al.*, 2011).

Although VTs are potent inhibitors of eukaryotic protein synthesis via their RNA (and DNA?; Brigotti *et al.*, 2002) glycanase activity, it is clear that inhibition of protein synthesis is not their only mechanism of action. In addition to the induction of various receptor binding dependent signal cascades (Tetaud *et al.*, 2003), A subunit mediated activation of select mRNA species (Petruzziello-Pellegrini *et al.*, 2012) also plays an important intermediary role in HUS pathogenesis.

The enteroaggregative property of the *E. coli* strain associated with the recent German outbreak has been suggested to be a key property responsible for the increased pathology associated with this outbreak. The enteroaggregative property of these *E. coli* has been shown to reduce the efficacy of the gastrointestinal epithelial barrier in model cell studies (Strauman *et al.*, 2010). Such a property will promote the incidence of bloody diarrhoea in addition to HUS. Indeed, the fact that a bloody diarrhoea prodrome is observed indicates that the toxin must have relatively free access, if only for a short period, to the systemic circulation and therefore increased efficacy of gastroepithelial cell transit might be superfluous. Alternatively, the bloody diarrhoea is not a toxin portal. In this outbreak the incidence of HUS after haemorrhagic colitis was 20% (Frank *et al.*, 2011), more than double the 'normal' incidence, and this was primarily the adult, rather than paediatric population. It has been proposed that the toxin undergoes carrier-mediated transport in the systemic circulation. A subunit mediated neutrophil binding has been proposed as a likely mechanism but this remains controversial (Brigotti, 2012).

Receptor glycolipid

The glycosphingolipid (GSL) globotriaosyl ceramide Gb_3, which is also the CD77 antigen marker of human germinal centre B cells (Mangeney *et al.*, 1991), the pk antigen of the P blood group (Spitalnik and Spitalnik, 1995) and the Burkitt lymphoma antigen (Klein *et al.*, 1983; Murray *et al.*, 1985), is the functional receptor for this family of toxins. VT2e, which is the oedema disease toxin in pigs, binds Gb_4 in preference to Gb_3 (DeGrandis *et al.*, 1989), though Gb_3 can still mediate VT2e cell killing (Keusch *et al.*, 1995). Incorporation of Gb_3 (Waddell *et al.*, 1990) or a Gb_3 analogue (Saito *et al.*, 2012) into Gb_3 negative cells induces their VT sensitivity and deletion of Gb_3 synthase completely protects cells and mice from VT pathology (Okuda *et al.*, 2006).

Gb_3 is the major GSL of the human kidney. The renal Gb_3 content increases as a function of age (Boyd and Lingwood, 1989) but paediatric renal glomeruli are more VT reactive than those of adults (Lingwood, 1994). This point has been challenged (Ergonul *et al.*, 2003), but verified under more physiological binding conditions (Khan *et al.*, 2009). These studies illustrate, however, that the binding of VT to renal glomerular epithelial cells is not just a question of whether the Gb_3 is present or not. Other factors, e.g. cholesterol (Khan *et al.*, 2009), regulate the availability of Gb_3 for binding and, therefore, constitute undefined risk factors for the development of HUS.

B subunit receptor-binding sites

The pentameric B subunit X-ray structure suggested an intrasubunit Gb_3 binding site (Sixma *et al.*, 1993). This was supported by molecular modelling which also suggested a second lower affinity Gb_3 binding site in a shallow groove as opposed to the membrane proximal surface of the B subunit monomer (Nyholm *et al.*, 1995, 1996). Co-crystallography using the Gb_3 oligosaccharide, indicated that this second site was the primary Gb_3 binding site and a third Gb_3 binding site B subunit monomer was also defined primarily via an interaction with tryptophan 34 (Ling *et al.*, 1998). Only site 2 was confirmed by NMR studies of the B subunit–Gb_3 oligosaccharide complex. Site 3 occupancy was not seen for co-crystal structures of another VT family member (Ling *et al.*, 2000). Site 3 was blocked by the A subunit C-terminus in VT2 (Fraser *et al.*, 2004). In addition, in VT2, the configuration of site 2 was distinct from that of VT1 such that Gb_3 oligosaccharide could not bind in the same way as for VT1. Nevertheless, mutation of amino acids in the sites 1, 2 and 3 of VT1 have all been shown to disrupt toxin-receptor binding and cytotoxicity and therefore the relationship between these postulated binding sites remains problematic. Certainly carbohydrate structures tailored to bind into site 2 are effective inhibitors of VT cytotoxicity (Kitov *et al.*, 2000, 2008). In contrast, mutations of amino acids in site 1 and 3 which have a marked reduction in holotoxin cell binding and cytotoxicity, had no effect on site 2 Gb_3 oligosaccharide binding (Soltyk *et al.*, 2002). Site 2 was proposed as an initial kinetic binding site and site 1 as the thermodynamically stable Gb_3 binding site (Nyholm *et al.*, 1996). Thus, while a potential 15 binding sites per pentamer have been proposed, the actual number of Gb_3 binding sites may be much less and has yet to be determined experimentally.

Mutation of amino acids primarily assigned in site 1 of VT1 results in a change of GSL binding specificity from Gb_3 to $Gb_3 + Gb_4$. An additional fact that still requires explanation is the observation that aminoGb$_4$ (N-deacetyl globoside), which could only dock in site 1, was preferentially bound by all VT family members tested (Nyholm *et al.*, 1996). Gb$_4$ does not bind to the majority of VTs. Removal of the acetyl moiety of the terminal N-acetylgalactosamine generated the preferred receptor species. In addition, more recent studies suggested that a (non-physiological) terminal GalNAc, rather than αGal residue in the 'globotriaose' moiety was a preferred receptor for VT2 (Flagler *et al.*, 2010).

VT signalling and internalization

VT binding to cell surface Gb$_3$ induces several transmembrane signalling pathways. VT binding to cell surface Gb$_3$ in lipid rafts up-regulates Src family kinases yes (Katagiri *et al.*, 1999), lyn (Mori *et al.*, 2000) and syk (Walchli *et al.*, 2009; Utskarpen *et al.*, 2010). The VT activation of syk increases clathrin heavy chain phosphorylation and is required for VT internalization (Lauvrak *et al.*, 2005). The A subunit of the holotoxin has been shown to increase the rate of clathrin-mediated toxin internalization into Gb$_3$-containing cells (Torgersen *et al.*, 2005). VT-induced signalling for apoptosis (Tesh, 2010) is complex and comprises both Gb$_3$ mediated (Tetaud *et al.*, 2003; Takenouchi *et al.*, 2004) and protein synthesis inhibition dependent pathways (Fujii *et al.*, 2008; Garibal *et al.*, 2010).

Cholesterol masking of VT receptors

We propose a new model of VTB/Gb$_3$ receptor binding which provides a potential explanation of the relationship between site 1 and site 2 binding in relation to cell cytotoxicity. We and others, have shown in many studies, that membrane GSL receptor function depends not only on the carbohydrate sequence of the GSL but also the chemical structure of the lipid moiety and the membrane environment (Lingwood, 1996a). The requirement for membrane Gb$_3$ to be located in more ordered, cholesterol enriched cell surface domains termed 'lipid rafts' in order to mediate cell cytotoxicity has been documented (Falguieres *et al.*, 2001). We have shown that VT1 and VT2c bind different fatty acid isoforms of Gb$_3$ (Kiarash *et al.*, 1994) and that the binding of VT1 to Gb$_3$ cholesterol-containing membranes is fatty acid dependent for VT1 but not VT2 (Mahfoud *et al.*, 2009). Different hydroxyl groups within the GSL carbohydrate sequence can be involved in the binding of different specific ligands to the same GSL. Thus, the same carbohydrate sequence can serve as a differential receptor for different specific ligands (Chark *et al.*, 2004). We have used the term 'aglycone modulation' to describe this phenomenon Lingwood, 1996a). In cell culture, VT1 and VT2 showed both coincident and differential binding sites on the cell surface which were differentially internalized to coalesce via retrograde transport to the Golgi and ER (Tam *et al.*, 2008). In human renal frozen sections, we observed that VT1 and VT2 binding most often overlapped. However, for some glomeruli, VT2 binding was apparent but the binding of VT1 was not (Khan *et al.*, 2009). Thus, in these glomeruli, Gb$_3$ was present but unavailable for VT1 recognition. We found that treatment of such glomeruli, initially with acetone (Chark *et al.*, 2004), but later with methyl-β-cyclodextrin (MCD) to specifically deplete cholesterol, resulted in the induction of VT1 binding within these glomeruli (Khan *et al.*, 2009), that is, cholesterol removal, now allowed VT1 to bind to the Gb$_3$ present in these glomeruli. Gb$_3$ within renal tubule epithelial cells is liable to detergent extraction and is thus not present in lipid rafts. However, Gb$_3$ in glomeruli, bound by both VT1 and VT2, is resistant to Triton X-100 detergent extraction, indicating that the glomerular Gb$_3$ is present in lipid rafts (Khan *et al.*, 2009). Even after cholesterol depletion with MCD, VT1 and VT2 glomerular binding was not liable to Triton extraction.

This explains the selective glomerular pathology associated with HUS and why renal tubular cells cultured *in vitro* are VT1 and VT2 sensitive but renal tubular damage is a secondary pathology of HUS. The fact that, for some glomeruli, cholesterol depletion results in the induction of VT1 binding suggests that cholesterol can in some way mask Gb$_3$ to prevent VT1 binding, and that high glomerular membrane cholesterol may be a negative risk factor for HUS.

The potential of cholesterol to mask membrane GSL antigens from binding their appropriate ligands was investigated initially in a model membrane vesicle system. Plasma membrane-derived cellular lipid rafts are typically isolated on a discontinuous sucrose gradient after detergent extraction in the cold. The lighter lipid raft fraction enriched in cholesterol and GSLs, floats to the 30/5% sucrose interface. Using this procedure, artificial Gb_3/cholesterol lipid rafts binding VT1 and VT2 were generated (Nutikka and Lingwood, 2004). In this model system, increasing cholesterol had an increasing inhibitory effect on VT1 binding. However, following increased gradient resolution and prolonged ultracentrifugation, the VT1 and VT2 Gb_3 binding fraction initially present at the 30/5% sucrose interface migrated to the top of the gradient, whereas the bulk of the Gb_3 and cholesterol remained at the 30/5% sucrose interface (Mahfoud *et al.*, 2010). Thus, VT1 and VT2 actually only bound a small fraction (~ 5%) of the total Gb_3 which could be separated from the bulk Gb_3/cholesterol-containing vesicles. Surprising, the same observations were found to be true for vesicles prepared from living Gb_3-containing cells or indeed, vesicles prepared from any cultured cell type. The GSL available for ligand binding was only a small fraction of the total membrane GSL fraction that could be separated via prolonged high-resolution ultracentrifugation from the bulk GSL fraction at the 30/5% sucrose interface. Thus, the plasma membrane GSLs of cells could be divided into two fractions, a small fraction available for exogenous ligand binding and a large fraction unavailable for binding of the same ligand. Treatment of cells with MCD or latrunculin, prior to vesicle preparation resulted in the unmasking of the so-called 'invisible' GSL fraction at the 30/5% sucrose interface (Mahfoud *et al.*, 2010). Thus, cholesterol in the membrane of tissue culture cells also masks a large fraction of GSLs to prevent the binding of an appropriate exogenous ligand.

The molecular explanation for this phenomenon was provided by the finding that in thermodynamic simulations, the carbohydrate of a glycosphingolipid complexed with cholesterol showed a different preferential conformation relative to the plane of the membrane, as compared to the GSL in a simple phospholipid membrane environment (Lingwood *et al.*, 2011). In the GSL/cholesterol complex, the steroid hydroxyl group formed a hydrogen bonding network with the anomeric oxygen of the first sugar, and the nitrogen of the ceramide, to bend the carbohydrate to a conformation approximately parallel to the plane of the plasma membrane, whereas in the phospholipid alone membrane environment, the carbohydrate orientation was essentially perpendicular to the membrane (Fig. 1.1).

New model for VTB subunit pentamer binding plasma membrane Gb_3

These data provide new insight into the potential mechanism of VT binding to cell membrane Gb_3. Binding of Gb_3 carbohydrate into site 2 of the VTB subunit pentamer places the terminal Galα1–4Gal disaccharide of Gb_3 essentially at right angles to the central pore axis of the B subunit pentamer, whereas the Gb_3 oligosaccharide docked in site 1,

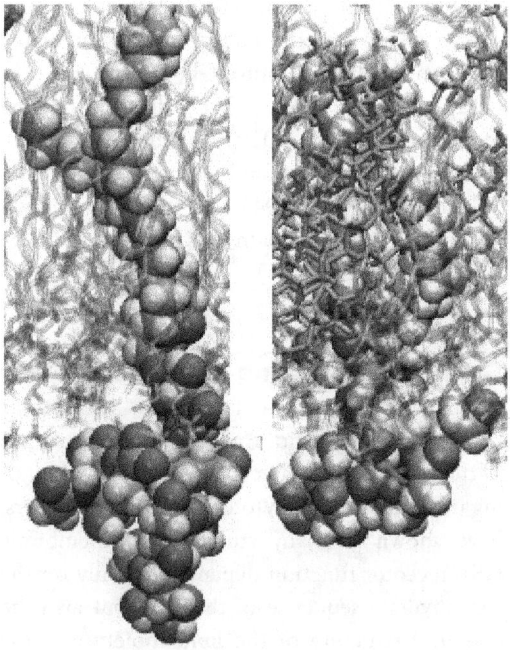

Figure 1.1 Conformation of GM1 ganglioside in phospholipid bilayer±cholesterol. Molecular simulation from Lingwood *et al.* (2011) by Vattulainen laboratory. The GM1 pentasaccharide is parallel to the membrane acyl chains in the absence (left), but rotated by 90° to become perpendicular to the acyl chains in the presence of cholesterol.

positions the terminal disaccharide parallel to the axis of the central pore (Fig. 1.2). Thus, when the B subunit pentamer binds to the Gb_3-containing target cell membrane, Gb_3 oligosaccharides oriented perpendicular to the plane of the plasma membrane are in a preferred conformation for binding in site 1, within the intersubunit cleft, whereas Gb_3 oligosaccharides oriented parallel to the membrane are more optimally positioned to bind in the shallow groove in the B subunit membrane-proximal surface that comprises site 2. Thus, binding to cholesterol complexed Gb_3 could be preferentially accommodated in site 2 whereas binding to the native Gb_3 oriented perpendicular to the plasma membrane could be preferentially accommodated in site 1. This could imply that initial cell binding is mediated by non-cholesterol associated Gb_3 (i.e. non-raft Gb_3) but that cell surface clustering of the multivalent ligand is mediated via lipid raft associated, cholesterol reoriented Gb_3.

The reduced efficacy of cholesterol to inhibit VT2 binding to Gb_3, both in renal tissue (Khan *et al.*, 2009) and in a model membrane vesicle system (Nutikka and Lingwood, 2004; Mahfoud *et al.*, 2009), could provide new insight into the increased pathology associated with VT2 as opposed to VT1 producing *E. coli* infection. For model and cell-derived vesicles containing excess cholesterol, however, neither VT1 nor VT2

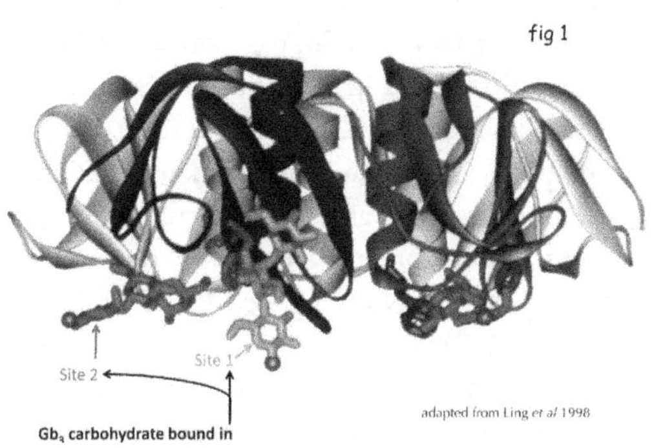

Figure 1.2 Co-crystal structure of VT1 B subunit pentamer and Gb_3 oligosaccharide. Three trisaccharide binding sites on the membrane adjacent side of the B subunit pentamer were defined by Ling *et al.* (1998) from the co-crystal structure with the 8-(methoxycarbonyl) octyl-globotriaose. The position of the trisaccharide in site 1 and site 2 is indicated.

binding to Gb_3 was observed (Mahfoud *et al.*, 2010), indicating that VT binding to membrane-parallel, cholesterol-associated Gb_3 in site 2 must occur subsequent to, and dependent on, site 1 binding of membrane-perpendicular Gb_3. This would imply that site 1 is primarily involved in tissue and cell membrane targeting of VT, whereas subsequent site 2 Gb_3/cholesterol binding may be more important in internalization and subsequent retrograde trafficking of the toxin receptor complex to the Golgi and ER. The difference in Gb_3 site 2 docking of VT2 as compare to VT1 (Fraser *et al.*, 2004) could then relate to the differential initial intracellular traffic of these toxins (Tam *et al.*, 2008).

Cholesterol is key for VT-Gb_3 retrograde transport

The interaction of membrane cholesterol with GSLs may also control the retrograde transport of internalized VT1 from endosomes to the Golgi. Although knockdown experiments have shown the depletion of many factors can compromise this endosome/trans-Golgi network transit (Sandvig *et al.*, 2009), the structure of the lipid moiety of Gb_3 and associated GSLs plays an important role in the signal transduction and molecular interactions mediated by and dependent upon VT internalization. The presence of Gb_3 within detergent resistant lipid rafts is required for targeting the internalized toxin for retrograde transport and cytotoxicity (Falguieres *et al.*, 2001). VT binding to non-raft Gb_3 results in internalization and trafficking to lysosomes where the toxin is degraded (Hoey *et al.*, 2003). Reconstitution of receptor negative cells with adamantylGb_3, a Gb_3 mimic we show does not interact with cholesterol (Saito *et al.*, 2012), results in the induction of cellular VT1 and VT2 sensitivity but no Golgi or ER containing VT is detected. The toxin is trafficked to early endosomes and lost intracellularly within a few hours.

Depletion of cellular cholesterol results in the blockage of VT endosome/Golgi transition (Falguieres *et al.*, 2001) and Gb_3 can be displaced from cell surface DRMs by selective depletion of glucosyl ceramide (Smith *et al.*, 2006). Moreover, the endosome/Golgi transition of VT can be reduced by the selective degradation of the C_{16} fatty acid isoform of glucosyl ceramide (Raa *et al.*, 2009). Long-chain fatty acid (C_{24}) containing Gb_3 is less effective to mediate Golgi/ER retrograde transport of VT1 than is short chain (C_{16}, C_{18}) Gb_3 (Arab and Lingwood, 1998). Gb_3 interaction with cholesterol is dependent on the fatty acid chain length (Mahfoud *et al.*, 2009). C_{16} Gb_3 was effectively bound but C_{18} and C_{20} Gb_3 were unavailable for VT1 binding. C_{22} and C_{24} Gb_3s also showed effective receptor function for VT1. This binding pattern was also seen for HIV gp120 binding to these Gb_3/cholesterol vesicles. C_{18} and C_{20} fatty acid chain lengths most closely match the dimensions of the membrane steroid, suggesting that hydrocarbon mismatching may reduce the efficacy of cholesterol-mediated carbohydrate reorientation in the GSL/cholesterol complex. However, the binding of VT2 to the ligand-available Gb_3 fraction was essentially unaffected by Gb_3 fatty acid composition. This could relate to the different site 2 occupancy reported for VT2 (Fraser *et al.*, 2004).

The Gb_3/cholesterol interaction, as monitored by changing the masking of Gb_3 carbohydrate, is reduced in the presence of glucosyl ceramide (Mahfoud *et al.*, 2010). Thus, GSL/GSL interaction plays an important role in the maintenance of Gb_3 within lipid rafts necessary for endosome/TGN Golgi transition and the availability for ligand binding. VT binding to Gb_3 varies for fatty acid isoforms (Kiarash *et al.*, 1994) and is increased for Gb_3 isoform mixtures (Pellizzari *et al.*, 1992). Indeed, in cholesterol-containing membranes, Gb_3 isoforms not bound by VT1, are bound when these isoforms are mixed together (Mahfoud *et al.*, 2009), suggesting that the interaction between different Gb_3 fatty acid isoforms can alter the presentation of Gb_3 carbohydrate for VT binding, providing a novel means for the regulation of Gb_3 receptor function (Lingwood, 2011).

Addition of adamantylGb_3 to Gb_3-containing cells resulted in a loss of distinct cell surface VT1/VT2 binding domains and the inhibition of VT1/VT2 retrograde trafficking beyond early endosomes. In contrast, reconstitution of Gb_3-containing cells with hydroxyl-ethyl-adamantylGb_3 had minimal effect on VT trafficking within the reconstituted cells. Inhibition of retrograde transport had a significant effect on VT1 but not VT2 cytotoxicity in adamantylGb_3 reconstituted cells. VT cytotoxicity became Brefeldin A resistant consistent with a loss of Golgi/ER retrograde transport. Unlike Gb_3, adamantylGb_3 could not protect liposomal cholesterol from MCD extraction indicating that adamantylGb_3 interacts poorly with cholesterol and thereby is resistant to the cholesterol medicated conformational change in Gb_3 carbohydrate. This could explain why the adamantylGb_3-mediated toxin trafficking to endosomes dominates the endogenous mediated retrograde transport to the Golgi and ER. Since hydroxy fatty acid in GSLs can also promote a membrane parallel carbohydrate conformation (Yahi *et al.*, 2010), the hydroxyl function of hydroxyl-ethyl-adamantylGb_3 may alter the conformation of the Gb_3 sugar such that this advantage is lost.

The interaction between GSLs and cholesterol may be crucial for the retrograde transport of Shiga toxin to the Golgi ER. The adamantylGSLs we have made have a bulky globular hydrophobic lipid moiety unsuitable for stacking against the hydrophobic rings of cholesterol. In a model membrane system the interaction of Gb_3 and cholesterol was sufficient to stabilize the cholesterol against MCD mediated cholesterol extraction. Membranes containing adamantylGb_3 and cholesterol were unable to achieve this. Moreover, inclusion of adamantylGb_3 together with Gb_3 and cholesterol resulted in a partial reversion of the protection against MCD cholesterol extraction afforded by Gb_3. Significantly adamantylGb_3 is able to mediate VT1 and VT2 internalization into previously receptor negative cells but only as far as early endosomes. No Golgi or ER VT trafficking was observed suggesting that, since cholesterol has been shown to be important in the transition of toxin trafficking between endosomes and Golgi, the interaction between Gb_3 and cholesterol is necessary to achieve this transition. AdamantylGSLs resistance to cholesterol-mediated carbohydrate conformation change may provide the explanation for the inability of adamantylGb_3 to mediate Golgi and ER retrograde traffic of VT1 and VT2.

Regulation of Gb_3 biosynthesis

Several signal transduction pathways have been shown to up-regulate cellular Gb_3 synthesis including cytokines αTNF (Louise and Obrig, 1991; Eisenhauer *et al.*, 2001) and IL1 (Louise and Obrig, 1991), particularly in combination. LPS can induce endothelial Gb_3 synthesis (Louise *et al.*, 1997). Thus, bacterial and inflammatory regulation of cellular Gb_3 levels is evident. The toxin itself increased neuronal Gb_3 in mice (Tironi-Farinati *et al.*, 2010). Such regulators therefore provide additional potential risk factors for HUS development.

To understand the molecular basis of VT tissue/cell targeting, cell susceptibility and intracellular VT traffic it is necessary to understand Gb_3 biosynthesis. The fatty acid heterogeneity of GSLs is due to a family of fatty acid selective ceramide synthases (Mizutani *et al.*, 2005; Stiban *et al.*, 2010). These are differentially distributed in different tissues (Mizutani *et al.*, 2006), so the same GSL from different tissues, even though made by the same glycosyl transferases, can have a different aglycone composition. Ceramide is made in the ER and transferred to the Golgi via the ceramide transport protein, CERT (Perry and Ridgway, 2005). GSLs for the most part, are synthesized from glucosyl ceramide. Glucosyl ceramide is synthesized on the cytosolic leaflet of the Golgi membrane by glucosyl ceramide synthase (Futerman and Pagano, 1991). Glucosyl ceramide therefore must be flipped into the lumen of the Golgi to access the glycosyl transferases responsible for further carbohydrate extension. The mechanism by which this occurs is not fully understood.

The drug efflux pump MDR1 is in part, responsible for glucosyl ceramide flipping into the Golgi (De Rosa and Lingwood, 2009). However, inhibition of MDR1 using cyclosporin A results in the depletion of neutral GSLs but not gangliosides (De Rosa *et al.*, 2004), suggesting that there could be separate pools of glucosyl ceramide and lactosyl ceramide for the synthesis of neutral versus acidic GSLs. siRNA knockdown of MDR1 is consistent with MDR1 glucosyl ceramide flippase function in some cell lines, but not others (studies in progress). Thus, MDR1 may serve as a glucosyl ceramide flippase only in a restricted subset of tissues. Depletion of the pleckstrin homology domain containing phosphatidylinositol-four-phosphate adapter protein 2 (FAPP2) has been shown to reduce GSL biosynthesis. FAPP2 transports glucosyl ceramide from the cytosolic face of the proximal to the distal Golgi stack (D'Angelo *et al.*, 2007). This suggests that glucosyl ceramide could be flipped in to distinct Golgi cisternae by different flippases.

The cholesterol induced conformational change in GSL carbohydrate will apply not only to cell surface GSLs but also to GSLs within the Golgi. Cholesterol is synthesized in the ER (Reinhart *et al.*, 1987) but the concentration in the ER membrane is very low and increases from the cis- to trans-Golgi and trans-Golgi network (Mukherjee *et al.*, 1998; Sato *et al.*, 2004). This increasing Golgi cholesterol concentration may differentially affect substrate GSL carbohydrate conformation during biosynthesis and thereby regulate GSL product formation (Fig. 1.3).

FAPP2 may deliver glucosyl ceramide to distal Golgi membrane cisternae of increased cholesterol content and the conformation of the glucose moiety is therefore potentially different from glucosyl ceramide of the proximal Golgi cisternae (Fig. 1.3). It is also likely that a glucosyl ceramide–cholesterol complex would flip less favourably than glucosyl ceramide alone, and therefore may require a different translocation mechanism in this location. Similarly, once translocated into the Golgi lumen, glucosyl ceramide complexed to cholesterol may show differential substrate activity for lactosyl ceramide synthase. Indeed, two lactosyl ceramide synthases have been described (Chatterjee and Pandey, 2008). Cholesterol may affect the carbohydrate conformation of lactosyl ceramide, which, in turn, by favouring the presentation of particular hydroxyl groups for example, may predispose substrate activity for one of the five glycosyl transferases which define the synthesis of the five different GSL series from the common lactosyl ceramide precursor, as a function of location within the Golgi.

Testosterone-dependent Gb_3 biosynthesis has been described in the kidney (McCluer *et al.*, 1983). This may relate to the unusual female bias for HUS incidence in the recent German outbreak (Frank *et al.*, 2011). Increased renal

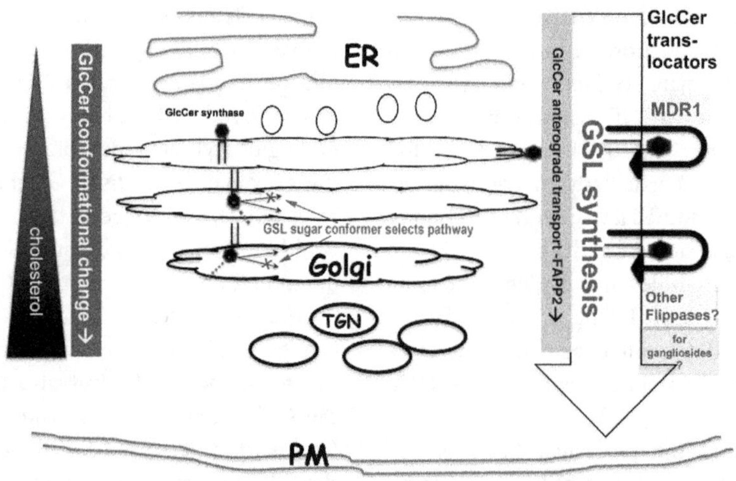

Figure 1.3 Potential role for cholesterol mediated carbohydrate conformational change in GSL biosynthesis. Varied Golgi/TGN membrane cholesterol content may alter conformation of GSL glycosyl transferase substrates as a function of anterograde transit. Translocation of glucosyl ceramide into the lumen at different positions within the Golgi stack could present different substrate conformers of glucosyl ceramide.

Gb_3 can actually protect from HUS since it have been shown that the Fabry mouse in which the αgalactosidase is deleted and Gb_3 accumulates due to abrogated breakdown, is less, not more, VT sensitive than the wild-type mouse (Cilmi *et al.*, 2006). Increased levels of non-raft Gb_3 (Falguieres *et al.*, 2001; Hoey *et al.*, 2003) could explain this observation.

Soluble GSL mimics

Adamantyl GSL mimics in which the ceramide fatty acid is replaced by an adamantane frame have been described (Lingwood *et al.*, 2006). These molecules are amphipathic, being both membrane and water-soluble and retain many of the receptor functions of membrane embedded GSLs in an aqueous environment. Adamantyl glucosyl and galactosyl ceramide can enter cells to subvert endogenous GSL synthetic pathways. At low concentrations, adamantyl glucosyl ceramide inhibits glucocerebrosidase and glucosyl ceramide and downstream GSLs accumulate. At higher concentrations lactosyl ceramide synthase is inhibited and all GSLs downstream of glucosyl ceramide are depleted (Kamani *et al.*, 2011). Adamantyl galactosyl ceramide provides an alternative substrate for Gb_3 synthase to generate adamantyl galabiosyl ceramide and competitively reduce Gb_3 synthesis (Kamani *et al.*, 2011). The adamantane frame of adamantylGSLs prevents these GSL mimics from forming a cholesterol complex (Saito *et al.*, 2012) and thereby adamantylGSLs may be resistant to cholesterol induced carbohydrate conformational change, providing adamantylGSLs an advantage over the parent GSL in terms of glycosyl transferase/glycohydrolase substrate activity. These soluble membrane GSL mimics will soon be commercially available.

Conclusions

The ability of Gb_3 glycosphingolipid to serve as the receptor mediating VT cytotoxicity is complex involving multiple Gb_3 binding sites, differential Gb_3 membrane presentation and intracellular trafficking, multiple lipid isoforms, lateral interaction with other membrane lipids.

Thus, the 'aglycone' effect, whereby the lipid moiety of GSL modulates ligand accessibility of the carbohydrate, is pervasive in both the receptor function of Gb_3 to bind verotoxins to the cells surface and the intracellular trafficking of the internalized VT–Gb_3 complex. Indeed, there may be separate receptor and trafficking conformations of Gb_3, bound by different sites within the B subunit pentamer. This mechanism of regulation may also apply to GSL synthesis during the biosynthetic anterograde transport of GSLs within the Golgi.

References

Arab, S., and Lingwood, C. (1998). Intracellular targeting of the endoplasmic reticulum/nuclear envelope by retrograde transport may determine cell hypersensitivity to Verotoxin: sodium butyrate or selection of drug resistance may induce nuclear toxin targeting via globotriaosyl ceramide fatty acid isoform traffic. J. Cell. Physiol. *177*, 646–660.

Boyd, B., and Lingwood, C.A. (1989). Verotoxin receptor glycolipid in human renal tissue. Nephron *51*, 207–210.

Brigotti, M. (2012). The interactions of human neutrophils with shiga toxins and related plant toxins: danger or safety? Toxins 4, 157–190.

Brigotti, M., Alfieri, R., Sestili, P., Bonelli, M., Petronini, P.G., Guidarelli, A., Barbieri, L., Stirpe, F., and Sperti, S. (2002). Damage to nuclear DNA induced by Shiga toxin 1 and ricin in human endothelial cells. FASEB J. *16*, 365–372.

Chark, D., Nutikka, A., Trusevych, N., Kuzmina, J., and Lingwood, C. (2004). Differential carbohydrate epitope recognition of globotriaosyl ceramide by verotoxins and monoclonal antibody: Role in human renal glomerular binding. Eur. J. Biochem. *271*, 405–417.

Chatterjee, S., and Pandey, A. (2008). The Yin and Yang of lactosylceramide metabolism: implications in cell function. Biochim. Biophys. Acta *1780*, 370–382.

Cilmi, S.A., Karalius, B.J., Choy, W., Smith, R.N., and Butterton, J.R. (2006). Fabry disease in mice protects against lethal disease caused by Shiga toxin-expressing enterohemorrhagic *Escherichia coli*. J. Infect. Dis. *194*, 1135–1140.

D'Angelo, G., Polishchuk, E., Di Tullio, G., Santoro, M., Di Campli, A., Godi, A., West, G., Bielawski, J., Chuang, C.C., van der Spoel, A.C., *et al.* (2007). Glycosphingolipid synthesis requires FAPP2 transfer of glucosylceramide. Nature *449*, 62–67.

De Rosa, M., and Lingwood, C. (2009). Multidrug resistance protein 1 (MDR1) and glycosphingolipid biosynthesis. In Glycobiology Reasech Trends, Powell, G., and McCabe, O., eds. (Nova Science Publishers Inc., New York), pp. 1–26.

De Rosa, M.F., Sillence, D., Ackerley, C., and Lingwood, C. (2004). Role of multiple drug resistance protein 1 in neutral but not acidic glycosphingolipid biosynthesis. J. Biol. Chem. *279*, 7867–7876.

DeGrandis, S., Law, H., Brunton, J., Gyles, C., and Lingwood, C.A. (1989). Globotetraosyl ceramide is recognized by the pig edema disease toxin. J. Biol. Chem. *264*, 12520–12525.

Eisenhauer, P.B., Chaturvedi, P., Fine, R.E., Richie, A.J., Pober, J.S., Cleary, T.G., and Newburg, D.S. (2001). Tumor necrosis factor alpha increases human cerebral endothelial cell Gb_3 and sensitivity to shiga toxin. Infect. Immun. *69*, 1889–1894.

Endo, Y., Tsurugi, K., Yutsudo, T., Takeda, Y., Ogasawara, K., and Igarashi, K. (1988). Site of the action of a vero toxin (VT2) from *Escherichia coli* O157:H7 and a Shiga toxin on eukaryotic ribosomes. Eur. J. Biochem. *171*, 45–50.

Ergonul, Z., Clayton, F., Fogo, A.B., and Kohan, D.E. (2003). Shigatoxin-1 binding and receptor expression in human kidneys do not change with age. Pediatr. Nephrol. *18*, 246–253.

Falguieres, T., Mallard, F., Baron, C., Hanau, D., Lingwood, C., Goud, B., Salamero, J., and Johannes, L. (2001). Targeting of Shiga toxin b-subunit to retrograde transport route in association with detergent-resistant membranes. Mol. Biol. Cell *12*, 2453–2468.

Flagler, M.J., Mahajan, S.S., Kulkarni, A.A., Iyer, S.S., and Weiss, A.A. (2010). Comparison of binding platforms yields insights into receptor binding differences between shiga toxins 1 and 2. Biochemistry *49*, 1649–1657.

Frank, C., Werber, D., Cramer, J.P., Askar, M., Faber, M., an der Heiden, M., Bernard, H., Fruth, A., Prager, R., Spode, A., *et al.* (2011). Epidemic profile of Shiga-toxin-producing *Escherichia coli* O104:H4 outbreak in Germany. N. Engl. J. Med. *365*, 1771–1780.

Fraser, M.E., Fujinaga, M., Cherney, M.M., Melton-Celsa, A.R., Twiddy, E.M., O'Brien, A.D., and James, M.N. (2004). Structure of shiga toxin type 2 (Stx2) from *Escherichia coli* O157:H7. J. Biol. Chem. *279*, 27511–27517.

Friedrich, A., Bielazewska, M., Zhang, W., Pulz, M., Kuczius, T., Ammon, A., and Karch, H. (2002). *Escherichia coli* harboring Shiga toxin 2 gene varients: frequency and association with clinical symptoms. J. Infect. Dis. *185*, 74–84.

Fujii, J., Wood, K., Matsuda, F., Carneiro-Filho, B.A., Schlegel, K.H., Yutsudo, T., Binnington-Boyd, B., Lingwood, C.A., Obata, F., Kim, K.S., *et al.* (2008). Shiga toxin 2 causes apoptosis in human brain microvascular endothelial cells via C/EBP homologous protein. Infect. Immun. *76*, 3679–3689.

Futerman, A.H., and Pagano, R.E. (1991). Determination of the intracellular sites and topology of glucosylceramide synthesis in rat liver. Biochem. J. *280(Pt 2)*, 295–302.

Garibal, J., Hollville, E., Renouf, B., Tetaud, C., and Wiels, J. (2010). Caspase-8-mediated cleavage of Bid and protein phosphatase 2A-mediated activation of Bax are necessary for Verotoxin-1-induced apoptosis in Burkitt's lymphoma cells. Cell Signal 22, 467–475.

Hoey, D.E.E., Currie, C., Lingwood, C.A., Gally, D.L., and Smith, D.G.E. (2003). Binding of verotoxin 1 to primary intestinal epithelial cells expressing Gb3 results in trafficking of toxin to lysosomal compartments. Cell. Microbiol. *5*, 85–97.

Kamani, M., Mylvaganum, M., Tian, R., Binnington, B., and Lingwood, C. (2011). Adamantyl glycosphingolipids provide a new approach to the selective regulation of cellular glycosphingolipid metabolism. J. Biol. Chem. *286*, 21413–21426.

Katagiri, Y., Mori, T., Nakajima, H., Katagiri, C., Taguchi, T., Takeda, T., Kiyokawa, N., and Fujimoto, J. (1999). Activation of Src family kinase induced by Shiga toxin binding to globotriaosyl ceramide (Gb3/CD77) in low density, detergent-insoluble microdomains. J. Biol. Chem. *274*, 35278–35282.

Keusch, G., Jacewicz, M., Acheson, D., Donohue-Rolfe, A., Kane, A., and McCluer, R. (1995). Globotriasylceramide,Gb3, is an alternative functional receptor for Shiga-like toxin 2e. Infect. Immun. *63*, 1138–1141.

Khan, F., Proulx, F., and Lingwood, C.A. (2009). Detergent-resistant globotriaosyl ceramide may define verotoxin/glomeruli-restricted hemolytic uremic syndrome pathology. Kidney Int. *75*, 1209–1216.

Kiarash, A., Boyd, B., and Lingwood, C.A. (1994). Glycosphingolipid receptor function is modified by fatty acid content: Verotoxin 1 and Verotoxin 2c preferentially recognize different globotriaosyl ceramide fatty acid homologues. J. Biol. Chem. *269*, 11138–11146.

Kitov, P.I., Sadowska, J.M., Mulvey, G., Armstrong, G.D., Lingaw, H., Pannu, N.S., Read, R.J., and Bundle, D.R. (2000). Shiga-like toxins are neutralized by tailored multivalent carbohydrate ligands. Nature *403*, 669–672.

Kitov, P.I., Mulvey, G.L., Griener, T.P., Lipinski, T., Solomon, D., Paszkiewicz, E., Jacobson, J.M., Sadowska, J.M., Suzuki, M., Yamamura, K., *et al.* (2008). *In vivo* supramolecular templating enhances the activity of multivalent ligands: a potential therapeutic against the *Escherichia coli* O157 AB5 toxins. Proc. Natl. Acad. Sci. U.S.A. *105*, 16837–16842.

Klein, G., Manneborg-Sandlund, A., Ehlin-Henriksson, B., Godal, T., Wiels, J., and Tursz, T. (1983). Expression of the BLA antigen, defined by the monoclonal 38.13 antibody, on Burkitt lymphoma lines, lymphoblastoid cell lines, their hybrids and other B-cell lymphomas and leukemias. Int. J. Cancer *31*, 535–542.

Lauvrak, S.U., Walchli, S., Iversen, T.G., Slagsvold, H.H., Torgersen, M.L., Spilsberg, B., and Sandvig, K. (2005). Shiga toxin regulates its entry in a Syk-dependent manner. Mol. Biol. Cell *17*, 1096–1109.

Ling, H., Boodhoo, A., Hazes, B., Cummings, M., Armstrong, G., Brunton, J., and Read, R. (1998). Structure of the Shiga toxin B-pentamer complexed with an analogue of its receptor Gb_3. Biochem *37*, 1777–1788.

Ling, H., Pannu, N.S., Boodhoo, A., Armstrong, G.D., Clark, C.G., Brunton, J.L., and Read, R.J. (2000). A mutant Shiga-like toxin IIe bound to its receptor Gb3: structure of a group II Shiga-like toxin with altered binding specificity. Structure *8*, 253–264.

Lingwood, C.A. (1994). Verotoxin-binding in human renal sections. Nephron *66*, 21–28.

Lingwood, C.A. (1996a). Aglycone modulation of glycolipid receptor function. Glycoconj. J. *13*, 495–503.

Lingwood, C.A. (1996b). Role of verotoxin receptors in pathogenesis. Trends Microbiol. *4*, 147–153.

Lingwood, C.A. (2011). Glycosphingolipid functions. Cold Spring Harb. Perspect. Biol. *3*, pii: a004788.

Lingwood, C.A., Sadacharan, S., Abul-Milh, A., Mylvaganam, M., and Peter, M. (2006). Soluble adamantyl glycosphingolipid analogs as probes of glycosphingolipid function. In Glycobiology Protocols, Brockhausen, I., ed. (Humana Press, Totowa, New Jersey), pp. 305–320.

Lingwood, D., Binnington, B., Róg, T., Vattulainen, I., Grzybek, M., Coskun, U., Lingwood, C., and Simons, K. (2011). Cholesterol modulates glycolipid conformation and receptor activity. Nat. Chem. Biol. *7*, 260–262.

Louise, C.B., and Obrig, T.G. (1991). Shiga toxin-associated hemolytic-uremic syndrome: combined cytotoxic effects of Shiga toxin, interleukin-1ß, and tumor necrosis factor alpha on human vascular endothelial cells *in vitro*. Infect. Immun. *59*, 4173–4179.

Louise, C.B., Tran, M.C., and Obrig, T.G. (1997). Sensitization of human umbilical vein endothelial cells to Shiga toxin: involvement of protein kinase C and NF-kappaB. Infect. Immun. *65*, 3337–3344.

Mahfoud, R., Manis, A., and Lingwood, C. (2009). Fatty acid-dependent globotriaosyl ceramide receptor function in detergent resistant model membranes. J. Lip. Res. *50*, 1744–1755.

Mahfoud, R., Manis, A., Binnington, B., Ackerley, C., and Lingwood, C.A. (2010). A major fraction of glycosphingolipids in model and cellular cholesterol containing membranes are undetectable by their binding proteins. J. Biol. Chem. *285*, 36049–36059.

Mangeney, M., Richard, Y., Coulaud, D., Tursz, T., and Wiels, J. (1991). CD77: an antigen of germinal center B cells entering apoptosis. Eur. J. Immunol. *21*, 1131–1140.

McCluer, R., Deutsch, C., and Gross, S. (1983). Testosterone-responsive mouse kidney glycosphi ngolipids:developmental and inbred strain effects. Endocrinol. *113*, 251–258.

Mizutani, Y., Kihara, A., and Igarashi, Y. (2005). Mammalian Lass6 and its related family members regulate synthesis of specific ceramides. Biochem. J. *390*, 263–271.

Mizutani, Y., Kihara, A., and Igarashi, Y. (2006). LASS3 (longevity assurance homologue 3) is a mainly testis-specific (dihydro)ceramide synthase with relatively broad substrate specificity. Biochem. J. *398*, 531–538.

Mori, T., Kiyokawa, N., Katagiri, Y.U., Taguchi, T., Suzuki, T., Sekino, T., Sato, N., Ohmi, K., Nakajima, H., Takeda, T., *et al.* (2000). Globotriaosyl ceramide (CD77/Gb3) in the glycolipid-enriched membrane domain participates in B-cell receptor-mediated apoptosis by regulating Lyn kinase activity in human B cells. Exp. Hemat. *28*, 1260–1268.

Mukherjee, S., Zha, X., Tabas, I., and Maxfield, F.R. (1998). Cholesterol distribution in living cells: fluorescence imaging using dehydroergosterol as a fluorescent cholesterol analog. Biophys. J. *75*, 1915–1925.

Murray, L.J., Habeshaw, J.A., Wiels, J., and Greaves, M.F. (1985). Expression of Burkitt lymphoma-associated antigen (defined by the monoclonal antibody 38.13) on both normal and malignant germinal-centre B cells. Int. J. Cancer *36*, 561–565.

Nutikka, A., and Lingwood, C. (2004). Generation of receptor active, globotriaosyl ceramide/cholesterol lipid 'rafts' *in vitro*: a new assay to define factors affecting glycosphingolipid receptor activity. Glycoconj. J. *20*, 33–38.

Nyholm, P.-G., Brunton, J.L., and Lingwood, C.A. (1995). Modelling of the interaction of verotoxin-1 (VT1) with its glycolipid receptor, globotriaosylceramide (Gb_3). Int. J. Biol. Macromol. *17*, 199–205.

Nyholm, P.G., Magnusson, G., Zheng, Z., Norel, R., Binnington-Boyd, B., and Lingwood, C.A. (1996). Two distinct binding sites for globotriaosyl ceramide on verotoxins: molecular modelling and confirmation by analogue studies and a new glycolipid receptor for all verotoxins. Chem. Biol. *3*, 263–275.

Okuda, T., Tokuda, N., Numata, S., Ito, M., Ohta, M., Kawamura, K., Wiels, J., Urano, T., Tajima, O., and Furukawa, K. (2006). Targeted disruption of Gb3/CD77 synthase gene resulted in the complete deletion of globo-series glycosphingolipids and loss of sensitivity to verotoxins. J. Biol. Chem. *281*, 10230–10235.

Pellizzari, A., Pang, H., and Lingwood, C.A. (1992). Binding of verocytotoxin 1 to its receptor is influenced by differences in receptor fatty acid content. Biochemistry *31*, 1363–1370.

Perry, R.J., and Ridgway, N.D. (2005). Molecular mechanisms and regulation of ceramide transport. Biochim. Biophys. Acta *1734*, 220–234.

Petruzziello-Pellegrini, T.N., Yuen, D.A., Page, A.V., Patel, S., Soltyk, A.M., Matouk, C.C., Wong, D.K., Turgeon, P.J., Fish, J.E., Ho, J.J., *et al.* (2012). The CXCR4/CXCR7/SDF-1 pathway contributes to the pathogenesis of Shiga toxin-associated hemolytic uremic syndrome in humans and mice. J. Clin. Invest. *122*, 759–776.

Raa, H., Grimmer, S., Schwudke, D., Bergan, J., Walchli, S., Skotland, T., Shevchenko, A., and Sandvig, K. (2009). Glycosphingolipid requirements for endosome-to-golgi transport of shiga toxin. Traffic *10*, 868–882.

Reinhart, M.P., Billheimer, J.T., Faust, J.R., and Gaylor, J.L. (1987). Subcellular localization of the enzymes of cholesterol biosynthesis and metabolism in rat liver. J. Biol. Chem. *262*, 9649–9655.

Saito, M., Mylvaganum, M., Tam, P., Binnington, B., and Lingwood, C. (2012). Structure-dependent pseudo-receptor intracellular traffic of adamantyl globotriaosyl ceramide mimics. J. Biol. Chem. *287*, 16073–16087.

Sandvig, K., Bergan, J., Dyve, A.B., Skotland, T., and Torgersen, M.L. (2009). Endocytosis and retrograde transport of Shiga toxin. Toxicon *56*, 1181–1185.

Sato, S.B., Ishii, K., Makino, A., Iwabuchi, K., Yamaji-Hasegawa, A., Senoh, Y., Nagaoka, I., Sakuraba, H., and Kobayashi, T. (2004). Distribution and transport of cholesterol-rich membrane domains monitored by a membrane-impermeant fluorescent polyethylene glycol-derivatized cholesterol. J. Biol. Chem. *279*, 23790–23796.

Sixma, T.K., Stein, P.E., Hol, W.G.J., and Read, R.J. (1993). Comparison of the B-pentamers of heat-labile enterotoxin and verotoxin-1: Two structures with remarkable similarity and dissimilarity. Biochemistry *32*, 191–198.

Smith, D.C., Sillence, D.J., Falguieres, T., Jarvis, R.M., Johannes, L., Lord, J.M., Platt, F.M., and Roberts, L.M. (2006). The association of Shiga-like toxin with detergent-resistant membranes is modulated by glucosylceramide and is an essential requirement in the endoplasmic reticulum for a cytotoxic effect. Mol. Biol. Cell *17*, 1375–1387.

Soltyk, A.M., MacKenzie, C.R., Wolski, V.M., Hirama, T., Kitov, P.I., Bundle, D.R., and Brunton, J.L. (2002). A mutational analysis of the Globotriaosylceramide binding sites of Verotoxin VT1. J. Biol. Chem. *277*, 5351–5359.

Spitalnik, P., and Spitalnik, S. (1995). The P blood group system: Biochemical, serological and clinical aspects. Transfus. Med. Rev. *9*, 110–122.

Stiban, J., Tidhar, R., and Futerman, A.H. (2010). Ceramide synthases: roles in cell physiology and signaling. Adv. Exp. Med. Biol. *688*, 60–71.

Strauman, M.C., Harper, J.M., Harrington, S.M., Boll, E.J., and Nataro, J.P. (2010). Enteroaggregative *Escherichia coli* disrupts epithelial cell tight junctions. Infect. Immun. *78*, 4958–4964.

Takenouchi, H., Kiyokawa, N., Taguchi, T., Matsui, J., Katagiri, Y.U., Okita, H., Okuda, K., and Fujimoto, J. (2004). Shiga toxin binding to globotriaosyl ceramide induces intracellular signals that mediate cytoskeleton remodeling in human renal carcinoma-derived cells. J. Cell Sci. *117*, 3911–3922.

Tam, P., Mahfoud, R., Nutikka, A., Khine, A., Binnington, B., Paroutis, P., and Lingwood, C. (2008). Differential intracellular trafficking and binding of verotoxin 1 and verotoxin 2 to globotriaosylceramide-containing lipid assemblies. J. Cell. Physiol. *216*, 750–763.

Tesh, V.L. (2010). Induction of apoptosis by Shiga toxins. Future Microbiol. *5*, 431–453.

Tetaud, C., Falguieres, T., Carlier, K., Lecluse, Y., Garibal, J., Coulaud, D., Busson, P., Steffensen, R., Clausen, H., Johannes, L., *et al.* (2003). Two distinct Gb3/CD77 signaling pathways leading to apoptosis are triggered by anti-Gb3/CD77 mAb and verotoxin-1. J. Biol. Chem. *278*, 45200–45208.

Tironi-Farinati, C., Loidl, C.F., Boccoli, J., Parma, Y., Fernandez-Miyakawa, M.E., and Goldstein, J. (2010). Intracerebroventricular Shiga toxin 2 increases the expression of its receptor globotriaosylceramide and causes dendritic abnormalities. J. Neuroimmunol. *222*, 48–61.

Torgersen, M.L., Lauvrak, S.U., and Sandvig, K. (2005). The A-subunit of surface-bound Shiga toxin stimulates clathrin-dependent uptake of the toxin. FEBS J. *272*, 4103–4113.

Utskarpen, A., Massol, R., van Deurs, B., Lauvrak, S.U., Kirchhausen, T., and Sandvig, K. (2010). Shiga toxin increases formation of clathrin-coated pits through Syk kinase. PLoS One *5*, e10944.

Waddell, T., Cohen, A., and Lingwood, C.A. (1990). Induction of verotoxin sensitivity in receptor deficient cell lines using the receptor glycolipid globotriosyl ceramide. Proc. Natl. Acad. Sci. U.S.A. *87*, 7898–7901.

Walchli, S., Aasheim, H.C., Skanland, S.S., Spilsberg, B., Torgersen, M.L., Rosendal, K.R., and Sandvig, K. (2009). Characterization of clathrin and Syk interaction upon Shiga toxin binding. Cell Signal *21*, 1161–1168.

Yahi, N., Aulas, A., and Fantini, J. (2010). How cholesterol constrains glycolipid conformation for optimal recognition of Alzheimer's beta amyloid peptide (Abeta1–40). PLoS One *5*, e9079.vww

The *Helicobacter pylori* CagA Protein: A Multifunctional Bacterial Toxin Delivered by Type IV Secretion

2

Wolfgang Fischer and Benjamin Busch

Abstract

To deliver proteins with host cell-modulating activities, pathogenic bacteria either secrete protein toxins that are able to enter target cells autonomously, or use specialized secretion systems (type III, type IV or type VI secretion systems) to inject effector proteins directly into the host cell cytosol. The human gastric pathogen *Helicobacter pylori* secretes one major virulence determinant, the vacuolating cytotoxin VacA, to the extracellular environment, and translocates another, the cytotoxin-associated antigen CagA, by contact-dependent injection using the Cag type IV secretion system. Transfer of CagA into host cells is considered as a major risk factor for development of gastric cancer, and ectopically expressed CagA is sufficient to cause neoplastic transformations. Once injected, CagA becomes phosphorylated by cellular tyrosine kinases, and the subsequent interaction with a large number of host cell proteins results in cytoskeleton rearrangements and in deregulation of several signal transduction pathways that may lead to precancerous changes. Moreover, CagA binds to several interaction partners independently of tyrosine phosphorylation, and these interactions lead to a loss of cell polarity and to increased cell motility. This review summarizes the current knowledge of the molecular mechanisms of the CagA translocation process and of the diverse functions of CagA in target cells.

Introduction

The human gastric pathogen *H. pylori* is the principal cause of chronic active gastritis and peptic ulcer disease, and it is also involved in development of mucosa-associated lymphoid tissue (MALT) lymphoma and gastric cancer (Peek and Blaser, 2002; Suerbaum and Michetti, 2002). Gastric adenocarcinoma is the second most common cause of cancer-related death worldwide and thus represents a major health problem. *H. pylori* infection is the strongest known risk factor for the development of gastric malignancies (Polk and Peek, Jr., 2010), which has led to the classification of *H. pylori* as a definitive (class I) carcinogen. Whereas all infected individuals develop an active gastritis, characterized by massive infiltration of granulocytes and subsequently lymphocytes into the gastric submucosa, this condition remains asymptomatic in most cases. Only a subset of infected persons suffers from symptomatic gastritis or the more severe complications such as duodenal ulcer, gastric ulcer or gastric adenocarcinoma. Duodenal ulcer usually develops in patients with high gastric acid output and with predominant infection and gastritis in the antrum, the distal part of the stomach. Damage in the duodenal epithelium caused by gastric acid leakage results in gastric metaplasia in the duodenum, followed by *H. pylori* infection, inflammation and ulceration (Suerbaum and Michetti, 2002; Kusters *et al.*, 2006). In contrast, gastric ulcers develop in patients with normal or reduced acid output and corpus-dominant or pan-gastritis, mostly at strongly inflamed corpus–antrum junctions. Likewise, gastric adenocarcinoma develops in patients with corpus-dominant or pan-gastritis and low acid output, and involves in the more common intestinal type a progression from gastritis over atrophic gastritis, intestinal metaplasia and dysplasia to cancer (for reviews, see Suerbaum and

Michetti, 2002; Atherton, 2006; Atherton and Blaser, 2009).

The development of gastric diseases is based on both host and bacterial factors. The most prominent host factors for increased risk of gastric cancer are genetic polymorphisms resulting in high expression of proinflammatory cytokines such as IL-1β or TNF-α, and two of the major bacterial virulence factors are certain genotypes (termed s1/i1/m1) of the *vacA* gene encoding the vacuolating cytotoxin, and the presence of the cytotoxin-associated gene (*cag*) pathogenicity island (reviewed in Amieva and El-Omar, 2008). The vacuolating cytotoxin VacA causes a number of different effects on epithelial cells and on cells of the immune system (reviewed in Gebert *et al.*, 2004; Cover and Blanke, 2005), and it has mainly been associated with ulcer development (Ogura *et al.*, 2000; Fujikawa *et al.*, 2003; Basso *et al.*, 2008). The *cag* pathogenicity island (*cag*PAI) is a 37 kb genome island which represents one of the major variable genome regions of *H. pylori*. Strains carrying the *cag* pathogenicity island are also mostly equipped with the more virulent *vacA* s1/i1/m1 genotype, suggesting a common selective pressure (Gangwer *et al.*, 2010), and have been termed type I strains (in contrast to type II strains having *vacA* s2/i2/m2 genotypes and lacking the *cag* PAI) (Xiang *et al.*, 1995). This correlation has originally led to the designation *cagA* (for cytotoxin-associated gene A) (Covacci *et al.*, 1993; Tummuru *et al.*, 1993). The *cag* pathogenicity island encodes the immunodominant antigen CagA and several components forming a type IV protein secretion apparatus (for a recent review see Fischer, 2011). The Cag type IV secretion system is responsible for induction of a pronounced proinflammatory response, and for direct translocation of the CagA protein into various host cells. Although the exact influence of CagA protein translocation on *H. pylori* pathogenesis is still not clear, strains containing the *cag* pathogenicity island are clearly associated with an enhanced risk of developing gastric adenocarcinoma (Peek and Blaser, 2002).

Since *H. pylori* is a human-specific pathogen, suitable animal models to study *H. pylori*-induced pathogenesis have been difficult to establish. Several mouse models have been used for colonization and pathogenesis studies, but one drawback of these models is that the *cag* pathogenicity island is unstable or non-functional in mouse-colonizing strains (Philpott *et al.*, 2002). However, stability and functionality of the *cag* pathogenicity island depend on the age of mice upon infection; thus, it has been shown that neonatally infected mice develop a tolerance to the Cag system that does not occur in infected adult mice (Arnold *et al.*, 2011b). The best animal model (apart from primate models) with respect to reproducibility of human pathology is the Mongolian gerbil (*Meriones unguiculatus*) model. Mongolian gerbils develop not only gastritis, but also ulceration and precancerous lesions such as intestinal metaplasia and dysplasia, or even cancer (Watanabe *et al.*, 1998), and these pathologies are dependent on the presence of the *cag* pathogenicity island (Wirth *et al.*, 1998; Ogura *et al.*, 2000; Peek *et al.*, 2000). Interestingly, a Mongolian gerbil-adapted *H. pylori* strain showing enhanced CagA translocation was found to induce gastric cancer rapidly after infection (Franco *et al.*, 2005). Furthermore, CagA-translocating strains are able to colonize the gastric corpus in the gerbil model and to cause precancerous lesions, whereas CagA-negative strains are predominantly colonizing the antrum (Rieder *et al.*, 2005). While these data suggested a contribution of CagA to the development of gastric cancer, generation and examination of transgenic mice expressing CagA has shown that CagA itself is sufficient to induce gastric epithelial hyperplasia as well as gastric and intestinal carcinoma in the absence of a pronounced inflammatory response (Ohnishi *et al.*, 2008). Accordingly, CagA has been termed a bacterial oncoprotein (Hatakeyama, 2011).

The *cag* pathogenicity island and its effector protein CagA

Variations of the *cag* pathogenicity island

An important feature of the species *H. pylori* is its unusual genetic diversity, which has mostly been characterized as a microdiversity resulting from high mutation and recombination rates (Suerbaum and Josenhans, 2007). According to this microdiversity, *H. pylori* can be divided into several distinct subpopulations that reflect their geographical origin (Falush *et al.*, 2003). However, pronounced variations between strains are also evident in their gene content (Fischer *et al.*, 2010), or in host-interacting factors showing signs of positive selection (Ogura *et al.*, 2007).

In most patient isolates, the *cag* pathogenicity island is either completely present or absent; truncations or partial deletions occur, but are not very common (Gressmann *et al.*, 2005). However, the percentage of *cag*PAI-positive strains varies considerably between geographically distinct groups, ranging from almost universal presence in East Asian isolates to complete absence in certain African populations (Gressmann *et al.*, 2005). Specifically, the *cag*PAI is most common in Asian and some African strains (about 95% in *hp*Africa1, *hsp*EAsia or *hp*Asia2 populations or subpopulations), completely absent in other African strains (*hp*Africa2 population), and variably present among the other strains (58% in *hp*Europe, 81% in *hp*NEAfrica, 65% in *hp*Sahul, and 28% in *hsp*Amerind populations) (Olbermann *et al.*, 2010). Phylogenetic analyses indicate that the *cag*PAI was acquired only once about 60,000 years ago, prior to human migration out of Africa, since the microdiversity within *cag*PAI genes correlates with the microdiversity of housekeeping genes (Linz *et al.*, 2007; Olbermann *et al.*, 2010).

The *cag*PAI contains 27–30 genes in a mostly conserved gene arrangement, 15 of which are thought to encode essential components of the type IV secretion apparatus (Fischer, 2011). Variations in gene arrangement include a duplication of the *cagA* gene together with one or two flanking genes found in some strains of American Indian origin (Kersulyte *et al.*, 2010), and a larger chromosomal rearrangement uncoupling the *cagA* gene from the rest of the pathogenicity island (Fischer, 2011). Several genes within the *cag*PAI have probably evolved under diversifying selection. This includes genes encoding proteins that are exposed to the bacterial surface or are part of a type IV secretion pilus structure (*cagI*, *cagL* and *cagY*; see below), but also the *cagA* gene and genes encoding proteins with unknown function (Olbermann *et al.*, 2010). This observation suggests the presence of strong selective pressures due to interactions with host cells or the immune system (Aras *et al.*, 2003).

Originally, the Cag type IV secretion system has been associated with the ability of bacteria to induce a strong proinflammatory response in epithelial cells, represented by secretion of interleukin-8 (IL-8) and other chemokines and cytokines (Akopyants *et al.*, 1998), but, in fact, several hundred host cell genes might be up- or down-regulated in response to the Cag system, probably making it the most important host interaction factor of *H. pylori* (Guillemin *et al.*, 2002; El Etr *et al.*, 2004). Although the majority of differentially regulated host cell genes depends on translocation of the CagA protein (El Etr *et al.*, 2004), many important responses are caused by the secretion apparatus itself and can therefore be elicited by *cagA* mutants as well. Such CagA-independent, but type IV secretion system-dependent signal transduction pathways have been reported in numerous studies. We will not further discuss these signalling pathways here and refer the reader to recent reviews instead (Backert and Naumann, 2010; Smolka and Backert, 2012).

Variability in the *cagA* gene and disease association

The CagA protein has originally been identified as a dominant antigen of 128 kDa (Covacci *et al.*, 1993), but there are considerable size variations between strains. Variations have been described in the 5′ (van Doorn *et al.*, 1999) and 3′ regions (Yamaoka *et al.*, 1998, 1999) of the *cagA* gene, both of which are divergent between strains from different populations. The 3′ variable region encodes the so-called EPIYA region containing the CagA tyrosine phosphorylation Glu-Pro-Ile-Tyr-Ala (EPIYA) motifs (Fig. 2.1A). According to their flanking amino acid sequences, these motifs can be classified as EPIYA-A, B, C, and D motifs, respectively (Hatakeyama, 2003; Fig. 2.1A and B). Interestingly, 'Western' *H. pylori* strains (including *hp*Europe, *hp*NEAfrica, *hp*Africa1, *hp*Sahul and parts of *hp*Asia2 populations) produce CagA proteins with different combinations of EPIYA motifs A, B, and C, whereas East Asian strains produce CagA proteins with motif A, B, and D combinations (Higashi *et al.*, 2002a; Naito *et al.*, 2006). Most strains have single A and B motifs, but motifs

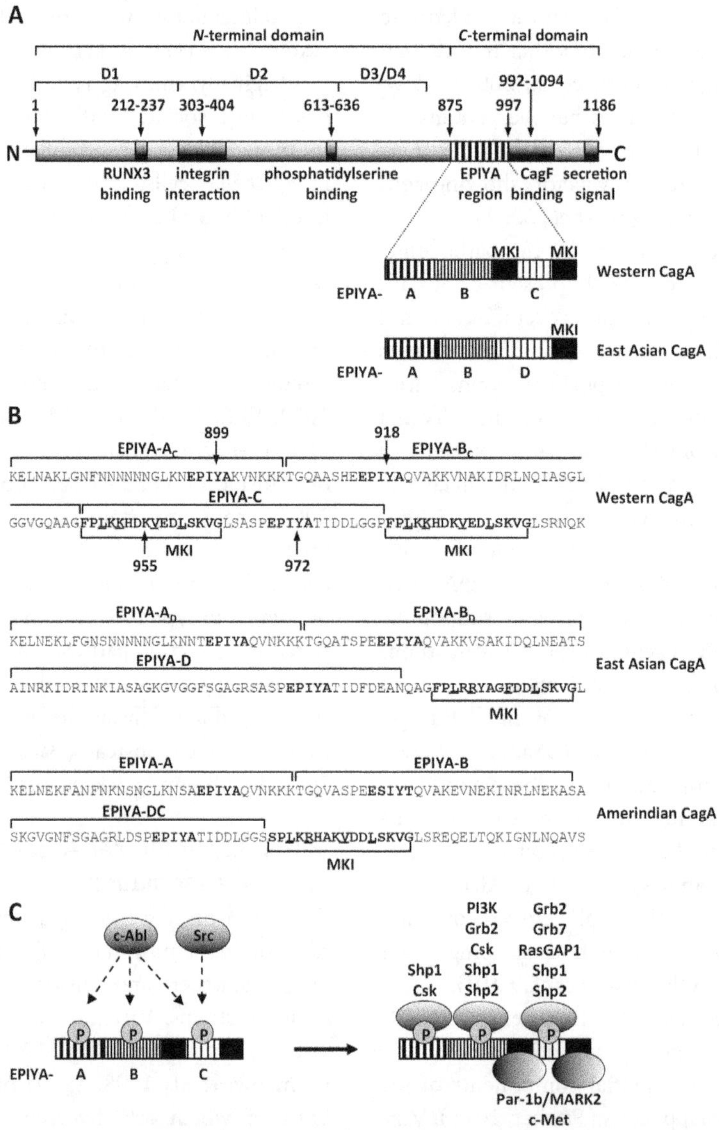

Figure 2.1 Functional domains and target protein interaction modules of the CagA protein. (A) Overall organization of the CagA protein. CagA consists of an N-terminal 100kDa part with known structure, and a C-terminal 30–50kDa part which is intrinsically disordered. Within the N-terminal part, several domains (D1, D2, D3/D4, or domains I to III) have been identified. The N-terminal part contains a region necessary for binding to RUNX3, a region required for interaction with β1 integrins, and a phosphatidylserine (PS)-binding region. The most important functional domain in the C-terminal part is the EPIYA region, consisting of different EPIYA motifs termed EPIYA-A, EPIYA-B and EPIYA-C motifs for Western *H. pylori* strains, and EPIYA-A, EPIYA-B and EPIYA-D motifs for East Asian strains, respectively. Additionally, the EPIYA region contains one or more MARK2 kinase inhibitor (MKI) motifs. Apart from the EPIYA region, the C-terminal part contains a region necessary for interaction with the secretion chaperone CagF, and the C-terminal type IV secretion signal. Amino acid positions refer to the CagA sequence of the Western reference strain 26695. (B) Representative amino acid sequences of the EPIYA regions from Western, East Asian, and Amerindian strains. Locations of the individual EPIYA motifs, labelled according to (Xia *et al.*, 2009), and MKI motifs are indicated. EPIYA sequences and MKI motif sequences are printed in boldface, and amino acids required for MARK2 binding are underlined. (C) Phosphorylation of EPIYA motifs and interaction partners in the EPIYA region. Specific kinase activities of Src and c-Abl kinases towards individual EPIYA motifs are shown on the left, and cellular proteins with reported interactions with the different phosphorylated EPIYA motifs, or with MKI motifs, on the right.

C are frequently duplicated in Western strains and may occur in one to five copies, whereas East Asian strains usually have only one D motif in addition to their A and B motifs (Argent *et al.*, 2008a). Although EPIYA motifs A and B are more similar between Western and East Asian strains, they can also be distinguished into A_C and B_C, or A_D and B_D motifs, respectively (Xia *et al.*, 2009; Fig. 2.1B). Immediately adjacent to the EPIYA motifs, a second conserved motif has been identified and termed MARK2 kinase inhibitor (MKI) motif due to its ability to bind to this kinase (see below). East Asian CagA molecules usually have single MKI motifs downstream of their EPIYA-D motifs, whereas Western CagA molecules contain one additional MKI motif within each EPIYA-C sequence, so that a CagA molecule with one EPIYA-C motif has two MKI motifs (Fig. 2.1B), a CagA molecule with two EPIYA-C motifs has three MKI motifs, and so on. The amino acid sequences of these motifs are also distinguishable between Western and East Asian strains, with clear signs of positive selection in East Asian strains (Olbermann *et al.*, 2010; Delgado-Rosado *et al.*, 2011; Furuta *et al.*, 2011; Kawai *et al.*, 2011).

Some notable exceptions in the primary structure of CagA were encountered in *H. pylori* strains isolated from South American Indian (Amerindian) and Japanese populations. Phylogenetic analysis of full-length CagA sequences revealed a clustering into four distinct clades (Duncan *et al.*, 2012). Group 1 CagA sequences contained most Western CagA variants with EPIYA-C motifs, and group 2 CagA sequences contained East Asian CagA types with EPIYA-D motifs. Group 3 comprised CagA sequences of strains originating from Okinawa that had previously been termed J-Western CagA variants (Truong *et al.*, 2009), and of isolates classified as *hp*Europe. Group 3 CagA variants possess a characteristic insertion of four amino acids at position 206 of reference strain 26695 CagA, with unknown function. Finally, CagA sequences of Amerindian *H. pylori* strains form a distinct group (Mane *et al.*, 2010; Camorlinga-Ponce *et al.*, 2011; Duncan *et al.*, 2012); these CagA variants are characterized by hybrid EPIYA-DC motifs composed of the left flanking region of EPIYA-D and the right flanking region of EPIYA-C (Mane *et al.*, 2010; Duncan *et al.*, 2012; Fig. 2.1B). A more detailed analysis of the corresponding nucleotide sequences suggested that East Asian EPIYA regions have evolved through several recombination events from Western EPIYA regions, with Amerindian EPIYA regions as intermediates (Mane *et al.*, 2010; Furuta *et al.*, 2011). CagA sequence analysis of more Amerindian strains showed that Amerindian CagA can be further divided into two subgroups termed AM-I and AM-II (Suzuki *et al.*, 2011), in which EPIYA-B motifs are usually altered (mostly to ESIYT for AM-I and GSIYD for AM-II variants, respectively). Furthermore, many AM-II CagA molecules contain variant MKI motifs and two deletions in the N-terminal CagA region with a total size of 180 amino acids (Suzuki *et al.*, 2011).

Despite the correlation between the CagA phylogeny and the multilocus sequence typing-based classification, there are considerable differences within the tree topologies (Duncan *et al.*, 2012). The strains classified as *hp*Europe according to the phylogeny of housekeeping genes are distributed among two groups (group 1 and group 3) in the CagA tree. Additionally, CagA phylogeny did not reveal a discrete African group of *H. pylori* strains, which might reflect horizontal gene transfer of the *cag*PAI after divergence of African and Non-African strains (Duncan *et al.*, 2012).

Different *cagA* genotypes, notably variations in the EPIYA region, are associated with the severity of diseases (Yamaoka *et al.*, 1999; Azuma *et al.*, 2002, 2004; Argent *et al.*, 2004; Basso *et al.*, 2008; Xia *et al.*, 2009; Ferreira *et al.*, 2012). Notably, the risk of developing gastric cancer increases in Western strains with increasing number of EPIYA-C motifs, and ABD-motif CagA variants are more often associated with gastric cancer than ABC-motif CagA variants.

Structure and functional domains of the CagA protein

As mentioned above, there are pronounced size variations between individual CagA proteins which are mostly due to variations in the EPIYA region. The prototypical CagA molecule of strain 26695 has 1186 amino acids (Fig. 2.1A) and a calculated molecular weight of 132 kDa, but other CagA variants may range in size between 130 and

150 kDa. As detailed below, proteolytic cleavage of native CagA after translocation into leucocytes, or of recombinant CagA produced in *E. coli*, results in an N-terminal fragment of 100 kDa and a C-terminal fragment of 30–50 kDa, suggesting that CagA can roughly be divided into corresponding N-terminal and C-terminal domains (Fig. 2.1A). Recent structural analysis has revealed that the C-terminal domain is intrinsically disordered and therefore not amenable to crystal structure determination, whereas the N-terminal domain (amino acids 1–876 or 1–884) has a defined structure with some short disordered regions (Hayashi *et al.*, 2012; Kaplan-Türköz *et al.*, 2012; Fig. 2.2). This N-terminal CagA fragment adopts a half moon shape consisting of three or four distinct domains (Figs. 2.1A and 2.2). The first domain comprises 10 α-helices and is connected to the central second domain by a disordered linker (Hayashi *et al.*, 2012). The second domain consists of a large β-sheet made of 11 antiparallel β-strands (residues 308–543) as well as an inserted subdomain (residues 370–446) and a helical region (residues 555–640), both of which contact the convex side of the β-sheet and presumably stabilize it (Hayashi *et al.*, 2012; Kaplan-Türköz *et al.*, 2012). The β-sheet itself is structurally similar to a particular region of the *Borrelia burgdorferi* outer membrane protein OspA (Kaplan-Türköz *et al.*, 2012). The second domain is followed by a long α helix which forms at its C-terminal end a four-helix bundle with three of the four α helices making up the last domain (Hayashi *et al.*, 2012; Kaplan-Türköz *et al.*, 2012; Fig. 2.2). Interestingly, the final two helices of this last domain have an interaction surface in the crystal with the same region of a second subunit, although recombinant CagA is monomeric in solution (Hayashi *et al.*, 2012; Kaplan-Türköz *et al.*, 2012). However, they comprise a sequence with significant similarity to a section of the C-terminal CagA domain, and N-terminal and C-terminal CagA fragments were shown to interact via hydrophobic interactions between these sequences (termed N-terminal and C-terminal binding sequences). Biochemical evidence suggests that this interaction induces folding in the otherwise unstructured C-terminal CagA domain and generates a lariat structure exposing different protein–protein interaction modules (Hayashi *et al.*, 2012).

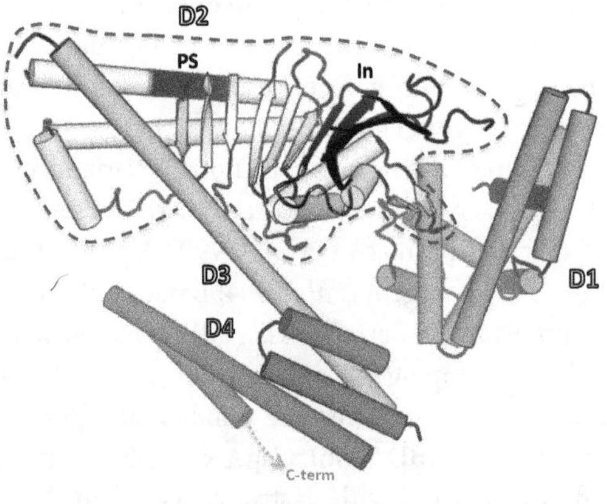

Figure 2.2 Structure of the N-terminal CagA part. A schematic structural view of an N-terminal CagA fragment (amino acids 1–884) is given, with α helices drawn as cylinders, and β-strands as arrows. The different domains (D1 to D4) are designated according to (Kaplan-Türköz *et al.*, 2012) and represented in different shades of grey; domain D2 is encircled by a broken line. Regions interacting with β1 integrins (In), or phosphatidylserine (PS), are depicted in black. Note that several loops connecting domains or domain substructures are not represented in the crystal structures due to high flexibility.

The Cag type IV secretion system as a toxin delivery system

Secretion versus translocation in bacterial type III and type IV secretion systems

Proteins delivered by bacterial type III or type IV secretion systems to eukaryotic cells are not usually considered as typical bacterial toxins, despite the fact that their roles in host cell manipulation are often very similar. For example, many type III-secreted effector proteins and many AB-type toxins modulate the activity of Rho-family GTPases (Aktories, 2011). A unifying feature of typical bacterial protein toxins is their ability to interact with and enter their target cells autonomously after secretion from the bacterial cell, whereas effector proteins of type III and type IV secretion systems require their cognate secretion apparatus not only for secretion, but also for translocation into the target cell. However, it is currently not well-understood how translocation of type III or type IV effector proteins across eukaryotic cell membranes is achieved. For type III secretion systems, a general opinion is that the secretion apparatus secretes up to three translocator proteins which subsequently form a translocon complex in the target cell lipid bilayer (Matteï *et al.*, 2011). Effector proteins are thought to be transported into target cells via this secretion system-delivered translocon. However, this does not necessarily mean that effector protein translocation is a one-step process. Recent data indicate that the type III effector YopH can also be delivered to target cells when supplied extracellularly, although this process still requires a functional type III secretion apparatus (Akopyan *et al.*, 2011).

In type IV secretion systems, a similar translocon complex has not been identified. Given that possible target cells of type IV secretion systems range from bacteria over yeast to mammalian and plant cells (Alvarez-Martinez and Christie, 2009), it is very likely that individual systems have adapted not only to their respective host bacteria, but also to these target cells. Although effector proteins of type IV secretion systems are also generally thought to be injected in a one-step process into target cells, several observations suggest that extracellular intermediates might exist as well. First, one particular type IV secretion system, the Ptl system of *Bordetella pertussis*, generally secretes its substrate, the pertussis toxin, to the extracellular environment, from where it enters target cells in the same way as other typical AB toxins do (Locht *et al.*, 2011). Second, in the well-characterized VirB type IV secretion system of the plant pathogen *Agrobacterium tumefaciens*, one effector protein (VirE2) has been suggested to act as a pore-forming protein in the plant cell membrane (Dumas *et al.*, 2001). Coinfection experiments, in which a VirE2-positive, but substrate (T-DNA)-deficient strain was able to rescue the T-DNA translocation defect of a *virE2* mutant (Binns *et al.*, 1995), suggested that extracellular delivery, at least of the VirE2 protein, occurs before transfer into the plant cell. However, more recent data suggested that the function of VirE2 might rather consist in pulling the T–DNA complex into the plant cell cytosol (Grange *et al.*, 2008). Nevertheless, the notion that extracellular intermediates may exist in this system is supported by identification of a secretion apparatus (*virB10*) mutant which secretes large amounts of VirE2 to the supernatant and is still able to translocate it into plant cells (Banta *et al.*, 2011). For the *H. pylori* Cag type IV secretion system, associated surface appendages (pilus structures) have been described, but they look very different from other type IV secretion-associated pili (Tanaka *et al.*, 2003; Rohde *et al.*, 2003), and their function might also be modulated by an unusual protein composition (see below). Intriguingly, the effector protein CagA has been visualized at the surface of these structures (Kwok *et al.*, 2007; Jiménez-Soto *et al.*, 2009), raising the possibility that extracellular CagA might indeed be a translocation intermediate. Moreover, CagA translocation could be reduced by addition of an anti-CagA antibody during infection, and it was shown that binding of surface-localized CagA to phosphatidylserine in the outer leaflet of the target cell lipid bilayer is important for its uptake into the cell (Murata-Kamiya *et al.*, 2010). However, the mechanism of CagA translocation across the host cell membrane is not well-understood.

CagA as a substrate for type IV secretion: signal sequence and signal recognition

CagA is so far the only known protein substrate of the Cag type IV secretion system. There is some evidence suggesting that peptidoglycan fragments can be transported by the Cag secretion apparatus to the host cell cytoplasm (Viala *et al.*, 2004), but how this transport is achieved is unclear. For the recognition of CagA as a substrate of the secretion apparatus and for translocation, the presence of its C-terminal 20 amino acids is crucial (Hohlfeld *et al.*, 2006). This is in agreement with other studies on type IV secretion signal sequences, which usually seem to be located at the C-terminus of the effector proteins, although they might be specific for the respective effector protein-secretion system combination (Alvarez-Martinez and Christie, 2009). For CagA, it was shown that exchange of 20 C-terminal amino acids for those of the conjugation effector protein MobA is possible without losing secretion capability (Hohlfeld *et al.*, 2006). However, the nature of the signal sequence is not clear in this system; whereas positively charged amino acids are important for the secretion signal of some type IV effector proteins, site-specific exchange of the corresponding basic amino acids of CagA did not result in a secretion defect (Hohlfeld *et al.*, 2006). Moreover, the CagA C-terminus is not sufficient for translocation, which might reflect the presence of either a non-linear signal recognized by the type IV secretion apparatus, or of a secondary signal recognition step. Such a secondary step might be necessary after CagA transport to the bacterial surface. Indeed, it has been shown that an N-terminal domain of CagA binds to integrin β1 chains (Jiménez-Soto *et al.*, 2009), and that a central region of the protein interacts with phosphatidylserine in the target cell membrane (Murata-Kamiya *et al.*, 2010).

CagA translocation into target cells depends on the presence of a functional type IV secretion apparatus, which has often been recognized by its ability to induce NF-κB activation or IL-8 secretion from epithelial cells (Fischer *et al.*, 2001). Although some translocated CagA variants are able to cause IL-8 release upon prolonged infection, these effects are mostly independent of CagA in the short-term infections usually employed (Backert and Naumann, 2010). Therefore, those 15 gene products of the *cag* pathogenicity island that are essential for both the IL-8 response and for CagA translocation are believed to be components of the secretion apparatus, whereas three further gene products (CagF, CagZ and Cagβ) were found to be essential for translocating CagA, but not for the IL-8 response, and have thus been considered as specific CagA translocation factors (reviewed in Fischer, 2011). In analogy to other type IV secretion systems, notably conjugation systems, where the so-called coupling proteins confer substrate specificity upon the secretion process by interacting both with substrates and the secretion apparatus (Llosa *et al.*, 2003), the coupling protein homologue Cagβ is likely to be the CagA signal recognition protein. It is anchored to the cytoplasmic membrane, and its cytoplasmic region has been shown to interact with CagA (Jurik *et al.*, 2010). Furthermore, Cagβ interacts with the second translocation factor CagZ, as identified in yeast two-hybrid screens and confirmed biochemically in *H. pylori* (Busler *et al.*, 2006; Jurik *et al.*, 2010). An isogenic *cagZ* deletion mutant was not only defective in CagA translocation, but also had strongly reduced Cagβ protein levels. Both defects could be restored by complementation of the mutant, suggesting that Cagβ and CagZ form a stable complex at the bacterial cytoplasmic membrane which might constitute the functional CagA signal recognition receptor (Jurik *et al.*, 2010).

The third translocation factor, CagF, was identified from immunoprecipitation experiments of bacterial lysates as a major component interacting with CagA (Couturier *et al.*, 2006; Pattis *et al.*, 2007). CagF is probably also located at the cytoplasmic membrane. Since the CagF-binding region is adjacent to, but distinct from the C-terminal secretion signal (Fig. 2.1A) and since these regions exert a dominant-negative effect on CagA translocation when fused to GFP, it has been proposed that CagF binding to the CagA C-terminus plays a role in recruitment to the secretion apparatus, similar to the function of secretion chaperones in type III secretion systems (Pattis *et al.*, 2007).

Pilus components and their interaction with integrins

Prior to injection of the CagA protein into target cells, adherence of the bacteria to the host cell surface is necessary. Binding of *H. pylori* to host cell receptors is generally mediated by outer membrane proteins of the Hop family and their cognate receptors such as Lewis^{b} or sialyl-Lewis^{x} molecules (see Fischer *et al.*, 2009; Backert *et al.*, 2011 for review). On the other hand, it has been found that the Cag secretion apparatus binds to β1 integrin molecules at the surface of host cells, and that this binding is required for translocation of the CagA protein (Kwok *et al.*, 2007; Jiménez-Soto *et al.*, 2009). It is not clear if and how integrin binding of the secretion apparatus and adherence mediated by Hop family adhesins depend on each other, but it has been shown that the blood group antigen-binding adhesin BabA is able to enhance CagA translocation and other type IV secretion-dependent effects in cells containing the BabA receptor Lewis^b (Ishijima *et al.*, 2011).

The Cag type IV secretion system elaborates sheathed surface appendages that are dissimilar from the pili commonly found in DNA-transporting type IV secretion systems. Nevertheless, these appendages are thought to be composed of the VirB2-like pilin subunit CagC (Andrzejewska *et al.*, 2006), although they can also be stained by immunogold labelling directed against CagY, CagT, CagX and CagL (Rohde *et al.*, 2003; Tanaka *et al.*, 2003; Kwok *et al.*, 2007). The CagL amino acid sequence contains an Arg-Gly-Asp (RGD) motif which is known from extracellular matrix proteins as an integrin-binding motif. Indeed, it was shown that purified CagL binds via this RGD motif to $\alpha_5\beta_1$ integrins, suggesting a VirB5-like tip adhesin function for CagL (Kwok *et al.*, 2007; Backert *et al.*, 2008), but conflicting results were obtained with respect to the requirement of this motif for CagA translocation (Kwok *et al.*, 2007; Jiménez-Soto *et al.*, 2009). CagL has also been implicated in activation of the gastrin promoter by interaction with $\alpha_v\beta_5$ integrin, which was also independent of its RGD motif (Wiedemann *et al.*, 2012). Binding of purified CagL to integrins has been shown to trigger several effects such as focal adhesion kinase (FAK) and Src kinase activation (Kwok *et al.*, 2007), as well as activation of epidermal growth factor receptor (EGFR), probably via heparin-binding epidermal growth factor (HB-EGF; see below). Apart from CagL, the secretion apparatus proteins CagY and CagI, and the effector protein CagA itself, have also been identified as bacterial interaction partners of β1 integrins (Jiménez-Soto *et al.*, 2009). Thus, interaction of different secretion system components with integrins seems to be an important prerequisite for Cag type IV secretion system function. However, the individual roles of these other integrin-binding Cag proteins are less clear.

With a size of about 220 kDa, CagY is the largest protein encoded on the *cag*PAI. It has a peculiar middle region containing numerous repeated motifs that mainly form α-helical structures (Liu *et al.*, 1999; Delahay *et al.*, 2008), and a significant sequence similarity to the *A. tumefaciens* secretion apparatus core component VirB10 which is confined to its 350 C-terminal amino acids only (Kutter *et al.*, 2008). Its predominant localization is in the bacterial cytoplasmic membrane, possibly with a bridging function between cytoplasmic and outer membranes, as shown for other VirB10 homologues (Chandran *et al.*, 2009), an assumption that is also supported by its interaction with a secretion apparatus subcomplex localized in the outer membrane (Kutter *et al.*, 2008). Antigenic variation in the CagY middle region has been suggested to promote immune evasion (Aras *et al.*, 2003), but it is currently unclear how CagY or its mid region are transported to the bacterial surface or assembled onto the pilus-like surface appendages. Interactions of CagY with other components of the type IV secretion pili have not been demonstrated so far.

The other putative integrin ligand CagI has recently been identified as an interaction partner of CagL (Shaffer *et al.*, 2011; Pham *et al.*, 2012). Because the *cagI* gene is located in an operon structure upstream of *cagL* and downstream of three further *cag* genes (*cagF*, *cagG* and *cagH*) (Sharma *et al.*, 2010), its function has been difficult to assess, but recent mutagenesis and complementation data clearly show that it is a component of the secretion apparatus probably required at later stages of secretion apparatus assembly (Pham *et al.*, 2012). In contrast to most other Cag secretion apparatus components, CagI does not have any homologues in other type IV secretion systems, and there is not much information about its properties (Fischer, 2011). In a recent survey of *cag* gene expression levels in gastric biopsies obtained from *H. pylori*-infected patients, *cagI* was among the most weakly transcribed genes (Avilés-Jiménez *et al.*, 2012). Interestingly, CagI is exposed to the bacterial surface and required to form type IV secretion pili (Shaffer *et al.*, 2011). Deletion of a common hexapeptide motif at the C-terminus of CagI and CagL results in loss of pilus formation and a defect in CagA translocation. A similar C-terminal hexapeptide motif in the secretion apparatus protein CagH seems to have a role in regulation of pilus length and diameter (Shaffer *et al.*, 2011).

CagA transport across the host cell membrane

As for other type IV secretion systems, it is not well understood how the transport of CagA across the host cell membrane is achieved mechanistically. It has been shown that cholesterol depletion of host cell membranes results in a significant reduction of CagA translocation, suggesting that lipid rafts play a role in transport into the cell (Lai *et al.*, 2008). This observation might be explained by a requirement for integrin clustering and recruitment to lipid rafts prior to translocation. Interestingly, however, it has also been shown that conversion of host cell cholesterol to cholesteryl glucosides by a bacterial cholesterol-α-glycosyltransferase (HP0421 or CapJ) is necessary for translocation (Wang *et al.*, 2012). Since the CagA translocation defect in a *capJ* deletion mutant could be partially complemented by external addition of cholesteryl glucosides and since CapJ-produced cholesteryl glucosides partition into detergent-resistant membrane microdomains of infected cells, it has been speculated that *H. pylori* actively alters the host cell membrane structure at translocation sites (Wang *et al.*, 2012). It has also been reported that binding of *H. pylori* to epithelial cells induces externalization of phosphatidylserine molecules from the inner to the outer leaflet of the target cell lipid bilayer, and that binding of CagA to phosphatidylserine is required for its uptake (Murata-Kamiya *et al.*, 2010). The interaction of CagA with phosphatidylserine is dependent on an R-X-R motif and on several other basic amino acids in a particular α-helix in the central region of the protein (Murata-Kamiya *et al.*, 2010; Hayashi *et al.*, 2012; Figs. 2.1A and 2.2). This observation, and the finding that CagA can indeed be found at the bacterial surface and type IV secretion pilus tips (Kwok *et al.*, 2007; Jiménez-Soto *et al.*, 2009), suggested that surface-localized CagA might be a translocation intermediate. Moreover, CagA interacts with extracellular domains of β1 integrin molecules, but since the binding affinity is very high, it is not clear whether these integrin-binding CagA molecules, or rather molecules from a distinct CagA pool, are actually transported into the cell (Jiménez-Soto *et al.*, 2009). In a yeast two-hybrid screen, the CagA region interacting with β1 integrins could be narrowed down to amino acids 303–404 (Kaplan-Türköz *et al.*, 2012; Figs. 2.1A and 2.2). Interestingly, recombinant CagA fragments corresponding to this region were sufficient to inhibit CagA translocation into target cells when added to the extracellular medium before infection with *H. pylori*. This suggests that binding of CagA to integrin molecules is not merely an epiphenomenon, but an important step during CagA translocation (Kaplan-Türköz *et al.*, 2012). Because CagA translocation can also be blocked by an antibody (9EG7) binding to the extended conformation of integrins only, it has been proposed that a conformational change between the extended and the bent integrin conformation triggers CagA injection (Jiménez-Soto *et al.*, 2009). However, it remains to be determined how the interplay between membrane lipids, membrane microdomains and integrin receptors eventually results in CagA translocation across the target cell membrane.

CagA interaction partners and associated effects in host cells

CagA translocation has mostly been studied in epithelial cells (for example, the gastric epithelial cell line AGS), but translocation into other cells, notably leucocytes, has also been reported (Lin *et al.*, 2010; Moese *et al.*, 2001; Odenbreit *et al.*, 2001). Conflicting results have often been obtained with respect to CagA interactions and their functional consequences, sometimes depending on the target cells used (for example, polarized or non-polarized epithelial cell lines), or on different experimental approaches for CagA administration (*H. pylori* infection or cell transfection with *cagA* expression constructs). Because of this, it should be noted that some of the results described below may be specific for the cells examined or the experimental conditions employed. This situation is further complicated by the fact that CagA induces up- or down-regulation of a large number of cellular genes (El Etr *et al.*, 2004), and also by CagA's unique ability to interact with more than 20 target cell proteins (Backert *et al.*, 2010).

CagA membrane localization, dimerization and processing

Upon arrival in the target cell, CagA is localized at the cell membrane close to bacterial attachment sites (Odenbreit *et al.*, 2000; Mimuro *et al.*, 2002; Fig. 2.3A), where it associates mainly with detergent-resistant membrane microdomains (Lai *et al.*, 2008). The molecular mechanisms that determine CagA membrane localization may depend on cell polarization, but are not entirely understood. When CagA is produced ectopically in canine kidney (MDCK) cells, which have often been used to study CagA effects on cell–cell junctions or on cell polarity, membrane targeting is determined by a central protein region ranging from amino acids 400 to 800 (Bagnoli *et al.*, 2005). In contrast, membrane association in (non-polarized) AGS cells after transfection depends on the EPIYA region (Higashi *et al.*, 2005). In line with this, an ectopically expressed C-terminal CagA fragment containing the EPIYA region, but lacking amino acids 400 to 800, was found at the membrane in non-polarized MDCK cells, and in the cytoplasm in polarized MDCK cells (Murata-Kamiya *et al.*, 2010). Membrane localization in polarized MDCK cells was shown to be mediated by the central phosphatidylserine-binding motif (Fig. 2.1A; Murata-Kamiya *et al.*, 2010; Hayashi *et al.*, 2012), but it is not clear why this positively charged motif is only functional in polarized cells. In another study, transfected CagA fragments comprising the N-terminal 200 amino acids, or amino acids 200–1216, were both found at the membrane in polarized MDCK cells and in non-polarized cells, but at different membrane regions in the latter case (Pelz *et al.*, 2011). Transfected CagA fragments lacking the N-terminal 200 amino acids, or only 26 N-terminal amino acids, had stronger effects on cell–cell junction disruption (see below) than full-length CagA, suggesting that targeting of the CagA N-terminus to corresponding membrane regions might suppress excessive activities of the C-terminal region (Pelz *et al.*, 2011). Interestingly, certain CagA variants from American Indian strains (termed AM-II) naturally contain two in-frame N-terminal deletions (amino acids 23–74 and 94–217) that lead to reduced CagA membrane localization, a phenotype that could also be observed in cells transfected with Western CagA constructs lacking the same N-terminal regions (Suzuki *et al.*, 2011).

Since Flag-tagged full-length CagA can be co-immunoprecipitated with HA-tagged full-length CagA in co-transfected cells, it has been suggested that CagA forms dimers or multimers (Ren *et al.*, 2006). The CagA region responsible for co-precipitation is identical to the region required for interaction with Par-1b/MARK2 kinase (see below), therefore it has been assumed that this interaction might be responsible for CagA dimerization (Saadat *et al.*, 2007; Fig. 2.3A). Although MARK kinase dimerization has not been shown *in vivo* (Marx *et al.*, 2010), it could indeed be demonstrated in overproducing AGS cells (Nagase *et al.*, 2011), and overexpression of Par-1b/MARK2 in COS-7 kidney fibroblast cells resulted in increased efficiency of CagA-CagA coprecipitation (Saadat *et al.*, 2007). On the other hand, biochemical evidence suggested that one Western CagA molecule (containing two MKI motifs; Fig. 2.1B) interacts with two MARK2 molecules (Nešić *et al.*, 2010). Therefore, it is currently unclear which stoichiometry CagA–MARK2 complexes have *in vivo*. In support of the notion that CagA dimerizes during *H. pylori* infection, an inhibitory effect of Amerindian CagA on Western CagA proinflammatory signalling has been shown upon coinfection with two corresponding strains (Suzuki *et al.*, 2011).

Apart from dimerization and phosphorylation (see below), CagA may also be processed. This has been shown after translocation into different leucocytes (mouse and human macrophage cell lines, primary human granulocytes, murine dendritic cells), where CagA is proteolytically cleaved into an N-terminal 100 kDa fragment and a tyrosine-phosphorylated C-terminal 35 kDa fragment (Moese *et al.*, 2001; Odenbreit *et al.*, 2001). The function of this proteolysis, which can also be observed to a minor extent in epithelial cells (Torruellas-Garcia *et al.*, 2006), is currently unknown. As mentioned before, a similar degradation can be observed with recombinant CagA fragments produced in *E. coli*, possibly involving a

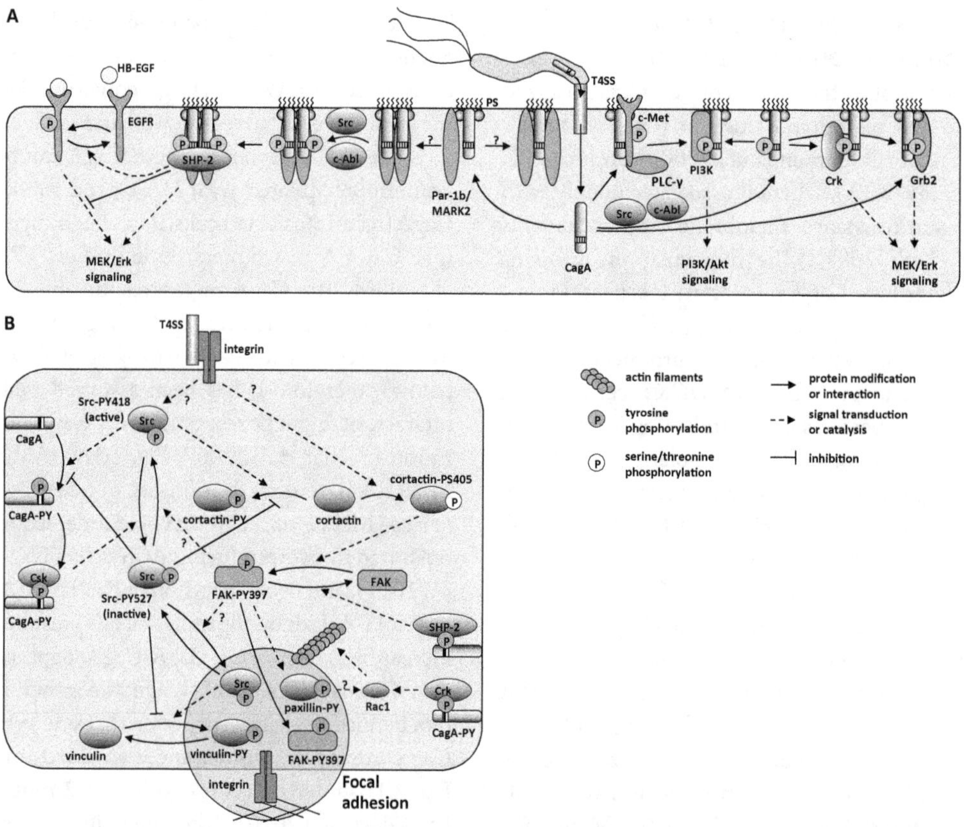

Figure 2.3 CagA interactions and associated effects in target cells. (A) CagA membrane association, phosphorylation and interactions. Upon arrival in the target cell (centre), CagA associates with the cell membrane, either by binding to phosphatidylserine (PS) in polarized cells, or by other unknown mechanisms. This step is independent of tyrosine phosphorylation and is therefore depicted with a non-phosphorylated CagA molecule; however, the order of membrane recruitment and phosphorylation is not known. Interaction with MARK2 may occur with a 1:2 or with a 2:2 stoichiometry, possibly depending on the CagA type. After tyrosine phosphorylation by Src or c-Abl kinases, CagA can interact with SHP-2, probably in a dimeric form. The SHP–2-CagA complex may stimulate MEK-Erk signalling, which can also be induced after epidermal growth factor receptor (EGFR) activation, for example by HB-EGF. However, SHP-2 activated by binding to phospho-CagA dephosphorylates EGFR and thus inhibits EGFR-dependent MEK/Erk signalling. Non-phosphorylated CagA interacts with phosphorylated c-Met and with phospholipase C-γ, resulting in activation of phosphatidylinositol-3-kinase (PI3K)-Akt signalling. CagA phosphorylated by Src and/or c-Abl kinases can also interact directly with PI3K, which may also induce Akt signalling. Binding of phosphorylated or unphosphorylated CagA to Crk or Grb2 adaptor proteins induces MEK/Erk signalling. Note that interactions that have been reported between non-phosphorylated CagA and TAK1 (Lamb *et al.*, 2009), RUNX3 (Tsang *et al.*, 2010), and α-Pix (Lim *et al.*, 2009), or between phosphorylated CagA and SHP-1, Grb7, and RasGAP1 (Selbach *et al.*, 2009) are not depicted. (B) Molecular mechanisms that might be involved in cytoskeletal rearrangements. A non-polarized epithelial cell is depicted. Injected CagA is first phosphorylated at the EPIYA motifs by Src kinase which has been activated by Tyr-418 phosphorylation due to an unknown signal provided by Cag secretion apparatus (or CagL) binding to integrin receptors. Kinase activity of c-Abl results in sustained CagA phosphorylation (not depicted). Phosphorylated CagA may bind to C-terminal Src kinase (Csk), which subsequently inactivates Src by phosphorylation at Tyr-527, resulting in an inhibition of phosphorylation of the Src substrates cortactin or vinculin, and possibly in reduction of focal adhesion formation. Cortactin becomes phosphorylated on Ser-405 by an unknown *cag*PAI-dependent mechanism and subsequently interacts with focal adhesion kinase (FAK), promoting FAK activation by Tyr-397 phosphorylation and thus enhanced focal adhesion formation. Activated FAK may also phosphorylate the focal adhesion protein paxillin, or Src kinase. Phosphorylated paxillin, or phospho-CagA bound to Crk adaptor proteins, may activate Rac1 to induce actin filament formation. SHP-2 activated by binding to phosphorylated CagA is able to dephosphorylate FAK.

poly-asparagine stretch within the EPIYA-A motif (Angelini *et al.*, 2009), or after protease treatment of purified CagA (Hayashi *et al.*, 2012), indicating that this region forms an exposed loop accessible for different proteases.

Tyrosine phosphorylation and interaction with SH2 domain-containing proteins

Tyrosine phosphorylation of CagA is a hallmark of its translocation into eukaryotic cells. In epithelial or macrophage cell lines, it increases with time and can be found as early as 20–30 minutes after infection (Backert *et al.*, 2001; Odenbreit *et al.*, 2001). Early work demonstrated that CagA is phosphorylated by Src-family kinases (c-Src, Lyn) (Selbach *et al.*, 2002; Stein *et al.*, 2002), mainly at the tyrosine residues within the EPIYA-C motifs of Western CagA variants (Y972 in CagA of reference strain 26695; Fig. 2.1B; Backert *et al.*, 2001), or within the (usually single) EPIYA-D motifs in East Asian CagA molecules (Higashi *et al.*, 2002a). More recently, it was shown that CagA tyrosine phosphorylation can also be catalysed by Abl-family kinases (c-Abl and Arg) (Poppe *et al.*, 2007; Tammer *et al.*, 2007). Recent data also suggest that, while Src-family kinases specifically phosphorylate CagA at EPIYA-C in Western strains, c-Abl is capable of phosphorylating motifs A, B and C (Mueller *et al.*, 2012; Fig. 2.1C). Time-course experiments showed that *H. pylori* infection of AGS cells results in sustained activation of c-Abl kinase activity by phosphorylation at tyrosine residue 412 (Poppe *et al.*, 2007; Tammer *et al.*, 2007), whereas Src is only transiently activated by phosphorylation of tyrosine residue 418, but becomes inactivated within 2 hours due to phosphorylation of tyrosine residue 527 (Selbach *et al.*, 2003; Poppe *et al.*, 2007; Tammer *et al.*, 2007). Activation of Src is probably mediated by binding of the secretion apparatus and particularly of its surface-exposed component CagL to integrin $\alpha_5\beta_1$ at the cell surface (Fig. 2.3B). Type IV secretion apparatus interaction with integrins also accounts for activation of other cellular kinases, e.g. focal adhesion kinase (FAK) (Kwok *et al.*, 2007), or Jun N-terminal kinase (JNK) (Snider *et al.*, 2008), but it is not clear whether activation of c-Abl, which is also dependent on the presence of the type IV secretion system (Tammer *et al.*, 2007), is initiated by this pathway as well. There is some evidence that early Src kinase activation is involved in subsequent c-Abl activation (Mueller *et al.*, 2012). However, activated c-Abl is then capable of phosphorylating CagA even in Src-deficient cells (Poppe *et al.*, 2007). Inactivation of Src depends on CagA tyrosine phosphorylation and is mediated either directly by interaction of phosphorylated CagA with Src (Selbach *et al.*, 2003), or indirectly via binding of phosphorylated CagA to C-terminal Src kinase (Csk), which then stimulates Csk kinase activity and leads to Src Tyr-527 phosphorylation (Tsutsumi *et al.*, 2003; Fig. 2.3B). In the latter case, Csk interacts with phosphorylated EPIYA-A and EPIYA-B, but not EPIYA-C or EPIYA-D motifs (Naito *et al.*, 2006; Selbach *et al.*, 2009), suggesting that the Csk-mediated Src inactivation occurs after c-Abl-dependent CagA phosphorylation. Thus, early CagA tyrosine phosphorylation after infection of cell lines is mediated by Src and can accordingly be blocked with the Src-family kinase inhibitor PP2, while later phosphorylation depends on the activity of c-Abl and is thus insensitive to PP2 inhibition (Tammer *et al.*, 2007).

Tyrosine-phosphorylated proteins in eukaryotic cells are known to interact either with phosphotyrosine binding (PTB) domains or with Src homology 2 (SH2) domains, and this is also the case for phospho-CagA. Surprisingly, phospho-CagA can interact with several different SH2 domain-containing proteins (Fig. 2.1C). The first interaction partner to be described was the protein-tyrosine phosphatase SHP-2 (Higashi *et al.*, 2002b). SHP-2 contains two SH2 domains, both of which were shown to interact with phosphorylated EPIYA-C or EPIYA-D motifs (Higashi *et al.*, 2002a). Binding of SHP-2 and of Csk to phosphorylated CagA was found to be mutually exclusive and might therefore occur consecutively (Tsutsumi *et al.*, 2006). Another possible interaction partner of phosphorylated CagA is the SH2 domain-containing adaptor protein Grb2, which was originally found to bind CagA independently of its tyrosine phosphorylation (Mimuro *et al.*, 2002), but has recently been identified in a systematic screen as an interaction partner of tyrosine-phosphorylated peptides corresponding to EPIYA-B and C motifs (Selbach *et al.*, 2009). Further phosphorylation-dependent interaction partners are Crk adaptor proteins (Crk-I, Crk-II and Crk-L) (Suzuki *et al.*, 2005) as well as phosphatidylinositol-3-kinase (PI3K), the phosphatase SHP-1 and the adaptor protein Grb7 (Selbach *et al.*, 2009).

It has been noted that while the CagA EPIYA motifs are well-adapted substrates for eukaryotic tyrosine kinases such as Src and Abl (Backert *et al.*, 2010), binding of tyrosine-phosphorylated EPIYA motifs to SH2 domains of the different target proteins is less specific than for typical eukaryotic phosphotyrosine-SH2 domain interactions. Also, tyrosine phosphorylation motifs in translocated bacterial effector proteins are generally more similar to each other than to endogenous eukaryotic SH2 domain-binding regions (Selbach *et al.*, 2009). This has led to the suggestion that CagA may have evolved to become a promiscuous binding partner of SH2 domain-containing proteins and thus to act as a 'master key' for interference with different cellular processes (Selbach *et al.*, 2009; Backert *et al.*, 2010). Interestingly, the EPIYA motif and related motifs seem to be significantly depleted from mammalian proteomes (Selbach *et al.*, 2009). Nevertheless, a screen for eukaryotic proteins containing EPIYA-like motifs revealed one protein, pragmin, that can be tyrosine-phosphorylated in epithelial cells and subsequently binds to Csk, which prevents Csk from inactivating Src, in contrast to CagA (Safari *et al.*, 2011).

Actin cytoskeleton rearrangements and the hummingbird phenotype

Focal adhesion proteins and the protein tyrosine phosphatase SHP-2

CagA tyrosine phosphorylation is a prerequisite for inducing a characteristic cell elongation phenotype (known as the hummingbird phenotype) in some epithelial cell lines such as AGS (Segal *et al.*, 1999). However, this phenotype, which is characterized by formation of thin cell protrusions that contain actin filaments and may be up to 100 µm long, is absent in most other gastric epithelial cell lines (Schneider *et al.*, 2011) and is thus probably only indicative of more subtle cytoskeletal changes in these other cells or *in vivo*. The hummingbird phenotype is supposed to represent a retraction defect caused by inhibition of focal contact disassembly (Bourzac *et al.*, 2007), or a general deregulation of focal adhesion complex turnover. Binding of the Cag secretion apparatus or of purified CagL to integrins on epithelial cells or fibroblasts induces formation of focal adhesion complexes and cell spreading accompanied by vinculin and actin reorganization, which depend on Src and FAK activation (Tegtmeyer *et al.*, 2010; Fig. 2.3B). It has also been reported that the focal adhesion protein paxillin becomes phosphorylated on tyrosine 118 in a *cag*PAI-, β1 integrin- and JNK-dependent manner (Snider *et al.*, 2008). Another study showed that the *H. pylori* outer membrane protein OipA is involved in tyrosine phosphorylation of paxillin at Tyr-31 and Tyr-118, whereas the *cag*PAI influences only paxillin Tyr-118 phosphorylation (Tabassam *et al.*, 2011). Paxillin phosphorylation was dependent on epidermal growth factor receptor (EGFR), phosphatidylinositol-3-kinase (PI3K)-Akt signalling, and FAK, and resulted in relocation of paxillin to focal adhesions and in stress fibre formation (Tabassam *et al.*, 2011; Fig. 2.3B). These data suggest that signalling processes induced by infection may result in formation of focal adhesion complexes. On the other hand, the feedback loop leading to inhibition of Src kinase results in dephosphorylation of the focal adhesion protein vinculin (Fig. 2.3B), and was found to result in a strong reduction and redistribution of focal adhesion complexes, and thus in a loss of cell matrix adhesion, which might also contribute to the cytoskeletal changes observed (Moese *et al.*, 2007). Specifically, vinculin dephosphorylation impairs its interaction with the p34Arc component of the Arp2/3 complex, which normally provides a link between integrins and the actin cytoskeleton (Moese *et al.*, 2007). Tyrosine dephosphorylation of cortactin, another protein associated with focal adhesions, is also mediated by feedback inhibition of Src kinase in *H. pylori*-infected cells (Selbach *et al.*, 2003). This dephosphorylation is a requirement for induction of the hummingbird phenotype, possibly due to its effect on FAK activation (see next section).

Induction of the hummingbird phenotype critically depends on phosphorylation of EPIYA motifs C or D for Western and East Asian CagA variants, respectively (Higashi *et al.*, 2004; Moese *et al.*, 2007), but additional phosphorylation of either EPIYA-A or EPIYA-B seems to be required at least for Western strains (Mueller *et al.*, 2012). Interestingly, the hummingbird phenotype could also be induced by CagA mutants in which EPIYA motif tyrosine residues had been exchanged for aspartate residues, thus mimicking phosphorylated tyrosine residues. Since this was only partially the case when EPIYA-A, B and C tyrosines were simultaneously exchanged for aspartates, it was suggested that tyrosine phosphorylation of all three motifs in one CagA molecule is not efficient or does not occur (Mueller *et al.*, 2012). Coinfection experiments showed that strains containing CagA variants with only single phosphorylatable motifs, which by themselves are not sufficient to induce the hummingbird phenotype, could functionally complement each other, suggesting that the individual EPIYA motifs act independently of each other after phosphorylation (Mueller *et al.*, 2012).

Because of the high number of possible phospho-CagA binding partners, it is difficult to estimate individual contributions for induction of cytoskeletal changes. Since Csk is a major interaction partner of phosphorylated EPIYA-A and EPIYA-B motifs, Csk-mediated Src inactivation and the above-mentioned consequences are likely to play an important role. On the other hand, the protein-tyrosine phosphatase SHP-2, which was originally identified as a CagA target molecule in gastric epithelial cells transfected with a *cagA* expression construct and which binds to phosphorylated EPIYA-C or EPIYA-D motifs, has also been shown to be required for inducing the hummingbird phenotype (Higashi *et al.*, 2002b). Binding of phosphorylated CagA stimulates SHP-2 phosphatase activity and recruits SHP-2 to the membrane, which alone might be sufficient to induce the hummingbird phenotype, even in the absence of CagA (Higashi *et al.*, 2002b). For the interaction with SHP-2, the above-described CagA dimerization seems to be a requirement (Higashi *et al.*, 2002a; Fig. 2.3A). In line with this, forced homodimerization of CagA has been shown to increase SHP-2 binding (Nagase *et al.*, 2011). An analogous interaction has been reported between CagA produced in a transgenic *Drosophila* model and the *Drosophila* SHP-2 homologue Corkscrew. As a consequence of this interaction, CagA is able to mimic and to partially substitute the *Drosophila* Gab adaptor protein Daughter of Sevenless (Botham *et al.*, 2008). Since SHP-2 acts downstream of several receptor tyrosine kinases with oncogenic potential and mutations of SHP-2 with increased activities are involved in many leukaemias (Östman *et al.*, 2006), the CagA–SHP-2 interaction has been suggested to be important for gastric cancer development (Hatakeyama, 2004). This is also supported by the fact that CagA variants which confer a higher risk of developing gastric cancer are more capable of interacting with SHP-2 (Naito *et al.*, 2006).

Role of focal adhesion kinase in cytoskeletal rearrangements

In *cagA*-transfected cells, one substrate of SHP-2 which becomes dephosphorylated as a consequence of the CagA–SHP-2 interaction is focal adhesion kinase (Tsutsumi *et al.*, 2006; Fig. 2.3B). FAK dephosphorylation at tyrosine residues 397 and 576 results in inhibition of FAK kinase activity (Tsutsumi *et al.*, 2006). However, the results of different studies suggest that *H. pylori* infection results in activation, rather than inactivation, of FAK (Kwok *et al.*, 2007; Tabassam *et al.*, 2008; Tegtmeyer *et al.*, 2010). Several mechanisms might be involved; apart from *cag*-dependent effects, the outer membrane protein OipA has been shown to induce Src activation and FAK tyrosine 397 phosphorylation via EGFR signalling (Tabassam *et al.*, 2008), finally resulting in activation of MEK-Erk signalling. Interestingly, *H. pylori* infection of AGS cells induces serine phosphorylation of cortactin at Ser-405, besides cortactin tyrosine dephosphorylation due to feedback inhibition of Src kinase. Cortactin serine phosphorylation, in turn, results in interaction of cortactin with FAK, promoting FAK kinase activity via phosphorylation at tyrosine residues 397, 407 and 576 (Tegtmeyer *et al.*, 2011; Fig. 2.3B). Interaction between cortactin and FAK was found to be dependent on the cortactin SH3 domain and a distinct PXXP motif in FAK. Serine 405 phosphorylation of cortactin was independent of CagA (Tegtmeyer *et al.*, 2011), which might explain why different results were obtained with ectopic expression of CagA after transfection. Nevertheless, in both cases the observed effects were involved in eliciting the hummingbird phenotype. When cells were treated with a FAK inhibitor during infection, the hummingbird phenotype was significantly impaired (Tegtmeyer *et al.*, 2011). On the other hand, development of the hummingbird phenotype could also be blocked in *cagA*-transfected cells by using a constitutively active FAK mutant (Tsutsumi *et al.*, 2006). Furthermore, overexpression of FAK in *H. pylori*-infected cells resulted in inhibition of cell elongation (Suzuki *et al.*, 2005), collectively suggesting that different ways of deregulating cellular signalling pathways might result in the same phenotype. Apart from the cellular proteins described above, the hummingbird phenotype requires the activity of vasodilator-stimulated phosphoprotein (VASP) (Knauer *et al.*, 2008), and an activation of protein kinase C-δ which is partly mediated by CagA and which results in signalling via the MEK-Erk pathway (Brandt *et al.*, 2009). It is not clear how activation of Erk contributes to cytoskeletal rearrangements, but one possibility is Erk-mediated phosphorylation of FAK or paxillin, as described above. A further pathway that is probably involved depends on binding of phosphorylated CagA to Crk adaptor proteins (see below), which has been shown to activate the small GTPase Rac1 (Suzuki *et al.*, 2005; Fig. 2.3B). Rac1 activation, which has also been shown to be required for induction of the hummingbird phenotype (Brandt *et al.*, 2007), may in turn stimulate actin polymerization via Wiskott-Aldrich syndrome protein family verprolin homologue (WAVE) and the Arp2/3 complex.

Interdependence of CagA and the vacuolating cytotoxin VacA

Because of the differential influence of CagA and other bacterial factors on activation of host cell kinases, and possibly on other cellular processes, it is not too surprising that controversial results have been obtained in *cagA*-transfected versus *H. pylori*-infected cells. Such differences may also account for the finding that interaction of phosphorylated CagA with SHP-2 was very strong in cells transfected with *cagA* expression vectors (Higashi *et al.*, 2002b), albeit dependent on the number and characteristics of the EPIYA motifs (Naito *et al.*, 2006), but comparably weak during infection of cells with *H. pylori* (Selbach *et al.*, 2003; Kikuchi *et al.*, 2012). Another bacterial virulence factor, which may have a major impact on CagA-associated phenotypes, is the vacuolating cytotoxin VacA. It has been realized for a long time that *H. pylori* strains producing highly active VacA molecules (*vacA* genotype s1/i1/m1) usually carry the *cag*PAI, whereas strains with less active VacA types are mostly *cag*PAI-deficient, suggesting a functional interdependence between both factors. Although early studies suggested that *cagA* is not required for VacA-induced vacuolation (Schmitt and Haas, 1994; Tummuru *et al.*, 1994), more recent studies showed that, on the contrary, *cagA* mutants elicit an even stronger VacA-induced vacuolation, and, vice versa, *vacA* mutants cause a more pronounced hummingbird phenotype (Argent *et al.*, 2008b). A transfected C-terminal CagA fragment including the EPIYA region was able to inhibit VacA-induced apoptosis and vacuolation (Oldani *et al.*, 2009), possibly due to phosphorylation-dependent inactivation of Src-family kinases and a concomitant inhibition of VacA endocytosis (Oldani *et al.*, 2009). Thus, infection of MKN28 or AGS cells with *cagA* mutants resulted in significantly increased apoptosis rates as compared to wild-type infections (Oldani *et al.*, 2009; Akada *et al.*, 2010). Generally, CagA seems to prevent clathrin-independent internalization mechanisms such as the uptake of the cholera toxin B subunit (Akada *et al.*, 2010). Conversely, it has been shown that VacA and CagA have an inverse influence on activation of the transcription factor nuclear factor of activated T cells (NFAT). Whereas VacA treatment of T cells results in inhibition of NFAT nuclear translocation (Gebert *et al.*, 2003), transfected CagA clearly induces NFAT activity in epithelial cells (Yokoyama *et al.*, 2005). Although this effect could be reproduced in *H. pylori*-infected cells, this was only possible using *vacA* mutants (Yokoyama *et al.*, 2005), again indicating differences between transfected and infected cells. The inhibitory effect of VacA on induction of the hummingbird phenotype was not dependent on vacuole formation, but required the presence of the VacA receptor RPTP-α (Tegtmeyer *et al.*, 2009). For this effect, reduced activation of the receptor tyrosine kinases EGFR and HER2/Neu in the presence of VacA, and a concomitant decrease in MEK-Erk signalling, were made responsible (Tegtmeyer *et al.*, 2009). It is not clear, however, what determines the ability of certain *H. pylori* strains, but not of others, to produce sufficient amounts of active VacA to suppress the hummingbird phenotype. Taken together, these observations suggest that VacA and CagA are able to reduce each other's effects on target cells, possibly to prevent too much damage.

Disruption of cell–cell junctions and cell migration

Induction of cell motility and disruption of tight junctions

Apart from eliciting the hummingbird phenotype, epithelial cell lines infected with *H. pylori* display a scattering phenotype similar to the one induced by hepatocyte growth factor (HGF) (Segal *et al.*, 1999). This phenotype involves a dissociation of cell–cell contacts and increased cell motility. In early studies with AGS cells, the motility phenotype was found to be concomitant with, but independent of, the hummingbird phenotype (Moese *et al.*, 2004). Moreover, it was shown to be *cag*PAI-dependent, but only partially CagA-dependent (Churin *et al.*, 2001; Al-Ghoul *et al.*, 2004), suggesting that both interactions of translocated CagA and signalling pathways induced by binding of the Cag secretion apparatus play a role. Motility induction in infected AGS cells depends on activation of the c-Met receptor tyrosine kinase and of phospholipase C-γ (PLC-γ) and involves downstream MEK-Erk signalling (Churin *et al.*, 2003). Moreover, non-phosphorylated CagA was found to form a complex with c-Met and PLC-γ (Fig. 2.3A).

However, AGS cells do not produce E-cadherin and do not form tight junctions or adherens junctions, both of which would have to be disrupted in order to render cells in polarized epithelial monolayers motile. Nevertheless, *H. pylori* is able to disrupt tight junctions and adherens junctions in polarized cells as well. Several tight junction proteins, including occludin and claudins (Fig. 2.4A), are either degraded or mislocalized to the cytosol upon *H. pylori* infection, resulting in a significant decrease of the transepithelial resistance. For these effects, bacterial urease production and an activation of myosin light-chain kinase (Lytton *et al.*, 2005; Fedwick *et al.*, 2005; Wroblewski *et al.*, 2009), or the vacuolating cytotoxin VacA (Papini *et al.*, 1998) have been reported to play a role (Fig. 2.4B). However, when *H. pylori* is used to infect polarized canine kidney (MDCK) cells, bacteria are preferably associated with apical tight junctions, and this association depends on CagA translocation (Amieva *et al.*, 2003). It involves a colocalization of CagA with the tight junction proteins zonula occludens protein 1 (ZO-1) and junctional adhesion molecules (JAMs; Fig. 2.4B), and a subsequent disruption of tight junctions, but is independent of CagA tyrosine phosphorylation (Amieva *et al.*, 2003). This effect, which can be reproduced by transfection of MDCK cells with *cagA* expression vectors, results in a loss of cell polarity, an extrusion of cells from the monolayer, enhanced cell migration, and an invasive behaviour of the migrating cells, reminiscent of cells that have undergone an epithelial–mesenchymal transition, as observed during development or cancer progression (Bagnoli *et al.*, 2005). Similar morphological changes including redistribution of the tight junction protein ZO-1 have also been observed in polarized T84 cells (El Etr *et al.*, 2004) and in primary gastric epithelial cells (Krueger *et al.*, 2007) infected with *H. pylori*. Although targeting of CagA to the membrane and to cell junctions is determined by N-terminal regions of the protein, an N-terminal fragment alone was not sufficient to induce a loss of cell polarity and cell migration (Bagnoli *et al.*, 2005).

The requirement of a C-terminal CagA region for eliciting cell motility could be explained by the finding that CagA interacts with partitioning-defective/microtubule affinity-regulating kinase (Par-1b/MARK2) (Saadat *et al.*, 2007; Zeaiter *et al.*, 2007). In *H. pylori*-infected AGS cells, MARK2 is recruited to the membranes close to bacterial attachment sites in a CagA-dependent manner (Zeaiter *et al.*, 2007). The interaction between CagA and MARK2 results in inhibition of MARK2 kinase activity, in its mislocalization

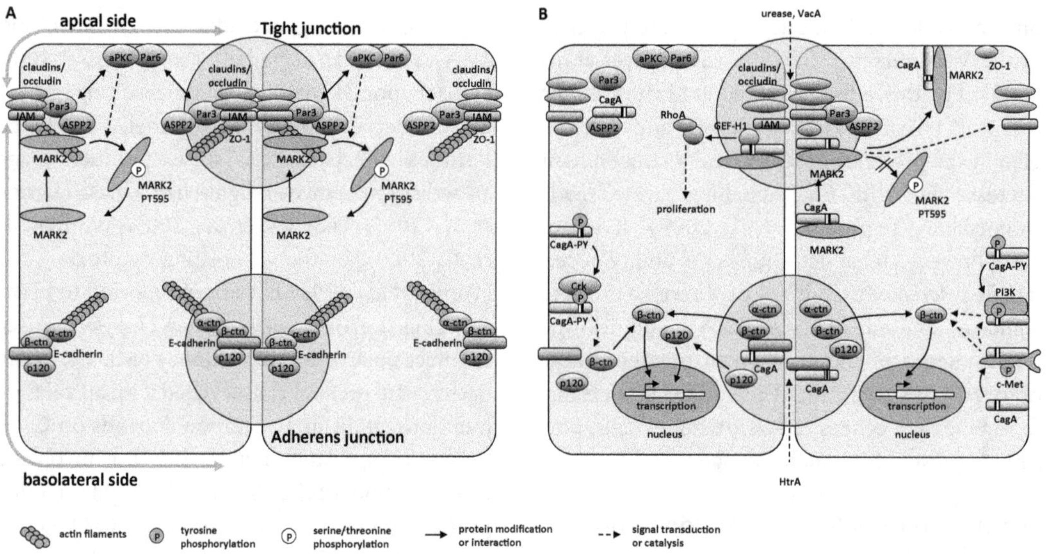

Figure 2.4 Disruption of cell–cell junctions by *H. pylori*. (A) Two polarized cells forming tight junctions and adherens junctions are depicted. Tight junctions are composed of the junctional proteins claudin, occludin and junctional adhesion molecule (JAM) as well as associated proteins such as *zonula occludens* Protein 1 (ZO-1), and are connected to the actin cytoskeleton. Cell polarity depends on a complex of Par proteins (Par3 and Par6) with atypical protein kinase C (aPKC) and possibly ASPP2. MARK2 kinase is excluded from apical membranes by aPKC-dependent phosphorylation at Thr-595, which causes its relocation to the cytosol. Adherens junctions consist of E-cadherin and the catenins (ctn) α-catenin, β-catenin and p120-catenin, and are also connected to the actin cytoskeleton. Cytoplasmic pools of β-catenin are rapidly phosphorylated by a cytoplasmic destruction complex and subsequently degraded by the proteasome (not depicted). (B) In *H. pylori*-infected polarized cells, non-phosphorylated CagA interacts with MARK2 and causes inactivation of MARK2 kinase activity as well as an inhibition of MARK2 Thr-595 phosphorylation, resulting in mislocalization to the apical membrane and general disruption of cell polarity. The CagA–MARK2 complex may interact at tight junctions with the guanine-nucleotide exchange factor GEF-H1, causing RhoA activation and cell proliferation. CagA also binds to ASPP2, but the influence of this interaction on cell polarity is unknown. Phosphorylation-independent binding of CagA to E-cadherin or phosphorylation-dependent signalling induced by CagA–Crk interaction causes adherens junction disruption and release of β-catenin and p120-catenin. Binding of CagA via its MKI motifs to phosphorylated c-Met and/or of phosphorylated CagA to PI3K results in PI3K/Akt signalling and thus accumulation of cytosolic β-catenin, which is then able to translocate to the nucleus and activate gene expression. Tight junctions may also be disrupted by urease or the vacuolating cytotoxin VacA, and adherens junctions by HtrA-mediated proteolysis and shedding of E-cadherin.

to apical membranes (Fig. 2.4B) and in a general loss of cell polarity (Saadat *et al.*, 2007). Loss of cell polarity is also reflected in mislocalization of the tight junction protein ZO-1 (Zeaiter *et al.*, 2007), and in a reduced transepithelial resistance of polarized MDCK monolayers (Lu *et al.*, 2008). MARK2 is usually located at the basolateral membrane and actively excluded from apical membrane regions due to phosphorylation at Thr-595 when reaching tight junctions. This phosphorylation is catalysed by atypical protein kinase C (aPKC) variants such as PKC-ζ located apically or at the junctions in a complex with several Par proteins, and results in relocation of MARK2 to the cytoplasm (Fig. 2.4A). In CagA-expressing cells (Fig. 2.4B), MARK2 phosphorylation at Thr-595 is reduced, suggesting that aPKC is unable to phosphorylate MARK2 in complex with CagA and thus to exclude it from the apical membrane, which might be sufficient for the observed loss of cell polarity (Saadat *et al.*, 2007).

Interestingly, CagA was also shown to interact with the tumour suppressor protein ASPP2 (Buti *et al.*, 2011), which is also predominantly located at tight junctions (McCaffrey and Macara, 2011; Fig. 2.4B). This interaction leads to binding of

ASPP2 to the tumour suppressor p53, which is subsequently degraded in the proteasome, resulting in an antiapoptotic phenotype (Buti *et al.*, 2011). However, an impact of the CagA–ASPP2 interaction on the function of tight junctions has not been examined.

CagA MKI motifs as phosphorylation-independent interaction modules

The interaction of CagA with MARK2 was found to be independent of CagA phosphorylation, but nevertheless dependent on the presence of the EPIYA region, which led to the discovery of the MKI motifs (Saadat *et al.*, 2007; Figs. 2.1A,B). The designation MKI (MARK2 kinase inhibitor) motifs derives from the fact that binding of CagA inhibits the kinase activity of MARK2 by mimicking kinase substrates in a similar way as eukaryotic kinase inhibitors do (Nešić *et al.*, 2010). Alternative names are the original designation CagA multimerization (CM) motifs (Ren *et al.*, 2006; Saadat *et al.*, 2007), or conserved repeats responsible for phosphorylation-independent activity (CRPIA) motifs (Suzuki *et al.*, 2009). Recently, the crystal structure of a C-terminal CagA fragment, ranging from amino acids 885 to 1005, bound to MARK2 has been solved (Nešić *et al.*, 2010). The major part of this CagA fragment including the EPIYA motifs was unstructured, only the 14 residues of the MKI motif had an interpretable electron density, forming an extended coil in the substrate-binding site of MARK2. Interestingly, the binding kinetics between CagA and MARK2, and in turn the binding efficiency of CagA to SHP-2, are influenced by the intramolecular interaction between the N-terminal and C-terminal binding sequences, suggesting that the putative C-terminal lariat structure described earlier is crucial for exposing the MKI and/or EPIYA motifs (Hayashi *et al.*, 2012). Another interesting observation was that each of the two MKI motifs of a single CagA fragment interacted with one MARK2 molecule (Nešić *et al.*, 2010; Fig. 2.3A). This seems to be contradictory to the previously mentioned finding that CagA dimerizes due to interaction with dimeric MARK2, and that this complex might be necessary, after CagA phosphorylation, for a productive interaction with the two SH2 domains of SHP-2 (Saadat *et al.*, 2007). Therefore, it has to be clarified how East Asian CagA variants, which usually have only one MKI motif, differ from Western CagA variants, which have at least two MKI motifs, with respect to MARK2 complex formation and dimerization. Differential binding of Western and East Asian MKI sequences to MARK2, and, as a consequence, also to SHP-2, has indeed been described (Lu *et al.*, 2008). Intriguingly, MKI motif amino acids show clear signs of positive selection in East Asian CagA variants (Torres-Morquecho *et al.*, 2010), suggesting that different interaction characteristics might indeed be relevant for Western and East Asian MKI sequences. Particularly, one amino acid substitution (K955G; Fig. 2.1B) which shows a better inhibition of MARK2 kinase activity (Nešić *et al.*, 2010), underlies positive selection in East Asian strains (Furuta *et al.*, 2011).

As detailed below, the CagA MKI motifs do not only mediate binding to Par-1b/MARK2 and to other Par-1 isoforms (Lu *et al.*, 2009), but also to c-Met receptor tyrosine kinase (Suzuki *et al.*, 2009). Thus, these motifs represent a second, phosphorylation-independent protein–protein interaction module in addition to the EPIYA motifs (Fig. 2.1C). Given that both interaction modules are located in close proximity on the CagA primary sequence, a challenging question for future research will be how the potential binding partners compete with each other, or how sequential binding of different proteins influences the functional outcome *in vivo*.

Consequences of the CagA–MARK2 interaction

Although the interaction between CagA and Par-1b/MARK2 or its isoforms is well-documented, the significance of MARK kinase inhibition is less clear. In accordance with the fact that MARK2 activity affects microtubule dynamics by phosphorylation of several microtubule-associated proteins, the CagA interaction could be shown to have a destabilizing effect on microtubules (Zeaiter *et al.*, 2007; Lu *et al.*, 2009), which may in turn influence the actin cytoskeleton. Interestingly, overexpression of MARK2 together with CagA suppressed induction of the hummingbird phenotype in AGS cells in a MARK2 kinase activity-dependent manner, despite the fact that this increased binding to SHP-2 (Saadat *et al.*, 2007). On the other hand, knockdown of MARK2 resulted in enhanced hummingbird formation (Lu *et al.*, 2009). Therefore, the CagA–MARK2 interaction might also play a role in the cytoskeletal rearrangements described above, either by interference with microtubule dynamics or by modulating binding to further contributing factors such as SHP-2. However, since the central phosphatidylserine-binding motif of CagA is required for MARK2 inhibition and disruption of polarity in polarized cells, but not for induction of the hummingbird phenotype in (non-polarized) AGS cells (Murata-Kamiya *et al.*, 2010), it is not clear if interaction of CagA with MARK2 might have different consequences in non-polarized cells. Considerable differences between non-polarized and polarized cells have also been described with respect to cell proliferation. Non-polarized MKN28 gastric epithelial cells transfected with *cagA* expression constructs showed an up-regulation of the cell cycle inhibitor $p21^{WAF1/Cip1}$ and were restricted in their proliferation, whereas polarized MDCK cells expressing CagA showed the opposite behaviour (Saito *et al.*, 2010). At the tight junctions of these cells, the MARK2–CagA complex was shown to interact with the RhoA guanine nucleotide exchange factor GEF-H1 (Fig. 2.4B). This interaction resulted in activation of RhoA, which in turn led to suppression of $p21^{WAF1/Cip1}$ and thus to enhanced cell proliferation of the migrating cells (Saito *et al.*, 2010).

CagA has thus a massive influence on host cell functions and epithelial integrity, but the significance of these changes for the bacteria has been less clear, apart from the assumption that loss of epithelial integrity may provide the bacteria with nutrients (Wessler and Backert, 2008). Recent studies have experimentally addressed this question by examination of *H. pylori*-epithelia cocultures. *H. pylori* cells adherent to cell–cell junctions at the apical side of polarized MDCK cells replicate with a mean doubling time of about 4 hours and form microcolonies (Tan *et al.*, 2009). In comparison with wild-type bacteria, *cagA* mutants display a 100-fold lower ability to replicate under these conditions. Interestingly, this growth defect was not observable at the basolateral side, or when cell polarity was impaired by blocking atypical protein kinase C-ζ or MARK2 (Tan *et al.*, 2009), suggesting that disruption of cell polarity due to the CagA–MARK2 interaction has a direct impact on bacterial survival. An important nutrient gained by disruption of cell polarity seems to be iron in the form of fully iron-loaded (holo-) transferrin bound to transferrin receptor, which was internalized more at the basolateral side and aberrantly transcytosed to the apical side upon infection (Tan *et al.*, 2011). However, this phenomenon seems to be significantly dependent on VacA, raising the question if it accounts for the reduced ability of *cagA* mutants to grow at the apical surface. Nevertheless, an *H. pylori cagA* mutant had a severe colonization defect in Mongolian gerbils under conditions of iron depletion (Tan *et al.*, 2011).

Modulation of adherens junctions and β-catenin signalling

Aside from tight junctions, epithelial integrity depends on adherens junctions, protein complexes composed mainly of E-cadherin and several catenins (α-catenin, β-catenin, p120-catenin), in which homophilic interactions between the E-cadherin extracellular domains mediate cell–cell attachment (Harris and Tepass, 2010; Fig. 2.4A). Intriguingly, *H. pylori* infection leads to disruption of adherens junctions as well (Conlin *et al.*, 2004; Suzuki *et al.*, 2005). The first mechanism suggested to be responsible for adherens junction disruption involved binding of CagA to Crk adaptor proteins (Suzuki *et al.*, 2005). The Crk adaptor proteins Crk-I, Crk-II and Crk-L could be co-precipitated with CagA from infected cells, depending on CagA phosphorylation (Suzuki *et al.*, 2005). These interactions may include the kinase c-Abl, which was shown to form a complex with phosphorylated CagA and Crk proteins at later time points of infection, resulting in Crk phosphorylation and c-Abl relocalization to focal adhesion complexes (Poppe *et al.*, 2007; Tammer *et al.*, 2007). Interaction of CagA with Crk adaptor proteins, and a subsequent activation of the MEK-Erk pathway and the GTPase Rac1 are required for the motility phenotype induced by *H. pylori* (Suzuki *et al.*, 2005). Concomitantly, a diffuse cytoplasmic staining of β-catenin and of

E-cadherin is observed, indicating a disruption of adherens junctions. Moreover, β-catenin also translocates to the nucleus (Fig. 2.4B), an observation that was also made in gastric epithelia of patients infected with *cag*-positive *H. pylori* strains (Franco *et al.*, 2005). In line with this, an adapted *H. pylori* strain reisolated after Mongolian gerbil infection, which was found to translocate higher amounts of CagA, also activated translocation of β-catenin into the nucleus very efficiently (Franco *et al.*, 2005). However, it should be noted that CagA does not seem to be the only factor involved, since β-catenin signalling independent of CagA could be observed in MDCK cells (Sokolova *et al.*, 2008), and both *oipA* mutants and *vacA* mutants were also defective in inducing nuclear translocation of β-catenin in AGS or AZ-521 gastric epithelial cells, respectively (Franco *et al.*, 2008; Nakayama *et al.*, 2009).

CagA expressed in cells after transfection has furthermore been shown to interact directly with E-cadherin in a tyrosine phosphorylation-independent manner and to release β-catenin due to this interaction (Murata-Kamiya *et al.*, 2007). Release of β-catenin from adherens junctions could also be observed in AGS gastric epithelial cells stably transfected with an E-cadherin expression construct (Suzuki *et al.*, 2009). However, β-catenin activation was also found upon infection of non-transfected AGS cells, or of other cells which do not express E-cadherin and consequently do not form adherens junctions (Suzuki *et al.*, 2009), suggesting that further pathways of β-catenin activation exist. This adherens junction-independent β-catenin translocation to the nucleus does not require CagA phosphorylation. Surprisingly, it involves binding of CagA to the phosphorylated c-Met receptor tyrosine kinase, which is also mediated by CagA's MKI motifs (Figs. 2.1C, 2.3A and 2.4B), and which results in activation of a phosphatidylinositol-3-kinase (PI3K)-Akt signalling pathway (Suzuki *et al.*, 2009). Direct binding of tyrosine-phosphorylated CagA to PI3K was also observed and might contribute to activation of Akt signalling as well (Figs. 2.1C, 2.3A and 2.4B). Activation of PI3K/Akt signalling by CagA is thus similar to Wnt signalling which is known to prevent phosphorylation of cytoplasmic β-catenin and its subsequent targeting to the proteasome (Clevers, 2006). Unphosphorylated β-catenin accumulating due to the CagA effects can then translocate to the nucleus and activate T cell factor/lymphoid enhancer factor (TCF/LEF) transcription factors (Franco *et al.*, 2005; Suzuki *et al.*, 2009). CagA-dependent β-catenin signalling might be further enhanced by binding of β-catenin to parafibromin, which occurs after tyrosine dephosphorylation of parafibromin by active SHP-2, and also prevents β-catenin from being degraded (Takahashi *et al.*, 2011).

An important feature of cells that have undergone an epithelial-mesenchymal transition is their ability to become invasive through the extracellular matrix. This phenotype, which has been demonstrated for *H. pylori*-infected epithelial cells (Bagnoli *et al.*, 2005), involves degradation of matrix components by matrix metalloproteinases (MMP). Such an invasive phenotype of AGS cells has been shown to depend on c-Met activation by CagA, and an associated up-regulation of MMP-2 and MMP-9 (Oliveira *et al.*, 2006). However, in transgenic AGS cells producing E-cadherin, the invasive phenotype was suppressed, although a complex of CagA, c-Met, E-cadherin and p120-catenin could be found in these cells (Oliveira *et al.*, 2009). Cag-dependent signalling is probably also involved in MMP-7 induction (Wroblewski *et al.*, 2003), which seems to be partly based on aberrant activation of p120-catenin from adherens junctions, although CagA plays only a minor role in this process (Ogden *et al.*, 2008). Furthermore, p120-catenin is involved together with β-catenin in CagA-dependent activation of the transcription factor peroxisome proliferator-activated receptor δ (PPAR-δ) (Nagy *et al.*, 2011), indicating that both β-catenin and p120-catenin play a role in adherens junction-dependent gene expression induced by *H. pylori* (Fig. 2.4B).

Apart from these mechanisms, *H. pylori* causes a disruption of adherens junctions which is not dependent on the *cag*PAI (Weydig *et al.*, 2007). This process, which does not induce β-catenin signalling, is mediated by the secreted protease HtrA and characterized by proteolytic shedding of E-cadherin extracellular domains (Hoy *et al.*, 2010). Changes in the epithelial barrier function might be favourable for the bacteria with respect to nutrient acquisition, but might also enable the transport of bacterial virulence factors such as VacA, or even of whole bacteria into the gastric submucosa (Necchi *et al.*, 2007; Dubois and Borén, 2007). Finally, opening of intercellular junctions might be a means to provide access for the type IV secretion apparatus to integrin receptors, which are usually located at the basolateral side of polarized cells (Wessler and Backert, 2008).

Other CagA interactions and associated signalling pathways

As mentioned above, both the Cag type IV secretion apparatus (via binding to integrin receptors) and translocated CagA can induce multiple signalling pathways which may be interdependent. One example for *cag*PAI- and CagA-dependent convergence of pathways is proinflammatory signalling leading to NF-κB activation and IL-8 secretion. Although the major part of NF-κB activation in *H. pylori*-infected cells is mediated by the (functional) secretion apparatus and is independent of CagA translocation, the presence of CagA in the cell after transfection or CagA accumulated by prolonged infection has an additional NF-κB activation effect which is not based on CagA tyrosine phosphorylation (Brandt *et al.*, 2005; Kim *et al.*, 2006). In transfected cells, this activation depends on a Ras-Raf-MEK-Erk signalling pathway and thus probably on Grb2 interaction (see below), but is independent of c-Met or SHP-2 (Brandt *et al.*, 2005). However, activation of transforming growth factor β-activated kinase 1 (TAK1) by binding to CagA (Lamb *et al.*, 2009), or the c-Met interaction mediated by the CagA MKI motifs (Suzuki *et al.*, 2009), might also play a role. Apart from TCF/LEF and NF-κB activation, c-Met signalling also leads to activation of the transcription factors SRE, SRF, NFAT and AP-1 (Hirata *et al.*, 2002; Suzuki *et al.*, 2009). Interestingly, all these signalling pathways were found to be much less induced by American Indian strains, particularly by strains containing AM-II-type CagA variants (Suzuki *et al.*, 2011).

The adaptor protein Grb2 was originally shown to interact with the CagA EPIYA region independently of tyrosine phosphorylation (Mimuro *et al.*, 2002); this interaction results in sustained activation of the MEK/Erk pathway (Fig. 2.3A). However, Grb2 was also identified as an interaction partner of phosphorylated EPIYA motifs B and C (Selbach *et al.*, 2009; Fig. 2.1C), suggesting that different Grb2-dependent responses are possible. In the Mongolian gerbil model, the presence of CagA suppresses apoptosis of epithelial cells induced by the drugs etoposide or staurosporine. This suppression is dependent on Grb2, the MEK-Erk pathway and partially on Crk adaptors, and leads to increased expression of the antiapoptotic protein MCL1 (Mimuro *et al.*, 2007). Interestingly, this antiapoptotic activity of CagA results in increased thickness of gastric pits and in higher colonization rates (Mimuro *et al.*, 2007). Gastric pit cells of *H. pylori*-infected gerbils also showed increased PI3K-Akt signalling (Suzuki *et al.*, 2009). In addition, antiapoptosis might be mediated by several other pathways (Choi *et al.*, 2003; Yanai *et al.*, 2003; Zhu *et al.*, 2007; Yan *et al.*, 2009). In contrast to these antiapoptotic effects, CagA-dependent up-regulation of spermine oxidase, which has been shown *in vitro* and *in vivo* in human biopsies and infected animals, is concomitant with an increase of apoptosis and also of DNA damage (Chaturvedi *et al.*, 2011). Nevertheless, a concurrent increase in a cell subpopulation from Mongolian gerbils showing increased DNA damage, but low apoptotic activity, was taken as an indication that spermine oxidase activity might lead to precancerous changes (Chaturvedi *et al.*, 2011).

EGF receptor is another important mediator of *H. pylori*-induced signalling pathways. It can be transactivated during *H. pylori* infection by different mechanisms, which lead to activation of MEK-Erk signalling (Keates *et al.*, 2001; Wallasch *et al.*, 2002; Ashktorab *et al.*, 2007; Basu *et al.*, 2008; Beswick and Reyes, 2008). One mechanism involves activation of the ADAM-17 (A disintegrin and metalloprotease) protein and subsequent secretion of the EGFR ligand HB-EGF (Keates *et al.*, 2001; Wallasch *et al.*, 2002) and can be induced by the Cag secretion apparatus or by purified CagL (Saha *et al.*, 2010; Tegtmeyer *et al.*, 2010). This ADAM-17-mediated pathway was also shown to induce PI3K-Akt signalling and to have an antiapoptotic effect (Yan *et al.*, 2009). Transactivation of EGFR and MEK-Erk signalling results in an up-regulation of the antimicrobial peptide human β-defensin 3 (Boughan *et al.*, 2006; Bauer *et al.*, 2012). However, upon prolonged infection of epithelial cell lines, CagA induces increased EGFR surface localization (Bauer *et al.*, 2008), but reduced EGFR activation (Bauer *et al.*, 2012). This decrease in EGFR activation and signalling is concomitant with EGFR dephosphorylation and accounts for an inhibition of expression of human β-defensin 3 which may contribute to bacterial survival (Bauer *et al.*, 2012), although the ability of human β-defensin 3 to kill *H. pylori* is controversial (Grubman *et al.*, 2010). Dephosphorylation of EGF receptor does not depend on Src inhibition, but rather on CagA-dependent SHP-2 activation (Bauer *et al.*, 2012; Fig. 2.3A). On the other hand, SHP-2 interacting with CagA has also been found to induce sustained MEK-Erk activation which is independent of Grb2 or Ras (Higashi *et al.*, 2004).

CagA interference with signalling pathways has also an influence on cell cycle progression. As mentioned before, transfected CagA stimulates nuclear translocation of the transcription factor NFAT, resulting in up-regulation of the cell cycle inhibitor $p21^{WAF1/Cip1}$ and other proteins (Yokoyama *et al.*, 2005). Since this activity is dependent on the presence of the CagA EPIYA region, but not on tyrosine phosphorylation, NFAT activation might be induced by interaction with c-Met, as described above. On the other hand, CagA transfected into MKN28 gastric epithelial cells induces cell cycle progression factors such as cyclin D1 by β-catenin-dependent activation of the TCF/LEF transcription factors (Murata-Kamiya *et al.*, 2007). However, cell cycle inhibition was found to be dominant. In contrast, a similar pathway was responsible for induction of peroxisome proliferator-activated receptor δ (PPAR-δ) in MKN28 cells, which promoted cell proliferation by activation of cyclin D1 (Nagy *et al.*, 2011). As mentioned before, these effects may depend on cell polarity, with $p21^{WAF1/Cip1}$-dependent cell cycle inhibition occurring in non-polarized cells, and proliferation due to of $p21^{WAF1/Cip1}$ suppression in polarized cells (Saito *et al.*, 2010). Cell cycle inhibition has also been found in B cells transfected with a *cagA* expression construct, in this case by suppression of JAK-STAT signalling (Umehara *et al.*, 2003). In the Mongolian gerbil model, however, the JAK-STAT pathway has been shown to be activated in a CagA-dependent manner (Bronte-Tinkew *et al.*, 2009). The concurrent activation of the STAT3 transcription factor may provide a selective advantage to the bacteria by inducing the lectin REG3γ which is bactericidal for Gram-positive bacteria (Lee *et al.*, 2012), but JAK-STAT activation might also contribute to carcinogenesis. Finally, another CagA-interacting molecule is the tumour-suppressing transcription factor RUNX3, which becomes ubiquitinated and degraded after binding to CagA. This interaction involves an N-terminal region of CagA with similarities to a WW domain, and a proline-rich motif in the C-terminal region of RUNX3 (Tsang *et al.*, 2010; Fig. 2.1A). Transfected CagA is also able to down-regulate RUNX3 expression via MEK-Erk or p38 signalling (Liu *et al.*, 2012). Since RUNX3 also induces $p21^{WAF1/Cip1}$, these mechanisms may have a further carcinogenic potential.

Conclusions

In marked contrast to other protein-translocating type IV secretion systems, which have evolved to transport from several to over 100 different effector proteins (Alvarez-Martinez and Christie, 2009), CagA is still the only known protein substrate of the Cag type IV secretion system. However, CagA seems to contain numerous protein–protein interaction modules, particularly in its EPIYA region, allowing it to target many interaction partners and giving rise to a huge number of potential cellular effects. Thus, CagA differs from typical bacterial protein toxins, which often have highly specific enzymatic activities, in that it is able to deregulate many crucial cellular processes in a given target cell. It will be a major challenge for future studies to define the simultaneous or sequential use of different CagA interactions and their relative contributions to pathogenesis *in vivo*. Because CagA is a major virulence factor of *H. pylori* and significantly enhances cancer risk, it is also a promising target for specific pharmacological inhibitors. In this regard, it has already been shown that a peptide corresponding to the MKI motif was able to block corresponding CagA-mediated proinflammatory responses (Suzuki *et al.*, 2009). It is likely that the recently determined CagA structures can be used as a basis for more rational inhibitor design.

Originally, the CagA protein was recognized as an immunodominant antigen in *H. pylori*-infected patients (Covacci *et al.*, 1993; Tummuru *et al.*, 1993). A significant fraction of CD4-positive T cells isolated from infected patients with peptic ulcer disease was also shown to be CagA-specific (D'Elios *et al.*, 1997). MHC class II-restricted T cell epitopes could be mapped to four distinct central and C-terminal CagA regions (including a part of the EPIYA-B motif) in a C57BL/6 mouse model (Arnold *et al.*, 2011a). Interestingly, immunization of adult C57BL/6 mice with recombinant CagA did not protect against *H. pylori* infection, but caused strong T cell-mediated immune responses which were sufficient to elicit gastric pathology similar to that observed during *H. pylori* infection. In contrast, mice were protected from infection-induced gastric pathology when they had been immunized as neonates with bacterial sonicates or recombinant CagA, suggesting that tolerance induction may be more useful than immune activation as a protective measure against gastric cancer (Arnold *et al.*, 2011a). This also implies that gastric pathology, including development of cancer, might only partially be elicited directly by CagA or other bacterial factors, and that the various activities of CagA found in host cells might be rather extreme manifestations of the more subtle modulating effects that occur *in vivo*. Again, it will be an important task for future research to understand the role of individual CagA interactions and associated effects for *H. pylori* persistence and pathogenesis.

Acknowledgements

This work was supported by a grant from the Deutsche Forschungsgemeinschaft (SFB 914, project B5). We are grateful to Laurent Terradot for preparing Fig. 2.2 and for sharing data prior to publication, to Rainer Haas for helpful discussions and critical reading of the manuscript, and to Beate Kern, Verena Königer and Carolin Spicher for critical comments. We apologize to all authors whose work could not be cited.

References

Akada, J.K., Aoki, H., Torigoe, Y., Kitagawa, T., Kurazono, H., Hoshida, H., Nishikawa, J., Terai, S., Matsuzaki, M., Hirayama, T., *et al.* (2010). *Helicobacter pylori* CagA inhibits endocytosis of cytotoxin VacA in host cells. Dis. Model. Mech. *3*, 605–617.

Akopyan, K., Edgren, T., Wang-Edgren, H., Rosqvist, R., Fahlgren, A., Wolf-Watz, H., and Fallman, M. (2011). Translocation of surface-localized effectors in type III secretion. Proc. Natl. Acad. Sci. U.S.A. *108*, 1639–1644.

Akopyants, N.S., Clifton, S.W., Kersulyte, D., Crabtree, J.E., Youree, B.E., Reece, C.A., Bukanov, N.O., Drazek, E.S., Roe, B.A., and Berg, D.E. (1998). Analyses of the *cag* pathogenicity island of *Helicobacter pylori*. Mol. Microbiol. *28*, 37–53.

Aktories, K. (2011). Bacterial protein toxins that modify host regulatory GTPases. Nat. Rev. Microbiol. *9*, 487–498.

Al-Ghoul, L., Wessler, S., Hundertmark, T., Krüger, S., Fischer, W., Wunder, C., Haas, R., Roessner, A., and Naumann, M. (2004). Analysis of the type IV secretion system-dependent cell motility of *Helicobacter pylori*-infected epithelial cells. Biochem. Biophys. Res. Commun. *322*, 860–866.

Alvarez-Martinez, C.E., and Christie, P.J. (2009). Biological diversity of prokaryotic type IV secretion systems. Microbiol. Mol. Biol. Rev. *73*, 775–808.

Amieva, M.R., and El-Omar, E.M. (2008). Host–bacterial interactions in *Helicobacter pylori* infection. Gastroenterol. *134*, 306–323.

Amieva, M.R., Vogelmann, R., Covacci, A., Tompkins, L.S., Nelson, W.J., and Falkow, S. (2003). Disruption of the epithelial apical–junctional complex by *Helicobacter pylori* CagA. Science *300*, 1430–1434.

Andrzejewska, J., Lee, S.K., Olbermann, P., Lotzing, N., Katzowitsch, E., Linz, B., Achtman, M., Kado, C.I., Suerbaum, S., and Josenhans, C. (2006). Characterization of the pilin ortholog of the *Helicobacter pylori* type IV *cag* pathogenicity apparatus, a surface-associated protein expressed during infection. J. Bacteriol. *188*, 5865–5877.

Angelini, A., Tosi, T., Mas, P., Acajjaoui, S., Zanotti, G., Terradot, L., and Hart, D.J. (2009). Expression of *Helicobacter pylori* CagA domains by library-based construct screening. FEBS J. *276*, 816–824.

Aras, R.A., Fischer, W., Perez-Perez, G.I., Crosatti, M., Ando, T., Haas, R., and Blaser, M.J. (2003). Plasticity of repetitive DNA sequences within a bacterial (Type IV) secretion system component. J. Exp. Med. *198*, 1349–1360.

Argent, R.H., Kidd, M., Owen, R.J., Thomas, R.J., Limb, M.C., and Atherton, J.C. (2004). Determinants and consequences of different levels of CagA phosphorylation for clinical isolates of *Helicobacter pylori*. Gastroenterol. *127*, 514–523.

Argent, R.H., Hale, J.L., El-Omar, E.M., and Atherton, J.C. (2008a). Differences in *Helicobacter pylori* CagA tyrosine phosphorylation motif patterns between Western and East Asian strains, and influences on interleukin-8 secretion. J. Med. Microbiol. *57*, 1062–1067.

Argent, R.H., Thomas, R.J., Letley, D.P., Rittig, M.G., Hardie, K.R., and Atherton, J.C. (2008b). Functional association between the *Helicobacter pylori* virulence factors VacA and CagA. J. Med. Microbiol. *57*, 145–150.

Arnold, I.C., Hitzler, I., Engler, D., Oertli, M., Agger, E.M., and Müller, A. (2011a). The C-terminally encoded, MHC class II-restricted T cell antigenicity of the *Helicobacter pylori* virulence factor CagA promotes gastric preneoplasia. J. Immunol. *186*, 6165–6172.

Arnold, I.C., Lee, J.Y., Amieva, M.R., Roers, A., Flavell, R.A., Sparwasser, T., and Müller, A. (2011b). Tolerance rather than immunity protects from *Helicobacter pylori*-induced gastric preneoplasia. Gastroenterol. *140*, 199–209.

Ashktorab, H., Daremipouran, M., Wilson, M., Siddiqi, S., Lee, E.L., Rakhshani, N., Malekzadeh, R., Johnson, A.C., Hewitt, S.M., and Smoot, D.T. (2007). Transactivation of the EGFR by AP-1 is induced by *Helicobacter pylori* in gastric cancer. Am. J. Gastroenterol. *102*, 2135–2146.

Atherton, J.C. (2006). The pathogenesis of *Helicobacter pylori*-induced gastro-duodenal diseases. Annu. Rev. Pathol. *1*, 63–96.

Atherton, J.C., and Blaser, M.J. (2009). Coadaptation of *Helicobacter pylori* and humans: ancient history, modern implications. J. Clin. Invest. *119*, 2475–2487.

Avilés-Jiménez, F., Reyes-Leon, A., Nieto-Patlán, E., Hansen, L.M., Burgueño, J., Ramos, I.P., Camorlinga-Ponce, M., Bermúdez, H., Blancas, J.M., Cabrera, L., *et al.* (2012). *In vivo* expression of *Helicobacter pylori* virulence genes in patients with gastritis, ulcer, and gastric cancer. Infect. Immun. *80*, 594–601.

Azuma, T., Yamakawa, A., Yamazaki, S., Fukuta, K., Ohtani, M., Ito, Y., Dojo, M., Yamazaki, Y., and Kuriyama, M. (2002). Correlation between variation of the 3′ region of the *cagA* gene in *Helicobacter pylori* and disease outcome in Japan. J. Infect. Dis. *186*, 1621–1630.

Azuma, T., Yamazaki, S., Yamakawa, A., Ohtani, M., Muramatsu, A., Suto, H., Ito, Y., Dojo, M., Yamazaki, Y., Kuriyama, M., *et al.* (2004). Association between diversity in the Src homology 2 domain-containing tyrosine phosphatase binding site of *Helicobacter pylori* CagA protein and gastric atrophy and cancer. J. Infect. Dis. *189*, 820–827.

Backert, S., and Naumann, M. (2010). What a disorder: proinflammatory signaling pathways induced by *Helicobacter pylori*. Trends Microbiol. *18*, 479–486.

Backert, S., Moese, S., Selbach, M., Brinkmann, V., and Meyer, T.F. (2001). Phosphorylation of tyrosine 972 of the *Helicobacter pylori* CagA protein is essential for induction of a scattering phenotype in gastric epithelial cells. Mol. Microbiol. *42*, 631–644.

Backert, S., Fronzes, R., and Waksman, G. (2008). VirB2 and VirB5 proteins: specialized adhesins in bacterial type-IV secretion systems? Trends Microbiol. *16*, 409–413.

Backert, S., Tegtmeyer, N., and Selbach, M. (2010). The versatility of *Helicobacter pylori* CagA effector protein functions: The master key hypothesis. Helicobacter *15*, 163–176.

Backert, S., Clyne, M., and Tegtmeyer, N. (2011). Molecular mechanisms of gastric epithelial cell adhesion and injection of CagA by *Helicobacter pylori*. Cell Commun. Signal. *9*, 28.

Bagnoli, F., Buti, L., Tompkins, L., Covacci, A., and Amieva, M.R. (2005). *Helicobacter pylori* CagA induces a transition from polarized to invasive phenotypes in MDCK cells. Proc. Natl. Acad. Sci. U.S.A. *102*, 16339–16344.

Banta, L.M., Kerr, J.E., Cascales, E., Giuliano, M.E., Bailey, M.E., McKay, C., Chandran, V., Waksman, G., and Christie, P.J. (2011). An *Agrobacterium* VirB10 mutation conferring a type IV secretion system gating defect. J. Bacteriol. *193*, 2566–2574.

Basso, D., Zambon, C.F., Letley, D.P., Stranges, A., Marchet, A., Rhead, J.L., Schiavon, S., Guariso, G., Ceroti, M., Nitti, D., *et al.* (2008). Clinical relevance of *Helicobacter pylori cagA* and *vacA* gene polymorphisms. Gastroenterol. *135*, 91–99.

Basu, S., Pathak, S.K., Chatterjee, G., Pathak, S., Basu, J., and Kundu, M. (2008). *Helicobacter pylori* protein HP0175 transactivates epidermal growth factor receptor through TLR4 in gastric epithelial cells. J. Biol. Chem. *283*, 32369–32376.

Bauer, B., Bartfeld, S., and Meyer, T.F. (2008). *H. pylori* selectively blocks EGFR endocytosis via the non-receptor kinase c-Abl and CagA. Cell. Microbiol. *11*, 156–169.

Bauer, B., Pang, E., Holland, C., Kessler, M., Bartfeld, S., and Meyer, T.F. (2012). The *Helicobacter pylori* virulence effector CagA abrogates human β-defensin 3 expression via inactivation of EGFR signaling. Cell Host Microbe *11*, 576–586.

Beswick, E.J., and Reyes, V.E. (2008). Macrophage migration inhibitory factor and interleukin-8 produced by gastric epithelial cells during *Helicobacter pylori* exposure induce expression and activation of the epidermal growth factor receptor. Infect. Immun. *76*, 3233–3240.

Binns, A.N., Beaupré, C.E., and Dale, E.M. (1995). Inhibition of VirB-mediated transfer of diverse substrates from *Agrobacterium tumefaciens* by the IncQ plasmid RSF1010. J. Bacteriol. *177*, 4890–4899.

Botham, C.M., Wandler, A.M., and Guillemin, K. (2008). A transgenic *Drosophila* model demonstrates that the *Helicobacter pylori* CagA protein functions as a eukaryotic Gab adaptor. PLoS Pathog. *4*, e1000064.

Boughan, P.K., Argent, R.H., Body-Malapel, M., Park, J.H., Ewings, K.E., Bowie, A.G., Ong, S.J., Cook, S.J., Sorensen, O.E., Manzo, B.A., *et al.* (2006). Nucleotide-binding oligomerization domain-1 and epidermal growth factor receptor: critical regulators of β-defensins during *Helicobacter pylori* infection. J. Biol. Chem. *281*, 11637–11648.

Bourzac, K.M., Botham, C.M., and Guillemin, K. (2007). *Helicobacter pylori* CagA induces AGS cell elongation through a cell retraction defect that is independent of Cdc42, Rac1, and Arp2/3. Infect. Immun. *75*, 1203–1213.

Brandt, S., Kwok, T., Hartig, R., König, W., and Backert, S. (2005). NF-κB activation and potentiation of proinflammatory responses by the *Helicobacter pylori* CagA protein. Proc. Natl. Acad. Sci. U.S.A. *102*, 9300–9305.

Brandt, S., Shafikhani, S., Balachandran, P., Jin, S., Hartig, R., König, W., Engel, J., and Backert, S. (2007). Use of a novel coinfection system reveals a role for Rac1, H-Ras, and CrkII phosphorylation in *Helicobacter pylori*-induced host cell actin cytoskeletal rearrangements. FEMS Immunol. Med. Microbiol. *50*, 190–205.

Brandt, S., Wessler, S., Hartig, R., and Backert, S. (2009). *Helicobacter pylori* activates protein kinase C delta to control Raf in MAP kinase signalling: role in AGS epithelial cell scattering and elongation. Cell Motil. Cytoskeleton *66*, 874–892.

Bronte-Tinkew, D.M., Terebiznik, M., Franco, A., Ang, M., Ahn, D., Mimuro, H., Sasakawa, C., Ropeleski, M.J., Peek, Jr, R.M.., and Jones, N.L. (2009). *Helicobacter pylori* cytotoxin-associated gene A activates the signal transducer and activator of transcription 3 pathway *in vitro* and *in vivo*. Cancer Res. *69*, 632–639.

Busler, V.J., Torres, V.J., McClain, M.S., Tirado, O., Friedman, D.B., and Cover, T.L. (2006). Protein-protein interactions among *Helicobacter pylori* cag proteins. J. Bacteriol. *188*, 4787–4800.

Buti, L., Spooner, E., van der Veen, A.G., Rappuoli, R., Covacci, A., and Ploegh, H.L. (2011). *Helicobacter pylori* cytotoxin-associated gene A (CagA) subverts the apoptosis-stimulating protein of p53 (ASPP2) tumor suppressor pathway of the host. Proc. Natl. Acad. Sci. U.S.A. *108*, 9238–9243.

Camorlinga-Ponce, M., Perez-Perez, G., Gonzalez-Valencia, G., Mendoza, I., Peñaloza-Espinosa, R., Ramos, I., Kersulyte, D., Reyes-Leon, A., Romo, C., Granados, J., *et al.* (2011). *Helicobacter pylori* genotyping from American indigenous groups shows novel Amerindian *vacA* and *cagA* alleles and Asian, African and European admixture. PLoS One *6*, e27212.

Chandran, V., Fronzes, R., Duquerroy, S., Cronin, N., Navaza, J., and Waksman, G. (2009). Structure of the outer membrane complex of a type IV secretion system. Nature *462*, 1011–1015.

Chaturvedi, R., Asim, M., Romero-Gallo, J., Barry, D.P., Hoge, S., de Sablet, T., Delgado, A.G., Wroblewski, L.E., Piazuelo, M.B., Yan, F., *et al.* (2011). Spermine oxidase mediates the gastric cancer risk associated with *Helicobacter pylori* CagA. Gastroenterology *141*, 1696–1708.

Choi, I.J., Kim, J.S., Kim, J.M., Jung, H.C., and Song, I.S. (2003). Effect of inhibition of extracellular signal-regulated kinase 1 and 2 pathway on apoptosis and *bcl-2* expression in *Helicobacter pylori*-infected AGS cells. Infect. Immun. *71*, 830–837.

Churin, Y., Kardalinou, E., Meyer, T.F., and Naumann, M. (2001). Pathogenicity island-dependent activation of Rho GTPases Rac1 and Cdc42 in *Helicobacter pylori* infection. Mol. Microbiol. *40*, 815–823.

Churin, Y., Al-Ghoul, L., Kepp, O., Meyer, T.F., Birchmeier, W., and Naumann, M. (2003). *Helicobacter pylori* CagA protein targets the c-Met receptor and enhances the motogenic response. J. Cell Biol. *161*, 249–255.

Clevers, H. (2006). Wnt/β-catenin signaling in development and disease. Cell *127*, 469–480.

Conlin, V.S., Curtis, S.B., Zhao, Y., Moore, E.D., Smith, V.C., Meloche, R.M., Finlay, B.B., and Buchan, A.M. (2004). *Helicobacter pylori* infection targets adherens junction regulatory proteins and results in increased rates of migration in human gastric epithelial cells. Infect. Immun. *72*, 5181–5192.

Couturier, M.R., Tasca, E., Montecucco, C., and Stein, M. (2006). Interaction with CagF is required for translocation of CagA into the host via the *Helicobacter pylori* type IV secretion system. Infect. Immun. *74*, 273–281.

Covacci, A., Censini, S., Bugnoli, M., Petracca, R., Burroni, D., Macchia, G., Massone, A., Papini, E., Xiang, Z., Figura, N., *et al.* (1993). Molecular characterization of the 128-kDa immunodominant antigen of *Helicobacter pylori* associated with cytotoxicity and duodenal ulcer. Proc. Natl. Acad. Sci. U.S.A. *90*, 5791–5795.

Cover, T.L., and Blanke, S.R. (2005). *Helicobacter pylori* VacA, a paradigm for toxin multifunctionality. Nat. Rev. Microbiol. *3*, 320–332.

D'Elios, M.M., Manghetti, M., Almerigogna, F., Amedei, A., Costa, F., Burroni, D., Baldari, C.T., Romagnani, S., Telford, J.L., and Del Prete, G. (1997). Different cytokine profile and antigen-specificity repertoire in *Helicobacter pylori*-specific T cell clones from the antrum of chronic gastritis patients with or without peptic ulcer. Eur. J. Immunol. *27*, 1751–1755.

Delahay, R.M., Balkwill, G.D., Bunting, K.A., Edwards, W., Atherton, J.C., and Searle, M.S. (2008). The highly repetitive region of the *Helicobacter pylori* CagY protein comprises tandem arrays of an α-helical repeat module. J. Mol. Biol. *377*, 956–971.

Delgado-Rosado, G., Dominguez-Bello, M.G., and Massey, S.E. (2011). Positive selection on a bacterial oncoprotein associated with gastric cancer. Gut Pathog. *3*, 18.

Dubois, A., and Borén, T. (2007). *Helicobacter pylori* is invasive and it may be a facultative intracellular organism. Cell. Microbiol. *9*, 1108–1116.

Dumas, F., Duckely, M., Pelczar, P., Van Gelder, P., and Hohn, B. (2001). An *Agrobacterium* VirE2 channel for transferred-DNA transport into plant cells. Proc. Natl. Acad. Sci. U.S.A. *98*, 485–490.

Duncan, S.S., Valk, P.L., Shaffer, C.L., Bordenstein, S.R., and Cover, T.L. (2012). J-Western forms of *Helicobacter pylori cagA* constitute a distinct phylogenetic group with a widespread geographic distribution. J. Bacteriol. *194*, 1593–1604.

El Etr, S.H., Mueller, A., Tompkins, L.S., Falkow, S., and Merrell, D.S. (2004). Phosphorylation-independent effects of CagA during interaction between *Helicobacter pylori* and T84 polarized monolayers. J. Infect. Dis. *190*, 1516–1523.

Falush, D., Wirth, T., Linz, B., Pritchard, J.K., Stephens, M., Kidd, M., Blaser, M.J., Graham, D.Y., Vacher, S., Perez-Perez, G.I., *et al.* (2003). Traces of human migrations in *Helicobacter pylori* populations. Science *299*, 1582–1585.

Fedwick, J.P., Lapointe, T.K., Meddings, J.B., Sherman, P.M., and Buret, A.G. (2005). *Helicobacter pylori* activates myosin light-chain kinase to disrupt claudin-4 and claudin-5 and increase epithelial permeability. Infect. Immun. *73*, 7844–7852.

Ferreira, R.M., Machado, J.C., Leite, M., Carneiro, F., and Figueiredo, C. (2012). The number of *Helicobacter pylori* CagA EPIYA C tyrosine phosphorylation motifs influences the pattern of gastritis and the development of gastric carcinoma. Histopathol. *60*, 992–998.

Fischer, W. (2011). Assembly and molecular mode of action of the *Helicobacter pylori* Cag type IV secretion apparatus. FEBS J. *278*, 1203–1212.

Fischer, W., Püls, J., Buhrdorf, R., Gebert, B., Odenbreit, S., and Haas, R. (2001). Systematic mutagenesis of the *Helicobacter pylori cag* pathogenicity island: essential genes for CagA translocation in host cells and induction of interleukin-8. Mol. Microbiol. *42*, 1337–1348. Erratum in: Mol. Microbiol. *47*, 1759 (2003).

Fischer, W., Prassl, S., and Haas, R. (2009). Virulence mechanisms and persistence strategies of the human gastric pathogen *Helicobacter pylori*. In Molecular Mechanisms of Bacterial Infection via the Gut, Sasakawa, C., ed. (Berlin, Heidelberg, Springer Verlag), pp. 129–171.

Fischer, W., Windhager, L., Rohrer, S., Zeiller, M., Karnholz, A., Hoffmann, R., Zimmer, R., and Haas, R. (2010). Strain-specific genes of *Helicobacter pylori*: genome evolution driven by a novel type IV secretion system and genomic island transfer. Nucleic Acids Res. *38*, 6089–6101.

Franco, A.T., Israel, D.A., Washington, M.K., Krishna, U., Fox, J.G., Rogers, A.B., Neish, A.S., Collier-Hyams, L., Perez-Perez, G.I., Hatakeyama, M., *et al.* (2005). Activation of β-catenin by carcinogenic *Helicobacter pylori*. Proc. Natl. Acad. Sci. U.S.A. *102*, 10646–10651.

Franco, A.T., Johnston, E., Krishna, U., Yamaoka, Y., Israel, D.A., Nagy, T.A., Wroblewski, L.E., Piazuelo, M.B., Correa, P., and Peek, Jr, R.M.. (2008). Regulation of gastric carcinogenesis by *Helicobacter pylori* virulence factors. Cancer Res. *68*, 379–387.

Fujikawa, A., Shirasaka, D., Yamamoto, S., Ota, H., Yahiro, K., Fukada, M., Shintani, T., Wada, A., Aoyama, N., Hirayama, T., *et al.* (2003). Mice deficient in protein tyrosine phosphatase receptor type Z are resistant to gastric ulcer induction by VacA of *Helicobacter pylori*. Nat. Genet. *33*, 375–381.

Furuta, Y., Yahara, K., Hatakeyama, M., and Kobayashi, I. (2011). Evolution of *cagA* oncogene of *Helicobacter pylori* through recombination. PLoS One *6*, e23499.

Gangwer, K.A., Shaffer, C.L., Suerbaum, S., Lacy, D.B., Cover, T.L., and Bordenstein, S.R. (2010). Molecular evolution of the *Helicobacter pylori* vacuolating toxin gene *vacA*. J. Bacteriol. *192*, 6126–6135.

Gebert, B., Fischer, W., Weiss, E., Hoffmann, R., and Haas, R. (2003). *Helicobacter pylori* vacuolating cytotoxin inhibts T lymphocyte activation. Science *301*, 1099–1102.

Gebert, B., Fischer, W., and Haas, R. (2004). The *Helicobacter pylori* vacuolating cytotoxin: from cellular vacuolation to immunosuppressive activities. Rev. Physiol. Biochem. Pharmacol. *152*, 205–220.

Grange, W., Duckely, M., Husale, S., Jacob, S., Engel, A., and Hegner, M. (2008). VirE2: a unique ssDNA-compacting molecular machine. PLoS Biol. *6*, e44.

Gressmann, H., Linz, B., Ghai, R., Pleissner, K.P., Schlapbach, R., Yamaoka, Y., Kraft, C., Suerbaum, S., Meyer, T.F., and Achtman, M. (2005). Gain and loss of multiple genes during the evolution of *Helicobacter pylori*. PLoS Genet. *1*, e43.

Grubman, A., Kaparakis, M., Viala, J., Allison, C., Badea, L., Karrar, A., Boneca, I.G., Le Bourhis, L., Reeve, S., Smith, I.A., *et al.* (2010). The innate immune molecule, NOD1, regulates direct killing of *Helicobacter pylori* by antimicrobial peptides. Cell. Microbiol. *12*, 626–639.

Guillemin, K., Salama, N.R., Tompkins, L.S., and Falkow, S. (2002). Cag pathogenicity island-specific responses of gastric epithelial cells to *Helicobacter pylori* infection. Proc. Natl. Acad. Sci. U.S.A. *99*, 15136–15141.

Harris, T.J., and Tepass, U. (2010). Adherens junctions: from molecules to morphogenesis. Nat. Rev. Mol. Cell Biol. *11*, 502–514.

Hatakeyama, M. (2003). *Helicobacter pylori* CagA – a potential bacterial oncoprotein that functionally mimics the mammalian Gab family of adaptor proteins. Microbes Infect. *5*, 143–150.

Hatakeyama, M. (2004). Oncogenic mechanisms of the *Helicobacter pylori* CagA protein. Nat. Rev. Cancer *4*, 688–694.

Hatakeyama, M. (2011). Anthropological and clinical implications for the structural diversity of the *Helicobacter pylori* CagA oncoprotein. Cancer Sci. *102*, 36–43.

Hayashi, T., Senda, M., Morohashi, H., Higashi, H., Horio, M., Kashiba, Y., Nagase, L., Sasaya, D., Shimizu, T., Venugopalan, N., *et al.* (2012). Tertiary structure–function analysis reveals the pathogenic signaling potentiation mechanism of *Helicobacter pylori* oncogenic effector CagA. Cell Host Microbe *12*, 20–33.

Higashi, H., Tsutsumi, R., Fujita, A., Yamazaki, S., Asaka, M., Azuma, T., and Hatakeyama, M. (2002a). Biological activity of the *Helicobacter pylori* virulence factor CagA is determined by variation in the tyrosine phosphorylation sites. Proc. Natl. Acad. Sci. U.S.A. *99*, 14428–14433.

Higashi, H., Tsutsumi, R., Muto, S., Sugiyama, T., Azuma, T., Asaka, M., and Hatakeyama, M. (2002b). SHP-2 tyrosine phosphatase as an intracellular target of *Helicobacter pylori* CagA protein. Science *295*, 683–686.

Higashi, H., Nakaya, A., Tsutsumi, R., Yokoyama, K., Fujii, Y., Ishikawa, S., Higuchi, M., Takahashi, A., Kurashima, Y., Teishikata, Y., *et al.* (2004). *Helicobacter pylori* CagA induces Ras-independent morphogenetic response through SHP-2 recruitment and activation. J. Biol. Chem. *279*, 17205–17216.

Higashi, H., Yokoyama, K., Fujii, Y., Ren, S., Yuasa, H., Saadat, I., Murata-Kamiya, N., Azuma, T., and Hatakeyama, M. (2005). EPIYA motif is a membrane-targeting signal of *Helicobacter pylori* virulence factor CagA in mammalian cells. J. Biol. Chem. *280*, 23130–23137.

Hirata, Y., Maeda, S., Mitsuno, Y., Tateishi, K., Yanai, A., Akanuma, M., Yoshida, H., Kawabe, T., Shiratori, Y., and Omata, M. (2002). *Helicobacter pylori* CagA protein activates serum response element-driven transcription independently of tyrosine phosphorylation. Gastroenterology *123*, 1962–1971.

Hohlfeld, S., Pattis, I., Püls, J., Plano, G.V., Haas, R., and Fischer, W. (2006). A C-terminal secretion signal is necessary, but not sufficient for type IV secretion of the *Helicobacter pylori* CagA protein. Mol. Microbiol. *59*, 1624–1637.

Hoy, B., Löwer, M., Weydig, C., Carra, G., Tegtmeyer, N., Geppert, T., Schröder, P., Sewald, N., Backert, S., Schneider, G., *et al.* (2010). *Helicobacter pylori* HtrA is a new secreted virulence factor that cleaves E-cadherin to disrupt intercellular adhesion. EMBO Rep. *11*, 798–804.

Ishijima, N., Suzuki, M., Ashida, H., Ichikawa, Y., Kanegae, Y., Saito, I., Borén, T., Haas, R., Sasakawa, C., and Mimuro, H. (2011). BabA-mediated adherence is a potentiator of the *Helicobacter pylori* type IV secretion system activity. J. Biol. Chem. *286*, 25256–25264.

Jiménez-Soto, L.F., Kutter, S., Sewald, X., Ertl, C., Weiss, E., Kapp, U., Rohde, M., Pirch, T., Jung, K., Retta, S.F., *et al.* (2009). *Helicobacter pylori* type IV secretion apparatus exploits β1 integrin in a novel RGD-independent manner. PLoS Pathog. *5*, e1000684.

Jurik, A., Haußer, E., Kutter, S., Pattis, I., Praßl, S., Weiss, E., and Fischer, W. (2010). The coupling protein Cagβ and its interaction partner CagZ are required for type IV secretion of the *Helicobacter pylori* CagA protein. Infect. Immun. *78*, 5244–5251.

Kaplan-Türköz, B., Jiménez-Soto, L.J., Dian, C., Ertl, C., Remaut, H., Louche, A., Tosi, T., Haas, R., and Terradot, L. (2012). Structural insights into *Helicobacter pylori* oncoprotein CagA interaction with β1 integrin. Proc. Natl. Acad. Sci. U.S.A. *109*, 14640–14645.

Kawai, M., Furuta, Y., Yahara, K., Tsuru, T., Oshima, K., Handa, N., Takahashi, N., Yoshida, M., Azuma, T., Hattori, M., *et al.* (2011). Evolution in an oncogenic bacterial species with extreme genome plasticity: *Helicobacter pylori* East Asian genomes. BMC Microbiol. *11*, 104.

Keates, S., Sougioultzis, S., Keates, A.C., Zhao, D., Peek, R.M.J., Shaw, L.M., and Kelly, C.P. (2001). *cag+ Helicobacter pylori* induce transactivation of the epidermal growth factor receptor in AGS gastric epithelial cells. J. Biol. Chem. *276*, 48127–48134.

Kersulyte, D., Kalia, A., Gilman, R.H., Mendez, M., Herrera, P., Cabrera, L., Velapatiño, B., Balqui, J., Paredes Puente de la Vega, F., Rodriguez Ulloa, C.A., *et al.* (2010). *Helicobacter pylori* from Peruvian amerindians: traces of human migrations in strains from remote Amazon, and genome sequence of an Amerind strain. PLoS One *5*, e15076.

Kikuchi, K., Murata-Kamiya, N., Kondo, S., and Hatakeyama, M. (2012). *Helicobacter pylori* stimulates epithelial cell migration via CagA-mediated perturbation of host cell signaling. Microbes Infect. *14*, 470–476.

Kim, S.Y., Lee, Y.C., Kim, H.K., and Blaser, M.J. (2006). *Helicobacter pylori* CagA transfection of gastric epithelial cells induces interleukin-8. Cell. Microbiol. *8*, 97–106.

Knauer, O., Binai, N.A., Carra, G., Beckhaus, T., Hanschmann, K.M., Renné, T., Backert, S., Karas, M., and Wessler, S. (2008). Differential phosphoproteome profiling reveals a functional role for VASP in *Helicobacter pylori*-induced cytoskeleton turnover in gastric epithelial cells. Cell. Microbiol. *10*, 2285–2296.

Krueger, S., Hundertmark, T., Kuester, D., Kalinski, T., Peitz, U., and Roessner, A. (2007). *Helicobacter pylori* alters the distribution of ZO-1 and p120ctn in primary human gastric epithelial cells. Pathol. Res. Pract. *203*, 433–444.

Kusters, J.G., van Vliet, A.H., and Kuipers, E.J. (2006). Pathogenesis of *Helicobacter pylori* infection. Clin. Microbiol. Rev. *19*, 449–490.

Kutter, S., Buhrdorf, R., Haas, J., Schneider-Brachert, W., Haas, R., and Fischer, W. (2008). Protein subassemblies of the *Helicobacter pylori* Cag type IV secretion system revealed by localization and interaction studies. J. Bacteriol. *190*, 2161–2171.

Kwok, T., Zabler, D., Urman, S., Rohde, M., Hartig, R., Wessler, S., Misselwitz, R., Berger, J., Sewald, N., König, W., *et al.* (2007). *Helicobacter* exploits integrin for type IV secretion and kinase activation. Nature *449*, 862–866.

Lai, C.H., Chang, Y.C., Du, S.Y., Wang, H.J., Kuo, C.H., Fang, S.H., Fu, H.W., Lin, H.H., Chiang, A.S., and Wang, W.C. (2008). Cholesterol depletion reduces *Helicobacter pylori* CagA translocation and CagA-induced responses in AGS cells. Infect. Immun. *76*, 3293–3303.

Lamb, A., Yang, X.D., Tsang, Y.H., Li, J.D., Higashi, H., Hatakeyama, M., Peek, R.M., Blanke, S.R., and Chen, L.F. (2009). *Helicobacter pylori* CagA activates NF-κB by targeting TAK1 for TRAF6-mediated Lys 63 ubiquitination. EMBO Rep. *10*, 1242–1249.

Lee, K.S., Kalantzis, A., Jackson, C.B., O'Connor, L., Murata-Kamiya, N., Hatakeyama, M., Judd, L.M., Giraud, A.S., and Menheniott, T.R. (2012). *Helicobacter pylori* CagA triggers expression of the bactericidal lectin REG3γ via gastric STAT3 activation. PLoS One 7, e30786.

Lim, J.W., Kim, K.H., and Kim, H. (2009). αPix interacts with *Helicobacter pylori* CagA to induce IL-8 expression in gastric epithelial cells. Scand. J. Gastroenterol. *44*, 1166–1172.

Lin, W.C., Tsai, H.F., Kuo, S.H., Wu, M.S., Lin, C.W., Hsu, P.I., Cheng, A.L., and Hsu, P.N. (2010). Translocation of *Helicobacter pylori* CagA into Human B lymphocytes, the origin of mucosa-associated lymphoid tissue lymphoma. Cancer Res. *70*, 5740–5748.

Linz, B., Balloux, F., Moodley, Y., Manica, A., Liu, H., Roumagnac, P., Falush, D., Stamer, C., Prugnolle, F., van der Merwe, S.W., *et al.* (2007). An African origin for the intimate association between humans and *Helicobacter pylori*. Nature *445*, 915–918.

Liu, G., McDaniel, T.K., Falkow, S., and Karlin, S. (1999). Sequence anomalies in the *cag7* gene of the *Helicobacter pylori* pathogenicity island. Proc. Natl. Acad. Sci. U.S.A. *96*, 7011–7016.

Liu, Z., Xu, X., Chen, L., Li, W., Sun, Y., Zeng, J., Yu, H., Chen, C., and Jia, J. (2012). *Helicobacter pylori* CagA inhibits the expression of Runx3 via Src/MEK/ERK and p38 MAPK pathways in gastric epithelial cell. J. Cell Biochem. *113*, 1080–1086.

Llosa, M., Zunzunegui, S., and De La Cruz, F. (2003). Conjugative coupling proteins interact with cognate and heterologous VirB10-like proteins while exhibiting specificity for cognate relaxosomes. Proc. Natl. Acad. Sci. U.S.A. *100*, 10465–10470.

Locht, C., Coutte, L., and Mielcarek, N. (2011). The ins and outs of pertussis toxin. FEBS J. *278*, 4668–4682.

Lu, H., Murata-Kamiya, N., Saito, Y., and Hatakeyama, M. (2009). Role of partitioning-defective 1/microtubule affinity-regulating kinases in the morphogenetic activity of *Helicobacter pylori* CagA. J. Biol. Chem. *284*, 23024–23036.

Lu, H.S., Saito, Y., Umeda, M., Murata-Kamiya, N., Zhang, H.M., Higashi, H., and Hatakeyama, M. (2008). Structural and functional diversity in the PAR1b/MARK2-binding region of *Helicobacter pylori* CagA. Cancer Sci. *99*, 2004–2011.

Lytton, S.D., Fischer, W., Nagel, W., Haas, R., and Beck, F.X. (2005). Production of ammonium by *Helicobacter pylori* mediates occludin processing and disruption of tight junctions in Caco-2 cells. Microbiology *151*, 3267–3276.

Mane, S.P., Dominguez-Bello, M.G., Blaser, M.J., Sobral, B.W., Hontecillas, R., Skoneczka, J., Mohapatra, S.K., Crasta, O.R., Evans, C., Modise, T., *et al.* (2010). Host-interactive genes in Amerindian *Helicobacter pylori* diverge from their Old World homologs and mediate inflammatory responses. J. Bacteriol. *192*, 3078–3092.

Marx, A., Nugoor, C., Panneerselvam, S., and Mandelkow, E. (2010). Structure and function of polarity-inducing kinase family MARK/Par-1 within the branch of AMPK/Snf1-related kinases. FASEB J. *24*, 1637–1648.

Matteï, P.J., Faudry, E., Job, V., Izoré, T., Attree, I., and Dessen, A. (2011). Membrane targeting and pore formation by the type III secretion system translocon. FEBS J. *278*, 414–426.

McCaffrey, L.M., and Macara, I.G. (2011). Epithelial organization, cell polarity and tumorigenesis. Trends Cell Biol. *21*, 727–735.

Mimuro, H., Suzuki, T., Tanaka, J., Asahi, M., Haas, R., and Sasakawa, C. (2002). Grb2 is a key mediator of *Helicobacter pylori* CagA protein activities. Mol. Cell *10*, 745–755.

Mimuro, H., Suzuki, T., Nagai, S., Rieder, G., Suzuki, M., Nagai, T., Fujita, Y., Nagamatsu, K., Ishijima, N., Koyasu, S., *et al.* (2007). *Helicobacter pylori* dampens gut epithelial self-renewal by inhibiting apoptosis, a bacterial strategy to enhance colonization of the stomach. Cell Host Microbe *2*, 250–263.

Moese, S., Selbach, M., Zimny-Arndt, U., Jungblut, P.R., Meyer, T.F., and Backert, S. (2001). Identification of a tyrosine-phosphorylated 35 kDa carboxy-terminal fragment ($p35^{CagA}$) of the *Helicobacter pylori* CagA protein in phagocytic cells: processing or breakage? Proteomics *1*, 618–629.

Moese, S., Selbach, M., Kwok, T., Brinkmann, V., König, W., Meyer, T.F., and Backert, S. (2004). *Helicobacter pylori induces* AGS cell motility and elongation via independent signaling pathways. Infect. Immun. *72*, 3646–3649.

Moese, S., Selbach, M., Brinkmann, V., Karlas, A., Haimovich, B., Backert, S., and Meyer, T.F. (2007). The *Helicobacter pylori* CagA protein disrupts matrix adhesion of gastric epithelial cells by dephosphorylation of vinculin. Cell. Microbiol. *9*, 1148–1161.

Mueller, D., Tegtmeyer, N., Brandt, S., Yamaoka, Y., De Poire, E., Sgouras, D., Wessler, S., Torres, J., Smolka, A., and Backert, S. (2012). c-Src and c-Abl kinases control hierarchic phosphorylation and function of the CagA effector protein in Western and East Asian *Helicobacter pylori* strains. J. Clin. Invest. *122*, 1553–1566.

Murata-Kamiya, N., Kurashima, Y., Teishikata, Y., Yamahashi, Y., Saito, Y., Higashi, H., Aburatani, H., Akiyama, T., Peek, Jr, R.M.., Azuma, T., *et al.* (2007). *Helicobacter pylori* CagA interacts with E-cadherin and deregulates the β-catenin signal that promotes intestinal transdifferentiation in gastric epithelial cells. Oncogene 26, 4617–4626.

Murata-Kamiya, N., Kikuchi, K., Hayashi, T., Higashi, H., and Hatakeyama, M. (2010). *Helicobacter pylori* exploits host membrane phosphatidylserine for delivery, localization, and pathophysiological action of the CagA oncoprotein. Cell Host Microbe 7, 399–411.

Nagase, L., Murata-Kamiya, N., and Hatakeyama, M. (2011). Potentiation of *Helicobacter pylori* CagA protein virulence through homodimerization. J. Biol. Chem. *286*, 33622–33631.

Nagy, T.A., Wroblewski, L.E., Wang, D., Piazuelo, M.B., Delgado, A., Romero-Gallo, J., Noto, J., Israel, D.A., Ogden, S.R., Correa, P., *et al.* (2011). β-Catenin and p120 mediate PPARδ-dependent proliferation induced by *Helicobacter pylori* in human and rodent epithelia. Gastroenterology *141*, 553–564.

Naito, M., Yamazaki, T., Tsutsumi, R., Higashi, H., Onoe, K., Yamazaki, S., Azuma, T., and Hatakeyama, M. (2006). Influence of EPIYA-repeat polymorphism on the phosphorylation-dependent biological activity of *Helicobacter pylori* CagA. Gastroenterology *130*, 1181–1190.

Nakayama, M., Hisatsune, J., Yamasaki, E., Isomoto, H., Kurazono, H., Hatakeyama, M., Azuma, T., Yamaoka, Y., Yahiro, K., Moss, J., *et al.* (2009). *Helicobacter pylori* VacA-induced inhibition of GSK3 through the PI3K/Akt signaling pathway. J. Biol. Chem. *284*, 1612–1619.

Necchi, V., Candusso, M.E., Tava, F., Luinetti, O., Ventura, U., Fiocca, R., Ricci, V., and Solcia, E. (2007). Intracellular, intercellular, and stromal invasion of gastric mucosa, preneoplastic lesions, and cancer by *Helicobacter pylori*. Gastroenterology *132*, 1009–1023.

Nešić, D., Miller, M.C., Quinkert, Z.T., Stein, M., Chait, B.T., and Stebbins, C.E. (2010). *Helicobacter pylori* CagA inhibits PAR1-MARK family kinases by mimicking host substrates. Nat. Struct. Mol. Biol. *17*, 130–132.

Odenbreit, S., Püls, J., Sedlmaier, B., Gerland, E., Fischer, W., and Haas, R. (2000). Translocation of *Helicobacter pylori* CagA into gastric epithelial cells by type IV secretion. Science *287*, 1497–1500.

Odenbreit, S., Gebert, B., Püls, J., Fischer, W., and Haas, R. (2001). Interaction of *Helicobacter pylori* with professional phagocytes: role of the *cag* pathogenicity island and translocation, phosphorylation and processing of CagA. Cell. Microbiol. *3*, 21–31.

Ogden, S.R., Wroblewski, L.E., Weydig, C., Romero-Gallo, J., O'brien, D.P., Israel, D.A., Krishna, U.S., Fingleton, B., Reynolds, A.B., Wessler, S., *et al.* (2008). p120 and Kaiso regulate *Helicobacter pylori*-induced expression of matrix metalloproteinase-7. Mol. Biol. Cell *19*, 4110–4121.

Ogura, K., Maeda, S., Nakao, M., Watanabe, T., Tada, M., Kyutoku, T., Yoshida, H., Shiratori, Y., and Omata, M. (2000). Virulence factors of *Helicobacter pylori* responsible for gastric diseases in Mongolian gerbil. J. Exp. Med. *192*, 1601–1610.

Ogura, M., Perez, J.C., Mittl, P.R., Lee, H.K., Dailide, G., Tan, S., Ito, Y., Secka, O., Dailidiene, D., Putty, K., *et al.* (2007). *Helicobacter pylori* evolution: lineage-specific adaptations in homologs of eukaryotic Sel1-like genes. PLoS Comput. Biol. *3*, 1455–1467.

Ohnishi, N., Yuasa, H., Tanaka, S., Sawa, H., Miura, M., Matsui, A., Higashi, H., Musashi, M., Iwabuchi, K., Suzuki, M., *et al.* (2008). Transgenic expression of *Helicobacter pylori* CagA induces gastrointestinal and hematopoietic neoplasms in mouse. Proc. Natl. Acad. Sci. U.S.A. *105*, 1003–1008.

Olbermann, P., Josenhans, C., Moodley, Y., Uhr, M., Stamer, C., Vauterin, M., Suerbaum, S., Achtman, M., and Linz, B. (2010). A global overview of the genetic and functional diversity in the *Helicobacter pylori cag* pathogenicity island. PLoS Genet. *6*, e1001069.

Oldani, A., Cormont, M., Hofman, V., Chiozzi, V., Oregioni, O., Canonici, A., Sciullo, A., Sommi, P., Fabbri, A., Ricci, V., *et al.* (2009). *Helicobacter pylori* counteracts the apoptotic action of its VacA toxin by injecting the CagA protein into gastric epithelial cells. PLoS Pathog. *5*, e1000603.

Oliveira, M.J., Costa, A.C., Costa, A.M., Henriques, L., Suriano, G., Atherton, J.C., Machado, J.C., Carneiro, F., Seruca, R., Mareel, M., *et al.* (2006). *Helicobacter pylori* induces gastric epithelial cell invasion in a c-Met and type IV secretion system-dependent manner. J. Biol. Chem. *281*, 34888–34896.

Oliveira, M.J., Costa, A.M., Costa, A.C., Ferreira, R.M., Sampaio, P., Machado, J.C., Seruca, R., Mareel, M., and Figueiredo, C. (2009). CagA associates with c-Met, E-cadherin, and p120-catenin in a multiproteic complex that suppresses *Helicobacter pylori*-induced cell-invasive phenotype. J. Infect. Dis. *200*, 745–755.

Östman, A., Hellberg, C., and Böhmer, F.D. (2006). Protein-tyrosine phosphatases and cancer. Nat. Rev. Cancer *6*, 307–320.

Papini, E., Satin, B., Norais, N., De Bernard, M., Telford, J.L., Rappuoli, R., and Montecucco, C. (1998). Selective increase of the permeability of polarized epithelial cell monolayers by *Helicobacter pylori* vacuolating toxin. J. Clin. Invest. *102*, 813–820.

Pattis, I., Weiss, E., Laugks, R., Haas, R., and Fischer, W. (2007). The *Helicobacter pylori* CagF protein is a type IV secretion chaperone-like molecule that binds close to the C-terminal secretion signal of the CagA effector protein. Microbiology *153*, 2896–2909.

Peek, R.M.J., and Blaser, M.J. (2002). *Helicobacter pylori* and gastrointestinal tract adenocarcinomas. Nat. Rev. Cancer *2*, 28–37.

Peek, R.M.J., Wirth, H.P., Moss, S.F., Yang, M., Abdalla, A.M., Tham, K.T., Zhang, T., Tang, L.H., Modlin, I.M., and Blaser, M.J. (2000). *Helicobacter pylori* alters gastric epithelial cell cycle events and gastrin secretion in Mongolian gerbils. Gastroenterology *118*, 48–59.

Pelz, C., Steininger, S., Weiss, C., Coscia, F., and Vogelmann, R. (2011). A novel inhibitory domain of *Helicobacter pylori* protein CagA reduces CagA effects on host cell biology. J. Biol. Chem. *286*, 8999–9008.

Pham, K.T., Weiss, E., Jiménez-Soto, L.F., Breithaupt, U., Haas, R., and Fischer, W. (2012). CagI is an essential component of the *Helicobacter pylori* Cag type IV secretion system and forms a complex with CagL. PLoS One *7*, e35341.

Philpott, D.J., Belaid, D., Troubadour, P., Thiberge, J.M., Tankovic, J., Labigne, A., and Ferrero, R.L. (2002). Reduced activation of inflammatory responses in host cells by mouse-adapted *Helicobacter pylori* isolates. Cell. Microbiol. *4*, 285–296.

Polk, D.B., and Peek, Jr, R.M.. (2010). *Helicobacter pylori*: gastric cancer and beyond. Nat. Rev. Cancer *10*, 403–414.

Poppe, M., Feller, S.M., Römer, G., and Wessler, S. (2007). Phosphorylation of *Helicobacter pylori* CagA by c-Abl leads to cell motility. Oncogene *26*, 3462–3472.

Ren, S., Higashi, H., Lu, H., Azuma, T., and Hatakeyama, M. (2006). Structural basis and functional consequence of *Helicobacter pylori* CagA multimerization in cells. J. Biol. Chem. *281*, 32344–32352.

Rieder, G., Merchant, J.L., and Haas, R. (2005). *Helicobacter pylori cag*-type IV secretion system facilitates corpus colonization to induce precancerous conditions in Mongolian gerbils. Gastroenterology *128*, 1229–1242.

Rohde, M., Püls, J., Buhrdorf, R., Fischer, W., and Haas, R. (2003). A novel sheathed surface organelle of the *Helicobacter pylori cag* type IV secretion system. Mol. Microbiol. *49*, 219–234.

Saadat, I., Higashi, H., Obuse, C., Umeda, M., Murata-Kamiya, N., Saito, Y., Lu, H., Ohnishi, N., Azuma, T., Suzuki, A., *et al.* (2007). *Helicobacter pylori* CagA targets PAR1/MARK kinase to disrupt epithelial cell polarity. Nature *447*, 330–333.

Safari, F., Murata-Kamiya, N., Saito, Y., and Hatakeyama, M. (2011). Mammalian Pragmin regulates Src family kinases via the Glu-Pro-Ile-Tyr-Ala (EPIYA) motif that is exploited by bacterial effectors. Proc. Natl. Acad. Sci. U.S.A. *108*, 14938–14943.

Saha, A., Backert, S., Hammond, C.E., Gooz, M., and Smolka, A.J. (2010). *Helicobacter pylori* CagL activates ADAM17 to induce repression of the gastric H, K-ATPase α subunit. Gastroenterology *139*, 239–248.

Saito, Y., Murata-Kamiya, N., Hirayama, T., Ohba, Y., and Hatakeyama, M. (2010). Conversion of *Helicobacter pylori* CagA from senescence inducer to oncogenic driver through polarity-dependent regulation of p21. J. Exp. Med. *207*, 2157–2174.

Schmitt, W., and Haas, R. (1994). Genetic analysis of the *Helicobacter pylori* vacuolating cytotoxin: structural similarities with IgA protease type of exported protein. Mol. Microbiol. *12*, 307–319.

Schneider, S., Carra, G., Sahin, U., Hoy, B., Rieder, G., and Wessler, S. (2011). Complex cellular responses of *Helicobacter pylori*-colonized gastric adenocarcinoma cells. Infect. Immun. *79*, 2362–2371.

Segal, E.D., Cha, J., Lo, J., Falkow, S., and Tompkins, L.S. (1999). Altered states: involvement of phosphorylated CagA in the induction of host cellular growth changes by *Helicobacter pylori*. Proc. Natl. Acad. Sci. U.S.A. *96*, 14559–14564.

Selbach, M., Moese, S., Hauck, C.R., Meyer, T.F., and Backert, S. (2002). Src is the kinase of the *Helicobacter pylori* CagA protein *in vitro* and *in vivo*. J. Biol. Chem. *277*, 6775–6778.

Selbach, M., Moese, S., Hurwitz, R., Hauck, C.R., Meyer, T.F., and Backert, S. (2003). The *Helicobacter pylori* CagA protein induces cortactin dephosphorylation and actin rearrangement by c-Src inactivation. EMBO J. *22*, 515–528.

Selbach, M., Paul, F.E., Brandt, S., Guye, P., Daumke, O., Backert, S., Dehio, C., and Mann, M. (2009). Host cell interactome of tyrosine-phosphorylated bacterial proteins. Cell Host Microbe *5*, 397–403.

Shaffer, C.L., Gaddy, J.A., Loh, J.T., Johnson, E.M., Hill, S., Hennig, E.E., McClain, M.S., McDonald, W.H., and Cover, T.L. (2011). *Helicobacter pylori* exploits a unique repertoire of type IV secretion system components for pilus assembly at the bacteria-host cell interface. PLoS Pathog. *7*, e1002237.

Sharma, C.M., Hoffmann, S., Darfeuille, F., Reignier, J., Findeiss, S., Sittka, A., Chabas, S., Reiche, K., Hackermüller, J., Reinhardt, R., *et al.* (2010). The primary transcriptome of the major human pathogen *Helicobacter pylori*. Nature *464*, 250–255.

Smolka, A.J., and Backert, S. (2012). How *Helicobacter pylori* infection controls gastric acid secretion. J. Gastroenterol. *47*, 609–618.

Snider, J.L., Allison, C., Bellaire, B.H., Ferrero, R.L., and Cardelli, J.A. (2008). The β_1 integrin activates JNK independent of CagA, and JNK activation is required for *Helicobacter pylori* $CagA^+$-induced motility of gastric cancer cells. J. Biol. Chem. *283*, 13952–13963.

Sokolova, O., Bozko, P.M., and Naumann, M. (2008). *Helicobacter pylori* suppresses glycogen synthase kinase 3β to promote β-catenin activity. J. Biol. Chem. *283*, 29367–29374.

Stein, M., Bagnoli, F., Halenbeck, R., Rappuoli, R., Fantl, W.J., and Covacci, A. (2002). c-Src/Lyn kinases activate *Helicobacter pylori* CagA through tyrosine-phosphorylation of the EPIYA motifs. Mol. Microbiol. *43*, 971–980.

Suerbaum, S., and Josenhans, C. (2007). *Helicobacter pylori* evolution and phenotypic diversification in a changing host. Nat. Rev. Microbiol. *5*, 441–452.

Suerbaum, S., and Michetti, P. (2002). *Helicobacter pylori* infection. N. Engl. J. Med. *347*, 1175–1186.

Suzuki, M., Mimuro, H., Suzuki, T., Park, M., Yamamoto, T., and Sasakawa, C. (2005). Interaction of CagA with Crk plays an important role in *Helicobacter pylori*-induced loss of gastric epithelial cell adhesion. J. Exp. Med. *202*, 1235–1247.

Suzuki, M., Mimuro, H., Kiga, K., Fukumatsu, M., Ishijima, N., Morikawa, H., Nagai, S., Koyasu, S., Gilman, R.H., Kersulyte, D., *et al.* (2009). *Helicobacter pylori* CagA phosphorylation-independent function in epithelial proliferation and inflammation. Cell Host Microbe *5*, 23–34.

Suzuki, M., Kiga, K., Kersulyte, D., Cok, J., Hooper, C.C., Mimuro, H., Sanada, T., Suzuki, S., Oyama, M., Kozuka-Hata, H., *et al.* (2011). Attenuated CagA oncoprotein in *Helicobacter pylori* from Amerindians in Peruvian Amazon. J. Biol. Chem. *286*, 29964–29972.

Tabassam, F.H., Graham, D.Y., and Yamaoka, Y. (2008). OipA plays a role in *Helicobacter pylori*-induced focal adhesion kinase activation and cytoskeletal re-organization. Cell. Microbiol. *10*, 1008–1020.

Tabassam, F.H., Graham, D.Y., and Yamaoka, Y. (2011). Paxillin is a novel cellular target for converging *Helicobacter pylori*-induced cellular signaling. Am. J. Physiol. Gastrointest. Liver Physiol. *301*, G601-G611.

Takahashi, A., Tsutsumi, R., Kikuchi, I., Obuse, C., Saito, Y., Seidi, A., Karisch, R., Fernandez, M., Cho, T., Ohnishi, N., *et al.* (2011). SHP2 tyrosine phosphatase converts parafibromin/Cdc73 from a tumor suppressor to an oncogenic driver. Mol. Cell *43*, 45–56.

Tammer, I., Brandt, S., Hartig, R., König, W., and Backert, S. (2007). Activation of Abl by *Helicobacter pylori*: a novel kinase for CagA and crucial mediator of host cell scattering. Gastroenterology *132*, 1309–1319.

Tan, S., Tompkins, L.S., and Amieva, M.R. (2009). *Helicobacter pylori* usurps cell polarity to turn the cell surface into a replicative niche. PLoS Pathog. *5*, e1000407.

Tan, S., Noto, J.M., Romero-Gallo, J., Peek, Jr, R.M.., and Amieva, M.R. (2011). *Helicobacter pylori* perturbs iron trafficking in the epithelium to grow on the cell surface. PLoS Pathog. *7*, e1002050.

Tanaka, J., Suzuki, T., Mimuro, H., and Sasakawa, C. (2003). Structural definition on the surface of *Helicobacter pylori* type IV secretion apparatus. Cell. Microbiol. *5*, 395–404.

Tegtmeyer, N., Zabler, D., Schmidt, D., Hartig, R., Brandt, S., and Backert, S. (2009). Importance of EGF receptor, HER2/Neu and Erk1/2 kinase signalling for host cell elongation and scattering induced by the *Helicobacter pylori* CagA protein: antagonistic effects of the vacuolating cytotoxin VacA. Cell. Microbiol. *11*, 488–505.

Tegtmeyer, N., Hartig, R., Delahay, R.M., Rohde, M., Brandt, S., Conradi, J., Takahashi, S., Smolka, A.J., Sewald, N., and Backert, S. (2010). A small fibronectin-mimicking protein from bacteria induces cell spreading and focal adhesion formation. J. Biol. Chem. *285*, 23515–23526.

Tegtmeyer, N., Wittelsberger, R., Hartig, R., Wessler, S., Martinez-Quiles, N., and Backert, S. (2011). Serine phosphorylation of cortactin controls focal adhesion kinase activity and cell scattering induced by *Helicobacter pylori*. Cell Host Microbe *9*, 520–531.

Torres-Morquecho, A., Giono-Cerezo, S., Camorlinga-Ponce, M., Vargas-Mendoza, C.F., and Torres, J. (2010). Evolution of bacterial genes: evidences of positive Darwinian selection and fixation of base substitutions in virulence genes of *Helicobacter pylori*. Infect. Genet. Evol. *10*, 764–776.

Torruellas-Garcia, J., Ferracci, F., Jackson, M.W., Joseph, S., Pattis, I., Plano, L.R., Fischer, W., and Plano, G.V. (2006). Measurement of effector protein injection by type III and type IV secretion systems using a 13-residue phosphorylatable GSK tag. Infect. Immun. *74*, 5645–5657.

Truong, B.X., Mai, V.T., Tanaka, H., Ly le, T., Thong, T.M., Hai, H.H., Van, L.D., Furumatsu, K., Yoshida, M., Kutsumi, H., *et al.* (2009). Diverse characteristics of the *cagA* gene of *Helicobacter pylori* strains collected from patients from southern Vietnam with gastric cancer and peptic ulcer. J. Clin. Microbiol. *47*, 4021–4028.

Tsang, Y.H., Lamb, A., Romero-Gallo, J., Huang, B., Ito, K., Peek, Jr, R.M.., Ito, Y., and Chen, L.F. (2010). *Helicobacter pylori* CagA targets gastric tumor suppressor RUNX3 for proteasome-mediated degradation. Oncogene *29*, 5643–5650.

Tsutsumi, R., Higashi, H., Higuchi, M., Okada, M., and Hatakeyama, M. (2003). Attenuation of *Helicobacter pylori* CagA-SHP-2 signaling by interaction between CagA and C-terminal Src kinase. J. Biol. Chem. *278*, 3664–3670.

Tsutsumi, R., Takahashi, A., Azuma, T., Higashi, H., and Hatakeyama, M. (2006). Focal adhesion kinase is a substrate and downstream effector of SHP-2 complexed with *Helicobacter pylori* CagA. Mol. Cell Biol. *26*, 261–276.

Tummuru, M.K.R., Cover, T.L., and Blaser, M.J. (1993). Cloning and expression of a high-molecular-mass major antigen of *Helicobacter pylori*: Evidence of linkage to cytotoxin production. Infect. Immun. *61*, 1799–1809.

Tummuru, M.K.R., Cover, T.L., and Blaser, M.J. (1994). Mutation of the cytotoxin-associated *cagA* gene does not affect the vacuolating cytotoxin activity of *Helicobacter pylori*. Infect. Immun. *62*, 2609–2613.

Umehara, S., Higashi, H., Ohnishi, N., Asaka, M., and Hatakeyama, M. (2003). Effects of *Helicobacter pylori* CagA protein on the growth and survival of B lymphocytes, the origin of MALT lymphoma. Oncogene 22, 8337–8342.

van Doorn, L.J., Figueiredo, C., Sanna, R., Blaser, M.J., and Quint, W.G. (1999). Distinct variants of *Helicobacter pylori cagA* are associated with *vacA* subtypes. J. Clin. Microbiol. *37*, 2306–2311.

Viala, J., Chaput, C., Boneca, I.G., Cardona, A., Girardin, S.E., Moran, A.P., Athman, R., Memet, S., Huerre, M.R., Coyle, A.J., *et al.* (2004). Nod1 responds to peptidoglycan delivered by the *Helicobacter pylori cag* pathogenicity island. Nat. Immunol. *5*, 1166–1174.

Wallasch, C., Crabtree, J.E., Bevec, D., Robinson, P.A., Wagner, H., and Ullrich, A. (2002). *Helicobacter pylori*-stimulated EGF receptor transactivation requires metalloprotease cleavage of HB-EGF. Biochem. Biophys. Res. Commun. *295*, 695–701.

Wang, H.J., Cheng, W.C., Cheng, H.H., Lai, C.H., and Wang, W.C. (2012). *Helicobacter pylori* cholesteryl glucosides interfere with host membrane phase and affect type IV secretion system function during infection in AGS cells. Mol. Microbiol. *83*, 67–84.

Watanabe, T., Tada, M., Nagai, H., Sasaki, S., and Nakao, M. (1998). *Helicobacter pylori* infection induces gastric cancer in mongolian gerbils. Gastroenterology *115*, 642–648.

Wessler, S., and Backert, S. (2008). Molecular mechanisms of epithelial-barrier disruption by *Helicobacter pylori*. Trends Microbiol. *16*, 397–405.

Weydig, C., Starzinski-Powitz, A., Carra, G., Löwer, J., and Wessler, S. (2007). CagA-independent disruption of adherence junction complexes involves E-cadherin shedding and implies multiple steps in *Helicobacter pylori* pathogenicity. Exp. Cell Res. *313*, 3459–3471.

Wiedemann, T., Hofbaur, S., Tegtmeyer, N., Huber, S., Sewald, N., Wessler, S., Backert, S., and Rieder, G. (2012). *Helicobacter pylori* CagL dependent induction of gastrin expression via a novel $\alpha v \beta_5$-integrin-integrin linked kinase signalling complex. Gut *61*, 986–996.

Wirth, H.P., Beins, M.H., Yang, M., Tham, K.T., and Blaser, M.J. (1998). Experimental infection of Mongolian gerbils with wild-type and mutant *Helicobacter pylori* strains. Infect. Immun. *66*, 4856–4866.

Wroblewski, L.E., Noble, P.J., Pagliocca, A., Pritchard, D.M., Hart, C.A., Campbell, F., Dodson, A.R., Dockray, G.J., and Varro, A. (2003). Stimulation of MMP-7 (matrilysin) by *Helicobacter pylori* in human gastric epithelial cells: role in epithelial cell migration. J. Cell Sci. *116*, 3017–3026.

Wroblewski, L.E., Shen, L., Ogden, S., Romero-Gallo, J., Lapierre, L.A., Israel, D.A., Turner, J.R., and Peek, Jr, R.M.. (2009). *Helicobacter pylori* dysregulation of gastric epithelial tight junctions by urease-mediated myosin II activation. Gastroenterology *136*, 236–246.

Xia, Y., Yamaoka, Y., Zhu, Q., Matha, I., and Gao, X. (2009). A comprehensive sequence and disease correlation analyses for the C-terminal region of CagA protein of *Helicobacter pylori*. PLoS One *4*, e7736.

Xiang, Z., Censini, S., Bayeli, P.F., Telford, J.L., Figura, N., Rappuoli, R., and Covacci, A. (1995). Analysis of expression of CagA and VacA virulence factors in 43 strains of *Helicobacter pylori* reveals that clinical isolates can be divided into two major types and that

CagA is not necessary for expression of the vacuolating cytotoxin. Infect. Immun. *63*, 94–98.

Yamaoka, Y., Kodama, T., Kashima, K., Graham, D.Y., and Sepulveda, A.R. (1998). Variants of the 3′ region of the *cagA* gene in *Helicobacter pylori* isolates from patients with different *H. pylori*-associated diseases. J. Clin. Microbiol. *36*, 2258–2263.

Yamaoka, Y., El-Zimaity, H.M., Gutierrez, O., Figura, N., Kim, J.G., Kodama, T., Kashima, K., and Graham, D.Y. (1999). Relationship between the *cagA* 3′ repeat region of *Helicobacter pylori*, gastric histology, and susceptibility to low pH. Gastroenterol. *117*, 342–349.

Yan, F., Cao, H., Chaturvedi, R., Krishna, U., Hobbs, S.S., Dempsey, P.J., Peek, Jr, R.M.., Cover, T.L., Washington, M.K., Wilson, K.T., *et al.* (2009). Epidermal growth factor receptor activation protects gastric epithelial cells from *Helicobacter pylori*-induced apoptosis. Gastroenterology *136*, 1297–1307.

Yanai, A., Hirata, Y., Mitsuno, Y., Maeda, S., Shibata, W., Akanuma, M., Yoshida, H., Kawabe, T., and Omata, M. (2003). *Helicobacter pylori* induces antiapoptosis through nuclear factor-κB activation. J. Infect. Dis. *188*, 1741–1751.

Yokoyama, K., Higashi, H., Ishikawa, S., Fujii, Y., Kondo, S., Kato, H., Azuma, T., Wada, A., Hirayama, T., Aburatani, H., *et al.* (2005). Functional antagonism between *Helicobacter pylori* CagA and vacuolating toxin VacA in control of the NFAT signaling pathway in gastric epithelial cells. Proc. Natl. Acad. Sci. U.S.A. *102*, 9661–9666.

Zeaiter, Z., Cohen, D., Müsch, A., Bagnoli, F., Covacci, A., and Stein, M. (2007). Analysis of detergent-resistant membranes of *Helicobacter pylori* infected gastric adenocarcinoma cells reveals a role for MARK2/Par1b in CagA-mediated disruption of cellular polarity. Cell. Microbiol. *10*, 781–794.

Zhu, Y., Wang, C., Huang, J., Ge, Z., Dong, Q., Zhong, X., Su, Y., and Zheng, S. (2007). The *Helicobacter pylori* virulence factor CagA promotes Erk1/2-mediated Bad phosphorylation in lymphocytes: a mechanism of CagA-inhibited lymphocyte apoptosis. Cell. Microbiol. *9*, 952–961.

Pore-forming Toxins

Juliane Bubeck Wardenburg, James Whisstock and
Rodney K. Tweten

3

Abstract

Bacterial pore-forming protein toxins are widespread among bacterial pathogens and opportunistic pathogens. Their roles in disease progression are varied and complex. In this review the cholesterol dependent cytolysins (CDCs) and the related membrane attack complex/perforin (MACPF) protein families, as well as *Staphylococcus aureus* α-haemolysin will be discussed. These pore-forming proteins exhibit prominent structural and mechanistic features that are paradigms for other pore-forming proteins. In this review we will focus on their structure and mechanisms of action and how they relate to their contribution(s) to pathogenesis.

Introduction

Pore forming toxins (PFTs) are a ubiquitous family of bacterial toxins: they are found in bacterial pathogens and commensals, as well as fungi, protozoa, vertebrates, invertebrates and plants. Their primary target is the cell membrane: by opening a pore in the cell membrane these toxins can induce cell lysis, or at subcytolytic levels they can affect a multitude of cellular processes. Those cells that are targeted by PFTs have ways of neutralizing the activity of these toxins, underscoring the complex and dynamic interactions of these secreted proteins with their host (Idone *et al.*, 2008). Although PFTs are mostly thought of as toxins, many PFTs are produced by bacterial species that spend most of their existence as commensal organisms and only occasionally cause disease. PFTs may have therefore evolved to support a commensal existence.

The PFT pore can range in size from 1–2 nm up to 20–30 nm. This remarkable difference in the pore sizes implies that the dimensions of the pore may specify the function of the toxin. PFTs like the cholesterol-dependent cytolysins (CDCs) form very large pores comprised of many protein subunits, whereas *Staphylococcus aureus* α-haemolysin-like pore formers typically are comprised of 6–8 protein subunits. It remains unclear whether bacteria primarily use PFTs to induce direct cell lysis – a growing body of data would suggest that this is an overly simplistic view of the functional capabilities of these complex structures.

Most bacteria that express a PFT either engage in a commensal lifestyle with the host or occupy an environmental niche (e.g. soil or water) for the majority of their existence, only rarely causing serious infections. This raises the intriguing question of whether PFTs are more important for maintaining the commensal state by altering cell metabolism to suit their needs or whether their role as 'pathogenic factors' subserves the function of establishing and maintaining the commensal state. In this context, disease attributable to PFTs may represent the exaggerated effects of the PFT when the host–pathogen balance is skewed in marked favour of the bacterial organism. This can result from an increased bacterial load, altered regulation of toxin production, or more likely a host-dependent aberration where a defect in the immune system is unintentionally exploited by the bacterium or host genetic polymorphisms encode unique susceptibility to the toxin. Host susceptibility to a PFT may thus be viewed as the final measure of the complex interaction of the host with the toxin-secreting organism.

Membrane integrity is carefully preserved in eukaryotic cells. Interestingly, cellular aquaporins function as water channels in the plasma membrane and exhibit pore diameters of approximately 3 nm (Verkman, 2011) – slightly larger than those formed by the pore of *S. aureus* α-haemolysin (Song *et al.*, 1996). Therefore, the cell likely has means to compensate for the introduction of additional small pores to protect itself against simple colloid osmotic lysis. This fact may allow PFTs to utilize the pore structure to direct host cellular processes while avoiding the induction of host cell death. The recent work showing that *S. aureus* α-haemolysin activates its cellular receptor ADAM10 in a pore-dependent fashion to cause alterations of the host cytoskeleton and cell–cell interactions provides support for this notion (Wilke and Bubeck Wardenburg, 2010; Inoshima *et al.*, 2011). These studies (discussed below) suggest a new paradigm for PFT action in which local perturbations of the plasma membrane by the PFT alter cellular signalling pathways. Whether these alterations lead to cell death or the reprogramming of the cell for the benefit of the bacterium remains an unknown, though exciting concept.

It is possible that the PFTs may be used on a far larger scale to maintain the commensal state by reprogramming cell physiology or alternatively to protect the toxin-secreting organism from collateral damage when the host immune system is activated for other reasons. Many CDCs, which form extraordinarily large pores, have been shown to alter signalling pathways, often through perturbations of calcium levels (Cockeran *et al.*, 2001; Rose *et al.*, 2001; Braun *et al.*, 2002; Gekara *et al.*, 2007, 2008). It seems unlikely that the CDC pore initiates these changes simply by a pore-dependent alteration of calcium, since this could be achieved by the formation of a much smaller pore that requires significantly less energy to produce. There are also a plethora of reports that show the CDCs perturb many different signalling pathways, but the precise event that links these changes to the CDC pore (or the prepore oligomer) remains unclear. As will be evident from this review, we understand much more about how these pores are assembled in the membrane than how they are employed by the bacterial cell to promote either the commensal or pathogenic states. Emerging studies across the PFT field, however, are beginning to shed light on the diversity and specificity of cellular action of these toxins. It is quite reasonable to suggest that this knowledge will lead to a greater appreciation of the roles of PFTs in establishing the commensal-pathogen balance with the host.

The wide distribution of these toxins makes a review of all of the different families of pore-forming toxins impossible. This review will focus mainly on a few families of bacterial pore-forming toxins that illustrate many of the mechanistic and pathogenic paradigms that are observed in other families of pore-forming toxins. Two major families of pore-forming toxins will be the focus of this review, which are represented by *Staphylococcus aureus* α-haemolysin and the cholesterol-dependent cytolysins (CDCs), of which *Clostridium perfringens* perfringolysin O (PFO) is an archetype. The study of these two classes of pore-forming toxins has established several paradigms for the assembly of multimeric β-barrel pore-forming toxins, which have led to insights into the function of other pore-forming toxins. The aerolysin-like pore-forming toxins represent a third major class of pore-forming toxins that form a heptameric β-barrel pore structure similar to the *S. aureus* α-haemolysin class of toxins, but exhibit many features that set them apart. Aerolysin is the product of the Gram-negative bacteria *Aeromonas hydrophila* (Bernheimer and Avigad, 1974; Buckley *et al.*, 1981), but unlike other PFTs the aerolysin family members are also found in Gram-positive bacteria (Ballard *et al.*, 1995a; Cole *et al.*, 2004), plants (Fontes *et al.*, 1997) and mushrooms (Tateno and Goldstein, 2003). Humans also express a number of pore-forming proteins, such as perforin and the cytolytic membrane attack complex (MAC) of the complement system. These proteins can also be considered toxins, directing their toxic action towards invading bacterial pathogens, virus infected cells, and tumours. It is apparent from the literature on PFTs that unlike many other classes of toxins, the ability to form a pore in a target membrane is apparently useful to a vast array of organisms, from bacteria to humans, as well as replicative biological systems such as viruses (Gonzalez and Carrasco, 2003).

Introduction to *Staphylococcus aureus* α-haemolysin

Staphylococcus aureus α-haemolysin (α-toxin, Hla) is a membrane-damaging cytotoxin that serves as a prototype of small β-barrel pore-forming cytotoxins (PFTs) (Bhakdi and Tranum-Jensen, 1991). The study of α-toxin dates back to the late nineteenth century with the identification of a toxic activity associated with staphylococcal culture precipitates. Lethality in guinea pigs and rabbits, dermonecrosis, inflammation of the conjunctival epithelium, and haemolysis were among the first actions ascribed to α-toxin (De Christmas, 1888; von Leber, 1888; Breiger and Fraenkel, 1890; Rodet and Courmont, 1892; van de Velde, 1894; Neisser and Wechsberg, 1901; Kraus and Pribram, 1906). The subsequent identification of α-toxin as a single molecular entity was brought about in the 1920s and 1930s, most notably by the work of Glenny and Stevens who described that *S. aureus* produced two immunologically distinct toxins that displayed species-specific haemolytic activity, defining the rabbit-specific haemolysin as α-haemolysin (Glenny and Stevens, 1935). Through studies now spanning over a century, the cardinal activities of α-toxin documented in the late 1800s have proven to be among the core manifestations of its pathological activity.

The study of α-toxin has provided remarkable insight on the structure and function of the family of small β-barrel PFTs. Several defining features of these toxins include (1) secretion of toxin subunits in monomeric form by the bacterial organism, (2) selective recognition and binding to the host target cell, (3) assembly of an oligomeric structure on the host membrane following toxin binding and (4) perforation of the host lipid bilayer by amphipathic β-hairpins that collectively form a 1–3 nm solvent-filled β-barrel. Originally thought to function by simply causing irreversible membrane injury, it is now well appreciated that these toxins display a more subtle and refined action on nucleated target cells, altering specific cellular signalling pathways, eliciting immunological responses, modifying intracellular organelle function, and evoking specific functions in host cellular proteins (Fivaz *et al.*, 2001; Gonzalez *et al.*, 2008). Small β-barrel PFTs are utilized by Gram-positive and Gram-negative bacteria alike, with representative toxins including aerolysin from *Aeromonas hydrophila* (Bernheimer *et al.*, 1975; Buckley *et al.*, 1981), protective antigen (PA) from *Bacillus anthracis* (Vodkin and Leppla, 1983), *E. coli* α-haemolysin (Lovell and T.A., 1960; Bhakdi *et al.*, 1986), *V. cholerae* cytolysin (McCardell *et al.*, 1985; Krasilnikov *et al.*, 1992), and α-toxin of *Clostridium septicum* (Bernheimer, 1944; Ballard *et al.*, 1995b). The diverse array of bacteria that utilize this family of PFTs to generate pores in the target membrane highlights the exquisite ability of these toxins to provoke distinct and desirable responses in the host depending on the specific cellular target of the cytotoxin. Studies of the molecular and physiological effects of α-toxin on a number of human cell types including epithelial cells, platelets, T cells, endothelial cells and monocytes demonstrate this concept (Bhakdi and Tranum-Jensen, 1991; Bhakdi *et al.*, 2005).

The complex microenvironment of the host tissues, immunodefence strategies imposed by the host, and limited bacterial numbers during infection suggest the extraordinary potency and specificity ofbacterial cytotoxins. This has perhaps been best illustrated through the story of α-toxin, especially in light of significant recent advances in our knowledge of the molecular pathogenesis of toxin-mediated disease. In this chapter, we will discuss the structural and biological attributes of α-toxin as a model of the small β-barrel PFTs, highlighting features of other members of this toxin family that illustrate their dynamic interaction with the host. As this family of toxins, especially α-toxin, has been the subject of a number of excellent reviews (Bhakdi and Tranum-Jensen, 1991; van der Goot, 2001; Bhakdi *et al.*, 2005; Parker and Feil, 2005; Prevost *et al.*, 2005), we will recount the most salient historic studies that heralded the current, exciting findings in the field. In addition, we will note the impact that these findings may have on future studies in PFT biology and novel applications of this knowledge, both from the standpoint of *S. aureus* disease-modification and technology development.

Nomenclature and early observations

General PFT classification

Bacterial PFTs are classified as α-PFTs or β-PFTs, a structural notation that designates whether the toxin inserts an amphipathic pore comprised of an α-helix or β-hairpins, respectively (Heuck *et al.*, 2001). The β-PFTs are further classified as small or large, defined by the diameter of the membrane-spanning pore. Small β-PFTs such as α-toxin and aerolysin are generally <3 nm in diameter and thus only able to accommodate ions and small molecules. The large β-PFTs, exemplified by the cholesterol-dependent cytolysins such as streptolysin O, perfringolysin O and pneumolysin can span 20–30 nm (Gonzalez *et al.*, 2008). While the transmembrane β-barrel is utilized by a number of pathogenic bacteria as noted above, this structure is observed in outer membrane proteins of Gram-negative bacteria, mitochondria, and chloroplasts (Wimley, 2003; Fairman *et al.*, 2011). Distinct from the spontaneous insertion of β-PFTs into the host membrane, outer membrane β-barrel proteins rely on multiprotein–machinery complexes to facilitate transport, folding and insertion of the transmembrane channel (Fairman *et al.*, 2011). Further, these β-barrel proteins often act as transporters, adhesins, enzymes, and receptors, thus appearing distinct in function from the bacterial cytotoxins. The crystal structure of heptameric *S. aureus* α-toxin provided the first high-resolution view of the assembly of β-hairpins into a β-barrel pore (Song *et al.*, 1996). Subsequently, multiple additional structures of β-barrel proteins have been solved, including anthrax PA (Petosa *et al.*, 1997; Lacy *et al.*, 2004; Kintzer *et al.*, 2009), the *Yersinia pestis* adhesion protein Ail (Yamashita *et al.*, 2011), and the human mitochondrial transporter VDAC1 (Bayrhuber *et al.*, 2008) (Fig. 3.1), each revealing the ability of arrayed β-strands to demarcate a central pore of variable size that is formed by segregation of the membrane lipid from the hydrophilic channel interior. The diverse functions of these proteins illustrate the versatility of the β-barrel. In the context of infection with pathogenic bacteria, the β-barrel pore has been well studied as a vehicle of injury. The broad utilization of β-barrel-containing proteins in bacterial

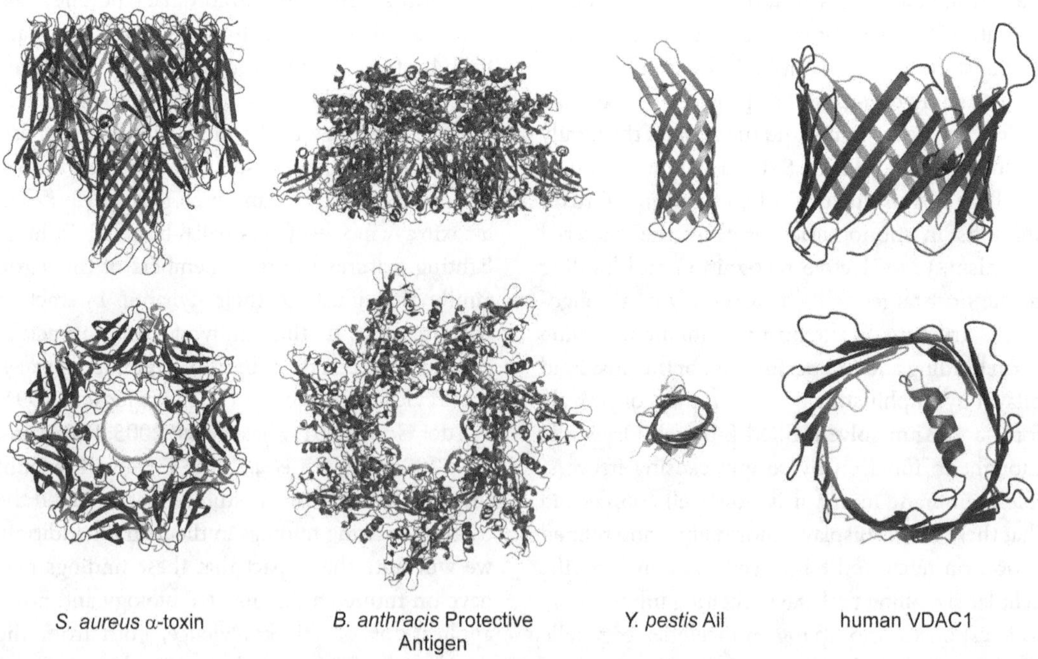

Figure 3.1 Structures of pore-forming proteins in which the β-strand motif is utilized to create the membrane-perforating pore. Structures were derived from the RCSB Protein Data Bank (PDB) and prepared using PYMOL (DeLano, 2002). *S. aureus* α-toxin, PDB 7AHL; *B. anthracis* Protective Antigen, PDB 1TZO; *Y. pestis* Ail, PDB 3QRC; human VDAC1, PDB 2JK4.

and eukaryotic homeostasis raises the intriguing possibility that injurious toxin pores may have the propensity to exhibit unique, non-cytotoxic functions if present at low concentrations in the setting of bacterial colonization (*Staphylococcus aureus*) or in the natural environment (*Aeromonas*).

Early studies on the biology of α-toxin

In the late 1920s, a keen focus on the injurious properties of *S. aureus* and its toxins arose following the death of 12 children in Bundaberg, Australia (Wiseman, 1975). These children were among a cohort of 21 that received an inoculation of a prophylactic diphtheria toxin–antitoxin preparation – within 2 days of receiving the immunization the children were dead. As early as a few hours after immunization, 16 children experienced vomiting, high fever, unconsciousness, convulsions, and cyanosis. All of the surviving children developed abscesses at the injection site. Investigation into the cause of this tragedy revealed contamination of the inoculum with a *S. aureus* strain capable of producing a soluble agent that led to haemolysis and lethal injury upon examination in rabbits (Kellaway *et al.*, 1928; Burnet, 1929; Elek, 1959). While the precise molecular attributes of this agent were not known at the time, a toxin was believed to be responsible.

The small number of studies on the toxic activity of staphylococcal supernatants conducted in the late 1800s and first two decades of the 1900s, together with the Bundaberg accident, drew the attention of F.M. Burnet to this area of investigation. He eloquently revisited the key findings of these studies, noting that multiple filtrates from distinct strains and preparations demonstrate haemolytic, dermonecrotic, and lethal activity (Burnet, 1929). Additional investigations were documented to reveal that antitoxin antiserum afforded protection against disease in rabbits and neutralized haemolytic and necrotic activity. Collectively, these seminal observations were communicated as five major conclusions (Fig. 3.2). Burnet's ideas not only captured the state of the field in his time, but foreshadowed molecular observations that would be made over the following 80 years that define α-toxin as a highly versatile exotoxin capable of eliciting distinct, cell-specific manifestations of injury. The formal designation of α-haemolysin as a distinct immunological and haemolytic entity followed from the studies of Glenny and Elek, respectively, a number of years after the Bundaberg accident and Burnet's work (Glenny and Stevens, 1935; Elek and Levy, 1950).

Despite the probable impurities of toxin preparations in the decades from 1930–1970, a number of significant observations on the nature of toxin-induced disease were compiled. Multiple small animal species were demonstrated to be susceptible to the lethal effects of the toxin (Wiseman, 1975). These comparative toxicity studies paralleled the exquisite sensitivity of

Conclusions.

(1) The agent obtained in staphylococcal culture filtrates that is capable of killing rabbits acutely on intravenous injection is a true exotoxin.

(2) The three activities of such filtrates in causing (*a*) hæmolysis *in vitro* (*b*) necrotic skin lesions on intradermal injection into rabbits and (*c*) acute death of rabbits after intravenous injections are manifestations of a single antigenic substance.

(3) A typical anatoxin can be obtained with full antigenic power but no toxicity and diminished hæmolytic power.

(4) The toxin is produced *in vivo* in rabbits infected with a toxigenic strain of staphylococci.

(5) If rabbits are actively or passively immunised to the toxin their survival time after intravenous injection of virulent staphylococci is increased.

JOURN. OF PATH.—VOL. XXXII. 3 B

Figure 3.2 F.M. Burnet's conclusions are shown as stated in 1929 on the properties and biological activity of the proposed *S. aureus* exotoxin responsible for the Bundaberg accident (Burnet, 1929). Burnet was the first to propose that distinct biological actions could be attributable to a single exotoxin.

rabbit erythrocytes to haemolysis (Cooper *et al.*, 1966; Bernheimer, 1968) – rabbits succumbed at an LD_{50} of 2 μg/kg α-toxin, the lowest LD_{50} of any species tested (Arbuthnott, 1970).

Properties of the α-haemolysin protein, including molecular weight and protein sequence determinations, were delineated in the mid-1900s, concomitant with the description of techniques to purify the toxin (Wiseman, 1975). In the 1960s, isolation of purified α-toxin from culture supernatants (Kumar *et al.*, 1962; Bernheimer and Schwartz, 1963; Lominski *et al.*, 1963) ushered in a new era of the study of this toxin, allowing for a vast array of biochemical, cellular biological and structural studies to be performed. Upon purification of the toxin, Bernheimer and Schwartz noted, 'In view of the rapidity with which it brings about cell damage and in view of its remarkable lytic action on red blood cells, as distinct from diphtheria, tetanus, and botulinum toxin which have neither of these properties, perhaps the best hypothesis is that it alters or disrupts cell membranes' (Bernheimer and Schwartz, 1963). In agreement with the observation that the toxin could cause systemic and local manifestations of disease in experimental animals, purified α-toxin was found to be cytotoxic to a wide array of cells in culture (Arbuthnott, 1970; Jeljaszewicz *et al.*, 1978). Several key lines of evidence emerged suggesting that the toxin caused direct injury to the host cell membrane: (1) low molecular weight markers of <1–2 nm in size leaked out of toxin-treated cells (Weissmann *et al.*, 1966; Thelestam *et al.*, 1973), (2) oligomeric toxin protein could be solubilized and recovered from red cell membrane preparations (Fussle *et al.*, 1981), and (3) membrane lesions were visualized in the plasma membrane of toxin-treated cells (Freer *et al.*, 1968). Importantly, studies on human platelets revealed that these cells become activated, aggregating in the presence of α-toxin (Siegel and Cohen, 1964). Thus, while the toxin was initially found to display haemolytic activity in a non-human species, it became clear that the toxin was capable of targeting a wider array of cells, most notably including those of a human. Multiple reviews provide a very detailed record of the many early studies that formed the basis for our knowledge of α-toxin; we recommend these to the reader (Arbuthnott, 1970; Wiseman, 1975; Bhakdi and Tranum-Jensen, 1991; Bhakdi *et al.*, 2005).

Primary structure and regulation of toxin production

While fundamental observations on the biological attributes and structure of α-toxin had been made in the early 1900s, identification of the encoding genetic locus was not reported until 1983 (Kehoe *et al.*, 1983). The full DNA coding and protein sequence of the toxin was reported shortly thereafter in 1984 (Grey and Kehoe, 1984). Utilizing a recombinant phage-based strategy paired with a species-specific screen for red blood cell haemolysis, Kehoe and colleagues successfully expressed α-toxin in *E. coli* as a cell-associated protein (Kehoe *et al.*, 1983). Originally harboured within a 7.6 kb region from the staphylococcal chromosome, transposon mutagenesis and subcloning enabled the isolation of a 1620 bp genomic DNA fragment that encoded the toxic activity as determined in a rabbit erythrocyte haemolysis assay. This cloned sequence facilitated the secretion of α-toxin from *B. subtilis* and *S. aureus* (Fairweather *et al.*, 1983).

The α-toxin gene is present in a single copy on the staphylococcal chromosome. Genomic sequencing has revealed that the toxin locus is nearly invariant across strains, with almost complete conservation of the primary amino acid sequence. While early reports of the protein sequence were conflicting in their delineation of the N-terminal (Kato and Watanabe, 1980; Grey and Kehoe, 1984) and C-terminal (Six and Harshman, 1973; Kato and Watanabe, 1980) amino acid residues, it is now accepted that the mature toxin is a 293 amino acid protein of ~34 kDa (Grey and Kehoe, 1984; Bhakdi and Tranum-Jensen, 1991; Song *et al.*, 1996). Analysis of the primary protein sequence reveals α-toxin to consist of a 26 amino acid signal sequence predicted to adopt an alpha helical confirmation (Grey and Kehoe, 1984), in agreement with biochemical studies demonstrating that the toxin is processed either before or during secretion to yield the mature toxin (Tweten *et al.*, 1983). Notably, the α-toxin polypeptide does not contain any cysteine residues. This fact has been capitalized on by a number of studies in which the introduction of cysteine residues has permitted both discrete site labelling for assessment of toxin structure and the generation of 'locked' mutants that are unable to form membrane-inserted pores (Valeva *et al.*, 1997a,b; Kawate and Gouaux, 2003). Circular dichroism analysis revealed that the toxin is principally composed of β-strands with very limited α-helical structure (Tobkes *et al.*, 1985).

Expression of the α-toxin monomer is subject to a complex regulatory system, enabling tight control of toxin production by *S. aureus* (Recsei *et al.*, 1986). The accessory gene regulator (*agr*) locus is a phosphorelay-based, quorum-sensing system that provides primary governance of toxin expression by a regulatory RNA molecule, RNAIII (Peng *et al.*, 1988; Novick *et al.*, 1993; Ji *et al.*, 1995). During the late-log and stationary phases of growth, the *agr* locus is transcriptionally active, driven by a two-promoter system that both enables the production of and response to the secreted autoinducer peptide, AIP. Peptide binding to its cognate cell surface receptor, AgrC, activates the response regulator AgrA (Ji *et al.*, 1995; Lina *et al.*, 1998; Lyon *et al.*, 2002); binding of this regulator to the promoter regions within the *agr* locus allows for the generation of the RNAIII transcript (Koenig *et al.*, 2004) and modulation of α-toxin expression (Novick *et al.*, 1993). These signals are modified by both the Sae and Sar regulatory circuits, providing an addition level of control on toxin expression (Xiong *et al.*, 2006; Reyes *et al.*, 2011). While it is challenging to objectively define precisely how these circuits function in a concerted manner *in vivo* to facilitate the host–pathogen interaction, it is straightforward to appreciate the relevance of tightly modulating the expression of such a toxin. Recent strategies targeting α-toxin-mediated *S. aureus* diseases have examined small molecule inhibitors of the Agr system as a means to down-modulate toxin expression, showing some degree of promise in preclinical studies (Leng *et al.*, 2011; Qiu *et al.*, 2011; Luo *et al.*, 2012).

Three-dimensional structure and function

The β-barrel PFTs share a similar overall mechanism of assembly that leads to the generation of a membrane-inserted oligomeric toxin pore (Heuck *et al.*, 2001). Toxin monomers are secreted from the host bacterial organism in a water-soluble state; these then associate with the host cell membrane, initially retaining their monomeric form. Host cell surface moieties such as lipids, sugars, or proteins govern toxin binding, imposing specificity of toxin action as will be discussed in more detail later in this review. Oligomerization of the toxin monomers occurs on the target cell membrane, likely facilitated by the lipid microenvironment and the clustering of toxin monomers. In the case of the small β-PFTs, the subsequent insertion of one β-hairpin motif from each monomer then occurs in a concerted fashion, displacing the membrane lipid as these hairpins form the cylindrical β-barrel structure that is the hallmark of this toxin family. Each step in the host cell assembly process has been explored through biochemical, biophysical, and structural studies for staphylococcal α-toxin, establishing a paradigm for understanding the mechanisms of cell binding, oligomerization and membrane insertion for related toxins.

Host cell binding

Foreshadowed by Glenny and Stevens' early observations on the specificity of α-toxin for rabbit erythrocytes, studies have noted that susceptibility to PFTs can vary considerably by cell type and species (Glenny and Stevens, 1935). The most dramatic illustration of this is seen in the dichotomy between the sensitivity of rabbit and human erythrocytes to α-toxin – the rabbit cell undergoes haemolysis at concentrations of toxin in the low nanomolar range, whereas the human erythrocyte resists injury into the high nanomolar to low micromolar range (Hildebrand *et al.*, 1991). Utilizing radioiodinated α-toxin, Cassidy and Harshman (1976) conducted a detailed analysis of host cell binding. They reported that the interaction of α-toxin with rabbit erythrocytes was saturable and time-dependent, calculating a dissociation constant (K_d) of 6×10^{-9} at 20°C. From these studies, 5000 discrete toxin-binding sites were estimated per red blood cell. These pioneering studies were the first to provide objective biochemical evidence for the presence of a high-affinity cell surface receptor, championed by these investigators to mediate the initial stage of toxin association with a susceptible host cell. An extension of these observations was provided by Hildebrand and colleagues, who established that a finite number of high-affinity 'receptors' (~ 1500–2000) were present on sensitive cells, resulting in half-maximal toxin binding at 2 nM. In contrast, non-susceptible cells were subject to the adsorptive binding of α-toxin that could yield lysis only at high toxin concentrations (Hildebrand *et al.*, 1991). Together, these observations suggested the presence of a proteinaceous cellular receptor for α-toxin, in stark contrast to contemporaneous observations demonstrating the ability of the toxin to perforate lipid vesicles and cause release of the intravesicular contents (Weissmann *et al.*, 1966; Arbuthnott *et al.*, 1973a; Menestrina, 1986; Belmonte *et al.*, 1987; Ikigai and Nakae, 1987a,b; Watanabe *et al.*, 1987; Forti and Menestrina, 1989). Paired with the elusive nature of the proteinaceous toxin receptor, the insertion of α-toxin pores into a purified lipid membrane cast considerable doubt on the necessity or relevance of a protein dock on susceptible cells.

To explain these conflicting observations, Valeva and colleagues hypothesized that clustered phosphocholine head groups in the membrane of susceptible cells functioned as the high-affinity α-toxin binding site (Valeva, 2006). While incapable of providing an explanation for species specificity in erythrocytes, this view went unchallenged in the field for several years until A disintegrin and metalloprotease 10 (ADAM10) was identified as the proteinaceous receptor for α-toxin (Wilke and Wardenburg, 2010). Identified through a biochemical approach taking advantage of the differential susceptibility of rabbit and human erythrocytes to lysis, ADAM10 was shown to be required for toxin binding and susceptibility to lysis in lung epithelial cells. This study documented species-specific red cell expression of ADAM10, revealed the necessity of the protein receptor at low toxin concentrations, and noted toxin-induced clustering of ADAM10 in caveolin-1-enriched membrane microdomains. The identification of ADAM10 as the α-toxin receptor thereby aligns an array of seemingly disparate observations spanning more than two decades, providing an opportunity to dissect the molecular underpinnings of toxin assembly on the host cell membrane.

Oligomerization

Electron micrographic images of cells or artificial liposomes treated with α-toxin revealed the presence of 10 nm ring-like structures estimated to contain six or seven arrayed subunits, suggesting that the toxin may require aggregation or the formation of a more complex structure for activity (Arbuthnott *et al.*, 1967, 1973a,b; Freer *et al.*, 1968, 1973; Remsen *et al.*, 1970; Fussle *et al.*, 1981). Subsequent biochemical studies were consistent with the formation of stable hexamers on the membrane that could be purified and visualized by SDS-PAGE analysis or quantified by differential ELISA-based methodology (Fussle *et al.*, 1981; Hugo *et al.*, 1987; Reichwein *et al.*, 1987). Gouaux and colleagues challenged the formation of toxin hexamers in 1994 with X-ray diffraction analysis that documented the toxin oligomer to be heptameric (Gouaux *et al.*, 1994). Shortly thereafter, a breakthrough in the field of PFT biology and the study of α-toxin followed with the solution of the crystal structure of the fully assembled toxin pore, confirming the heptameric structure (Song *et al.*, 1996).

Across the family of small β-PFTs, defining the precise mechanisms by which cell-bound toxin monomers assemble to form oligomers has proven challenging. Several rather plausible models have come to the forefront: (1) toxin-receptor binding allows for localization of the toxin on the cell surface, predicated by the distribution of the receptor or (2) membrane lipid sites provide a platform for the simultaneous binding of multiple toxin monomers. Aerolysin provides an interesting model wherein these strategies may be simultaneously applicable, and in which a molecular trigger of oligomerization has been defined. Aerolysin was the first small β-PFT resolved by X-ray crystallography, demonstrating a β-strand enriched four-domain structure (Parker *et al.*, 1994a). Proaerolysin recognizes GPI-anchored proteins as its cellular receptor through cooperative binding involving recognition of the N-linked sugar resides on the protein and an N-glycan moiety that decorates the GPI anchor by domains 1 and 2, respectively (Cowell *et al.*, 1997; Abrami *et al.*, 2002; Hong *et al.*, 2002). GPI anchored proteins predominate in membrane lipid rafts, thus the receptor and the membrane microenvironment are colocalized to favour monomer-monomer contact (Abrami and van Der Goot, 1999). Proaerolysin undergoes proteolytic cleavage at Arg_{432} by host cell furin to yield aerolysin (Abrami *et al.*, 1998), releasing a C-terminal peptide from Domain 4, the β-strand-enriched region of the toxin that inserts into the host cell membrane (Parker *et al.*, 1994a). Proteolysis results in significant structural changes in aerolysin, associated with increased β-sheet structure and the ability to oligomerize (Cabiaux *et al.*, 1997). Similar to aerolysin, *V. cholerae* cytolysin requires proteolysis by either bacterial or cellular proteases to enable the monomeric form of the toxin to transition to an oligomer (Valeva *et al.*, 2004).

In the case of α-toxin, strong evidence suggests that the toxin interacts with the membrane lipid –basic and aromatic residues in the so-called 'rim-stem crevice' of the toxin define sites of heptamer-lipid contact (Galdiero and Gouaux, 2004), the toxin can damage protein-free liposomes (Weissmann *et al.*, 1966; Arbuthnott *et al.*, 1973a; Menestrina, 1986; Belmonte *et al.*, 1987; Ikigai and Nakae, 1987a,b; Watanabe *et al.*, 1987; Forti and Menestrina, 1989), cholesterol extraction abrogates toxin binding (Valeva, 2006), and exogenous phosphocholine can antagonize toxin binding (Watanabe *et al.*, 1987; Valeva, 2006). The recent identification of ADAM10 as the cellular receptor for α-toxin provides the initial opportunity to examine the relative contribution of membrane-facilitated aggregation versus receptor binding for α-toxin oligomerization. ADAM10 is an intriguing proteinaceous receptor that may allow one to envisage another mode of oligomerization – binding of monomeric toxin to ADAM10 may be akin to a ligand–receptor interaction that induces receptor-mediated signal transduction events; these in turn may augment local toxin binding or nucleate oligomer formation.

Pore formation

The transforming structural events that result in perforation of the host lipid bilayer by PFTs have been the study of intensive investigation. Among the small β-PFTs, α-toxin and *B. anthracis* PA have been nicely characterized in this regard, taking advantage of biochemical and structural analytic tools to define the movements of discrete protein segments into the membrane to generate the pore. As this topic has been the subject of extensive review within the structural biology field, we refer the reader to these reviews for a detailed discussion of experimental data (Gouaux, 1997, 1998). In this section, we highlight several notable features of the α-toxin pore, as they have proven to be of considerable interest in light of recent observations on the toxin. α-toxin exhibits a well-defined pre-pore state, representing the fully assembled oligomeric structure on the host cell membrane. The transitions from monomeric toxin to the fully assembled pore are therefore discernible as distinct conformational entities. The pre-pore to pore transition observed in α-toxin is irreversible, and interestingly bears mechanistic similarity to the same transition made by *C. septicum* alpha toxin (Sellman *et al.*, 1997a) and the alpha-helical cytolysin A (ClyA) of *E. coli* and *S. enterica* (Mueller *et al.*, 2009).

The structure of the water-soluble α-toxin monomer has been reported (Meesters *et al.*, 2009), providing insight on the pre-pore mechanism. While the small angle X-ray scattering technique utilized to obtain this structure suffers from low resolution, this technique was combined with molecular modelling to reveal that the majority of the monomeric structure is geometrically conserved in the heptamer. Several significant movements or regions of flexibility are noted in the transition process: (1) the rather dramatic down-folding of the central glycine-rich β-hairpin stem (Lys_{100}-Tyr_{148}) away from the monomer to perforate the membrane, (2) extension of the N-terminal segment (Ala_1-Val_{20}) towards the neighbouring protomer in the assembled heptamer, stabilizing the structure, and (3) flexibility of the phosphocholine binding region at the rim–stem interface (Meesters *et al.*, 2009).

Valeva and colleagues rigorously examined the pre-pore to pore transition utilizing site-specific labelling of engineered cysteine mutants paired with biochemical and functional studies. They demonstrated that His_{35} moves into a hydrophobic environment at this transition point, precipitating the insertion of the stem domain into the membrane (Valeva *et al.*, 1997a). In a cooperative manner, the N-terminal region of the protomer undergoes a conformational change to 'latch' onto the immediately adjacent protomer, stabilizing the heptameric pore structure (Valeva *et al.*, 1997b). Consistent with these findings, truncation of 20 amino acids at the N-terminus yields inactive toxin oligomers (Walker *et al.*, 1992b). Early studies suggested that the toxin underwent proteolytic cleavage at the N-terminus to yield the active moiety (Wiseman and Caird, 1972); however, this has not been borne out in further studies and is inconsistent with a model in which this region stabilizes the active toxin conformation in the membrane (Freer *et al.*, 1973).

Paired with detailed analysis of the role of specific amino acids with the toxin for membrane binding, oligomerization, and pore formation (Jursch *et al.*, 1994; Menzies and Kernodle, 1994; Walker and Bayley, 1995), His_{35} has been defined as essential for progression of the toxin from the pre-pore to the pore state. Protein variants with substitutions at this residue are unable to form a stable oligomer, and as such, His_{35} mutants are incapable of assembly into a lytic pore in spite of their preservation of cellular binding ability. While fundamental to observations in the 1990s that informed our knowledge of the structural basis for α-toxin pore formation, mutant toxin molecules harbouring substitutions at His_{35} were readily appreciated to be potential vaccine candidates (Bhakdi *et al.*, 1994). Early studies in this regard demonstrated promising results in protecting mice against a lethal intraperitoneal challenge with *S. aureus* (Menzies and Kernodle, 1996).

Role in disease pathogenesis

In 2005, Bhakdi and colleagues stated, 'Inasmuch as the pathogenesis of staphylococcal disease is multifactorial, alpha-toxin probably assumes a dominant role in most cases' (Bhakdi *et al.*, 2005). Early studies alluded to the triad of lethal disease, haemolysis, and dermonecrosis as the chief manifestations of α-toxin-induced host injury. The identification of ADAM10 as the α-toxin receptor has allowed a rigorous examination of the role of the toxin in these disease states in the past two years (Inoshima *et al.*, 2011; Inoshima *et al.*, 2012; Powers *et al.*, 2012). Recent studies both confirm early observations and provide intriguing new molecular insight on the mechanism by which a single toxin can govern the pathogenic manifestations of disease in distinct organ systems and tissues. In addition, emerging observations on the ability of the toxin to manipulate other facets of host biology, particularly innate immune function, have greatly expanded our understanding of the diverse capabilities of the toxin (Yarovinsky *et al.*, 2008; Craven *et al.*, 2009; Hruz *et al.*, 2009; Frank *et al.*, 2012; Kebaier *et al.*, 2012; Lizak and Yarovinsky, 2012).

Advances in *S. aureus* genetic manipulation have allowed for perturbation of α-toxin expression and a rigorous analysis of its role in the pathogenesis of disease in experimental animals. Toxin-deficient strains display reduced virulence in pneumonia (Bubeck Wardenburg *et al.*, 2007a,b), severe dermonecrotic skin infection (Patel *et al.*, 1987; Kennedy *et al.*, 2010), lethal disease (Menzies and Kernodle, 1996; Powers *et al.*, 2012), peritonitis (O'Reilly *et al.*, 1986; Patel *et al.*, 1987), infection of the cornea (O'Callaghan *et al.*, 1997), central nervous system (Kielian *et al.*, 2001) and endocardium (Bayer *et al.*, 1997), as well as in the setting of mastitis (Bramley *et al.*, 1989).

It has been clear for many years that α-toxin elicits biological responses in host cells and tissues at concentrations not known to cause overt cellular lysis (Bhakdi and Tranum-Jensen, 1991; Bhakdi *et al.*, 2005). This finding is enigmatic, raising two questions: (1) Why form a membrane-active pore with the capability to kill cells if this is not strictly required for toxin action? and (2) Mechanistically, how does a small membrane pore elicit specific signal transduction pathways (biological responses) in an wide array of susceptible, yet very distinct cell and tissue types? Answers to these questions begin to emerge from substantial recent advances in our understanding of the host–pathogen interaction as it pertains to the interface of the toxin with target cells and the immune system – these will be the focus of this section of the chapter.

Mechanisms of cellular and tissue injury

The α-toxin receptor ADAM10 is a zinc-dependent metalloprotease expressed on a wide array of human cells as a type-I transmembrane protein (Seals, 2003; Edwards *et al.*, 2008). The extracellular domain of ADAM10 is comprised of the N-terminal enzymatic domain followed by the so-called 'disintegrin' and cysteine-rich domains, both of which facilitate protein–protein interactions. Within the short cytoplasmic tail of ADAM10 is a proline-rich sequence and a consensus binding site for calmodulin (Nagano *et al.*, 2004). The primary role of ADAM10 is as a cell-surface 'sheddase', responsible for the cleavage of an array of ectodomain-containing proteins proximal to the cell membrane (Tousseyn *et al.*, 2006; Reiss and Saftig, 2009). ADAM10 substrates include cytokines and chemokines, members of the cadherin family, amyloid precursor protein (APP), platelet glycoprotein VI, notch, and ephrin to name a few (Allinson *et al.*, 2003; Janes *et al.*, 2005; Maretzky *et al.*, 2005; Gardiner *et al.*, 2007; Schulte *et al.*, 2007; Schulz *et al.*, 2008; Gibb *et al.*, 2010). Substrate specificity varies depending on the cell type and the regulatory mechanisms in place in the specific cellular context to govern ADAM10 enzymatic action (Van Wart and Birkedal-Hansen, 1990; Janes *et al.*, 2005; Ludwig *et al.*, 2005; Escrevente *et al.*, 2008; Janes *et al.*, 2009; Endres and Fahrenholz, 2010). Ectodomain cleavage by ADAM10 results in the liberation of the substrate's extracellular domain and retention of a membrane-bound fragment, which often is rapidly subjected to further proteolytic cleavage events. ADAM10-mediated cleavage yields discrete biological outcomes, as many extracellular and intracellular cleavage products are active signalling moieties that facilitate both physiological and pathophysiological processes.

The requirement for ADAM10 as the cellular receptor for α-toxin in *S. aureus* pathogenesis was recently demonstrated utilizing conditional knockout approaches in the alveolar epithelium and the mature epidermis (Inoshima *et al.*, 2011, 2012). Germline deletion of ADAM10 results in embryonic lethality at E9.5 (Hartmann *et al.*, 2002), and is associated with cell-specific abnormalities upon conditional knockout in T cell precursors and in the developing epidermis (Tian *et al.*, 2008; Weber *et al.*, 2011). The loss of ADAM10 expression in the lung led to a marked improvement in the outcome of *S. aureus* pneumonia (Inoshima *et al.*, 2011), minimizing lethal disease compared with control mice. Similarly, ADAM10 knockout in the skin was associated with a reduction in the size of *S. aureus* skin lesions and abrogation of the severe dermonecrotic tissue injury that is a hallmark of the action of α-toxin (Inoshima *et al.*, 2012).

In the context of epithelial cells (such as pneumocytes and keratinocytes) that are key targets of α-toxin, E-cadherin is a principal substrate for ADAM10 (Maretzky *et al.*, 2005, 2008). Cleavage of E-cadherin by ADAM10 results in a loss of the homotypic interaction of the cadherin molecules on adjacent cells at the adherens junction, thereby injuring the epithelial tissue barrier function. *In vitro* studies demonstrated the surprising finding that treatment of epithelial cells with sublytic concentrations of α-toxin leads to a rapid up-regulation of the metalloprotease activity of the receptor, which in turn dismantled the adherens junction through cleavage of E-cadherin (Inoshima *et al.*, 2011, 2012). Extension of these studies in the context of infection revealed that α-toxin caused a primary disturbance of the epithelial barrier in the lung and the skin, manifesting as proteinaceous pulmonary oedema and dermonecrotic injury, respectively.

More broadly considering barrier functions within the host as a target of the toxin, studies of the vascular endothelial barrier and staphylococcal sepsis demonstrated that toxin binding to ADAM10 on primary endothelial cells leads to rapid activation of the metalloprotease and VE-cadherin cleavage (Powers *et al.*, 2012). Toxin-deficient *S. aureus* strains display a virulence defect in a mouse model of lethal sepsis induced by intravenous inoculation of the pathogen; this correlates with toxin-mediated induction of endovascular injury and increased vascular permeability as documented by dye extravasation studies (Powers *et al.*, 2012).

These observations hearken back to the early studies on the role of Hla in lethal disease and dermonecrotic injury, now providing molecular detail of α-toxin action in these states of host injury. It becomes clear that the functional outcome of the toxin–receptor interaction is predicated on the tissue-specific actions of ADAM10, precipitating the central pathophysiological consequences of infection. The pleiotropic effects of a single toxin can therefore now be explained by two principles – (1) targeting of multiple host cell types through widespread expression of the toxin receptor ADAM10 and (2) diversity of cellular effects of the toxin dependent on the both the level of ADAM10 expression and the native functions of ADAM10 in the cell/tissue. Importantly, these outcomes are absolutely dependent on the assembly and patency of the α-toxin pore (Inoshima *et al.*, 2011), demonstrating the primacy of the β-PFT structure for the full range of the toxin's biological activity.

Immunomodulation

Multiple studies have indicated that immune cells are targets of α-toxin. While of interest, these historic studies have not yet led to an appreciation for the role of the toxin in manipulating the immune response *in vivo* to facilitate pathogenesis. In recent years, several key studies have shed light on this topic, illustrating that α-toxin targets both innate and adaptive immune cells, altering the host response to staphylococcal infection and again demonstrating the diverse capabilities of the toxin.

Pro-inflammatory signalling in the host in response to infection is a double-edged sword, affording protection from pathogens yet contributing to self-injury when overly robust. Inflammation is a key feature of *S. aureus* infection, most readily appreciated in the lungs and skin wherein the rapid infiltration of innate immune cells is observed in both human and murine hosts (McElroy *et al.*, 1999; Miller and Cho, 2011; Parker and Prince, 2011). In murine pneumonia, α-toxin is required to generate a gradient of keratinocyte-derived chemokine (KC) and macrophage inflammatory protein-2 (MIP-2), CXC chemokines that facilitate neutrophil recruitment to the lung (Bartlett *et al.*, 2008). α-toxin induces inflammatory responses in multiple cells, resulting in the release of cytokines and vasoactive agents (Suttorp *et al.*, 1985, 1992, 1993; Bhakdi *et al.*, 1989; Grimminger *et al.*, 1997; Onogawa, 2002; Rose *et al.*, 2002). One hallmark of innate immune cell activation is the secretion of the inflammatory cytokine interleukin 1β (IL-1β), a consequence of inflammasome activation and caspase-mediated cleavage of pro-IL-1β to yield the active form of this molecule. Intoxication with α-toxin induces IL-1β secretion in macrophages and monocytes, implicating this lineage as a target of the toxin during infection (Bhakdi *et al.*, 1989; Craven *et al.*, 2009; Kebaier *et al.*, 2012). Craven and colleagues demonstrated that the nucleotide binding domain and leucine-rich repeat-containing gene family, pyrin domain-containing protein (NLRP3) inflammasome was activated in monocytic cells following exposure to α-toxin, resulting in caspase-1 activation and IL-1β secretion (Craven *et al.*, 2009). While antagonism of the toxin *in vivo* through toxin-neutralizing antibodies blunted IL-1β secretion during *S. aureus* pneumonia (Bubeck Wardenburg and Schneewind, 2008), the molecular mechanisms underlying this response *in vivo* had not been investigated until recently. Following on the work of Craven, Kebaier *et al.* (2012) demonstrated toxin-dependent activation of the NLRP3 inflammasome in *S. aureus* pneumonia, leading to necrotic tissue injury. Consistent with these findings, an attenuation of disease is observed in mice harbouring germline deletion of *Nlrp3*. Mice lacking expression of NLRP3 display increased survival following intra-tracheal instillation of purified α-toxin, and show decreased lung pathology, IL-1β secretion and neutrophil infiltrates upon infection with live staphylococci. These studies strongly implicate the role of NLRP3 and this inflammatory cascade as downstream effectors of α-toxin-mediated pathogenesis, linking this pathway to toxin-induced cellular necrosis. Interestingly, while antagonism of α-toxin by active vaccination, passive immunization, and direct small molecule inhibitors decreases the bacterial load in the lung during infection (Bubeck Wardenburg and Schneewind, 2008; Ragle and Bubeck Wardenburg, 2009; Ragle *et al.*, 2010), mice harbouring a deletion of either *Nlrp3* or conditional deletion of *Adam10* in the alveolar epithelium do not display a decreased bacterial burden following infection (Inoshima *et al.*, 2011; Kebaier *et al.*, 2012). Together, these results further highlight the multiple cellular and tissue pathways that are toxin responsive, and the complexity of the host–pathogen interaction centred on this PFT.

In addition to these effects on host innate immunity, a number of emerging studies indicate that α-toxin modulates the crosstalk between innate and adaptive immunity and patterns the adaptive immune response. Treatment of human monocytes with subcytolytic concentrations of α-toxin potently stimulates the secretion of interleukin 17A (IL-17A) (Niebuhr *et al.*, 2011). This cytokine polarizes the helper T cell response towards the induction of Th17 cells, a subset of $CD4^+$ T cells that both respond to and express IL-17A, and contribute most notably to immunoprotection of the epithelium. The Th17 response has been shown to be important in the context of *S. aureus* skin infection, implicated both in the immunopathogenesis of toxin-enhanced inflammatory skin disease and in protection against acute infection (Cho *et al.*, 2010; Miller and Cho, 2011; Niebuhr *et al.*, 2011). Most recently, a whole-transcriptome approach was utilized as a tool to discern the impact of α-toxin on the host response to infection. This study compared responses in mice infected with wild-type *S. aureus* to those infected with toxin-deficient *S. aureus* in a murine model of pneumonia. Toxin expression was strongly associated with induction of the IL-17A response (Frank *et al.*, 2012). Mice infected with α-toxin-expressing *S. aureus* generate a T cell repertoire characterized by a greater number of $IL\text{-}17A^+$ cells, illustrating a direct impact of the toxin on the adaptive immune system. While the molecular mechanisms that underlie the induction of the polarized Th17 response are not yet known, the nucleotide-binding oligomerization domain-containing 2 (NOD2) has been shown to play an important role in mediating innate immunodefence against *S. aureus*, dependent on the expression of α-toxin (Hruz *et al.*, 2009). As NOD2 signalling can promote Th17 differentiation (van Beelen *et al.*, 2007), this represents a plausible pathway for further investigation.

α-toxin induces apoptotic cell death in monocytes, B cells and T cells (Nygaard *et al.*, 2012), further indicative of the broad impact of this virulence factor on innate and adaptive immunity. While many of these observations suggest that the toxin's harmful actions are unopposed by the host, epithelial cells are able to repair toxin-induced membrane injury and eliminate the toxin through a linked process of endocytosis from the membrane and exocytosis of the toxin in so-called 'toxosomes', or exosome-like vesicles (Valeva *et al.*, 2000; Husmann *et al.*, 2006, 2009). Further, type-I interferon (IFN) produced by the host can afford protection against toxin-induced injury (Yarovinsky *et al.*, 2008). This effect is dependent on the presence of phospholipid scramblase I (PLSCR1), which provides protection against cellular leakage of ATP (Lizak and Yarovinsky, 2012). While the precise mechanisms of action of PLSCR1 are not yet elucidated, strong support for the role of this protein in host protection against α-toxin is observed in PLSCR1 knockout mice that demonstrate increased susceptibility to the injurious actions of the toxin in the lung (Lizak and Yarovinsky, 2012).

These concepts are depicted in Fig. 3.3, highlighting the toxin–receptor interaction on the membrane, the wide array of cells targeted by α-toxin, and the cellular responses that vary not only by the cell that is being intoxicated, but by the relative concentration of toxin to which the cell is exposed.

Summary and future perspectives – α-toxin

Together, recent studies illuminate two notable features of α-toxin biology in the mammalian host: (1) pleiotropic actions of the toxin on multiple cell and tissue types, predicated by expression of the toxin receptor ADAM10 and the native, cell-specific functions of the receptor, and (2) versatility of the toxin in the setting of the complex microenvironment of the host tissue – α-toxin is capable of causing both overt, severe lytic injury to cells yet is equipotent at directing the host–pathogen interaction via more subtle mechanisms that become apparent at low toxin concentrations. Discoveries on the biology of α-toxin as compiled

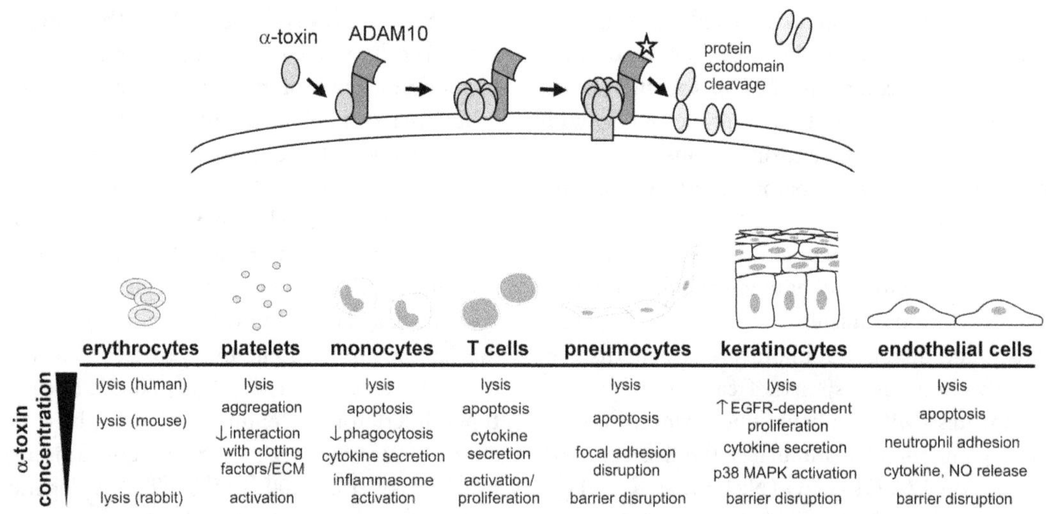

Figure 3.3 Shown is a diagram illustrating the formation of the α-toxin–ADAM10 complex, facilitating membrane binding of the toxin, oligomerization, and pore formation (upper panel). The formation of the toxin pore leads to rapid up-regulation of the metalloprotease activity of ADAM10 (denoted by a star). ADAM10, in turn, acts in a cell-specific manner to cleave ectodomain-containing proteins that may represent biological mediators of α-toxin action. Multiple cell types are targeted by α-toxin, each displaying unique effects that are dependent on the relative concentration of toxin to which the cell is exposed (lower panel).

over the past 120 years inform our understanding of the entire family of small β-PFTs, define innovative strategies for human disease modification, and reveal the critical questions that remain unanswered in the field thereby highlighting novel areas for future investigation.

A number of technological advances have been explored based on knowledge of the structural and functional biology of α-toxin. Among the most prominent of these is targeting the toxin to afford protection against *S. aureus* disease. While the concept of toxin antagonism to ameliorate infectious disease dates back to the days of Emil von Behring and Paul Ehrlich, a dramatic increase in the number of efforts to target α-toxin has been witnessed in recent years. Vaccines, passive immunization strategies, small molecule inhibitors of the toxin, and most recently small molecule-based targeting of host ADAM10 have all demonstrated a degree of efficacy in combatting *S. aureus* disease (Bubeck Wardenburg and Schneewind, 2008; Ragle and Bubeck Wardenburg, 2009; Kennedy *et al.*, 2010; Ragle *et al.*, 2010; Inoshima *et al.*, 2011, 2012; Powers *et al.*, 2012; Spaulding *et al.*, 2012; Tkaczyk *et al.*, 2012). As such, many of these modalities are being developed for and examined in human clinical trials.

A second avenue of technology development around small β-PFTs has capitalized on the biophysical and structural attributes of the toxin to elicit controlled permeabilization and biosensor capability. The addition of adapter molecules to the α-toxin pore allows this structure to function as a sensor element in detecting organic molecules, highlighting the adaptability of this bacterial toxin as a useful machine (Gu *et al.*, 1999). As a 'nanopore', fully assembled α-toxin permits the translocation of single stand nucleic acids, allowing for the discrimination of single base pairs via electrical recordings of the pore's ionic current as nucleotides traverse the channel (Branton *et al.*, 2008; Clarke *et al.*, 2009; Stoddart *et al.*, 2009, 2010; Japrung *et al.*, 2010). This technology may be applied for DNA sequencing applications, both lessening expense and eliminating the need for primary amplification of nucleotides prior to identification.

Perhaps the most exciting aspect of small β-PFT biology relates to the many unanswered questions that arise in light of recent remarkable advances in our understanding of these toxins. The field is now poised to define how host sensitivity and resistance to intoxication are governed. While much is already known in this regard, this knowledge must be ultimately expanded to the human host – a challenging task given the genetic heterogeneity of the population. Further, it will be essential to think more broadly about the potential role of these toxins in the non-pathogenic state. While many of the small β-PFTs have been implicated in virulence, the bacterial organisms that secrete these toxins quite likely utilize these 'weapons' as tools to facilitate survival. Survival may not arise from overt injury to the host, but rather from a more subtle manipulation of the environment to facilitate niche establishment, colonization, or transmission. In this regard, the coming years may illustrate common threads of functionality shared by the small β-PFTs and the large PFTs, which will be discussed in the next section of this chapter.

Introduction to the cholesterol-dependent cytolysins and membrane attack complex/perforin (MACPF) family proteins

The cholesterol-dependent cytolysins (CDCs) are a family of PFTs that are found in many Gram-positive pathogens, but are conspicuously absent in Gram-negative bacterial pathogens. CDC-expressing species are present in species from *Streptococcus, Listeria, Clostridium, Bacillus, Trueperella* (formerly *Arcanobacterium* and before that *Actinomyces*), *Gardnerella, Brevibacillus, Lactobacillus* and *Arcanobacterium*, and with time more CDCs will likely be identified in other genera. The CDCs contribute to pathogenesis in these organisms in a variety of ways, which range from auxiliary to primary pathogenic factors. For instance, the CDC of *Listeria monocytogenes*, listeriolysin O (LLO), is absolutely required for pathogenesis (Portnoy *et al.*, 1988; Cossart *et al.*, 1989) whereas perfringolysin O (PFO) contributes a supportive role in *Clostridium perfringens* gas gangrene (Awad *et al.*, 1995, 2001; Hickey *et al.*, 2008). Nearly a century ago Wuth (1923) described the haemolytic properties of a toxin from *C. perfringens*, which was consistent with the CDC PFO. The CDCs have since been classified as β-haemolysins, yet the attack of erythrocytes by the CDCs is primarily diagnostic and likely has little to do with their roles in the pathogenesis of these various bacterial species.

The wide distribution of the CDCs in the Gram-positive bacteria suggests that the ability to form a large pore in a eukaryotic membrane is useful to their lifestyles. Although the CDCs likely contribute to the pathogenesis of many of these species, our understanding of their contributions remains limited to a few species. Virtually nothing is understood, however, about the role of CDCs in bacterial survival in the commensal state, or in environmental niches outside of humans and/or animals. Many CDC-producing species infrequently cause disease in humans or animals, whereas they may spend all or the great majority of their existence as commensals in the human or animal, or as free-living species in the soil or water environments. In the soil and water environments CDCs may be used to ameliorate the impact of eukaryotic predation on bacteria (Habte and Alexander, 1975; Sherr and Sherr, 2002; Pernthaler, 2005), which may exhibit parallels to those roles in humans and animals where they provide protection from the phagocytic cells of the immune system. As suggested in the section on α-haemolysin: have these pore-forming toxins really evolved to promote disease or simply to maintain a commensal lifestyle in humans and animals (or the environment outside of a host)? Of those bacterial species that infect humans or animals, many exist primarily as commensals and only upon breakdown in a natural mucosal barrier or the immune system do they infrequently cause disease. Hence, is the primary role of the CDCs (as well as other PFTs) mainly to maintain the commensal (beneficial?) state of some bacteria? Although investigators have made inroads into understanding the contribution of the CDCs to the pathogenic state for a few organisms, a significant void in our knowledge exists concerning their contribution to survival in non-pathogenic states of bacterial existence, which may be their primary function.

In contrast to the complexities that confront our understanding of the *in vivo* roles of the CDCs, the study of the assembly mechanism of the CDC pore-forming complex has rapidly expanded in the last 15 years. This was due, in part, to solution of the crystal structure of *C. perfringens* PFO in 1997 and the identification of is membrane-spanning structures (Rossjohn *et al.*, 1997; Shepard *et al.*, 1998; Shatursky *et al.*, 1999). These two discoveries not only allowed investigators to probe deeply into the mechanism of the CDC pore assembly, but have recently allowed others to hypothesize that the large family of membrane attack complex/perforin (MACPF) proteins utilize a pore-forming mechanism that exhibits some key features of the CDC mechanism (Hadders *et al.*, 2007; Rosado *et al.*, 2007; Slade *et al.*, 2008; Law *et al.*, 2010; Lovelace *et al.*, 2011).

Unlike the α-toxin class of PFTs, the soluble CDC monomers undergo a significant number of structural changes to form the pore. The CDC pore is also about 20 times the diameter and 100 times the area of the α-toxin pore. The CDC pore-forming mechanism exhibits several hallmark features: the first of these to be discovered was the cholesterol dependence of the CDC pore-forming mechanism. The cholesterol-dependence was identified shortly after the beginning of the twentieth century (reviewed in (Bernheimer, 1976). Until recently, however, the role of cholesterol in the mechanism of the CDCs was not entirely clear, thus they were renamed cholesterol-dependent cytolysins due to this ambiguity (Shatursky *et al.*, 1999). As will be discussed, the role of cholesterol in the CDC mechanism is not the same for all CDCs, although they all bind cholesterol at the same site. The second hallmark feature of the CDCs is that they form large oligomeric pore complexes, which are assembled from up to 40 (or possibly more) soluble CDC monomers and represent the largest pores generated in a eukaryotic cell by a bacterial toxin. Most of the structure–function studies on the CDCs have focused on the assembly of this large membrane-embedded pore complex from the soluble monomeric CDC proteins. As will be seen, this transition involves a complex choreography of intra- and intermolecular interactions and secondary and tertiary structural changes in the CDC monomers in order to assemble the CDC pore complex.

The crystal structures of several members of the MACPF family of proteins have been solved recently and have a revealed that they contain a protein fold that is strikingly similar to that of domain 3 of the CDCs (Fig. 3.4). As will be described in this section, the function of domain 3 of the CDCs has been worked out in some detail, which showed that it is involved in oligomerization and formation of the β-barrel pore of the CDCs. Therefore, it was hypothesized that like the CDCs the MACPF pore-forming mechanism may also exhibit features of the CDC mechanism (Hadders *et al.*, 2007; Rosado *et al.*, 2007; Slade *et al.*, 2008). Although studies are only beginning on the MACPF pore-forming mechanism, it is already apparent that their mechanism exhibits some important fundamental differences from the CDC mechanism.

In this section the main features of the pore-forming mechanism of the CDCs will be examined and how these studies have led to insights into the pore-forming mechanism of the membrane attack complex/perforin (MACPF) family of proteins. Some members of the MACPF protein family contribute to microbial (prokaryotic and eukaryotic microbial pathogens) pathogenesis whereas others are mediators of immune defence against microbes. As will be seen below our understanding of the mechanism and contribution to pathogenesis of the microbial MACPF proteins is limited. We will also explore, on a limited basis, the role of the CDCs in pathogenesis with respect to certain structural and mechanistic features of the CDCs. For extensive reviews of the roles of CDCs in pathogenesis the reader is directed to some relatively recent reviews on the contribution of CDCs to microbial pathogenesis (Kayal and Charbit, 2006; Pizarro-Cerda and Cossart, 2006; Marriott *et al.*, 2008; Hamon *et al.*, 2012).

General features of the CDC primary structure

The CDCs are typically secreted as soluble proteins that exhibit a mass of 50–70 kDa, although the cytolysin core structure is typically about 50 kDa. The additional mass observed in some CDCs is always found as an amino-terminal extension, which can range from a short peptide to a 15 kDa protein. The CDC core structure is typically hydrophilic with no extended stretches of

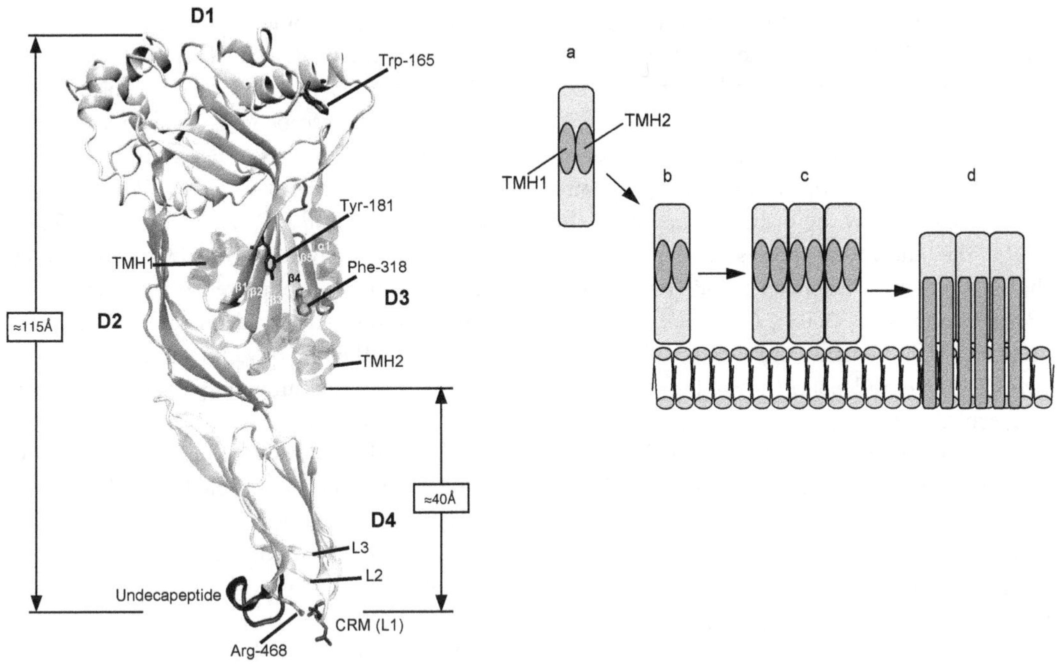

Figure 3.4 The left panel is the crystal structure of the soluble PFO monomer (Rossjohn *et al.*, 1997) (1PFO) with the various structures and amino acids residues identified in this review. The α-helical bundles that unravel and form the transmembrane β-hairpins are designated TMH1 and 2. The locations of the undecapeptide and cholesterol recognition/binding motif (CRM) are shown. The monomer binds to the membrane via the structures at the tip of domain 4 (CRM, loops L1-L1, undecapeptide) (Ramachandran *et al.*, 2002; Soltani *et al.*, 2007a; Farrand *et al.*, 2010) in a perpendicular orientation to the membrane as shown (Czajkowsky *et al.*, 2004; Ramachandran *et al.*, 2005; Tilley *et al.*, 2005). Shown in the right panel is a schematic overview of the basic stages of the CDC pore assembly. After secretion by the bacterial cell the soluble CDCs (a) monomers bind to the membrane receptor (cholesterol or in some cases human CD59) (b). The bound monomers begin to oligomerize into the prepore complex (c) and then upon conversion of the prepore to pore complex (d) the TMHs insert into the bilayer forming the β-barrel pore. During this transition the vertical height of the monomer collapses from about 115Å by about 40Å (Czajkowsky *et al.*, 2004; Ramachandran *et al.*, 2005) in order to bring the domain 3 TMHs into position so that the hairpins can cross the bilayer thereby forming the β-barrel pore. β-strands 1–5 are designated β1-β5. As described in the text the small loop comprised of β5 and α1 must swing out of the way of β4 to allow it to form edge-on backbone hydrogen bonds with β1 of a second monomer during oligomerization. The structure was prepared using VMD (Humphrey *et al.*, 1996).

hydrophobic residues (except for the signal peptides for secretion by a type II secretion system). PFO exhibited a highly soluble nature and could easily be concentrated to 15–20 mg/ml during crystallization trials (Feil *et al.*, 1996), although not all CDCs exhibit high solubility for reasons that remain unclear (Tweten, unpublished data).

The undecapeptide signature motif

The first primary structure of a CDC was revealed upon the cloning and sequencing of the streptolysin O (SLO) gene from *Streptococcus pyogenes* (Kehoe *et al.*, 1987), which was quickly followed by the gene sequences for pneumolysin (Walker *et al.*, 1987), listeriolysin O (LLO) (Mengaud *et al.*, 1988) and perfringolysin O (PFO) (Tweten, 1988). The DNA-derived primary structures showed that these CDCs exhibited a high degree of identity (≈40–70% identical residues) and that they contained a conserved 11-residue peptide near the C-terminus, which has been termed the tryptophan-rich or undecapeptide motif. This sequence (ECTGLAWEWWR) has been designated the signature motif of the CDCs. The undecapeptide is mostly conserved within the family of CDCs: some CDCs exhibit variations on this theme, but they are still recognizable as the signature motif and so it is likely that if a protein exhibits this motif it is a member of the CDC family.

As is evident from its sequence, the undecapeptide is rich in tryptophan and usually contains the only cysteine in the mature structure of the CDC. The known exceptions to that rule are the CDCs from *Clostridium tetani* (tetanolysin) (Bruggemann *et al.*, 2003), *C. botulinum* (botulinolysin) (Bruggemann and Gottschalk, 2004), two CDCs from *C. novyi* (novyiolysin I and II) (Bettegowda *et al.*, 2006) and *Listeria ivanovii* (ivanolysin) (Haas *et al.*, 1992), which contain 1–3 additional cysteines in their secreted forms. Specific function(s), if any, of these additional cysteines have not been demonstrated. Two cysteines are conserved in position between tetanolysin, botulinolysin and the two novyiolysins, but their locations in molecular models, which are based on the PFO crystal structure (Rossjohn *et al.*, 1997), are not predicted to form either intra- or intermolecular disulfides in the oligomer or monomer. The function of the conserved cysteine in the undecapeptide also remains unclear, although its chemical modification or mutation to certain amino acids can dramatically decrease the cytolytic activity of the CDCs (Iwamoto *et al.*, 1987; Pinkney *et al.*, 1989; Saunders *et al.*, 1989). Early studies showed that chemical modification of this cysteine inactivated the pore-forming mechanism of PFO by affecting both cell binding and the subsequent formation of the pore (Iwamoto *et al.*, 1987). The cytolytic activity of PFO could be regained, however, if the thiol was modified with a reagent that was reversible by reduction of a disulfide. The substitution of the cysteine with alanine does not have a significant effect on its activity, although serine is not well tolerated (Pinkney *et al.*, 1989; Saunders *et al.*, 1989; Shepard *et al.*, 1998). Therefore, the cysteine is located at a sensitive site that apparently does not allow modification of the thiol, but the thiol group alone is not absolutely required for cytolytic activity.

The reason that the cysteine has been maintained in the undecapeptide structure of most CDCs remains unclear, although Singh *et al.* (2008) provided some insight into its possible function in the CDC from *L. monocytogenes*, listeriolysin O (LLO). LLO has been shown to be essential for the escape of *L. monocytogenes* from the phagosome during cellular invasion (Gaillard *et al.*, 1987; Berche *et al.*, 1988; Portnoy *et al.*, 1988; Cossart *et al.*, 1989). Singh *et al.* (2008) showed that activation of LLO pore-forming activity required the presence of γ-interferon-inducible lysosomal thiol reductase (GILT) in the phagosome. Their observations implied that after LLO secretion the thiol group of the undecapeptide cysteine was oxidized by disulfide formation, which, as indicated above, would inactivate LLO. Therefore, this disulfide would have to be reduced by GILT in the phagosome. In the absence of GILT, escape of *L. monocytogenes* from the phagosome of cells was compromised, as was its growth in mice deficient in GILT. When the cysteine in LLO was replaced with alanine, which makes LLO resistant to oxidative inactivation, GILT was no longer required for phagosomal escape. Hence, it is possible that the oxidative state of the cysteine sulfhydryl may be a mechanism to sense the environment of the organism and regulate its activity.

Although the undecapeptide is rich in tryptophan residues, the reason for their conspicuous presence is not understood. In PFO at least two of the three tryptophans partition into the bilayer when it binds to the membrane (Nakamura *et al.*, 1995, 1998), though they do not penetrate deeply into the bilayer (Heuck *et al.*, 2003). Aromatic residues are often present at the lipid–water interface of the membrane-spanning β-strands of bacterial porins, thereby forming rings of aromatic residues that flank the transmembrane β-barrel. The reason for their presence in the poring β-barrel structure is not understood, as stacking interactions cannot occur between the non-planar and non-aromatic rings of cholesterol and tryptophan. The entry of the CDC tryptophans into the bilayer can be monitored by an increase in their fluorescence emission, which results from the exclusion of water that quenches the tryptophan fluorescence in the soluble monomer (Nakamura *et al.*, 1995). The change in the fluorescence emission of the tryptophans has been a convenient method to monitor membrane binding of the CDC monomers to synthetic lipid bilayers.

The cytolytic activity of PFO, and other related CDCs, is highly sensitive to mutations within the undecapeptide (Sekino-Suzuki *et al.*, 1996; Nakamura *et al.*, 1998; Jacobs *et al.*, 1999; Weis and Palmer, 2001; Polekhina *et al.*, 2005). Except for the first two residues (Glu and Cys) none of the other residues can be substituted with alanine in PFO without having a detrimental effect on the activity (Polekhina *et al.*, 2005). Hence, the undecapeptide is clearly a key element in the activity of the CDCs, yet its function in the pore-forming mechanism of the CDCs has remained enigmatic for many years. Recent studies, however, have revealed that the role of the undecapeptide in the CDC mechanism is complex and is addressed later in this chapter.

The amino terminal modifications of the CDCs

The core CDC primary structure, which is typified by the secreted form of PFO (Fig. 3.4), is about 470 amino acids in length (approximately 53 kDa) and is conserved in all CDCs. This core structure is the pore-forming structure. As indicated above, some CDCs exhibit an extra peptide sequence at their amino terminus, whereas extra peptide sequences have not been found at the C-terminus. Lectinolysin (LLY) (Farrand *et al.*, 2008), the CDC from some strains of *Streptococcus mitis* and *S. pseudopneumoniae*, exhibits a primary structure of about 628 amino acids or about 70 kDa. The extra amino-terminal sequence in LLY constitutes a fucose-binding lectin specific for Lewis blood group Le^y and Le^b antigens (Farrand *et al.*, 2008; Feil *et al.*, 2012). The 3D crystal structure of this lectin domain (Feil *et al.*, 2012) is similar to an agglutinin produced by the European eel *Anguilla anguilla* (Bianchet *et al.*, 2002), as well as the carbohydrate-binding module (CBM) of the family 98 glycoside hydrolase (SpX-1) of *Streptococcus pneumonia* (Boraston *et al.*, 2006). Why LLY contains this lectin domain remains unknown, although there is some evidence that may increase LLY cytolysin activity on cells expressing carbohydrates recognized by the lectin (Farrand *et al.*, 2008).

Another example of an extra amino terminal peptide is found in *Streptococcus pyogenes* SLO, which is about 70 amino acids in length (Kehoe *et al.*, 1987) and is involved in the SLO-mediated translocation of an NAD-glycohydrolase enzyme (SPN) into eukaryotic cells (Madden *et al.*, 2001). *Listeria monocytogenes* listeriolysin O (LLO) also exhibits an amino-terminal PEST-like sequence (Decatur and Portnoy, 2000), which does not act as a PEST sequence to increase proteolytic degradation of LLO, but instead appears to affect LLO expression (Schnupf *et al.*, 2006). There are other CDCs that exhibit extra amino terminal peptide additions, but generally their function, if any, remains unknown and these peptides do not exhibit any sequence similarity to other known peptides or proteins.

The CDC monomer crystal structure

Rossjohn *et al.* (1997) solved the first CDC crystal structure, which was the soluble monomer structure of PFO. PFO exhibited a β-sheet-rich, elongated structure that contained an IgG-like domain 4 (Fig. 3.4). Since that time the structures for three additional CDCs have been solved and include *Streptococcus intermedius* intermedilysin (ILY) (Polekhina *et al.*, 2005), *Bacillus anthracis* anthrolysin O (ALO) (Bourdeau *et al.*, 2009) and *Streptococcus suis* suilysin (SLY) (Xu *et al.*, 2010), all of which exhibit the same overall shape of the PFO structure. When the PFO structure was revealed comparatively little was understood about the CDC mechanism and the structure of PFO did not reveal any clear insights into how it formed a pore. Furthermore, its structure was not related to the previously solved structures for *Aeromonas hydrophila* aerolysin monomer (Parker *et al.*, 1994b) or the monomer structure observed within the *Staphylococcus aureus* α-haemolysin heptameric pore structure (Song *et al.*, 1996). Nevertheless, the solution of the PFO structure was a watershed event in the study of the CDCs: as described in the following sections it provided an essential basis for many subsequent biochemical and biophysical studies of CDC pore-forming mechanism that revealed new paradigms in the assembly of a β-barrel pore.

Secretion of the CDCs

The CDCs are expressed and secreted by a type II secretion system, as all but one exhibit the classic amino terminal signal peptide for this secretion system. The only known exception to this rule is the CDC from *Streptococcus pneumoniae*, pneumolysin (PLY) (Walker *et al.*, 1987), which is turning into intriguing story. It had been suggested in an early study that PLY was released from the bacterial cell by lysis (Berry *et al.*, 1989), whereas a later study suggested that some strains actively secreted PLY (Benton *et al.*, 1997). More recent studies by Andrew Camilli's group have discovered that a fraction of the PLY is localized to the *S. pneumoniae* cell wall (Price and Camilli, 2009), which is eventually released from the cell. Their study suggested that *S. pneumoniae* actively secretes PLY. Importantly, they showed that PLY could not re-associate with the cell wall from the extracellular milieu, which showed that if PLY was released by lysis of some cells it could not populate the cell surface of intact cells. Remarkably, they found that adding a canonical type II signal peptide to PLY *inhibits* its secretion, suggesting that the *S. pneumoniae* type II secretion was not able to handle PLY with the signal peptide. They have also shown that the secretion system that secretes PLY in *S. pneumoniae* also appears to operate in *Bacillus subtilis*, as a fraction of PLY is localized to the cell wall of *B. subtilis* when expressed in that system (Price *et al.*, 2012).

In contrast to what was observed in *S. pneumoniae*, the *B. subtilis* type II system secreted PLY that was fused to a canonical *S. pneumoniae* secretion signal (Price *et al.*, 2012). They also found that the Sec-machinery of *S. pneumoniae* appears to be intolerant of other CDCs with a signal peptide, as the gene for PFO with its native signal peptide was lethal to *S. pneumoniae*. The only viable *S. pneumoniae* recombinants isolated after transformation with the gene for the full-length PFO exhibited frame-shift mutations within the signal peptide or the region of the gene that encodes the signal was deleted. PFO without a signal peptide was expressed by *S. pneumoniae* and a small fraction appeared to be localized to the cell wall. These data suggested that the *S. pneumoniae* Sec-machinery is incompatible with CDCs that have a type II secretion signal and that an alternative, yet unidentified secretion pathway is responsible for the secretion of PLY. Why the *S. pneumoniae* Sec-machinery is not compatible with signal peptide bearing CDCs is not known, but it suggests that there are fundamental differences in the protein repertoire that can be handled by this system in different bacterial species.

Cellular recognition

Following secretion the soluble CDC monomers bind to a cellular receptor, which is mediated by domain 4 of the CDC structure, which exhibits an immunoglobulin-like fold (Fig. 3.4). For many years the only known CDC receptor appeared to be membrane cholesterol; however, a second class of CDCs was identified in 2004 that used another receptor, human CD59 (Table 3.1). *Streptococcus intermedius* intermedilysin (ILY) (Giddings *et al.*, 2004) was the first of the human CD59 binding CDCs to be identified (Giddings *et al.*, 2004), which was followed by identification of human CD59 binding CDCs in *Gardnerella vaginalis* (vaginolysin or VLY) (Gelber *et al.*, 2008) and a CDC identified in some isolates of *Streptococcus mitis* and *S. pseudopneumoniae* termed lectinolysin (LLY) (Farrand *et al.*, 2008; Wickham *et al.*, 2011). The discovery that ILY only bound the human form of CD59 (Giddings *et al.*, 2004) explained the previous studies that showed it was cytolytic only on human cells (Nagamune *et al.*, 1996). More importantly, these studies showed that some CDCs had evolved to bind a receptor other than cholesterol. This discovery, however, also presented a new problem: the ILY cytolytic mechanism remained cholesterol-dependent, but cholesterol clearly was not the receptor (Giddings *et al.*, 2003). It was this incongruity that drove the investigation into the basis of its cholesterol dependence, which eventually led to the discovery of the cholesterol-binding motif of the CDCs, like PFO, that use cholesterol as their receptor and also explained why ILY remained cholesterol dependent.

The discovery of the ILY receptor (Giddings *et al.*, 2004) showed that CD59 was required for ILY binding to cells: cells that lacked human CD59 were not bound by ILY. Yet, they had shown earlier (Giddings *et al.*, 2003) that depletion of cholesterol

Table 3.1 Examples of CDCs, their bacterial species of origin and receptor specificity. The list of CDCs shows most members of the CDC family, but it is not an exhaustive list of all members. At this time the receptor for Arcanolysin is unknown, as it has been shown that it exhibits species-specific cytolytic activity, which is different from the human-specific CDCs, which bind to the human form of CD59 and are active on human cells (Giddings *et al.*, 2004; Gelber *et al.*, 2008; Wickham *et al.*, 2011). The receptor for pyolysin is uncertain, as it exhibits significantly less cytolytic activity than most CDCs that utilize cholesterol as their receptor (Jost *et al.*, 2011), which could also be explained if it uses a receptor other than cholesterol that was not present or in low levels on cells

CDC name (abbreviation)	Bacterial species	Receptor
Perfringolysin O (PFO)	Clostridium perfringens	Cholesterol
Botulinolysin	Clostridium botulinum	Cholesterol
Tetanolysin	Clostridium tetani	Cholesterol
Novyiolysin I	Clostridium novyi	Cholesterol
Novyiolysin II	Clostridium novyi	Cholesterol
Streptolysin O (SLO)	Streptococcus pyogenes	Cholesterol
Pneumolysin (PLY)	Streptococcus pneumoniae	Cholesterol
Caniolysin	Streptococcus canis	Cholesterol
Equisimilysin	*Streptococcus dysgalactiae* subsp. *equisimilis*	Cholesterol
Intermedilysin (ILY)	Streptococcus intermedius	Human CD59
Suilysin (SLY)	Streptococcus suis	Cholesterol
Lectinolysin (LLY)[a]	Streptococcus mitis, S. pseudopneumoniae	Human CD59[b]
Anthrolysin O (ALO)	Bacillus anthracis	Cholesterol
Cereolysin O (CLO)	Bacillus cereus	Cholesterol
Sphaericolysin	Bacillus sphaericus	Cholesterol
Brevilysin	Brevibacillus brevis	Cholesterol
Alveolysin	Paenibacillus alvei	Cholesterol
Inerolysin (INY)	Lactobacillus iners	Cholesterol
Ivanolysin O (IVO)	Listeria ivanovii	Cholesterol
Listeriolysin O (LLO)	Listeria monocytogenes	Cholesterol
Seeligeriolysin O (LSO)	Listeria seeligeri	Cholesterol
Vaginolysin (VLY)	Gardnerella vaginalis	Human CD59
Pyolysin	*Trueperella* (formerly *Arcanobacterium*) *pyogenes*	Cholesterol?
Arcanolysin	Arcanobacterium haemolyticum	?

[a]Not all isolates of *S. mitis* and *S. pseudopneumoniae* express an LLY-like CDC.
[b]It has not been rigorously demonstrated whether LLY binds to only the human form of CD59.

from cell membranes blocked the pore-forming activity of ILY, but not binding. Hence, CD59 was required for cell binding, but cholesterol remained necessary for its pore-forming activity. These results suggested that cholesterol played different roles in the pore-forming mechanisms of the CD59 binding ILY-like CDCs and that of the PFO-like CDCs, which used cholesterol as their receptor. As described below, the basis for the cholesterol-dependency of these two toxin families was resolved in two studies by Farrand *et al.* (2010) and LaChapelle (now Wickham) *et al.* (LaChapelle *et al.*, 2009).

There may be at least one more CDC that exhibits a cellular specificity that implies a receptor that is different than either cholesterol or human CD59, which is Arcanolysin from *Arcanobacterium haemolyticum* (Jost *et al.*, 2011). This CDC is cytolytically active on human, rabbit, pig and horse cells whereas it is 10–60 times less active on ovine, bovine, murine, and canine cells. Pyolysin (PLO) is also significantly less active on cells than the cholesterol binding PFO suggesting that it too may utilize an alternative receptor, but PLO has not been shown to exhibit a species selective activity.

The cholesterol recognition/binding motif (CRM)

The CRM of the PFO-like CDCs had been generally accepted to be the conserved undecapeptide motif since changes within this structure affects binding (Iwamoto *et al.*, 1987; Sekino-Suzuki *et al.*, 1996; Polekhina *et al.*, 2005). Furthermore, the ILY undecapeptide exhibited significant changes in its primary structure (GATGLAWEPWR vs. the consensus ECTGLAWEWWR) (Nagamune *et al.*, 2000), suggesting that through evolution ILY had lost the ability to bind cholesterol (Tweten, 2005). This scenario turned out to be wrong, as studies by Soltani *et al.* (2007b) uncoupled membrane binding from the membrane insertion of the PFO undecapeptide. Hence, the undecapeptide did not appear to be the cholesterol-binding motif for the CDCs that used cholesterol as their receptor. The only other structures within domain 4 of the PFO membrane-binding interface (Fig. 3.4) that interacted with the membrane were the nearby loops L1-L3 (Ramachandran *et al.*, 2004; Soltani *et al.*, 2007a), which suggested that the CRM of PFO was located somewhere within these loops.

Farrand *et al.* (2010) subsequently performed an alanine scan of the L1–L3 loops and identified two residues in the L1 loop (Fig. 3.4), which had the greatest effect on membrane binding and cytolytic activity of PFO. The two residues are a Thr–Leu pair (490 and 491 in PFO) that is conserved in all known CDCs identified to date. Knocking out both residues completely eliminated binding of PFO, as well as the related CDCs SLO and PLY, to cholesterol-rich liposomes and cells. Hence, the Thr–Leu pair formed the CRM, as all other mutations within L1-L3 exhibited much smaller effects on binding. The CRM was also shown to be relatively intolerant to conservative substitutions, which is consistent with its strict conservation in the CDCs. The small size of the CRM was at first surprising, but it made sense when the structure of cholesterol at the membrane surface was considered. The headgroup of cholesterol exposed at the membrane surface is comparatively small, comprising only a 3-β-hydroxyl group and perhaps parts of the A-ring of cholesterol that may be exposed at the membrane surface. Previous studies had shown that the 3-hydroxyl in the β-configuration is absolutely essential for CDC binding (Prigent and Alouf, 1976), which suggested a stereospecific interaction of the CRM with this group.

Not all membrane cholesterol is bound by the CDCs: the lipid environment of the cholesterol has a significant effect on CDC binding to membranes (Zitzer *et al.*, 2001; Heuck *et al.*, 2003; Nelson *et al.*, 2008; Flanagan *et al.*, 2009). The lipid environment of the cholesterol apparently affects its exposure at the membrane surface. In liposomes comprised of only phosphatidylcholine and cholesterol Heuck *et al.* (2003) showed that PFO binding went from 0% binding to 100% binding as the cholesterol concentration was increased from about 45 mol% to 55 mol%, which suggested that at concentrations of less than 45 mol% cholesterol was not accessible due to its lipid environment. However, binding at lower concentrations of cholesterol could be achieved by simply changing the structure of the lipids, thus showing that the lipid environment exerts a significant effect on the accessibility of cholesterol to the CDC CRM. The structure of loops 2 and 3 also influences the binding interaction (Farrand *et al.*, 2010; Johnson *et al.*, 2012). Although the exact nature of this effect is not understood, it is probably related to their ability to insert into specific lipid environments after the CRM recognizes cholesterol, thereby firming up the monomer interaction with the membrane.

Interestingly, the Thr–Leu pair is also conserved in the ILY-like CDCs that use human CD59 as their receptor. The presence of the CRM suggested that they were capable of binding to cholesterol, but cholesterol-mediated binding clearly did not occur on cholesterol-rich cell membranes that lacked human CD59 (Giddings *et al.*, 2004). Yet, in the absence of cholesterol, CD59 binding alone was not sufficient to form the pore, although it did trigger formation of the ILY oligomeric complex (Giddings *et al.*, 2003). As indicated above, the studies of Lachapelle *et al.* (2009) and Farrand *et al.* (2010) resolved this enigma. These two studies showed that ILY disengaged from CD59 as it made the conversion from the prepore state to the inserted pore complex and that the CRM of ILY had to establish a cholesterol-dependent interaction of loops L1-L3 with the membrane to maintain the contact between the prepore complex and the membrane surface. If the CRM was knocked out in ILY it disengaged from the surface upon prepore to pore conversion since it lacked the second anchor point mediated by the CRM-dependent interaction of the loops with the membrane. Therefore, although the CRM does not mediate the interaction of ILY with the membrane, it must still form a cholesterol-dependent interaction with the membrane, after CD59 binding, in order to anchor the oligomeric complex to the membrane as it disengages from CD59 during the critical prepore to pore transition.

The ILY-CD59 binding interaction

The binding interface of the ILY–CD59 interaction (Wickham *et al.*, 2011) deserves special attention, as it exhibits several interesting features. The normal function of CD59 on human cells is as a late stage inhibitor of the complement membrane attack complex or MAC. CD59 interacts with MAC components C8α and C9, which are the primary proteins involved in the formation of the MAC membrane pore (reviewed in Plumb and Sodetz, 1998). Hence, CD59 prevents the formation of the MAC pore by preventing the polymerization of C8α and C9 (interestingly, as described later in this chapter, the CDCs and MAC proteins may be distant ancestors that share features of their pore-forming mechanisms). The complement MAC is used to directly attack the bacterial membrane of Gram-negative bacteria by pore formation, but can also insert pores into eukaryotic cell membranes. Therefore, CD59, along with other MAC inhibitors, minimizes MAC-mediated collateral damage to the host cells during complement activation. Interestingly, ILY binds to and uses human CD59 to initiate the formation of its own oligomeric pore complex. During the assembly of the ILY pore complex on the eukaryotic cell those cells become more sensitive to lysis by their own complement (LaChapelle *et al.*, 2009), which implied that the binding sites for ILY, C8α and C9 were similar and that ILY binding blocked the protective effect of CD59 to MAC-mediated damage.

Wickham *et al.* (2011) mapped the residues in both human CD59 and ILY that contributed to the binding interaction of these two proteins. These studies and previous studies of the CD59 binding site for C8α and C9 (Huang *et al.*, 2005) revealed that the binding sites for all three proteins were remarkably similar and that ILY bound to a subset of residues in CD59 that form its binding interface with C8α and C9. The sites were so closely related that mutants previously shown to increase CD59 mediated protection to MACPF mediated lysis, presumably by increasing the affinity of the interaction with C8α and/or C9, also increased the affinity of the ILY–CD59 interaction. Interestingly, the increased affinity of the ILY-CD59 binding interaction decreased the haemolytic activity of ILY by slowing the prepore to pore transition. This was consistent with the fact that ILY disengages from CD59 during the prepore to pore transition: if the affinity of this interaction increases it slows this transition, which slows the rate of pore formation. Hence, the evolution of the CD59 binding site on ILY did not occur without some constraints: the binding affinity of the interaction had to be carefully balanced to maintain specificity, but could not be so high as to interfere with the disengagement of the receptor, which is necessary for prepore to pore conversion.

Although the CD59 binding site for ILY and the MAC proteins was nearly identical, the respective binding sites of these three proteins for CD59 exhibit no similarity (Huang *et al.*, 2006; Wickham *et al.*, 2011). A consensus signature motif was identified for the CD59-binding CDCs, Tyr-X-Tyr-X_{14}-Ser-Arg, which is conserved in all three CD59-binding CDCs (Wickham *et al.*, 2011), whereas the binding sites of C8α and C9 were narrowed to peptides with the sequences of LGIQYE and VSLAFS, respectively (Huang *et al.*, 2006). Presumably, specific contact residues are oriented similarly for ILY and the MAC proteins that contact the same site on CD59, but as is apparent, these contact residues do not exhibit any significant similarity.

The biological basis for the evolution of the CD59 binding specificity remains a mystery: CD59 is expressed to some extent on almost all cell types, therefore it is difficult to rationalize that this binding specificity allowed these CDCs to narrow their specificity to certain cell types. So far, the CD59-specific CDCs have been identified in mucosal commensals that rarely cause disease. Although these organisms can cause disease, which is primarily abscess formation by the streptococcal species and bacterial vaginosis by *G. vaginalis*, it seems unlikely that these CD59-specific CDCs evolved to support the occasional opportunistic infection. It is also interesting to note that ILY and VLY appear to be specific for the human-derived CD59 (human specificity has not been confirmed unambiguously for LLY). CD59 is largely homologously restricted (Rollins *et al.*, 1991): each species CD59 specifically binds to and inhibits its homologous MAC. Therefore, although the human CD59 binding specificity did not allow ILY to target specific cell types, it did restrict its activity at the organism level to humans. This restriction to human cells suggests that these bacterial species and their CDCs evolved with humans. The likelihood of coevolution with humans suggests that the CD59-binding CDCs evolved to not cause disease, but to maintain their commensal state in some way.

The CDC membrane anchoring system

In addition to the interaction of the monomer with its receptor, cholesterol or CD59, and the membrane insertion of the undecapeptide, two short hydrophobic loops, L2 and L3 at the tip of domain 4 (Fig. 3.4), also contribute to anchoring the monomer to the membrane surface (Ramachandran *et al.*, 2002; Soltani *et al.*, 2007a,b). If aspartate is substituted into the core of L2 or L3, or the Leu of the CRM in L1, the interaction of all three loops is abolished (Soltani *et al.*, 2007a). If just the core alanine of loop L3 (Ala-437 of PFO) is changed to glycine binding is significantly decreased. Therefore, in the absence of membrane insertion of L2 and L3, or L3 alone, the binding strength of the CRM–cholesterol interaction is apparently not sufficient to anchor the monomer to the surface. These observations suggest the CRM directs binding to cholesterol, but the insertion of these loops is essential to increase the binding strength of this interaction. Upon oligomerization of the monomers into the prepore the avidity effects of the multiple contacts of the oligomerized monomers then significantly increase the strength of the binding interaction, thereby making it essentially irreversible.

The combination of the CRM interaction mediated by L1 and the subsequent membrane insertion of L2, L3 and the undecapeptide anchor the monomers in an upright position in which the long axis of the CDC molecule is perpendicular to the membrane surface (Czajkowsky *et al.*, 2004; Ramachandran *et al.*, 2005). This anchoring system therefore maintains a stable platform for the subsequent conformation changes in the monomer that lead to oligomerization and insertion of the β-barrel pore.

The CDC allosteric pathway

The CDCs do not assemble an oligomeric complex in the absence of membrane binding: the interaction of the monomers with the cell surface is required to initiate assembly of the oligomeric pore complex. Receptor binding (cholesterol or human CD59) initiates structural changes that are required for assembly of the pore complex. As will be described below, most of the major structural changes associated with monomer oligomerization and formation of the β-barrel pore occur within domain 3, which is nearly 60 Å distal from the binding site of the monomer at the tip of domain 4 (Fig. 3.4). Domains 3 and 4 are not in direct physical contact, but are connected by domain 2, which forms contacts with both domains. Therefore, how does membrane binding by the CDC monomer initiate oligomerization and pore formation?

Palmer *et al.* (1998b) initially proposed that structural changes within domain 3 were allosterically linked to membrane binding. Heuck *et al.* (2000) subsequently demonstrated that domains 3 and 4 were conformationally coupled by demonstrating that mutations within the domain 3 amphipathic hairpins that altered their rate of membrane insertion correspondingly altered the rate of the membrane insertion of the undecapeptide tryptophans. The undecapeptide tryptophans insert into the membrane and contribute to the binding interaction and so this event precedes the insertion of the β-barrel pore, which is the final step of in the pore-forming mechanism (Heuck *et al.*, 2000). The fact that mutations in the domain 3 β-hairpins could correspondingly altered the rate of membrane insertion of the undecapeptide and β-barrel pore could only be explained by the conformational coupling of these two events. Hence, the undecapeptide is apparently involved in an allosteric pathway that couples the membrane interaction of the monomer to structural events that are necessary to initiate the assembly of the oligomeric pore.

As indicated above, the pore-forming mechanism of PFO is sensitive to mutations within its undecapeptide structure whereas the ILY pore-forming mechanism is much less sensitive to mutations within its undecapeptide (Polekhina *et al.*, 2005). This observation suggested that the undecapeptide played a lesser role in the pore-forming mechanism of ILY than in PFO. Furthermore, cholesterol binding by PFO initiates the process of oligomerization of the membrane bound monomers whereas binding to human CD59 initiates these changes in ILY. Therefore, the allosteric pathways coupling receptor binding to the domain 3 structural changes are likely to exhibit some differences in these two receptor-binding classes of CDCs. These observations suggested that the undecapeptide played an important role in coupling binding to structural changes necessary for pore formation in PFO-like cholesterol binding proteins, but it was not linked to the CRM in the CD59-binding ILY, thereby transferring control of the initiation of oligomerization to the CD59 binding site.

Consistent with this scenario, Dowd *et al.* (2012) showed that mutation of the last residue of the PFO undecapeptide, Arg-468, could block the conversion of monomers to an oligomeric pore complex, whereas the mutation of the analogous residue in ILY had been previously shown to have little effect on its pore-forming mechanism (Polekhina *et al.*, 2005). Arg-468 only forms two contacts in the crystal structure of the PFO monomer: its NH1 forms hydrogen bonds to the backbone carbonyls of the CRM Thr–Leu pair. This contact is missing in the ILY structure (Polekhina *et al.*, 2005), as the analogous arginine is pulled out of position and cannot form the same interaction with the CRM. Substitution of Arg-468 in PFO with alanine (PFO^{R468A}) effectively prevented all domain 3 structural changes and also prevented the intermolecular interaction of the membrane bound monomers: in essence the membrane bound monomers remained inert. This study suggested that the undecapeptide of PFO is an essential component in the allosteric pathway that couples binding to the initiation of events necessary for the oligomerization of PFO.

As discussed above, ILY still forms a cholesterol-dependent interaction with the membrane surface via its CRM after it binds to human CD59 and also inserts loops L2 and L3 into the membrane, which is necessary to maintain its anchor to the membrane as it disengages from CD59 during prepore to pore conversion (Farrand *et al.*, 2010; Wickham *et al.*, 2011). Although ILY cannot directly bind to cholesterol in natural membranes that lack human CD59 (Giddings *et al.*, 2004), Dowd *et al.* (2012) showed that it could bind to cholesterol-rich liposomes in the absence of CD59. Interestingly, ILY bound better than PFO to these liposomes in a CRM-dependent fashion, yet this binding did not trigger any domain 3 conformational changes in ILY or the formation of a pore. Therefore, cholesterol-bound ILY remained inert on the membrane, similar to PFO^{R468A}. These data showed that in PFO the undecapeptide forms an integral element of an allosteric pathway that couples cholesterol binding to the initiation of changes in the monomers that allow them to interact and drive the formation of the oligomeric pore complex. In ILY, however, this pathway appears to have been disrupted during the evolution of the CD59 binding site so that the allosteric pathway could be re-routed to the CD59 binding site.

The allosteric pathway likely induces some changes in the domain 2–3 interface (Fig. 3.4), which alters the monomer structure sufficiently so that membrane-bound monomers can interact. Several different crystal structures of the PFO monomer show that torsion around the domain 2–4 interface is associated with the disruption of several contacts in the domain 2–3 interface (Rossjohn *et al.*, 2007). Therefore, it is likely that binding is allosterically signalled via the domain 2–4 interface, which then may alter the domain

2–3 interface in a way that facilitates the interaction of the monomers. Binding and the allosteric pathway alone, however, are probably not directly responsible for driving the major structural transitions in domain 3 that are necessary for formation of the pore. As described in the next section evidence now suggests that the monomer–monomer interactions drive the major structural changes in the CDC monomers that are required for establishing the geometry of the oligomer and the formation of the β-barrel pore.

Formation of the prepore oligomer

Upon binding the CDC monomers oligomerize with each other to form what is termed the prepore complex (Shepard *et al.*, 2000). The prepore is essentially the state of the oligomer prior to the membrane insertion of the β-barrel pore, although there can be more than one prepore state. The prepore stage for several small pore-forming toxins, such as *Staphylococcus aureus* α-haemolysin (Walker *et al.*, 1992a), *Clostridium septicum* α-toxin (Sellman *et al.*, 1997) and *Aeromonas hydrophila* aerolysin (Garland and Buckley, 1988) were identified prior to the studies with the CDCs. In contrast to these toxins, which form oligomeric pore complexes of 6–8 monomers, the CDCs form a much larger prepore, which entails complexities that are not encountered in the assembly of smaller oligomeric complexes. As will be evident in the following discussion the CDC monomer structure undergoes significantly more changes than does α-haemolysin during formation of the oligomeric prepore and pore. Furthermore, the prepore state is not defined by a single structural state of the monomer in the CDCs, as PFO has been trapped in at least three different prepore states, which include a SDS-sensitive linear oligomeric complex that is in equilibrium with the monomer state (Hotze *et al.*, 2012), a SDS-sensitive circular oligomeric complex (Hotze *et al.*, 2002) and an SDS-stable circular oligomeric complex (Shepard *et al.*, 2000; Hotze *et al.*, 2001).

An early study on the oligomerization of PFO using fluorescence resonance energy transfer (FRET) suggested that oligomerization preceded pore formation (Harris *et al.*, 1991). Later Shepard *et al.* (Shepard *et al.*, 2000) used low temperature to separate the oligomerization and pore states of PFO: low temperature facilitated the oligomerization of PFO into a SDS-stable prepore complex, but prevented insertion of the β-barrel pore. This low temperature oligomer, however, could be triggered to form a pore by raising the temperature. The prepore intermediate for PFO was originally a controversial topic: it had been suggested that the insertion of such a large β-barrel from a prepore state was energetically unfavourable and an alternative model was proposed in which a dimer inserted and formed a small pore to which monomers were added to increase the size of the pore (Palmer *et al.*, 1998a). Energetically, however, quite the opposite is true since membrane insertion of the naked hairpins of monomers or dimers would incur a large energetic cost due to the free energy required to insert hairpins with unsatisfied hydrogen bound potential (White and Wimley, 1999). Furthermore, Hotze *et al.* (2001) used a disulfide-trapped prepore mutant of PFO to show that membrane insertion of the β-barrel pore occurred rapidly from the prepore state. Therefore, formation of the CDC prepore may be necessary to generate a partially or fully hydrogen-bonded pre-β-barrel in order to reduce the energetic cost of its membrane insertion.

Upon interaction of monomers with the membrane specific structural changes are observed in domain 3 of PFO, which are required to establish a SDS-stable prepore. Hotze *et al.* initially describe a mutant of PFO in which the substitution of an alanine for Tyr-181 in β-strand 1 (Fig. 3.4) of the core β-sheet blocked formation of the pore and trapped PFO in an SDS-sensitive, but stable ring-shaped oligomeric complex (Hotze *et al.*, 2002). A tyrosine is conserved in all known CDC structures at the equivalent position, which suggested that it was important to the activity of the CDCs. They further showed that native PFO, when mixed with this mutant, could drive the necessary structural changes so that it formed an SDS-stable oligomeric pore complex. These studies were the first to suggest that monomer–monomer interaction could drive the necessary structural changes for pore formation.

Ramachandran *et al.* (2004) subsequently showed that a corresponding aromatic residue in β-strand 4 of PFO, Phe-318, formed an intermolecular π-stacking interaction with Tyr-181 of a second monomer. Substituting either residue with alanine or moving one aromatic out of register with the other disrupted this stacking interaction and trapped PFO in a SDS-sensitive prepore state. It was proposed that this stacking interaction maintained the correct in-register pairing of β-stands 1 and 4 of two monomers. It is evident, however, that this stacking interaction is also important for the conversion of the SDS-sensitive oligomer to an SDS-resistant oligomer. Whether the conversion from an SDS-sensitive to resistant oligomeric state is accomplished simply by ensuring the correct in-register pairing of the strands or something more complex remains unclear. The Tyr-181 analogue is present in all known CDCs, although the Phe-318 analogue is not: therefore the π-stacking function is likely specific to the CDCs that have this aromatic pair. The function(s) of this pair is apparently accomplished by other means in those CDCs that lack the Phe-318 analogue.

Ramachandran *et al.* also showed that the loop comprised of α-helix 1 and β-strand 5 (Fig. 3.4) must break its interaction with β-strand 4 (primarily backbone hydrogen bonds between β-strands 4 and 5) so it can rotate away and expose the edge of β-strand 4. Once β-strand 4 is exposed it becomes available to form edge-on hydrogen bonds with β-strand 1 of another monomer. This interaction may also be key to how the CDCs regulate the assembly of the oligomeric pore complex, which is described in detail below.

Monomer–monomer interactions and the assembly of the CDC oligomer

As indicated in the previous sections there is accumulating evidence that membrane binding initiates monomer–monomer interaction, but that it is the actual interaction of the monomers that drive the major conformational changes that lead to pore formation. Recently, Hotze *et al.* (2012) showed that monomer–monomer interactions drive the specific structural changes in domain 3 that lead to pore formation and that specific domain 3-mediated interactions are required to set the geometry of the pore complex. They showed that mutation of a conserved tryptophan in domain 1 of PFO to threonine (or any other non-aromatic residue), Trp-165 (PFO^{W165T}), blocked all major domain 3 structural transitions, but allowed the monomers to form SDS-sensitive linear oligomers in which the monomers were freely exchangeable in the oligomers. Cholesterol binding apparently triggered the allosteric pathway, as described above (Dowd *et al.*, 2012), but the mutation of Trp-165 introduced a barrier to the domain 3 structural changes. Functional PFO, however, formed chimeric oligomers with the PFO^{W165T} monomers and drove the specific disruption of the domain 2–3 interface and induced β-strand 5 to rotate away from β-strand 4, which resulted in the membrane insertion of the transmembrane β-hairpins of PFO^{W165T}.

These studies suggested that in a fraction of the monomers weak intermolecular interactions, which result from conformational changes induced by membrane binding via the allosteric pathway, lead to stronger intermolecular interactions that induce the major structural changes in domain 3. Hotze *et al.* (2012) proposed that the conversion of the weakly interacting monomers to a stable dimer was the rate-limiting step in oligomerization. In essence, the weak, reversible interactions promoted by membrane binding, as typified by PFO^{W165T}, do not lead to the quantitative conversion to stable dimers. Instead, to form the stable dimer the β-strand 1 of one monomer has to displace β-strand 5 of a second monomer and form interstrand hydrogen bonds with β-strand 4. Under normal conditions the thermal fluctuations within domain 3 would likely partially disrupt the β-strand 4–5 interactions, which form only 4 backbone hydrogen bonds in the monomer of PFO. Hence, in a fraction of the weakly interacting dimers the β-strand 4–5 interaction will be in this transient open state due to the thermal fluctuations, which then can allow β-strand 1 of the second monomer of the dimer to displace β-strand 5 in the first monomer and form interstrand hydrogen bonds with β-strand 4 of the first monomer. Since β-strands 1 and 4 can form up to eight intermolecular hydrogen bonds the reversal of this interaction is significantly reduced, thereby stabilizing the dimer.

Hotze *et al.* (2012) further showed that the interaction of β-strand 4 of one monomer with β-strand 1 of the second monomer could displace β-strand 5 from its interaction with β-strand 4 of the second monomer, thereby increasing the probability that β-strand 4 of the second monomer will be in the open state. Hence, the interaction of β-strand 1 of free membrane-bound monomers with the β-strand 4 of the dimer, or higher order oligomer, would not be subjected to the same energetic constraints (i.e. the disruption of the β-strand 4–5 interaction by thermal fluctuations) that slow the initial formation of the dimer. Therefore, the extension of established dimers or higher order oligomers would be favoured over the establishment of new dimers, thus increasing the efficiency of assembling full-length functional oligomeric pore complexes. This mechanism is also consistent with the non-stochastic assembly of the PFO oligomer observed by Shepard *et al.* (2000). When they followed the assembly kinetics of the PFO oligomer at low temperature the only SDS stable species that were identified by SDS-agarose gel electrophoresis, which separates monomer and oligomer, were the monomer and full-length oligomer: intermediate sized oligomers were not observed. Even at the very earliest stages of assembly, when oligomerization was slowed by low temperature and stabilized by a cross-linker, the monomer, dimer and full-length oligomer were the only major species observed. If oligomerization were stochastic then at the earliest stages of assembly the concentration of the different species would be monomer > dimer >trimer >>> full-length oligomer in the samples stabilized by cross-linking. These data suggested that the extension of established oligomers into full-length oligomers was preferred over establishment of new oligomers, consistent with the above studies and the non-stochastic mechanism of CDC assembly (Hotze *et al.*, 2012).

A non-stochastic mechanism of assembly is also energetically more efficient that a stochastic mechanism of assembly. For a single oligomeric pore of 36 monomers more than 50,000 ATP would be required to build and secrete the necessary monomers. In a stochastic system of assembly many more oligomers would be initiated and the probability of depleting the available supply monomers before their conversion to functional complexes would be higher than for the non-stochastic model. The efficient assembly of oligomers on eukaryotic cells may not be important near the bacterial cell where the concentration of monomers are high, but as the monomers diffuse away from the bacteria cell their concentration rapidly decreases. In the case of *C. perfringens* gas gangrene PFO has been shown to act distally from the site of infection in the muscle to affect the infiltration of neutrophils and to cause coagulative necrosis (Awad *et al.*, 2001). Hence, the more efficient non-stochastic model of assembly would increase the probability that functional oligomeric pores would assemble on the target cells.

Formation of the pore complex

Shortly after the solution of the monomer crystal structure for PFO the first major aspect of the PFO pore mechanism to be revealed was the identity of its membrane-penetrating domain. Using a combination of fluorescence spectroscopic approaches Shepard *et al.* (1998) and Shatursky *et al.* (1999) revealed that the two domain 3 α-helical bundles (Fig. 3.4) unravelled to form two amphipathic β-hairpins that penetrated and crossed the membrane bilayer (transmembrane β-hairpins or TMHs). These studies established that the CDCs employ a β-barrel pore to penetrate the bilayer rather than an amphipathic α-helical pore, as had been suggested earlier (Palmer *et al.*, 1996). The PFO oligomeric complex is comprised of about 36 monomers (there is a small variability in the final size) (Czajkowsky *et al.*, 2004), therefore its β-barrel pore contains approximately 144 amphipathic β-strands. At the time of its discovery PFO was the only pore-forming toxin that employed two β-hairpins per monomer, which were combined to form the β-barrel pore. The heptameric pore formers, such as *S. aureus* α-haemolysin, employ a single β-hairpin per monomer (Song *et al.*, 1996; Melton *et al.*, 2004; Iacovache *et al.*, 2006). The use of two β-hairpins per monomer was in part due to the size of the oligomeric pore and its radius of curvature: the width of a single β-hairpin per monomer is not sufficient to form the interstrand backbone hydrogen bonds necessary to form a contiguous β-barrel pore whereas a single hairpin is sufficient to form a contiguous β-barrel for the heptameric pore-forming toxins.

The fact that the monomers are anchored in a perpendicular manner to the membrane surface along their long axis (Fig. 3.4) posed a problem for the membrane insertion of the β-barrel: when the α-helical bundles were unravelled in the β-hairpins the length of the β-hairpins was only sufficient to reach the bilayer surface, but not cross the membrane bilayer. The long axis of the PFO monomers is approximately ≈115 Å, which positions domain 4 near the surface and domain 1 furthest from the membrane surface, as depicted in Fig. 3.4. This orientation positions the domain 3 α-helical bundles, which ultimately form the transmembrane β-hairpins, about 40 Å above the membrane surface. When they are extended the hairpins are only long enough to reach the membrane surface, yet the studies of Shepard *et al.* (1998) and Shatursky *et al.* (1999) showed that they did cross the bilayer. This conundrum was resolved in two studies where atomic force microscopy (AFM) and Förster resonance energy transfer (FRET) were used independently to demonstrate that a 35–40 Å collapse of the prepore structure occurred upon its conversion to the pore complex (Czajkowsky *et al.*, 2004; Ramachandran *et al.*, 2005) (Fig. 3.4). The AFM data showed that the ring radius did not significantly change upon the vertical collapse of the prepore to pore. Based on this observation they proposed that the domain 2 structure collapsed and allowed domains 1 and 3 to drop nearly straight down to the membrane surface, thereby bringing domain 3 sufficiently close to the membrane surface so that the extended β-hairpins could cross the bilayer. The structure of domain 2 is largely maintained through its interface with domain 3, which has to be disrupted to extend transmembrane hairpin 1 (TMH1, Fig. 3.4) and relieve the twist in the core β-sheet. It was unlikely that the structure of domain 2 would remain intact once domain 3 rotated away from its interface with domain 2, thus, it was predicted that domain 2 would collapse thereby bringing domains 1 and 3 closer to the membrane surface. These studies were further supported by the subsequent 3D reconstruction of the pneumolysin pore fitted with the PFO structure: a prominent bulge was observed on the outside of the oligomeric pore complex, which was consistent with domain 2 folded in half (Tilley *et al.*, 2005).

Hence, these studies showed that upon prepore to pore conversion the prepore complex undergoes about a large vertical collapse to plunge the β-barrel into the membrane bilayer. This process is unlike that seen for α-haemolysin, or currently any other pore-forming toxin. The α-haemolysin prepore oligomer undergoes no change in the vertical dimension of the prepore upon formation of the pore. The transmembrane β-hairpin of the α-haemolysin is folded in half in the monomer and unfolds to extend into the transmembrane β-hairpin of the β-barrel pore (Olson *et al.*, 1999).

Ring versus arc oligomers

There has been a continuing discussion about the presence and function of incomplete oligomers versus the complete ring-shaped complexes (Bhakdi *et al.*, 1985), as well as the presence of small pores induced by CDCs (Shaughnessy *et al.*, 2006). The evidence that suggests that incomplete rings, i.e. 'arcs' can insert into the bilayer is primarily derived from electron micrographs of the complexes. These incomplete rings suggest the presence of what appears to be a pore that is ostensibly lined with lipid on one side, as suggested by Palmer *et al.* (Palmer *et al.*, 1998a). As of now there are no unequivocal data that demonstrate that incomplete oligomers can insert and form pores. Furthermore, the examination of the fragments (i.e. incomplete rings) within the EM images often suggests that they result from the breakup of intact ring structures by mechanical stress. This is particularly evident in a recent publication on listeriolysin O where there are many examples of arc complexes that appear to be derived directly from ring fragmentation in the EM samples (Vadia *et al.*, 2011). This does not mean incomplete rings cannot insert into the membrane, but to date there has been no direct, convincing evidence for the membrane insertion of incomplete rings. Furthermore, it may be hard or impossible to distinguish between arcs that are simply fragments from the breakup of intact pore complexes during preparation for EM versus those that may have inserted prior to the formation of the complete ring.

The CDCs and bacterial pathogenesis

As described in several reviews (Kayal and Charbit, 2006; Pizarro-Cerda and Cossart, 2006; Marriott *et al.*, 2008; Hamon *et al.*, 2012) the CDCs from *S. pneumoniae* (PLY) and *L. monocytogenes* (LLO) can have a myriad of cellular effects on different cell types. Many of these effects are correlated with disturbances in cellular calcium or potassium levels induced by these toxins, although it is unclear if these disturbances are a direct or indirect effect of pore formation by the CDCs. Using a CDC pore to dysregulate calcium metabolism seems overkill, since a 25 nm pore is not necessary to allow passage of calcium ions. Many of the cellular effects reported for CDCs may simply be the cellular response to the attack of the plasma membrane by a pore-forming toxin, whereas others may be a direct result of the CDC action on a specific cell type. CDC binding is highly sensitive to the lipid environment of the cholesterol molecule (Heuck *et al.*, 2000; Nelson *et al.*, 2008; Flanagan *et al.*, 2009); therefore, it is possible that CDCs can target specific microdomains on the cell surface, which might interfere with specific processes of the cell and may differ depending on the cell type. However, as indicated in the introduction, a comprehensive discussion of these cellular effects is beyond the scope of this review. Instead, we will describe a few fundamental pathogenesis-related features of the CDCs that have been known for years, yet they are still not well understood. Three examples will be discussed where CDCs contribute to pathogenesis, but how they accomplish these feats is not always apparent.

Listeriolysin O

The difficulty in elucidating the direct effect of a CDC in pathogenesis is illustrated by LLO. An essential phase of *L. monocytogenes* infection is the invasion of the eukaryotic cell. Once the bacterial cell successfully invades it must escape from the phagosome before it fuses with the lysosome, which destroys the bacteria cell. Once in the cytosol *L. monocytogenes* is then free to replicate without being exposed to the immune system and can spread cell to cell without an extracellular stage. It has been known since 1987 that LLO production is essential to the intracellular replication of *L. monocytogenes* (Gaillard *et al.*, 1987; Berche *et al.*, 1988; Portnoy *et al.*, 1988; Cossart *et al.*, 1989) by mediating its escape from the phagosome. Initially, the concept was that LLO lysed the phagosome, yet this mechanism is not feasible since unlike a cell, where the cytoplasmic contents would flow out, the osmotic pressures would be reversed between the cell cytoplasm and vacuole. Furthermore, once inserted into the phagosomal membrane the size of the LLO pore would simply allow the contents of the two compartments to mix. The LLO pore, although large, is also much too small to allow passage of the bacteria cell. Some studies suggest that LLO facilitates vacuolar escape by perturbing the rate of phagosome–lysosome fusion by pH and calcium disturbances, which provide the bacterial cell more time to escape (Shaughnessy *et al.*, 2006). How the bacterial cells escape still remains unclear, although the studies of Burrack *et al.* (2009) suggests that LLO perforation of the phagosome slows its maturation sufficiently to allow the bacterial phospholipases to degrade the liposomal membrane. LLO also appears to be essential to escaping the double membrane during cell-to-cell spread. In this case, however, its function appears to be to provide access for the bacterial phospholipases to the second membrane of the dual membrane endosome. These studies suggest that the large CDC pore facilitates the transfer of proteins across the membrane, which makes sense considering its large size and the significant investment in energy that is necessary to produce it. Although this scenario is consistent with the large CDC pore and passage of proteins through this pore, many studies suggest that the CDCs trigger cellular ion perturbations (as is evident from the reviews of CDC involvement in pathogenesis described in the introduction to this section). However, it is not yet clear in all cases if the CDC pore alone facilitates the passage of the ions or it is an indirect effect of pore formation. Protein translocation through the CDC pore seemed to be a reasonable theory to explain why the CDC pore is so large, but then biology threw a curve ball at us in the form of streptolysin O (SLO) where the pore may not be required for protein translocation.

Streptolysin O

Streptococcus pyogenes causes a wide variety of diseases, but one of the most invasive and serious is necrotizing fasciitis (Stevens, 1996). Like many other CDCs its CDC, SLO, has been shown to induce a number of cellular effects (e.g. Ruiz *et al.*, 1998; Limbago *et al.*, 2000; Cywes Bentley *et al.*, 2005; Nilsson *et al.*, 2006; Harder *et al.*, 2009; Timmer *et al.*, 2009; Keyel *et al.*, 2011; Logsdon *et al.*, 2011; Usmani *et al.*, 2011). From a structural viewpoint one of the most interesting, but puzzling aspects of SLO activity is associated with its specific transport of an NAD glycohydrolase enzyme (SPN) into the keratinocyte (Madden *et al.*, 2001). SLO mediates the translocation of SPN into cells, which then hydrolyses NAD^+ into ADP-ribose and nicotinamide. Diphtheria and cholera toxins also exhibit NAD glycohydrolase activity, but in the presence of a protein acceptor these toxins will transfer the ADP-ribose moiety to that protein and alter its activity (Collier and Pappenheimer, 1964; Collier, 1967; Cassel and Pfeuffer, 1978; Gill and Meren, 1978). These toxins are termed A-B toxins, in which the A fragment is the catalytic fragment and the B fragment is the receptor binding and translocation subunit that facilitates the entry of the A-subunit into cells. SPN is secreted by the *S. pyogenes* type II secretion system, but for years it was unclear whether it entered cells, as there was no apparent translocation or 'B' subunit. Madden *et al.* (Madden *et al.*, 2001) showed that SLO was responsible for the translocation of SPN into eukaryotic cells. It was subsequently shown that this translocation system is apparently highly specific, as no other *S. pyogenes* protein has been identified that is translocated by SLO (Meehl and Caparon, 2004). It was assumed that the SLO pore, which like other CDC pores, is large and could easily facilitate the passage of quite large proteins, was necessary to translocate SPN into cells. It was proposed that this system was the analogue to the Type III secretion found in Gram-negative bacteria (Madden *et al.*, 2001), which directly translocates effectors from the bacterial cytosol to the eukaryotic cytosol (Blocker *et al.*, 2003). However, nearly a decade later it was shown by the same group that the SLO pore is not necessary for the transfer of SPN since SLO trapped at various non-pore-forming stages translocates SPN as well or better than wild-type SLO (Magassa *et al.*, 2010)! Hence, this SLO-based translocation system shows that the CDCs can function in ways that do not require formation of the pore, but the mechanism by which SLO effects SPN translocation remains unknown.

Perfringolysin O

PFO contributes to *C. perfringens* gas gangrene, but in a completely unexpected way. The CDCs are highly cytolytic and a hallmark of *C. perfringens* gas gangrene is massive necrosis of the muscle tissue (myonecrosis). Yet, PFO does not contribute extensively to the development of myonecrosis, but instead contributes to effects at the periphery of the infection (Awad *et al.*, 1995, 2001; Ellemor *et al.*, 1999). It contributes to leukostasis, coagulative necrosis and vascular disruption, which would suggest that it can diffuse to the periphery of the infection to cause these effects. The primary mediator of the muscle necrosis is *C. perfringens* α-toxin, a phospholipase/sphingomyelinase enzyme. It remains unclear why PFO does not exhibit a greater contribution to the destruction of the muscle tissue, or why it alone is not employed by the bacterial cell to effect myonecrosis. Curiously, Kennedy *et al.* (2009) showed that the pore-forming α-toxin from *Clostridium septicum* (Ballard *et al.*, 1992), another myonecrotic clostridial species, could effectively replace the phospholipase/sphingomyelinase α-toxin (Titball *et al.*, 1989) of *C. perfringens*. *C. septicum* α-toxin is a heptameric pore former, similar, but unrelated to α-haemolysin from *S. aureus*. Interestingly, the *C. septicum* α-toxin in *C. perfringens* caused significant myonecrosis, but did not induce the characteristic leukostasis, which is a hallmark feature of myonecrosis and in *C. perfringens* requires the synergistic action of PFO and the phospholipase/sphingomyelinase α-toxin (Awad *et al.*, 1995, 2001; Ellemor *et al.*, 1999). Hence, these studies show that a pore-forming toxin can induce significant myonecrosis, but for reasons unknown *C. perfringens* has not employed PFO for this purpose. The fact that PFO does not exhibit significant cytolytic activity on the muscle tissue during gangrene suggests that the lipid environment of the cholesterol in the muscle cells plasma membrane does not readily support efficient binding of PFO and so PFO can diffuse to and act at the periphery of the infection.

The membrane attack complex/ perforin (MACPF) proteins

The Membrane Attack Complex/Perforin-like (MACPF) protein family was originally discovered through the identification of sequence similarly between the pore-forming immunity proteins perforin and Complement C9 [a component of the complement MAC (Young *et al.*, 1986)]. Perforin is secreted by cytotoxic T-lymphocytes and is responsible for eliminating virus infected and malignant cells, whereas the complement MAC is responsible for destroying certain Gram-negative pathogens as well as foreign cells. Together C9 and perforin represented two important examples of what was thought to be a pore-forming protein family exclusive to higher eukaryotes. In 1999, however, the first bacterial MACPF protein, CT153 from *Chlamydia* spp., was identified (Ponting, 1999) and over subsequent years it became clear that representative MACPF proteins were widely distributed throughout most kingdoms of life. From a microbial perspective, representative family members could be identified in slime moulds (e.g. *Dictyostelium fasciculatum*), certain fungi (Stephens *et al.*, 1999), pathogenic apicomplexans (Kafsack *et al.*, 2009), as well as occasionally in prokaryotic organisms (Ponting, 1999; Rosado *et al.*, 2007). However, the distribution of microbial MACPF proteins is generally sporadic and, accordingly, the presence of MACPF genes in microbes (and in particular in prokaryote organisms) appear to be the exception rather than the rule.

The relationship between MACPF-like proteins and the microbial world was further illuminated through structural studies. The structure of the bacterial MACPF protein Plu-MACPF, from the insect pathogenic γ-proteobacteria *Photorhabdus luminescens* (Rosado *et al.*, 2007), together with the structure of the MACPF domain of Complement C8 (Hadders *et al.*, 2007; Slade *et al.*, 2008) revealed the surprising finding: the MACPF proteins are homologous to the CDCs (discussed earlier in this chapter; Fig. 3.5). The two branches of the family share exceptionally limited sequence identity, with only two glycine residues absolutely conserved in position in the structures of both MACPF proteins and CDCs. Crucially, like CDCs, MACPF proteins contain two clusters of helices that are structurally equivalent to TMH1 and TMH2 of the CDCs (Fig. 3.5). Analysis of pore-forming MACPF proteins further revealed that the sequence of TMH1 and TMH2 is consistent with the requirement to form an amphipathic, membrane embedded β-barrel. These data, together with additional structural and biochemical insights (Law *et al.*, 2010; Hadders *et al.*, 2012), thus provide strong support for the idea that MACPF proteins most likely form pores *via* a mechanism that is broadly CDC-like, but likely exhibits important differences. Taken together, these structural and mechanistic data also support the idea that CDCs and MACPF proteins share a common evolutionary ancestor.

At the time of writing, the MACPF proteins and CDCs together number well over 1000 members. While non-pore-forming family members

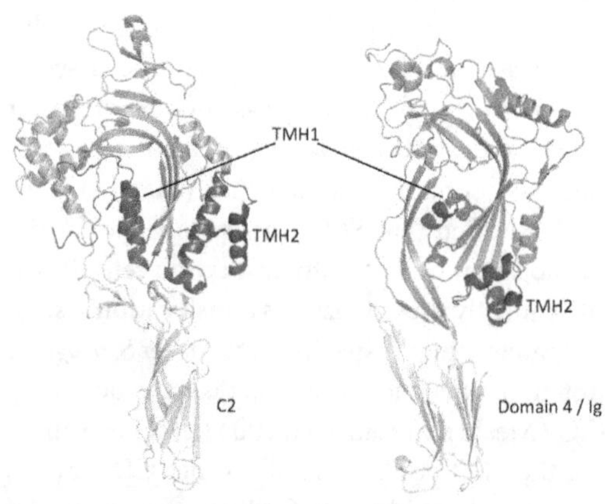

Figure 3.5 Ribbon representations are shown illustrating the structures of the MACPF proteins mouse perforin (3NSJ) (Law *et al.*, 2010) in the left panel and PFO (1PFO) (Rossjohn *et al.*, 1997) in the right panel. The structures are shown in a similar orientation, and the membrane penetrating regions TMH1 and TMH2 are labelled. Perforin and PFO deploy a C2 domain and Ig-like fold (domain IV), respectively, to interact with membranes. In both molecules these membrane interacting domains are positioned directly underneath the MACPF/CDC membrane penetrating domain.

have been identified (for example complement C6 and C7 function as scaffolding proteins in the assembly of the MAC; Hadders *et al.*, 2012), the majority of MACPF proteins characterized to date perform a role in membrane insertion and/or pore formation. Accordingly, the MACPF/CDC superfamily is currently the largest and most diverse group of pore-forming toxins.

Most functional and biochemical studies published to date are focused on mammalian MACPF proteins such as perforin and members of the complement MAC. In contrast, the role of microbial MACPF proteins generally remains obscure. Many of the microbes that contain MACPF proteins are, however noted pathogens or are organisms that interact closely with mammalian or other eukaryote or bacterial hosts. For example analysis of genomic data reveal members of the MACPF superfamily can be found in *Plasmodium* spp., *Toxoplasma* spp., *Theileria parva* (a parasitic protozoan that causes East Coast fever in livestock), pathogenic fungi such as *Fusarium oxysporum* and ciliates such as *Paramecium* spp. In addition to the CDC branch of the family, bacterial MACPF genes can be identified in *Chlamydial* spp., several members of the Cytophaga–Flavobacteria–Bacteroides (CFB) group and enterobacteria such as *Photorhabdus luminescens*.

Better-characterized microbial MACPF proteins include those produced by the protozoan parasites that cause Malaria and Toxoplasmosis. In malaria a MACPF protein termed SPECT-2 is essential for the traversal of the sinusoidal cell layer as a prelude to hepatocyte infection (Ishino *et al.*, 2005). It was suggested that this molecule functions to damage host cell membranes, thus permitting parasite invasion. In addition, a second malaria MACPF protein, MAOP, is required to transverse the mosquito midgut (Kadota *et al.*, 2004). MACPF proteins in *Toxoplasma gondii* also play an important role in invasion – in this instance a perforin-like protein, TGPLP1, is required to damage the parasitophorous vacuole and host cell plasma membrane and parasites lacking TgPLP1 are unable to exit the host cell (Kafsack *et al.*, 2009). Finally, however, it is perhaps worth reiterating that the best-characterized members of the MACPF/CDC superfamily are the CDCs themselves, which are important virulence factors, discussed elsewhere in this chapter. Given what we know to date, it is thus tempting to broadly label microbial MACPF proteins as possible toxins or virulence factors. However, it is important to note that certain MACPF proteins perform an important role distinct from defence or attack. Most notably the Drosophila MACPF protein Torso-like is a maternally secreted factor essential for proper embryonic patterning (Stevens *et al.*, 1990) and certain fungal MACPF proteins such as SpoC1-C1C are postulated to be involved in reproduction (Stephens *et al.*, 1999). Accordingly, it is quite possible that in addition to potential roles in invasion or defence, microbial MACPF proteins may also be more generally involved in specialized housekeeping or reproductive functions.

Summary and future perspectives – CDCs and MACPF proteins

The CDCs are widespread in Gram-positive bacteria, yet a greater understanding of their contribution to bacterial survival during infections, or in their more frequent commensal states, is required. These studies will be aided by a better understanding of how CDCs target cells and subcellular membrane domains. It is clear that the lipid environment of the cholesterol receptor has a significant influence on binding; hence, can different CDCs recognize and bind to different cholesterol-rich microdomains, thereby preferentially targeting different cell types with the 'right' lipid environment for the cholesterol? If they target discrete microdomains it may result in microdomain-dependent effects on cells: i.e. the formation of a large pore complex may disrupt specific cellular signalling pathways localized to specific lipid environments. We also do not understand the basis for the shift in receptor specificity in the human CD59 binding CDCs. Since CD59 is comparatively ubiquitous on cells it does not appear to provide any apparent advantage in cellular specificity for these CDCs. Furthermore, this receptor specificity restricts the activity of these CDCs to human cells, which suggests these organisms coevolved with humans. Yet we do not understand why these bacterial species, which are mucosal commensals, evolved a CDC to utilize this specific receptor and what advantages this change in receptor specificity confers to their survival in the human host.

Although we have made significant progress in understanding the pore-forming mechanism of the CDCs, several facets of this process remain unclear. One of the outstanding questions is how SLO mediates translocation of SPN in a pore-independent manner. We also do not understand the fate of the lipid and the associated membrane proteins that may be trapped within the interior of the β-barrel pore of the CDCs. The CDC β-hairpins are not long enough to sweep the lipids out of the way, which probably occurs with the smaller pore-forming toxins like *S. aureus* α-haemolysin. Hence, when CDCs attack cells large sections of the plasma membrane may be displaced or lost. The area of a CDC pore is on the order of 700 nm^2, which could encompass several large protein complexes or many individual proteins. If specific microdomains are targeted by the CDCs then the cell may lose proteins or protein complexes that are important to the cell's normal function. The possible loss of signalling complexes could be critical if the cells are associated with the early immune response to these infections. The assembly of the β-barrel pore and the forces that drive its insertion into the bilayer also remain unclear: do the CDCs assemble a pre-β-barrel in the prepore complex and then plunge this into the bilayer like a cookie cutter and if so what supplies the energy to initiate the insertion and what triggers the insertion process? The answer to these questions will undoubtedly reveal new paradigms in the mechanism of pore formation and role of the CDCs in pathogenesis.

Finally, we can now begin to explore the pore-forming mechanism of the MACPF proteins in earnest, now that we have structures for several members of this family, the ability to manipulate purified recombinant forms and a basis for the pore-forming mechanism from the structural similarities between the CDCs and MACPF proteins. It is likely that the MACPF proteins mechanism will form a β-barrel pore like the CDCs, but the assembly of the β-barrel and its insertion mechanism will likely hold some surprises for us. The studies on perforin already suggest that its oligomer structure does not undergo a large vertical collapse like the CDCs to form its pore (Law *et al.*, 2010). Hence, the study of the MACPF proteins will likely lead us to new insights into how they function and contribute to cellular survival of microbes or protection of the host.

In the last 20 years the study of the pore-forming toxins has shown us that their mechanisms are far more complex than we originally envisaged. Furthermore, more recent studies have changed our view of these proteins as simple pore formers that lyse cells: its is now clear that they are far more complex and sophisticated molecules whose real functions are only now becoming apparent.

References

Abrami, L., and van Der Goot, F.G. (1999). Plasma membrane microdomains act as concentration platforms to facilitate intoxication by aerolysin. J. Cell Biol. *147*, 175–184.

Abrami, L., Fivaz, M., Decroly, E., Seidah, N.G., Jean, F., Thomas, G., Leppla, S.H., Buckley, J.T., and van der Goot, F.G. (1998). The pore-forming toxin proaerolysin is activated by furin. J. Biol. Chem. *273*, 32656–32661.

Abrami, L., Velluz, M.C., Hong, Y., Ohishi, K., Mehlert, A., Ferguson, M., Kinoshita, T., and Gisou van der Goot, F. (2002). The glycan core of GPI-anchored proteins modulates aerolysin binding but is not sufficient: the polypeptide moiety is required for the toxin–receptor interaction. FEBS Lett. *512*, 249–254.

Allinson, T.M., Parkin, E.T., Turner, A.J., and Hooper, N.M. (2003). ADAMs family members as amyloid precursor protein alpha-secretases. J. Neurosci. Res. *74*, 342–352.

Arbuthnott, J.P. (1970). Staphylococcal alpha-toxin. In Microbial Toxins, Montie, T.C., Kadis, S., and Ajl, S.I., eds. (New York, Academic Press), pp. 189–236.

Arbuthnott, J.P., Freer, J.H., and Bernheimer, A.W. (1967). Physical states of staphylococcal alpha-toxin. J. Bacteriol. *94*, 1170–1177.

Arbuthnott, J.P., Freer, J.H., and Billcliffe, B. (1973a). Lipid-induced polymerization of staphylococcal -toxin. J. Gen. Microbiol. *75*, 309–319.

Arbuthnott, J.P., Freer, J.H., and McNiven, A.C. (1973b). Physical properties of staphylococcal alpha-toxin and aspects of alpha-toxin membrane interactions. Contrib. Microbiol. Immunol. *1*, 285–297.

Awad, M.M., Bryant, A.E., Stevens, D.L., and Rood, J.I. (1995). Virulence studies on chromosomal alpha-toxin and theta-toxin mutants constructed by allelic exchange provide genetic evidence for the essential role of alpha-toxin in Clostridium perfringens-mediated gas gangrene. Mol. Microbiol. *15*, 191–202.

Awad, M.M., Ellemor, D.M., Boyd, R.L., Emmins, J.J., and Rood, J.I. (2001). Synergistic effects of alpha-toxin and perfringolysin O in Clostridium perfringens-mediated gas gangrene. Infect. Immun. *69*, 7904–7910.

Ballard, J., Bryant, A., Stevens, D., and Tweten, R.K. (1992). Purification and characterization of the lethal toxin (alpha-toxin) of *Clostridium septicum*. Infect. Immun. *60*, 784–790.

Ballard, J., Crabtree, J., Roe, B.A., and Tweten, R.K. (1995a). The primary structure of *Clostridium septicum* alpha toxin exhibits similarity with *Aeromonas hydrophila* aerolysin. Infect. Immun. *63*, 340–344.

Ballard, J., Crabtree, J., Roe, B.A., and Tweten, R.K. (1995b). The primary structure of *Clostridium septicum* alpha-toxin exhibits similarity with that of Aeromonas hydrophila aerolysin. Infect. Immun. *63*, 340–344.

Bartlett, A.H., Foster, T.J., Hayashida, A., and Park, P.W. (2008). Alpha-toxin facilitates the generation of CXC chemokine gradients and stimulates neutrophil homing in *Staphylococcus aureus* pneumonia. J. Infect. Dis. *198*, 1529–1535.

Bayer, A.S., Ramos, M.D., Menzies, B.E., Yeaman, M.R., Shen, A.J., and Cheung, A.L. (1997). Hyperproduction of alpha-toxin by *Staphylococcus aureus* results in paradoxically reduced virulence in experimental endocarditis: a host defense role for platelet microbicidal proteins. Infect. Immun. *65*, 4652–4660.

Bayrhuber, M., Meins, T., Habeck, M., Becker, S., Giller, K., Villinger, S., Vonrhein, C., Griesinger, C., Zweckstetter, M., and Zeth, K. (2008). Structure of the human voltage-dependent anion channel. Proc. Natl. Acad. Sci. U.S.A. *105*, 15370–15375.

Belmonte, G., Cescatti, L., Ferrari, B., Nicolussi, T., Ropele, M., and Menestrina, G. (1987). Pore formation by *Staphylococcus aureus* alpha-toxin in lipid bilayers. Dependence upon temperature and toxin concentration. Eur. Biophys. J. *14*, 349–358.

Benton, K.A., Paton, J.C., and Briles, D.E. (1997). Differences in virulence for mice among *Streptococcus pneumoniae* strains of capsular types 2, 3, 4, 5, and 6 are not attributable to differences in pneumolysin production. Infect. Immun. *65*, 1237–1244.

Berche, P., Gaillard, J.L., and Richard, S. (1988). Invasiveness and intracellular growth of *Listeria monocytogenes*. Infection. *16* (Suppl.2), S145–S148.

Bernheimer, A.W. (1944). Parallelism in the lethal and hemolytic activity of the toxin of *Clostridium septicum*. J. Exp. Med. *80*, 309–320.

Bernheimer, A.W. (1968). Cytolytic toxins of bacterial origin. The nature and properties of cytolytic proteins are discussed with emphasis on staphylococcal alpha-toxin. Science *159*, 847–851.

Bernheimer, A.W. (1976). Sulfhydryl activated toxins. In Mechanism in Bacterial Toxinology, Berheimer, A.W., ed. (New York, John Wiley and Sons), pp. 85–97.

Bernheimer, A.W., and Avigad, L.S. (1974). Partial characterization of aerolysin, a lytic exotoxin from Aeromonas hydrophila. Infect. Immun. *9*, 1016–1021.

Bernheimer, A.W., and Schwartz, L.L. (1963). Isolation and composition of staphylococcal alpha toxin. J. Gen. Microbiol. *30*, 455–468.

Bernheimer, A.W., Avigad, L.S., and Avigad, G. (1975). Interactions between aerolysin, erythrocytes, and erythrocyte membranes. Infect. Immun. *11*, 1312–1319.

Berry, A.M., Lock, R.A., Hansman, D., and Paton, J.C. (1989). Contribution of autolysin to virulence of *Streptococcus pneumoniae*. Infect. Immun. *57*, 2324–2330.

Bettegowda, C., Huang, X., Lin, J., Cheong, I., Kohli, M., Szabo, S.A., Zhang, X., Diaz, Jr, L.A.., Velculescu, V.E., Parmigiani, G., *et al.* (2006). The genome and transcriptomes of the anti-tumor agent *Clostridium novyi*-NT. Nat. Biotechnol. *24*, 1573–1580.

Bhakdi, S., and Tranum-Jensen, J. (1991). Alpha-toxin of *Staphylococcus aureus*. Microbiol. Rev. *55*, 733–751.

Bhakdi, S., Tranum-Jensen, J., and Sziegoleit, A. (1985). Mechanism of membrane damage by streptolysin-O. Infect. Immun. *47*, 52–60.

Bhakdi, S., Mackman, N., Nicaud, J.M., and Holland, I.B. (1986). *Escherichia coli* hemolysin may damage target cell membranes by generating transmembrane pores. Infect. Immun. *52*, 63–69.

Bhakdi, S., Muhly, M., Korom, S., and Hugo, F. (1989). Release of interleukin-1 beta associated with potent cytocidal action of staphylococcal alpha-toxin on human monocytes. Infect. Immun. *57*, 3512–3519.

Bhakdi, S., Jursch, R., Broker, M., Ronneberger, H., and Hungerer, K.D. (1994). Functionally inactive S. aureus alpha-toxin containing a single amino acid substitution: potential usefulness as a vaccine. Behring Inst. Mitt. 80–84.

Bhakdi, S., Walev, I., Hussmann, M., and Valeva, A. (2005). Staphylococcal alpha-toxin. In Microbial Protein Toxins, Schmitt, M.J., and Schaffrath, R., eds. (Berlin, Springer-Verlag), pp. 91–110.

Bianchet, M.A., Odom, E.W., Vasta, G.R., and Amzel, L.M. (2002). A novel fucose recognition fold involved in innate immunity. Nat. Struct. Biol. *9*, 628–634.

Blocker, A., Komoriya, K., and Aizawa, S. (2003). Type III secretion systems and bacterial flagella: insights into their function from structural similarities. Proc. Natl. Acad. Sci. U.S.A. *100*, 3027–3030.

Boraston, A.B., Wang, D., and Burke, R.D. (2006). Blood group antigen recognition by a *Streptococcus pneumoniae* virulence factor. J. Biol. Chem. *281*, 35263–35271.

Bourdeau, R.W., Malito, E., Chenal, A., Bishop, B.L., Musch, M.W., Villereal, M.L., Chang, E.B., Mosser, E.M., Rest, R.F., and Tang, W.J. (2009). Cellular functions and X-ray structure of anthrolysin O, a cholesterol-dependent cytolysin secreted by *Bacillus anthracis*. J. Biol. Chem. *284*, 14645–14656.

Bramley, A.J., Patel, A.H., O'Reilly, M., Foster, R., and Foster, T.J. (1989). Roles of alpha-toxin and beta-toxin in virulence of *Staphylococcus aureus* for the mouse mammary gland. Infect. Immun. *57*, 2489–2494.

Branton, D., Deamer, D.W., Marziali, A., Bayley, H., Benner, S.A., Butler, T., Di Ventra, M., Garaj, S., Hibbs, A., Huang, X., *et al.* (2008). The potential and challenges of nanopore sequencing. Nat. Biotechnol. *26*, 1146–1153.

Braun, J.S., Sublett, J.E., Freyer, D., Mitchell, T.J., Cleveland, J.L., Tuomanen, E.I., and Weber, J.R. (2002). Pneumococcal pneumolysin and H(2)O(2) mediate brain cell apoptosis during meningitis. J. Clin. Invest. *109*, 19–27.

Breiger, L., and Fraenkel, C. (1890). Untersuchungen uber Bacteriengifte. Berlin Klin Wochschr *27*, 241.

Bruggemann, H., and Gottschalk, G. (2004). Insights in metabolism and toxin production from the complete genome sequence of *Clostridium tetani*. Anaerobe *10*, 53–68.

Bruggemann, H., Baumer, S., Fricke, W.F., Wiezer, A., Liesegang, H., Decker, I., Herzberg, C., Martinez-Arias, R., Merkl, R., Henne, A., *et al.* (2003). The genome sequence of *Clostridium tetani*, the causative agent of tetanus disease. Proc. Natl. Acad. Sci. U.S.A. *100*, 1316–1321.

Bubeck Wardenburg, J., and Schneewind, O. (2008). Vaccine protection against *Staphylococcus aureus* pneumonia. J. Exp. Med. *205*, 287–294.

Bubeck Wardenburg, J., Bae, T., Otto, M., Deleo, F.R., and Schneewind, O. (2007a). Poring over pores: alpha-hemolysin and Panton–Valentine leukocidin in *Staphylococcus aureus* pneumonia. Nat. Med. *13*, 1405–1406.

Bubeck Wardenburg, J., Patel, R.J., and Schneewind, O. (2007b). Surface Proteins and Exotoxins are Required for the Pathogenesis of *Staphylococcus aureus* Pneumonia. Infect. Immun. *75*, 1040–1044.

Buckley, J.T., Halasa, L.N., Lund, K.D., and MacIntyre, S. (1981). Purification and some properties of the hemolytic toxin aerolysin. Can. J. Biochem. *59*, 430–435.

Burnet, F.M. (1929). The exotoxins of *Staphylococcus pyogenes aureus*. J. Pathol. Bacteriol. *32*, 717–734.

Burrack, L.S., Harper, J.W., and Higgins, D.E. (2009). Perturbation of vacuolar maturation promotes listeriolysin O-independent vacuolar escape during *Listeria monocytogenes* infection of human cells. Cell. Microbiol. *11*, 1382–1398.

Cabiaux, V., Buckley, J.T., Wattiez, R., Ruysschaert, J.M., Parker, M.W., and van der Goot, F.G. (1997). Conformational changes in aerolysin during the transition from the water-soluble protoxin to the membrane channel. Biochemistry *36*, 15224–15232.

Cassel, D., and Pfeuffer, T. (1978). Mechanism of cholera toxin action: covalent modification of the guanyl nucleotide-binding protein of the adenylate cyclase system. Proc. Natl. Acad. Sci. U.S.A. *75*, 2669–2673.

Cassidy, P., and Harshman, S. (1976). Studies on the binding of staphylococcal 125I-labeled α-toxin to rabbit erythrocytes. Biochemistry *15*, 2348–2355.

Cho, J.S., Pietras, E.M., Garcia, N.C., Ramos, R.I., Farzam, D.M., Monroe, H.R., Magorien, J.E., Blauvelt, A., Kolls, J.K., Cheung, A.L., *et al.* (2010). IL-17 is essential for host defense against cutaneous *Staphylococcus aureus* infection in mice. J. Clin. Invest. *120*, 1762–1773.

Clarke, J., Wu, H.C., Jayasinghe, L., Patel, A., Reid, S., and Bayley, H. (2009). Continuous base identification for single-molecule nanopore DNA sequencing. Nat. Nanotechnol. *4*, 265–270.

Cockeran, R., Steel, H.C., Mitchell, T.J., Feldman, C., and Anderson, R. (2001). Pneumolysin potentiates production of prostaglandin E(2) and leukotriene B(4) by human neutrophils. Infect. Immun. *69*, 3494–3496.

Cole, A.R., Gibert, M., Popoff, M., Moss, D.S., Titball, R.W., and Basak, A.K. (2004). *Clostridium perfringens* epsilon-toxin shows structural similarity to the pore-forming toxin aerolysin. Nat. Struct. Mol. Biol. *11*, 797–798.

Collier, R.J. (1967). Effect of diphtheria toxin on protein synthesis: inactivation of one of the transfer factors. J. Mol. Biol. *25*, 83–98.

Collier, R.J., and Pappenheimer, Jr, A.M.. (1964). Studies on the mode of action of diphtheria toxin. II. Effect of toxin on amino acid incorporation in cell-free systems. J. Exp. Med. *120*, 1019–1039.

Cooper, L.Z., Madoff, M.A., and Weinstein, L. (1966). Heat stability and species range of purified staphylococcal alpha-toxin. J. Bacteriol. *91*, 1686–1692.

Cossart, P., Vincente, M.F., Mengaud, J., Baquero, F., Perez-Diaz, J.C., and Berche, P. (1989). Listeriolysin O is essential for the virulence of *Listeria monocytogenes*: direct evidence obtained by gene complementation. Infect. Immun. *57*, 3629–3639.

Cowell, S., Aschauer, W., Gruber, H.J., Nelson, K.L., and Buckley, J.T. (1997). The erythrocyte receptor for the channel-forming toxin aerolysin is a novel glycosylphosphatidylinositol-anchored protein. Mol. Microbiol. *25*, 343–350.

Craven, R.R., Gao, X., Allen, I.C., Gris, D., Bubeck Wardenburg, J., McElvania-Tekippe, E., Ting, J.P., and Duncan, J.A. (2009). *Staphylococcus aureus* alpha-hemolysin activates the NLRP3-inflammasome in human and mouse monocytic cells. PLoS One *4*, e7446.

Cywes Bentley, C., Hakansson, A., Christianson, J., and Wessels, M.R. (2005). Extracellular group A Streptococcus induces keratinocyte apoptosis by dysregulating calcium signalling. Cell. Microbiol. *7*, 945–955.

Czajkowsky, D.M., Hotze, E.M., Shao, Z., and Tweten, R.K. (2004). Vertical collapse of a cytolysin prepore moves its transmembrane β-hairpins to the membrane. EMBO J. *23*, 3206–3215.

De Christmas, M.J. (1888). Recherches Expérimentales sur la Suppuration. Ann. Inst. Pasteur *2*, 469.

Decatur, A.L., and Portnoy, D.A. (2000). A PEST-like sequence in listeriolysin O essential for *Listeria monocytogenes* pathogenicity. Science *290*, 992–995.

DeLano, W.L. (2002). The PyMOL Molecular Graphics System (San Carlos, CA, USA, DeLano Scientific).

Dowd, K.J., Farrand, A.J., and Tweten, R.K. (2012). The cholesterol-dependent cytolysin signature motif: a critical element in the allosteric pathway that couples membrane binding to pore assembly. PLoS Pathog. *8*, e1002787.

Edwards, D., Handsley, M., and Pennington, C. (2008). The ADAM metalloproteinases. Mol. Aspects Med. *29*, 258–289.

Elek, S.D. (1959). Staphylococcus Pyogenes (Edinburgh and London, E., and Livingstone, S., Ltd.).

Elek, S.D., and Levy, E. (1950). Distribution of hemolysis in pathogenic and non-pathogenic staphylococci. J. Pathol. Bacteriol. *62*, 541–554.

Ellemor, D.M., Baird, R.N., Awad, M.M., Boyd, R.L., Rood, J.I., and Emmins, J.J. (1999). Use of genetically manipulated strains of *Clostridium perfringens* reveals that both alpha-toxin and theta-toxin are required for vascular leukostasis to occur in experimental gas gangrene. Infect. Immun. *67*, 4902–4907.

Endres, K., and Fahrenholz, F. (2010). Upregulation of the α-secretase ADAM10 – risk or reason for hope? FEBS Journal *277*, 1585–1596.

Escrevente, C., Morais, V.A., Keller, S., Soares, C.M., Altevogt, P., and Costa, J. (2008). Functional role of N-glycosylation from ADAM10 in processing, localization and activity of the enzyme. Biochim. Biophys. Acta *1780*, 905–913.

Fairman, J.W., Noinaj, N., and Buchanan, S.K. (2011). The structural biology of beta-barrel membrane proteins: a summary of recent reports. Curr. Opin. Struct. Biol. *21*, 523–531.

Fairweather, N., Kennedy, S., Foster, T.J., Kehoe, M., and Dougan, G. (1983). Expression of a cloned *Staphylococcus aureus* alpha-hemolysin determinant in Bacillus subtilis and *Staphylococcus aureus*. Infect. Immun. *41*, 1112–1117.

Farrand, A.J., LaChapelle, S., Hotze, E.M., Johnson, A.E., and Tweten, R.K. (2010). Only two amino acids are essential for cytolytic toxin recognition of cholesterol at the membrane surface. Proc. Natl. Acad. Sci. U.S.A. *107*, 4341–4346.

Farrand, S., Hotze, E., Friese, P., Hollingshead, S.K., Smith, D.F., Cummings, R.D., Dale, G.L., and Tweten, R.K. (2008). Characterization of a streptococcal cholesterol-dependent cytolysin with a lewis y and b specific lectin domain. Biochemistry *47*, 7097–7107.

Feil, S.C., Rossjohn, J., Rohde, K., Tweten, R.K., and Parker, M.W. (1996). Crystallization and preliminary X-ray analysis of a thiol-activated cytolysin. FEBS Lett. *397*, 290–292.

Feil, S.C., Lawrence, S., Mulhern, T.D., Holien, J.K., Hotze, E.M., Farrand, S., Tweten, R.K., and Parker, M.W. (2012). Structure of the lectin regulatory domain of the cholesterol-dependent cytolysin lectinolysin reveals the basis for its Lewis antigen specificity. Structure *20*, 248–258.

Fivaz, M., Abrami, L., Tsitrin, Y., and van der Goot, F.G. (2001). Not as simple as just punching a hole. Toxicon *39*, 1637–1645.

Flanagan, J.J., Tweten, R.K., Johnson, A.E., and Heuck, A.P. (2009). Cholesterol exposure at the membrane surface is necessary and sufficient to trigger perfringolysin O binding. Biochemistry *48*, 3977–3987.

Fontes, W., Sousa, M.V., Aragao, J.B., and Morhy, L. (1997). Determination of the amino acid sequence of the plant cytolysin enterolobin. Arch. Biochem. Biophys. *347*, 201–207.

Forti, S., and Menestrina, G. (1989). Staphylococcal alpha-toxin increases the permeability of lipid vesicles by cholesterol- and pH-dependent assembly of oligomeric channels. Eur. J. Biochem. *181*, 767–773.

Frank, K.M., Zhou, T., Moreno-Vinasco, L., Hollett, B., Garcia, J.G., and Bubeck Wardenburg, J. (2012). Host response signature to *Staphylococcus aureus* alpha-Hemolysin implicates pulmonary TH17 response. Infect Immun. *80*, 3161–3169.

Freer, J.H., Arbuthnott, J.P., and Bernheimer, A.W. (1968). Interaction of staphylococcal alpha-toxin with artificial and natural membranes. J. Bacteriol. *95*, 1153–1168.

Freer, J.H., Arbuthnott, J.P., and Billcliffe, B. (1973). Effects of staphylococcal -toxin on the structure of erythrocyte membranes: a biochemical and freeze-etching study. J. Gen. Microbiol. *75*, 321–332.

Fussle, R., Bhakdi, S., Sziegoleit, A., Tranum-Jensen, J., Kranz, T., and Wellensiek, H.J. (1981). On the mechanism of membrane damage by *Staphylococcus aureus* alpha-toxin. J. Cell Biol. *91*, 83–94.

Gaillard, J.L., Berche, P., Mounier, J., Richard, S., and Sansonetti, P. (1987). *In vitro* model of penetration and intracellular growth of *Listeria monocytogenes* in the human enterocyte-like cell line Caco-2. Infect. Immun. *55*, 2822–2829.

Galdiero, S., and Gouaux, E. (2004). High resolution crystallographic studies of alpha–hemolysin-phospholipid complexes define heptamer-lipid head group interactions: implication for understanding protein–lipid interactions. Protein Sci. *13*, 1503–1511.

Gardiner, E.E., Karunakaran, D., Shen, Y., Arthur, J.F., Andrews, R.K., and Berndt, M.C. (2007). Controlled shedding of platelet glycoprotein (GP)VI and GPIb-IX-V by ADAM family metalloproteinases. J. Thromb. Haemost. *5*, 1530–1537.

Garland, W.J., and Buckley, J.T. (1988). The cytolytic toxin aerolysin must aggregate to disrupt erythrocytes, and aggregation is stimulated by human glycophorin. Infect. Immun. *56*, 1249–1253.

Gekara, N.O., Westphal, K., Ma, B., Rohde, M., Groebe, L., and Weiss, S. (2007). The multiple mechanisms of Ca^{2+} signalling by listeriolysin O, the cholesterol-dependent cytolysin of *Listeria monocytogenes*. Cell. Microbiol. *9*, 2008–2021.

Gekara, N.O., Groebe, L., Viegas, N., and Weiss, S. (2008). *Listeria monocytogenes* desensitizes immune cells to subsequent Ca^{2+} signaling via listeriolysin O-induced depletion of intracellular Ca^{2+} stores. Infect. Immun. *76*, 857–862.

Gelber, S.E., Aguilar, J.L., Lewis, K.L., and Ratner, A.J. (2008). Functional and phylogenetic characterization of Vaginolysin, the human-specific cytolysin from *Gardnerella vaginalis*. J. Bacteriol. *190*, 3896–3903.

Gibb, D.R., El Shikh, M., Kang, D.J., Rowe, W.J., El Sayed, R., Cichy, J., Yagita, H., Tew, J.G., Dempsey, P.J., Crawford, H.C., *et al.* (2010). ADAM10 is essential for Notch2-dependent marginal zone B cell development and CD23 cleavage *in vivo*. J. Exp. Med. *207*, 623–635.

Giddings, K.S., Johnson, A.E., and Tweten, R.K. (2003). Redefining cholesterol's role in the mechanism of the cholesterol-dependent cytolysins. Proc. Natl. Acad. Sci. U.S.A. *100*, 11315–11320.

Giddings, K.S., Zhao, J., Sims, P.J., and Tweten, R.K. (2004). Human CD59 is a receptor for the cholesterol-dependent cytolysin intermedilysin. Nat. Struct. Mol. Biol. *11*, 1173–1178.

Gill, D.M., and Meren, R. (1978). ADP-ribosylation of membrane proteins catalyzed by cholera toxin: basis of the activation of adenylate cyclase. Proc. Natl. Acad. Sci. U.S.A. *75*, 3050–3054.

Glenny, A.T., and Stevens, M.F. (1935). *Staphylococcus* toxins and antitoxins. J. Pathol. Bacteriol. *40*, 201–210.

Gonzalez, M.E., and Carrasco, L. (2003). Viroporins. FEBS Lett. *552*, 28–34.

Gonzalez, M.R., Bischofberger, M., Pernot, L., van der Goot, F.G., and Freche, B. (2008). Bacterial pore-forming toxins: the (w)hole story? Cell. Mol. Life Sci. *65*, 493–507.

Gouaux, E. (1997). Channel-forming toxins: tales of transformation. Curr. Opin. Struct. Biol. *7*, 566–573.

Gouaux, E. (1998). alpha-Hemolysin from *Staphylococcus aureus*: an archetype of beta-barrel, channel-forming toxins. J. Struct. Biol. *121*, 110–122.

Gouaux, J.E., Braha, O., Hobaugh, M.R., Song, L., Cheley, S., Shustak, C., and Bayley, H. (1994). Subunit stoichiometry of staphylococcal alpha-hemolysin in crystals and on membranes: a heptameric transmembrane pore. Proc. Natl. Acad. Sci. U.S.A. *91*, 12828–12831.

Gray, G.S., and Kehoe, M. (1984). Primary sequence of the alpha-toxin gene from *Staphylococcus aureus* wood 46. Infect. Immun. *46*, 615–618.

Grimminger, F., Rose, F., Sibelius, U., Meinhardt, M., Potzsch, B., Spriestersbach, R., Bhakdi, S., Suttorp, N., and Seeger, W. (1997). Human endothelial cell activation and mediator release in response to the bacterial exotoxins *Escherichia coli* hemolysin and staphylococcal alpha-toxin. J. Immunol. *159*, 1909–1916.

Gu, L.Q., Braha, O., Conlan, S., Cheley, S., and Bayley, H. (1999). Stochastic sensing of organic analytes by a pore-forming protein containing a molecular adapter. Nature *398*, 686–690.

Haas, A., Dumbsky, M., and Kreft, J. (1992). Listeriolysin genes: complete sequence of ILO from *Listeria ivanovii* and of ISO from *Listeria seeligeri*. Biochim. Biophys. Acta *1130*, 81–84.

Habte, M., and Alexander, M. (1975). Protozoa as agents responsible for the decline of *Xanthomonas campestris* in soil. Appl. Microbiol. *29*, 159–164.

Hadders, M.A., Beringer, D.X., and Gros, P. (2007). Structure of C8alpha-MACPF reveals mechanism of membrane attack in complement immune defense. Science *317*, 1552–1554.

Hadders, M.A., Bubeck, D., Roversi, P., Hakobyan, S., Forneris, F., Morgan, B.P., Pangburn, M.K., Llorca, O., Lea, S.M., and Gros, P. (2012). Assembly and regulation of the membrane attack complex based on structures of C5b6 and sC5b9. Cell Rep. *1*, 1–8.

Hamon, M.A., Ribet, D., Stavru, F., and Cossart, P. (2012). Listeriolysin O: the Swiss army knife of Listeria. Trends Microbiol. *20*, 360–368.

Harder, J., Franchi, L., Munoz-Planillo, R., Park, J.H., Reimer, T., and Nunez, G. (2009). Activation of the Nlrp3 inflammasome by Streptococcus pyogenes requires streptolysin O and NF-kappa B activation but proceeds independently of TLR signaling and P2X7 receptor. J. Immunol. *183*, 5823–5829.

Harris, R.W., Sims, P.J., and Tweten, R.K. (1991). Kinetic aspects of the aggregation of *Clostridium perfringens* theta toxin on erythrocyte membranes: a fluorescence energy transfer study. J. Biol. Chem. *266*, 6936–6941.

Hartmann, D., de Strooper, B., Serneels, L., Craessaerts, K., Herreman, A., Annaert, W., Umans, L., Lubke, T., Lena Illert, A., von Figura, K., *et al.* (2002). The disintegrin/metalloprotease ADAM 10 is essential for Notch signalling but not for alpha-secretase activity in fibroblasts. Hum. Mol. Genet. *11*, 2615–2624.

Heuck, A.P., Hotze, E., Tweten, R.K., and Johnson, A.E. (2000). Mechanism of membrane insertion of a multimeric b-barrel protein: Perfringolysin O creates a pore using ordered and coupled conformational changes. Mol. Cell *6*, 1233–1242.

Heuck, A.P., Tweten, R.K., and Johnson, A.E. (2001). Beta-barrel pore-forming toxins: intriguing dimorphic proteins. Biochemistry *40*, 9065–9073.

Heuck, A.P., Tweten, R.K., and Johnson, A.E. (2003). Assembly and topography of the prepore complex in cholesterol-dependent cytolysins. J. Biol. Chem. *278*, 31218–31225.

Hickey, M.J., Kwan, R.Y., Awad, M.M., Kennedy, C.L., Young, L.F., Hall, P., Cordner, L.M., Lyras, D., Emmins, J.J., and Rood, J.I. (2008). Molecular and cellular basis of microvascular perfusion deficits induced by *Clostridium perfringens* and *Clostridium septicum*. PLoS Pathog. *4*, e1000045.

Hildebrand, A., Pohl, M., and Bhakdi, S. (1991). *Staphylococcus aureus* alpha-toxin. Dual mechanism of binding to target cells. J. Biol. Chem. *266*, 17195–17200.

Hong, Y., Ohishi, K., Inoue, N., Kang, J.Y., Shime, H., Horiguchi, Y., van der Goot, F.G., Sugimoto, N., and Kinoshita, T. (2002). Requirement of N-glycan on GPI-anchored proteins for efficient binding of aerolysin but not *Clostridium septicum* alpha-toxin. EMBO J. *21*, 5047–5056.

Hotze, E.M., Wilson-Kubalek, E.M., Rossjohn, J., Parker, M.W., Johnson, A.E., and Tweten, R.K. (2001). Arresting pore formation of a cholesterol-dependent cytolysin by disulfide trapping synchronizes the insertion of the transmembrane beta-sheet from a prepore intermediate. J. Biol. Chem. *276*, 8261–8268.

Hotze, E.M., Heuck, A.P., Czajkowsky, D.M., Shao, Z., Johnson, A.E., and Tweten, R.K. (2002). Monomer–monomer interactions drive the prepore to pore conversion of a beta-barrel-forming cholesterol-dependent cytolysin. J. Biol. Chem. *277*, 11597–11605.

Hotze, E.M., Wilson-Kubalek, E., Farrand, A.J., Bentsen, L., Parker, M.W., Johnson, A.E., and Tweten, R.K. (2012). Monomer–monomer interactions propagate structural transitions necessary for pore-formation by the cholesterol-dependent cytolysins. J. Biol. Chem. *287*, 24534–24543.

Hruz, P., Zinkernagel, A.S., Jenikova, G., Botwin, G.J., Hugot, J.P., Karin, M., Nizet, V., and Eckmann, L. (2009). NOD2 contributes to cutaneous defense against *Staphylococcus aureus* through alpha-toxin-dependent innate immune activation. Proc. Natl. Acad. Sci. U.S.A. *106*, 12873–12878.

Huang, Y., Smith, C.A., Song, H., Morgan, B.P., Abagyan, R., and Tomlinson, S. (2005). Insights into the human CD59 complement binding interface toward engineering new therapeutics. J. Biol. Chem. *280*, 34073–34079.

Huang, Y., Qiao, F., Abagyan, R., Hazard, S., and Tomlinson, S. (2006). Defining the CD59-C9 binding interaction. J. Biol. Chem. *281*, 27398–27404.

Hugo, F., Sinner, A., Reichwein, J., and Bhakdi, S. (1987). Quantitation of monomeric and oligomeric forms of membrane-bound staphylococcal alpha-toxin by enzyme-linked immunosorbent assay with a neutralizing monoclonal antibody. Infect. Immun. *55*, 2933–2939.

Humphrey, W., Dalke, A., and Schulten, K. (1996). VMD: visual molecular dynamics. J. Mol. Graph. *14*, 33–38.

Husmann, M., Dersch, K., Bobkiewicz, W., Beckmann, E., Veerachato, G., and Bhakdi, S. (2006). Differential role of p38 mitogen activated protein kinase for cellular recovery from attack by pore-forming S. aureus alpha-toxin or streptolysin O. Biochem. Biophys. Res. Commun. *344*, 1128–1134.

Husmann, M., Beckmann, E., Boller, K., Kloft, N., Tenzer, S., Bobkiewicz, W., Neukirch, C., Bayley, H., and Bhakdi, S. (2009). Elimination of a bacterial pore-forming toxin by sequential endocytosis and exocytosis. FEBS Lett. *583*, 337–344.

Iacovache, I., Paumard, P., Scheib, H., Lesieur, C., Sakai, N., Matile, S., Parker, M.W., and van der Goot, F.G. (2006). A rivet model for channel formation by aerolysin-like pore-forming toxins. EMBO J. *25*, 457–466.

Idone, V., Tam, C., Goss, J.W., Toomre, D., Pypaert, M., and Andrews, N.W. (2008). Repair of injured plasma membrane by rapid Ca^{2+}-dependent endocytosis. J. Cell Biol. *180*, 905–914.

Ikigai, H., and Nakae, T. (1987a). Assembly of the alpha-toxin-hexamer of *Staphylococcus aureus* in the liposome membrane. J. Biol. Chem. 262, 2156–2160.

Ikigai, H., and Nakae, T. (1987b). Interaction of the alpha-toxin of *Staphylococcus aureus* with the liposome membrane. J. Biol. Chem. *262*, 2150–2155.

Inoshima, I., Inoshima, N., Wilke, G.A., Powers, M.E., Frank, K.M., Wang, Y., and Bubeck Wardenburg, J. (2011). A *Staphylococcus aureus* pore-forming toxin subverts the activity of ADAM10 to cause lethal infection in mice. Nat. Med. *17*, 1310–1314.

Inoshima, N., Wang, Y., and Wardenburg, J.B. (2012). Genetic requirement for ADAM10 in severe *Staphylococcus aureus* skin infection. J. Invest. Dermatol. *132*, 1513–1516.

Ishino, T., Chinzei, Y., and Yuda, M. (2005). A Plasmodium sporozoite protein with a membrane attack complex domain is required for breaching the liver sinusoidal cell layer prior to hepatocyte infection. Cell. Microbiol. *7*, 199–208.

Iwamoto, M., Ohno-Iwashita, Y., and Ando, S. (1987). Role of the essential thiol group in the thiol-activated cytolysin from Clostridium perfringens. Eur. J. Biochem. *167*, 425–430.

Jacobs, T., Cima-Cabal, M.D., Darji, A., Mendez, F.J., Vazquez, F., Jacobs, A.A., Shimada, Y., Ohno-Iwashita, Y., Weiss, S., and de los Toyos, J.R. (1999). The conserved undecapeptide shared by thiol-activated cytolysins is involved in membrane binding. FEBS Lett. *459*, 463–466.

Janes, P.W., Saha, N., Barton, W.A., Kolev, M.V., Wimmer-Kleikamp, S.H., Nievergall, E., Blobel, C.P., Himanen, J.-P., Lackmann, M., and Nikolov, D.B. (2005). Adam Meets Eph: an ADAM substrate recognition module acts as a Molecular switch for ephrin cleavage in trans. Cell *123*, 291–304.

Janes, P.W., Wimmer-Kleikamp, S.H., Frangakis, A.S., Treble, K., Griesshaber, B., Sabet, O., Grabenbauer, M., Ting, A.Y., Saftig, P., Bastiaens, P.I., *et al.* (2009). Cytoplasmic relaxation of active Eph controls ephrin shedding by ADAM10. PLoS Biol. *7*, e1000215.

Japrung, D., Henricus, M., Li, Q., Maglia, G., and Bayley, H. (2010). Urea facilitates the translocation of single-stranded DNA and RNA through the alpha-hemolysin nanopore. Biophys. J. *98*, 1856–1863.

Jeljaszewicz, J., Smiegielski, S., and Hryniewicz, P.J. (1978). Biological effects of staphylococcal and streptoccal toxins. In Bacterial Toxins and Cell Membranes, Jeljaszewicz, J., and Wadstrom, I., eds. (New York, Academic Press. Inc.), pp. 185–227.

Ji, G., Beavis, R.C., and Novick, R.P. (1995). Cell density control of staphylococcal virulence mediated by an octapeptide pheromone. Proc. Natl. Acad. Sci. U.S.A. *92*, 12055–12059.

Johnson, B.B., Moe, P.C., Wang, D., Rossi, K., Trigatti, B.L., and Heuck, A.P. (2012). Modifications in perfringolysin O domain 4 alter the cholesterol concentration threshold required for binding. Biochemistry. *51*, 3373–3382.

Jost, B.H., Lucas, E.A., Billington, S.J., Ratner, A.J., and McGee, D.J. (2011). Arcanolysin is a cholesterol-dependent cytolysin of the human pathogen Arcanobacterium haemolyticum. BMC Microbiol. *11*, 239.

Jursch, R., Hildebrand, A., Hobom, G., Tranum-Jensen, J., Ward, R., Kehoe, M., and Bhakdi, S. (1994). Histidine residues near the N terminus of staphylococcal alpha-toxin as reporters of regions that are critical for oligomerization and pore formation. Infect. Immun. *62*, 2249–2256.

Kadota, K., Ishino, T., Matsuyama, T., Chinzei, Y., and Yuda, M. (2004). Essential role of membrane-attack protein in malarial transmission to mosquito host. Proc. Natl. Acad. Sci. U.S.A. *101*, 16310–16315.

Kafsack, B.F., Pena, J.D., Coppens, I., Ravindran, S., Boothroyd, J.C., and Carruthers, V.B. (2009). Rapid membrane disruption by a perforin-like protein facilitates parasite exit from host cells. Science 323, 530–533.

Kato, I., and Watanabe, M. (1980). Chemical studies on staphylococcal alpha-toxin and its fragments. Toxicon *18*, 361–365.

Kawate, T., and Gouaux, E. (2003). Arresting and releasing staphylococcal α-hemolysin at intermediate stages of pore formation by engineered disulfide bonds. Protein Sci. *12*, 997–1006.

Kayal, S., and Charbit, A. (2006). Listeriolysin O: a key protein of *Listeria monocytogenes* with multiple functions. FEMS Microbiol. Rev. *30*, 514–529.

Kebaier, C., Chamberland, R.R., Allen, I.C., Gao, X., Broglie, P.M., Hall, J.D., Jania, C., Doerschuk, C.M., Tilley, S.L., and Duncan, J.A. (2012). *Staphylococcus aureus* alpha-hemolysin mediates virulence in a murine model of severe pneumonia through activation of the NLRP3 inflammasome. J. Infect. Dis. *205*, 807–817.

Kehoe, M., Duncan, J., Foster, T., Fairweather, N., and Dougan, G. (1983). Cloning, expression, and mapping of the *Staphylococcus aureus* alpha-hemolysin determinant in *Escherichia coli* K-12. Infect. Immun. *41*, 1105–1111.

Kehoe, M.A., Miller, L., Walker, J.A., and Boulnois, G.J. (1987). Nucleotide sequence of the streptolysin O (SLO) gene: structural homologies between SLO and other membrane-damaging, thiol-activated toxins. Infect. Immun. *55*, 3228–3232.

Kellaway, C.H., MacCallum, P., and Terbutt, A.H. (1928). Report of the Commission to the Governor of the Commonwealth of Australia (H.J. Green, Government Printer, Melbourne).

Kennedy, A.D., Bubeck Wardenburg, J., Gardner, D.J., Long, D., Whitney, A.R., Braughton, K.R., Schneewind, O., and DeLeo, F.R. (2010). Targeting of alpha-hemolysin by active or passive immunization decreases severity of USA300 skin infection in a mouse model. J. Infect. Dis. *202*, 1050–1058.

Kennedy, C.L., Lyras, D., Cheung, J.K., Hiscox, T.J., Emmins, J.J., and Rood, J.I. (2009). Cross-complementation of *Clostridium perfringens* PLC and *Clostridium septicum* alpha-toxin mutants reveals PLC is sufficient to mediate gas gangrene. Microbes and infection/Institut Pasteur *11*, 413–418.

Keyel, P.A., Heid, M.E., and Salter, R.D. (2011). Macrophage responses to bacterial toxins: a balance between activation and suppression. Immunol. Res. *50*, 118–123.

Kielian, T., Cheung, A., and Hickey, W.F. (2001). Diminished virulence of an alpha-toxin mutant of *Staphylococcus aureus* in experimental brain abscesses. Infect. Immun. *69*, 6902–6911.

Kintzer, A.F., Thoren, K.L., Sterling, H.J., Dong, K.C., Feld, G.K., Tang, I.I., Zhang, T.T., Williams, E.R., Berger, J.M., and Krantz, B.A. (2009). The protective antigen component of anthrax toxin forms functional octameric complexes. J. Mol. Biol. 392, 614–629.

Koenig, R.L., Ray, J.L., Maleki, S.J., Smeltzer, M.S., and Hurlburt, B.K. (2004). *Staphylococcus aureus* AgrA binding to the RNAIII-agr regulatory region. J. Bacteriol. *186*, 7549–7555.

Krasilnikov, O.V., Muratkhodjaev, J.N., and Zitzer, A.O. (1992). The mode of action of *Vibrio cholerae* cytolysin. The influences on both erythrocytes and planar lipid bilayers. Biochim. Biophys. Acta *1111*, 7–16.

Kraus, R., and Pribram, E. (1906). Ueber Staphylokokkentoxin und dessen Antitoxin. Wien Klin Wochschr *17*, 493.

Kumar, S., Loken, K.I., Kenyon, A.J., and Lindorfer, R.K. (1962). The characterization of staphylococcal toxins: II. The isolation and characterization of a homogeneous staphylococcal protein possessing alpha hemolytic, dermonecrotic, lethal, and leucocidal activities. J. Exp. Med. *115*, 1107–1115.

LaChapelle, S., Tweten, R.K., and Hotze, E.M. (2009). Intermedilysin–receptor interactions during assembly of the pore complex: assembly intermediates increase host cell susceptibility to complement-mediated lysis. J. Biol. Chem. *284*, 12719–12726.

Lacy, D.B., Wigelsworth, D.J., Melnyk, R.A., Harrison, S.C., and Collier, R.J. (2004). Structure of heptameric protective antigen bound to an anthrax toxin receptor: a role for receptor in pH-dependent pore formation. Proc. Natl. Acad. Sci. U.S.A. *101*, 13147–13151.

Law, R.H., Lukoyanova, N., Voskoboinik, I., Caradoc-Davies, T.T., Baran, K., Dunstone, M.A., D'Angelo, M.E., Orlova, E.V., Coulibaly, F., Verschoor, S., *et al.* (2010). The structural basis for membrane binding and pore formation by lymphocyte perforin. Nature *468*, 447–451.

Leng, B.F., Qiu, J.Z., Dai, X.H., Dong, J., Wang, J.F., Luo, M.J., Li, H.E., Niu, X.D., Zhang, Y., Ai, Y.X., *et al.* (2011). Allicin reduces the production of alpha-toxin by *Staphylococcus aureus*. Molecules *16*, 7958–7968.

Limbago, B., Penumalli, V., Weinrick, B., and Scott, J.R. (2000). Role of streptolysin O in a mouse model of invasive group A streptococcal disease. Infect. Immun. *68*, 6384–6390.

Lina, G., Jarraud, S., Ji, G., Greenland, T., Pedraza, A., Etienne, J., Novick, R.P., and Vandenesch, F. (1998). Transmembrane topology and histidine protein kinase activity of AgrC, the agr signal receptor in *Staphylococcus aureus*. Mol. Microbiol. *28*, 655–662.

Lizak, M., and Yarovinsky, T.O. (2012). Phospholipid scramblase 1 mediates type I interferon-induced protection against staphylococcal alpha-toxin. Cell Host Microbe *11*, 70–80.

Logsdon, L.K., Hakansson, A.P., Cortes, G., and Wessels, M.R. (2011). Streptolysin O inhibits clathrin-dependent internalization of group A *Streptococcus*. MBio. *2*, e00332–00310.

Lominski, L., Arbuthnott, J.P., and Spence, J.P. (1963). Purification of staphylococcus alpha-toxin. J. Pathol. Bacteriol. *86*, 258–262.

Lovelace, L.L., Cooper, C.L., Sodetz, J.M., and Lebioda, L. (2011). Structure of human C8 protein provides mechanistic insight into membrane pore formation by complement. J. Biol. Chem. *286*, 17585–17592.

Lovell, R., and T.A., R. (1960). A filterable hemolysin from *Escherichia coli*. Nature *188*, 755.

Ludwig, A., Hundhausen, C., Lambert, M.H., Broadway, N., Andrews, R.C., Bickett, D.M., Leesnitzer, M.A., and Becherer, J.D. (2005). Metalloproteinase inhibitors for the disintegrin-like metalloproteinases ADAM10 and ADAM17 that differentially block constitutive and phorbol ester-inducible shedding of cell surface molecules. Comb. Chem. High Throughput Screen. *8*, 161–171.

Luo, M., Qiu, J., Zhang, Y., Wang, J., Dong, J., Li, H., Leng, B., Zhang, Q., Dai, X., Niu, X., *et al.* (2012). alpha-Cyperone Alleviates Lung Cell Injury Caused by *Staphylococcus aureus* via Attenuation of alpha-Hemolysin Expression. Journal of microbiology and biotechnology *22*, 1170–1176.

Lyon, G.J., Wright, J.S., Muir, T.W., and Novick, R.P. (2002). Key determinants of receptor activation in the agr autoinducing peptides of *Staphylococcus aureus*. Biochemistry *41*, 10095–10104.

Madden, J.C., Ruiz, N., and Caparon, M. (2001). Cytolysin-mediated translocation (CMT): a functional equivalent of type III secretion in Gram-positive bacteria. Cell *104*, 143–152.

Magassa, N., Chandrasekaran, S., and Caparon, M.G. (2010). *Streptococcus pyogenes* cytolysin-mediated translocation does not require pore formation by streptolysin O. EMBO Rep. *5*, 400–405.

Maretzky, T., Reiss, K., Ludwig, A., Buchholz, J., Scholz, F., Proksch, E., de Strooper, B., Hartmann, D., and Saftig, P. (2005). ADAM10 mediates E-cadherin shedding and regulates epithelial cell–cell adhesion, migration, and beta-catenin translocation. Proc. Natl. Acad. Sci. U.S.A. *102*, 9182–9187.

Maretzky, T., Scholz, F., Koten, B., Proksch, E., Saftig, P., and Reiss, K. (2008). ADAM10-mediated E-cadherin release is regulated by proinflammatory cytokines and modulates keratinocyte cohesion in eczematous dermatitis. J. Invest. Dermatol. *128*, 1737–1746.

Marriott, H.M., Mitchell, T.J., and Dockrell, D.H. (2008). Pneumolysin: a double-edged sword during the host–pathogen interaction. Curr. Mol. Med. *8*, 497–509.

McCardell, B.A., Madden, J.M., and Shah, D.B. (1985). Isolation and characterization of a cytolysin produced by *Vibrio cholerae* serogroup non-O1. Can. J. Microbiol. *31*, 711–720.

McElroy, M.C., Harty, H.R., Hosford, G.E., Boylan, G.M., Pittet, J.F., and Foster, T.J. (1999). Alpha-toxin damages the air–blood barrier of the lung in a rat model of *Staphylococcus aureus*-induced pneumonia. Infect. Immun. *67*, 5541–5544.

Meehl, M.A., and Caparon, M.G. (2004). Specificity of streptolysin O in cytolysin-mediated translocation. Mol. Microbiol. *52*, 1665–1676.

Meesters, C., Brack, A., Hellmann, N., and Decker, H. (2009). Structural characterization of the alpha-hemolysin monomer from *Staphylococcus aureus*. Proteins *75*, 118–126.

Melton, J.A., Parker, M.W., Rossjohn, J., Buckley, J.T., and Tweten, R.K. (2004). The identification and structure of the membrane-spanning domain of the *Clostridium septicum* alpha toxin. J. Biol. Chem. *279*, 14315–14322.

Menestrina, G. (1986). Ionic channels formed by *Staphylococcus aureus* alpha-toxin: voltage-dependent inhibition by divalent and trivalent cations. J. Membr. Biol. *90*, 177–190.

Mengaud, J., Vicente, M.F., Chenevert, J., Pereira, J.M., Geoffroy, C., Gicquel-Sanzey, B., Baquero, F., Perez-Diaz, J.C., and Cossart, P. (1988). Expression in *Escherichia coli* and sequence analysis of the listeriolysin O determinant of *Listeria monocytogenes*. Infect. Immun. *56*, 766–772.

Menzies, B.E., and Kernodle, D.S. (1994). Site-directed mutagenesis of the alpha-toxin gene of *Staphylococcus aureus*: role of histidines in toxin activity *in vitro* and in a murine model. Infect. Immun. *62*, 1843–1847.

Menzies, B.E., and Kernodle, D.S. (1996). Passive immunization with antiserum to a nontoxic alpha-toxin mutant from *Staphylococcus aureus* is protective in a murine model. Infect. Immun. *64*, 1839–1841.

Miller, L.S., and Cho, J.S. (2011). Immunity against *Staphylococcus aureus* cutaneous infections. Nat. Rev. Immunol. *11*, 505–518.

Mueller, M., Grauschopf, U., Maier, T., Glockshuber, R., and Ban, N. (2009). The structure of a cytolytic alpha-helical toxin pore reveals its assembly mechanism. Nature *459*, 726–730.

Nagamune, H., Ohnishi, C., Katsuura, A., Fushitani, K., Whiley, R.A., Tsuji, A., and Matsuda, Y. (1996). Intermedilysin, a novel cytotoxin specific for human cells secreted by *Streptococcus intermedius* UNS46 isolated from a human liver abscess. Infect. Immun. *64*, 3093–3100.

Nagamune, H., Whiley, R.A., Goto, T., Inai, Y., Maeda, T., Hardie, J.M., and Kourai, H. (2000). Distribution of the intermedilysin gene among the anginosus group streptococci and correlation between intermedilysin production and deep-seated infection with *Streptococcus intermedius*. J. Clin. Microbiol. *38*, 220–226.

Nagano, O., Murakami, D., Hartmann, D., De Strooper, B., Saftig, P., Iwatsubo, T., Nakajima, M., Shinohara, M., and Saya, H. (2004). Cell–matrix interaction via CD44 is independently regulated by different metalloproteinases activated in response to extracellular $Ca^{(2+)}$ influx and PKC activation. J. Cell Biol. *165*, 893–902.

Nakamura, M., Sekino, N., Iwamoto, M., and Ohno-Iwashita, Y. (1995). Interaction of theta-toxin (perfringolysin O), a cholesterol-binding cytolysin, with liposomal membranes: change in the aromatic side chains upon binding and insertion. Biochemistry *34*, 6513–6520.

Nakamura, M., Sekino-Suzuki, N., Mitsui, K., and Ohno-Iwashita, Y. (1998). Contribution of tryptophan residues to the structural changes in perfringolysin O during interaction with liposomal membranes. Biochem, J.. (Tokyo) *123*, 1145–1155.

Neisser, M., and Wechsberg, F. (1901). Ueber das Staphylotoxin. Z. Hyg. Infektionskrankh *36*, 299.

Nelson, L.D., Johnson, A.E., and London, E. (2008). How interaction of perfringolysin O with membranes is controlled by sterol structure, lipid structure, and physiological low pH: insights into the origin of perfringolysin O-lipid raft interaction. J. Biol. Chem. *283*, 4632–4642.

Niebuhr, M., Gathmann, M., Scharonow, H., Mamerow, D., Mommert, S., Balaji, H., and Werfel, T. (2011). Staphylococcal alpha-toxin is a strong inducer of interleukin-17 in humans. Infect. Immun. *79*, 1615–1622.

Nilsson, M., Sorensen, O.E., Morgelin, M., Weineisen, M., Sjobring, U., and Herwald, H. (2006). Activation of human polymorphonuclear neutrophils by streptolysin O from Streptococcus pyogenes leads to the release of proinflammatory mediators. J. Thromb. Haemost. 95, 982–990.

Novick, R.P., Ross, H.F., Projan, S.J., Kornblum, J., Kreiswirth, B., and Moghazeh, S. (1993). Synthesis of staphylococcal virulence factors is controlled by a regulatory RNA molecule. EMBO J. *12*, 3967–3975.

Nygaard, T.K., Pallister, K.B., DuMont, A.L., DeWald, M., Watkins, R.L., Pallister, E.Q., Malone, C., Griffith, S., Horswill, A.R., Torres, V.J., *et al.* (2012). Alpha-toxin induces programmed cell death of human T cells, B cells, and monocytes during USA300 infection. PLoS One 7, e36532.

O'Callaghan, R.J., Callegan, M.C., Moreau, J.M., Green, L.C., Foster, T.J., Hartford, O.M., Engel, L.S., and Hill, J.M. (1997). Specific roles of alpha-toxin and beta-toxin during *Staphylococcus aureus* corneal infection. Infect. Immun. *65*, 1571–1578.

O'Reilly, M., de Azavedo, J.C., Kennedy, S., and Foster, T.J. (1986). Inactivation of the alpha-haemolysin gene of *Staphylococcus aureus* 8325-4 by site-directed mutagenesis and studies on the expression of its haemolysins. Microb. Pathog. *1*, 125–138.

Olson, R., Nariya, H., Yokota, K., Kamio, Y., and Gouaux, E. (1999). Crystal structure of Staphylococcal LukF delineates conformational changes accompanying formation of a transmembrane channel. Nat. Struct. Biol. *6*, 134–140.

Onogawa, T. (2002). Staphylococcal alpha-toxin synergistically enhances inflammation caused by bacterial components. FEMS Immunol. Med. Microbiol. *33*, 15–21.

Palmer, M., Saweljew, P., Vulicevic, I., Valeva, A., Kehoe, M., and Bhakdi, S. (1996). Membrane-penetrating domain of streptolysin O identified by cysteine scanning mutagenesis. J. Biol. Chem. *271*, 26664–26667.

Palmer, M., Harris, R., Freytag, C., Kehoe, M., Tranum-Jensen, J., and Bhakdi, S. (1998a). Assembly mechanism of the oligomeric streptolysin O pore: the early membrane lesion is lined by a free edge of the lipid membrane and is extended gradually during oligomerization. EMBO J. *17*, 1598–1605.

Palmer, M., Vulicevic, I., Saweljew, P., Valeva, A., Kehoe, M., and Bhakdi, S. (1998b). Streptolysin O: A proposed model of allosteric interaction between a pore-forming protein and its target lipid bilayer. Biochemistry *37*, 2378–2383.

Parker, D., and Prince, A. (2011). Immunopathogenesis of *Staphylococcus aureus* pulmonary infection. Semin Immunopathol. *34*, 281–297.

Parker, M.W., Buckley, J.T., Postma, J.P., Tucker, A.D., Leonard, K., Pattus, F., and Tsernoglou, D. (1994a). Structure of the Aeromonas toxin proaerolysin in its water-soluble and membrane-channel states. Nature *367*, 292–295.

Parker, M.W., Buckley, J.T., Postma, J.P.M., Tucker, A.D., Leonard, K., Pattus, F., and Tsernoglou, D. (1994b). Structure of the *Aeromonas* toxin proaerolysin in its water-soluble and membrane-channel states. Nature *367*, 292–295.

Parker, M.W., and Feil, S.C. (2005). Pore-forming protein toxins: from structure to function. Prog. Biophys. Mol. Biol. *88*, 91–142.

Patel, A.H., Nowlan, P., Weavers, E.D., and Foster, T. (1987). Virulence of protein A-deficient and alpha-toxin-deficient mutants of *Staphylococcus aureus* isolated by allele replacement. Infect. Immun. *55*, 3103–3110.

Peng, H.L., Novick, R.P., Kreiswirth, B., Kornblum, J., and Schlievert, P. (1988). Cloning, characterization, and sequencing of an accessory gene regulator (agr) in *Staphylococcus aureus*. J. Bacteriol. *170*, 4365–4372.

Pernthaler, J. (2005). Predation on prokaryotes in the water column and its ecological implications. Nat. Rev. Microbiol. 3, 537–546.

Petosa, C., Collier, R.J., Klimpel, K.R., Leppla, S.H., and Liddington, R.C. (1997). Crystal structure of the anthrax toxin protective antigen. Nature *385*, 833–838.

Pinkney, M., Beachey, E., and Kehoe, M. (1989). The thiol-activated toxin streptolysin O does not require a thiol group for activity. Infect. Immun. *57*, 2553–2558.

Pizarro-Cerda, J., and Cossart, P. (2006). Subversion of cellular functions by *Listeria monocytogenes*. J. Pathol. *208*, 215–223.

Plumb, M.E., and Sodetz, J.M. (1998). Proteins of the membrane attack complex. In The Human Complement System in Health and Disease, Volanakis, J.E., and Frank, M.M., eds. (New York, Marcel Dekker, Inc.), pp. 119–148.

Polekhina, G., Giddings, K.S., Tweten, R.K., and Parker, M.W. (2005). Insights into the action of the superfamily of cholesterol-dependent cytolysins from studies of intermedilysin. Proc. Natl. Acad. Sci. U.S.A. *102*, 600–605.

Ponting, C.P. (1999). Chlamydial homologues of the MACPF (MAC/perforin) domain. Curr. Biol. *9*, R911–913.

Portnoy, D.A., Jacks, P.S., and Hinrichs, D.J. (1988). Role of hemolysin for the intracellular growth of *Listeria monocytogenes*. J. Exp. Med. *167*, 1459–1471.

Powers, M.E., Kim, H.K., Wang, Y., and Bubeck Wardenburg, J. (2012). ADAM10 mediates vascular injury induced by *Staphylococcus aureus* alpha-hemolysin. J. Infect. Dis. *206*, 352–356.

Prevost, G., Mourey, L., Colin, D., Monteil, H., Dalla Serra, M., and Menestrina, G. (2005). Alpha-helix and beta-barrel pore-forming toxins (leucocidins, alpha-, gamma-, and delta-cytolysins) of *Staphylococcus aureus*. In The Comprehensive Sourcebook of Bacterial Toxins, Alouf, J.E., and Freer, J.H., eds. (London, Academic Press).

Price, K.E., and Camilli, A. (2009). Pneumolysin localizes to the cell wall of *Streptococcus pneumoniae*. J. Bacteriol. *191*, 2163–2168.

Price, K.E., Greene, N.G., and Camilli, A. (2012). Export requirements of pneumolysin in *Streptococcus pneumoniae*. J. Bacteriol. *194*, 3651–3660.

Prigent, D., and Alouf, J.E. (1976). Interaction of streptolysin-O with sterols. Biochim. Biophys. Acta *443*, 288–300.

Qiu, J., Luo, M., Wang, J., Dong, J., Li, H., Leng, B., Zhang, Q., Dai, X., Zhang, Y., Niu, X., *et al.* (2011). Isoalantolactone protects against *Staphylococcus aureus* pneumonia. FEMS Microbiol. Lett. *324*, 147–155.

Ragle, B.E., and Bubeck Wardenburg, J. (2009). Anti-alpha-hemolysin monoclonal antibodies mediate protection against *Staphylococcus aureus* pneumonia. Infect. Immun. *77*, 2712–2718.

Ragle, B.E., Karginov, V.A., and Bubeck Wardenburg, J. (2010). Prevention and treatment of *Staphylococcus aureus* pneumonia with a beta-cyclodextrin derivative. Antimicrob. Agents Chemother. *54*, 298–304.

Ramachandran, R., Heuck, A.P., Tweten, R.K., and Johnson, A.E. (2002). Structural insights into the membrane-anchoring mechanism of a cholesterol-dependent cytolysin. Nat. Struct. Biol. *9*, 823–827.

Ramachandran, R., Tweten, R.K., and Johnson, A.E. (2004). Membrane-dependent conformational changes initiate cholesterol-dependent cytolysin oligomerization and intersubunit β-strand alignment. Nat. Struct. Mol. Biol. *11*, 697–705.

Ramachandran, R., Tweten, R.K., and Johnson, A.E. (2005). The domains of a cholesterol-dependent cytolysin undergo a major FRET-detected rearrangement during pore formation. Proc. Natl. Acad. Sci. U.S.A. *102*, 7139–7144.

Recsei, P., Kreiswirth, B., O'Reilly, M., Schlievert, P., Gruss, A., and Novick, R.P. (1986). Regulation of exoprotein gene expression in *Staphylococcus aureus* by agr. Mol. Gen. Genet. *202*, 58–61.

Reichwein, J., Hugo, F., Roth, M., Sinner, A., and Bhakdi, S. (1987). Quantitative analysis of the binding and oligomerization of staphylococcal alpha-toxin in target erythrocyte membranes. Infect. Immun. *55*, 2940–2944.

Reiss, K., and Saftig, P. (2009). The "a disintegrin and metalloprotease" (ADAM) family of sheddases: physiological and cellular functions. Semin. Cell Dev. Biol. *20*, 126–137.

Remsen, C.C., Watson, S.W., and Bernheimer, A.W. (1970). Evidence for an ordered arrangement in erythrocyte membranes. Biochem. Biophys. Res. Commun. *40*, 1297–1304.

Reyes, D., Andrey, D.O., Monod, A., Kelley, W.L., Zhang, G., and Cheung, A.L. (2011). Coordinated regulation by AgrA, SarA, and SarR to control agr expression in *Staphylococcus aureus*. J. Bacteriol. *193*, 6020–6031.

Rodet, A., and Courmont, J. (1892). Produits du staphylocoque pyogène. Le Bulletin Med. 23, 84.

Rollins, S.A., Zhao, J., Ninomiya, H., and Sims, P.J. (1991). Inhibition of homologous complement by CD59 is mediated by a species-selective recognition conferred through binding to C8 within C5b-8 or C9 within C5b-9. J. Immunol. *146*, 2345–2351.

Rosado, C.J., Buckle, A.M., Law, R.H., Butcher, R.E., Kan, W.T., Bird, C.H., Ung, K., Browne, K.A., Baran, K., Bashtannyk-Puhalovich, T.A., *et al.* (2007). A common fold mediates vertebrate defense and bacterial attack. Science *317*, 1548–1551.

Rose, F., Zeller, S.A., Chakraborty, T., Domann, E., Machleidt, T., Kronke, M., Seeger, W., Grimminger, F., and Sibelius, U. (2001). Human endothelial cell activation and mediator release in response to *Listeria monocytogenes* virulence factors. Infect. Immun. *69*, 897–905.

Rose, F., Dahlem, G., Guthmann, B., Grimminger, F., Maus, U., Hanze, J., Duemmer, N., Grandel, U., Seeger, W., and Ghofrani, H.A. (2002). Mediator generation and signaling events in alveolar epithelial cells attacked by S. aureus alpha-toxin. Am. J. Physiol. Lung Cell Mol. Physiol. *282*, L207–214.

Rossjohn, J., Feil, S.C., McKinstry, W.J., Tweten, R.K., and Parker, M.W. (1997). Structure of a cholesterol-binding thiol-activated cytolysin and a model of its membrane form. Cell *89*, 685–692.

Rossjohn, J., Polekhina, G., Feil, S.C., Morton, C.J., Tweten, R.K., and Parker, M.W. (2007). Structures of perfringolysin O suggest a pathway for activation of cholesterol-dependent cytolysins. J.M.I. Biol. *367*, 1227–1236.

Ruiz, N., Wang, B., Pentland, A., and Caparon, M. (1998). Streptolysin O and adherence synergistically modulate proinflammatory responses of keratinocytes to group a streptococci. Mol. Microbiol. *27*, 337–346.

Saunders, F.K., Mitchell, T.J., Walker, J.A., Andrew, P.W., and Boulnois, G.J. (1989). Pneumolysin, the thiol-activated toxin of *Streptococcus pneumoniae*, does not require a thiol-group for *in vitro* activity. Infect. Immun. *57*, 2547–2552.

Schnupf, P., Portnoy, D.A., and Decatur, A.L. (2006). Phosphorylation, ubiquitination and degradation of listeriolysin O in mammalian cells: role of the PEST-like sequence. Cell. Microbiol. *8*, 353–364.

Schulte, A., Schulz, B., Andrzejewski, M., Hundhausen, C., Mletzko, S., Achilles, J., Reiss, K., Paliga, K., Weber, C., and Rosejohn, S. (2007). Sequential processing of the transmembrane chemokines CX3CL1 and CXCL16 by α- and γ-secretases. Biochem. Biophys. Res. Comm. *358*, 233–240.

Schulz, B., Pruessmeyer, J., Maretzky, T., Ludwig, A., Blobel, C.P., Saftig, P., and Reiss, K. (2008). ADAM10 regulates endothelial permeability and T-cell transmigration by proteolysis of vascular endothelial cadherin. Circ. Res. *102*, 1192–1201.

Seals, D.F. (2003). The ADAMs family of metalloproteases: multidomain proteins with multiple functions. Genes Dev. *17*, 7–30.

Sekino-Suzuki, N., Nakamura, M., Mitsui, K.I., and Ohno-Iwashita, Y. (1996). Contribution of individual tryptophan residues to the structure and activity of theta-toxin (perfringolysin o), a cholesterol-binding cytolysin. Eur. J. Biochem. *241*, 941–947.

Sellman, B.R., Kagan, B.L., and Tweten, R.K. (1997). Generation of a membrane-bound, oligomerized pre–pore complex is necessary for pore formation by

Clostridium septicum alpha toxin. Mol. Microbiol. *23*, 551–558.

Shatursky, O., Heuck, A.P., Shepard, L.A., Rossjohn, J., Parker, M.W., Johnson, A.E., and Tweten, R.K. (1999). The mechanism of membrane insertion for a cholesterol dependent cytolysin: A novel paradigm for pore-forming toxins. Cell *99*, 293–299.

Shaughnessy, L.M., Hoppe, A.D., Christensen, K.A., and Swanson, J.A. (2006). Membrane perforations inhibit lysosome fusion by altering pH and calcium in *Listeria monocytogenes* vacuoles. Cell. Microbiol. *8*, 781–792.

Shepard, L.A., Heuck, A.P., Hamman, B.D., Rossjohn, J., Parker, M.W., Ryan, K.R., Johnson, A.E., and Tweten, R.K. (1998). Identification of a membrane-spanning domain of the thiol-activated pore-forming toxin *Clostridium perfringens* perfringolysin O: an α-helical to β-sheet transition identified by fluorescence spectroscopy. Biochemistry *37*, 14563–14574.

Shepard, L.A., Shatursky, O., Johnson, A.E., and Tweten, R.K. (2000). The mechanism of assembly and insertion of the membrane complex of the cholesterol-dependent cytolysin perfringolysin O: formation of a large prepore complex. Biochemistry *39*, 10284–10293.

Sherr, E.B., and Sherr, B.F. (2002). Significance of predation by protists in aquatic microbial food webs. Antonie Van Leeuwenhoek *81*, 293–308.

Siegel, I., and Cohen, S. (1964). Action of Staphylococcal toxin on human platelets. J. Infect. Dis. *114*, 488–502.

Singh, R., Jamieson, A., and Cresswell, P. (2008). GILT is a critical host factor for *Listeria monocytogenes* infection. Nature *455*, 1244–1247.

Six, H.R., and Harshman, S. (1973). Purification and properties of two forms of staphylococcal toxin. Biochemistry *12*, 2672–2677.

Slade, D.J., Lovelace, L.L., Chruszcz, M., Minor, W., Lebioda, L., and Sodetz, J.M. (2008). Crystal structure of the MACPF domain of human complement protein C8 alpha in complex with the C8 gamma subunit. J. Mol. Biol. *379*, 331–342.

Soltani, C.E., Hotze, E.M., Johnson, A.E., and Tweten, R.K. (2007a). Specific protein-membrane contacts are required for prepore and pore assembly by a cholesterol-dependent cytolysin. J. Biol. Chem. *282*, 15709–15716.

Soltani, C.E., Hotze, E.M., Johnson, A.E., and Tweten, R.K. (2007b). Structural elements of the cholesterol-dependent cytolysins that are responsible for their cholesterol-sensitive membrane interactions. Proc. Natl. Acad. Sci. U.S.A. *104*, 20226–20231.

Song, L., Hobaugh, M.R., Shustak, C., Cheley, S., Bayley, H., and Gouaux, J.E. (1996). Structure of staphylococcal alpha-hemolysin, a heptameric transmembrane pore. Science *274*, 1859–1866.

Spaulding, A.R., Lin, Y.C., Merriman, J.A., Brosnahan, A.J., Peterson, M.L., and Schlievert, P.M. (2012). Immunity to *Staphylococcus aureus* secreted proteins protects rabbits from serious illnesses. Vaccine. *34*, 5099–5109.

Stephens, K.E., Miller, K.Y., and Miller, B.L. (1999). Functional analysis of DNA sequences required for conidium-specific expression of the SpoC1-C1C gene of *Aspergillus nidulans*. Fungal Genet. Biol. *27*, 231–242.

Stevens, D.L. (1996). Invasive group A streptococcal disease. Infect. Agents Dis. *5*, 157–166.

Stevens, L.M., Frohnhofer, H.G., Klingler, M., and Nusslein-Volhard, C. (1990). Localized requirement for torso-like expression in follicle cells for development of terminal anlagen of the Drosophila embryo. Nature *346*, 660–663.

Stoddart, D., Heron, A.J., Mikhailova, E., Maglia, G., and Bayley, H. (2009). Single-nucleotide discrimination in immobilized DNA oligonucleotides with a biological nanopore. Proc. Natl. Acad. Sci. U.S.A. *106*, 7702–7707.

Stoddart, D., Maglia, G., Mikhailova, E., Heron, A.J., and Bayley, H. (2010). Multiple base-recognition sites in a biological nanopore: two heads are better than one. Angew. Chem. Int. Ed. Engl. *49*, 556–559.

Suttorp, N., Seeger, W., Dewein, E., Bhakdi, S., and Roka, L. (1985). Staphylococcal alpha-toxin-induced PGI2 production in endothelial cells: role of calcium. Am. J. Physiol. *248*, C127–134.

Suttorp, N., Buerke, M., and Tannert-Otto, S. (1992). Stimulation of PAF-synthesis in pulmonary artery endothelial cells by *Staphylococcus aureus* alpha-toxin. Thromb. Res. *67*, 243–252.

Suttorp, N., Fuhrmann, M., Tannert-Otto, S., Grimminger, F., and Bhadki, S. (1993). Pore-forming bacterial toxins potently induce release of nitric oxide in porcine endothelial cells. J. Exp. Med. *178*, 337–341.

Tateno, H., and Goldstein, I.J. (2003). Molecular cloning, expression, and characterization of novel hemolytic lectins from the mushroom Laetiporus sulphureus, which show homology to bacterial toxins. J. Biol. Chem. *278*, 40455–40463.

Thelestam, M., Mollby, R., and Wadstrom, T. (1973). Effects of staphylococcal alpha-, beta-, delta-, and gamma-hemolysins on human diploid fibroblasts and HeLa cells: evaluation of a new quantitative as say for measuring cell damage. Infect. Immun. *8*, 938–946.

Tian, L., Wu, X., Chi, C., Han, M., Xu, T., and Zhuang, Y. (2008). ADAM10 is essential for proteolytic activation of Notch during thymocyte development. Int. Immunol. *20*, 1181–1187.

Tilley, S.J., Orlova, E.V., Gilbert, R.J., Andrew, P.W., and Saibil, H.R. (2005). Structural basis of pore formation by the bacterial toxin pneumolysin. Cell *121*, 247–256.

Timmer, A.M., Timmer, J.C., Pence, M.A., Hsu, L.C., Ghochani, M., Frey, T.G., Karin, M., Salvesen, G.S., and Nizet, V. (2009). Streptolysin O promotes group A Streptococcus immune evasion by accelerated macrophage apoptosis. J. Biol. Chem. *284*, 862–871.

Titball, R.W., Hunter, S.E.C., Martin, K.L., Morris, B.C., Shuttleworth, A.D., Rubridge, T., Anderson, D.W., and Kelly, D.C. (1989). Molecular cloning and nucleotide sequence of the alpha-toxin (phospholipase C) of *Clostridium perfringens*. Infect. Immun. *57*, 367–376.

Tkaczyk, C., Hua, L., Varkey, R., Shi, Y., Dettinger, L., Woods, R., Barnes, A., MacGill, R.S., Wilson, S., Chowdhury, P., *et al.* (2012). Identification of anti-alpha toxin monoclonal antibodies that reduce the severity of *Staphylococcus aureus* dermonecrosis and exhibit a correlation between affinity and potency. Clin. Vaccine Immunol. *19*, 377–385.

Tobkes, N., Wallace, B.A., and Bayley, H. (1985). Secondary structure and assembly mechanism of an oligomeric channel protein. Biochemistry *24*, 1915–1920.

Tousseyn, T., Jorissen, E., Reiss, K., and Hartmann, D. (2006). (Make) stick and cut loose – disintegrin metalloproteases in development and disease. Birth Defects Res. C Embryo Today *78*, 24–46.

Tweten, R.K. (1988). Nucleotide sequence of the gene for perfringolysin O (theta-toxin) from Clostridium perfringens: significant homology with the genes for streptolysin O and pneumolysin. Infect. Immun. *56*, 3235–3240.

Tweten, R.K. (2005). The cholesterol-dependent cytolysins; a family of versatile pore-forming toxins. Infect. Immun. *73*, 6199–6209.

Tweten, R.K., Christianson, K.K., and Iandolo, J.J. (1983). Transport and processing of staphylococcal alpha-toxin. J. Bacteriol. *156*, 524–528.

Usmani, S.M., Einem, J.V., Frick, M., Miklavc, P., Mayenburg, M., Husmann, M., Dietl, P., and Wittekindt, O.H. (2011). Molecular basis of early epithelial response to streptococcal exotoxin: Role of STIM1 and Orai1 proteins. Cell. Microbiol. *14*, 299–315.

Vadia, S., Arnett, E., Haghighat, A.C., Wilson-Kubalek, E.M., Tweten, R.K., and Seveau, S. (2011). The pore-forming toxin listeriolysin O mediates a novel entry pathway of *L. monocytogenes* into human hepatocytes. PLoS Pathog. *7*, e1002356.

Valeva, A. (2006). Evidence That Clustered Phosphocholine Head Groups Serve as Sites for Binding and Assembly of an Oligomeric Protein Pore. J. Biol. Chem. *281*, 26014–26021.

Valeva, A., Palmer, M., and Bhakdi, S. (1997a). Staphylococcal α-toxin: formation of the heptameric pore is partially cooperative and proceeds through multiple intermediate stages. Biochemistry *36*, 13298–13304.

Valeva, A., Pongs, J., Bhakdi, S., and Palmer, M. (1997b). Staphylococcal alpha-toxin: the role of the N-terminus in formation of the heptameric pore – a fluorescence study. Biochim. Biophys. Acta *1325*, 281–286.

Valeva, A., Walev, I., Gerber, A., Klein, J., Palmer, M., and Bhakdi, S. (2000). Staphylococcal alpha-toxin: repair of a calcium-impermeable pore in the target cell membrane. Mol. Microbiol. *36*, 467–476.

Valeva, A., Walev, I., Weis, S., Boukhallouk, F., Wassenaar, T.M., Endres, K., Fahrenholz, F., Bhakdi, S., and Zitzer, A. (2004). A cellular metalloproteinase activates *Vibrio cholerae* pro-cytolysin. J. Biol. Chem. *279*, 25143–25148.

van Beelen, A.J., Zelinkova, Z., Taanman-Kueter, E.W., Muller, F.J., Hommes, D.W., Zaat, S.A., Kapsenberg, M.L., and de Jong, E.C. (2007). Stimulation of the intracellular bacterial sensor NOD2 programs dendritic cells to promote interleukin-17 production in human memory T cells. Immunity *27*, 660–669.

van de Velde, H. (1894). Mécanisme de la virulence du staphylocoque pyogène. La Cellule *10*, 401–460.

van der Goot, F.G.E. (2001). Pore Forming Toxins (Berlin, Springer Verlag).

Van Wart, H.E., and Birkedal-Hansen, H. (1990). The cysteine switch: a principle of regulation of metalloproteinase activity with potential applicability to the entire matrix metalloproteinase gene family. Proc. Natl. Acad. Sci. U.S.A. *87*, 5578–5582.

Verkman, A.S. (2011). Aquaporins at a glance. J. Cell Sci. *124*, 2107–2112.

Vodkin, M.H., and Leppla, S.H. (1983). Cloning of the protective antigen gene of *Bacillus anthracis*. Cell *34*, 693–697.

von Leber, T. (1888). Uber die Entstehung der Entzundung und die Wirkung der entzundungserregenden Schadlichkeiten. Fortschr Med *6*, 460.

Walker, B., and Bayley, H. (1995). Key residues for membrane binding, oligomerization, and pore forming activity of staphylococcal alpha-hemolysin identified by cysteine scanning mutagenesis and targeted chemical modification. J. Biol. Chem. *270*, 23065–23071.

Walker, B., Krishnasastry, M., Zorn, L., and Bayley, H. (1992). Assembly of the oligomeric membrane pore formed by Staphylococcal alpha-hemolysin examined by truncation mutagenesis. J. Biol. Chem. *267*, 21782–21786.

Walker, J.A., Allen, R.L., Falmagne, P., Johnson, M.K., and Boulnois, G.J. (1987). Molecular cloning, characterization, and complete nucleotide sequence of the gene for pneumolysin, the sulfhydryl-activated toxin of *Streptococcus pneumoniae*. Infect. Immun. *55*, 1184–1189.

Watanabe, M., Tomita, T., and Yasuda, T. (1987). Membrane-damaging action of staphylococcal alpha-toxin on phospholipid-cholesterol liposomes. Biochim. Biophys. Acta *898*, 257–265.

Weber, S., Niessen, M.T., Prox, J., Lullmann-Rauch, R., Schmitz, A., Schwanbeck, R., Blobel, C.P., Jorissen, E., de Strooper, B., Niessen, C.M., *et al.* (2011). The disintegrin/metalloproteinase Adam10 is essential for epidermal integrity and Notch-mediated signaling. Development *138*, 495–505.

Weis, S., and Palmer, M. (2001). Streptolysin O: the C-terminal, tryptophan-rich domain carries functional sites for both membrane binding and self-interaction but not for stable oligomerization. Biochim. Biophys. Acta *1510*, 292–299.

Weissmann, G., Sessa, G., and Bernheimer, A.W. (1966). Staphylococcal alpha-toxin: effects on artificial lipid spherules. Science *154*, 772–774.

White, S.H., and Wimley, W.C. (1999). Membrane protein folding and stability: physical principles. Annu. Rev. Biophys. Biomol. Struct. *28*, 319–365.

Wickham, S.E., Hotze, E.M., Farrand, A.J., Polekhina, G., Nero, T.L., Tomlinson, S., Parker, M.W., and Tweten, R.K. (2011). Mapping the intermedilysin-human CD59 receptor interface reveals a deep correspondence with the binding site on CD59 for complement binding proteins C8alpha and C9. J. Biol. Chem. *286*, 20952–20962.

Wilke, G.A., and Bubeck Wardenburg, J. (2010). Role of a disintegrin and metalloprotease 10 in *Staphylococcus aureus* alpha-hemolysin-mediated cellular injury. Proc. Natl. Acad. Sci. U.S.A. *107*, 13473–13478.

Wimley, W.C. (2003). The versatile beta-barrel membrane protein. Curr. Opin. Struct. Biol. *13*, 404–411.

Wiseman, G.M. (1975). The hemolysins of *Staphylococcus aureus*. Bacteriol. Rev. *39*, 317–344.

Wiseman, G.M., and Caird, J.D. (1972). Further observations on the mode of action of the alpha toxin of *Staphylococcus aureus* "Wood-46". Can. J. Microbiol. *18*, 987–992.

Wuth, O. (1923). Serologishe und biochemische Studien uber das Hamolysin des *Frankelshen gasbazillus*. Biochem. Zeit. *142*, 19–28.

Xiong, Y.Q., Willard, J., Yeaman, M.R., Cheung, A.L., and Bayer, A.S. (2006). Regulation of *Staphylococcus aureus* alpha-toxin gene (hla) expression by agr, sarA, and sae *in vitro* and in experimental infective endocarditis. J. Infect. Dis. *194*, 1267–1275.

Xu, L., Huang, B., Du, H., Zhang, X.C., Xu, J., Li, X., and Rao, Z. (2010). Crystal structure of cytotoxin protein suilysin from Streptococcus suis. Protein Cell *1*, 96–105.

Yamashita, S., Lukacik, P., Barnard, T.J., Noinaj, N., Felek, S., Tsang, T.M., Krukonis, E.S., Hinnebusch, B.J., and Buchanan, S.K. (2011). Structural insights into Ail-mediated adhesion in *Yersinia pestis*. Structure *19*, 1672–1682.

Yarovinsky, T.O., Monick, M.M., Husmann, M., and Hunninghake, G.W. (2008). Interferons increase cell resistance to Staphylococcal alpha-toxin. Infect. Immun. *76*, 571–577.

Young, J.D., Cohn, Z.A., and Podack, E.R. (1986). The ninth component of complement and the pore-forming protein (perforin 1) from cytotoxic T cells: structural, immunological, and functional similarities. Science 233, 184–190.

Zitzer, A., Bittman, R., Verbicky, C.A., Erukulla, R.K., Bhakdi, S., Weis, S., Valeva, A., and Palmer, M. (2001). Coupling of cholesterol and cone-shaped lipids in bilayers augments membrane permeabilization by the cholesterol-specific toxins streptolysin O and *Vibrio cholerae* cytolysin. J. Biol. Chem. *276*, 14628–14633.

Bacterial Enterotoxins as Immunomodulators and Vaccine Adjuvants

4

Johan Mattsson and Nils Lycke

Abstract

The bacterial enterotoxins, cholera toxin (CT) and the closely related *E. coli* heat-labile toxin (LT) have been found to be the most potent mucosal immunoenhancers (adjuvants) we know of today. Hence, much research is focused on understanding the mechanism behind their potent augmenting function following mucosal immunizations and oral immunizations, in particular. These holotoxins consist of an AB5 structure, where the A1-subunit hosts ADP-ribosylating activity and the B-subunit is the receptor binding element, which exists as a pentamer and specifically binds to GM1 ganglioside on the membrane of most mammalian cells. The A1 and B-subunit pentamer are attached through the linker A2. Because of severe toxicity of the holotoxins following either oral or nasal administration clinical use of the holotoxins is precluded. Therefore, attempts to mutate the A-subunit so as to reduce enzymatic activity with retained augmenting effect have been successful. We have developed the CTA1-DD molecule which has retained the full enzymatic activity of CT, but without the toxic side effects of the holotoxin. In the present review we describe the mechanism of action for ADP-ribosylating holotoxins and we discuss the mechanistic benefits of mutant holotoxins or the unique CTA1-DD adjuvant for future prospects of developing effective mucosal vaccines in general and oral vaccines, in particular.

Introduction

As we have gained more information about how the immune system is organized and functions we have also explored avenues to modulate and impact on how it is regulated. This has proven very successful in several fields of clinical medicine. For example, organ transplantation would not have been possible without appropriate immunosuppression, nor would we have effective treatments against autoimmune diseases had we not understood how to ameliorate or block inflammatory responses (Blazar *et al.*, 2012; Schwartz, 2012). Vaccines and immune protection against infectious diseases is another area where knowledge about how the immune system works, has been successfully applied. By trial and error we have learnt how vaccines could be made effective by adding immunoenhancing components to the vaccine (Pulendran and Ahmed, 2011; Schijns and Lavelle, 2011). These components are termed, adjuvants, from the Latin *adjuvare,* meaning to help. Most of today's adjuvants or immunoenhancers are derived from microorganisms and have in common that they can be recognized by the immune system by special receptors, pattern recognition receptors (PRRs) (Baccala *et al.*, 2009). Most adjuvants use toll like receptors (TLRs) or NOD-like receptors (NLRs) to activate the immune system (Ebensen and Guzman, 2008). Thus, adjuvants are substances that have the ability to potentiate the immunogenicity of vaccines and they allow for lower doses of antigen as well as fewer immunizations to stimulate a strong immune response (Cox and Coulter, 1997). Most importantly, the adjuvant greatly improves the quality and the longevity of the immune response. The selection of adjuvant to be used in a vaccine is often as critical as which antigen or combination of antigens to include in the vaccine. The immunoenhancer can dramatically affect the long-term protective effect of a vaccine as the generation of long-lived memory B cells and plasma cells may directly be influenced by the choice of adjuvant (Tagliabue and Rappuoli, 2008; Galli *et al.*, 2009).

Perhaps the most effective immunomodulators we know of today are holotoxins from different bacterial species. A characteristic of these holotoxins is their ADP-ribosylating ability. The most used and best studied examples of holotoxin adjuvants are cholera toxin (CT) and *E. coli* heat-labile toxin (LT). These molecules, CT or LT, share considerable homology (Freytag *et al.*, 2010; Liang and Hajishengallis, 2010). The holotoxins are AB_5 complexes that consist of an ADP-ribosylating A1-subunit linked to a pentamer of B-subunits via an A2-moiety (Spangler, 1992; Fan *et al.*, 2004; Holmgren and Czerkinsky, 2005). The ability of CT to function as an adjuvant was first discovered by administrating the toxin intravenously (Northrup and Fauci, 1972). These pioneering studies by Fauci and colleagues showed a strong augmenting effect on the antibody response to sheep red blood cells following parenteral immunization, while later Elson and co-workers demonstrated that CT was an exceptionally strong adjuvant for enhancing immune responses after mucosal vaccination (Elson and Ealding, 1984). The holotoxin adjuvants bind with their B-subunits to distinct cell-membrane receptors present on most nucleated cells and cause ADP-ribosylation of membrane Gsα-protein, eventually resulting in cytoplasmic cAMP increases (Snider, 1995). CT binds the GM1 ganglioside receptor, while LT to some extent also binds other gangliosides (Fan *et al.*, 2001; Connell, 2007). In fact, LT has also been reported to bind to GD1b, GM2, asialo GM1 as well as some glycoproteins (Fukuta *et al.*, 1988; Karlsson *et al.*, 1996).

The receptor-binding is both a strength and a weakness of this adjuvant system. A clear benefit is the facilitated uptake of the adjuvant across mucosal membranes and, perhaps, the augmented up-take by follicle-associated epithelium (FAE), and accumulation in dendritic cells (DC) (Anosova *et al.*, 2008). On the other hand the lack of selectivity may be associated with unwanted side-effects. For example, the promiscuous distribution of the receptors also to nerve cells preclude the use of these adjuvants after intranasal (in) administration as seen with the virus-like particle (VLP) flu-vaccine using LT as the adjuvant, which gave facial paralysis (Bell's palsy) in a few cases and had to be taken off the market (Llewellyn-Smith *et al.*, 1990; van Ginkel *et al.*, 2000; Glueck, 2001; Fujihashi *et al.*, 2002; Mutsch *et al.*, 2004; Armstrong *et al.*, 2005; van Ginkel *et al.*, 2005). Furthermore, after oral vaccination using CT or LT adjuvants in the clinic overt diarrhoea was observed in some vaccinees (Levine *et al.*, 1984; Summerton *et al.*, 2010). However, notwithstanding the toxicity problem, there is still firm belief that CT and LT, through site-directed mutations in the enzymatically active A1/A2-part, will turn the holotoxins into less toxic and safer mucosal adjuvants (Pizza *et al.*, 2001). Indeed, a good example of this strategy is the LTR192G mutant developed by Clements and co-workers (Freytag *et al.*, 2010; Summerton *et al.*, 2010). Especially the recently developed double mutant LTR192G/L211A (dmLT) of LT has proven to be trypsin-resistant, i.e. resistant to degradation, and it exerts no enterotoxicity, while retaining strong adjuvant function. It is currently being tested in clinical trials. Also, Rappuoli and colleagues have very successfully pursued research and development of genetically detoxified LT mutants (Douce *et al.*, 1998; Pizza *et al.*, 2001; Brereton *et al.*, 2011). They showed that mutants with some ADP-ribosylating activity (LTR72) were more adjuvant active than mutants completely lacking enzymatic activity (LTK63) (Giuliani *et al.*, 1998). Unfortunately, the fact that also LTK63, devoid of ADP-ribosylating properties, was found to associate with Bell's palsy in a few vaccinees in two clinical trials, argue, though, for caution in the use of GM1-binding adjuvants given i.n. (Stephenson *et al.*, 2006; Lewis *et al.*, 2009). However, for oral or sublingual vaccination this neurotoxicity problem may not exist, although from a regulatory perspective it may still be problematic.

Because of the toxicity problems and the risk of side-effects when the B-subunits were part of the holotoxin adjuvant we developed a new family of targeted fusion proteins that explore the full enzymatic activity of CT, while excluding the B-subunit binding to the GM1-ganglioside receptor (Agren *et al.*, 1997). The CTA1-DD fusion protein was developed as a genetically encoded protein with the complete CTA1-moiety linked together with a dimer of the D-fragment from *Staphylococcus aureus* protein A (Agren *et al.*, 1997). This unique strategy turned out to be very successful and we have demonstrated in multiple infectious disease models the efficacy and safety of the CTA1-DD adjuvant (Belyakov *et al.*, 1998; Eriksson *et al.*, 2004; Akhiani *et al.*, 2006; McNeal *et al.*, 2007; Eliasson *et al.*, 2008; Cunningham *et al.*, 2009). While the D fragment targets Fc and Fab fragments on immunoglobulins, preferentially of the IgG subclasses, it was originally thought to target B cells *in vivo* (Agren *et al.*, 1997, 1999). Today, we also know that other cells, DCs, in particular, are targeted by CTA1-DD. These different binding properties of CTA1-DD, as compared to CT, renders the former completely non-toxic and safe as shown in mice, guinea pigs and non-human primates (Agren *et al.*, 1997; Lycke, 2007; Sundling *et al.*, 2008). This is because CTA1-DD cannot bind to GM1-gangliosides as it does not host the B-subunit and, therefore, the risk of developing Bell's palsy following intranasal delivery is eliminated (Agren *et al.*, 1997; Eriksson *et al.*, 2004). In fact, CTA1-DD has been shown to be safe and effective when using a number of different immunization routes, including mucosal as well as parenteral immunizations using doses that are even 10× the expected clinical dose in humans (Sundling *et al.*, 2008).

In the present chapter we will discuss the holotoxins and their derivatives as immunomodulators and vaccine adjuvants in greater detail. We will focus on their mechanisms of action and make comparisons between the different adjuvants, although very few studies have, in fact, attempted to compare the toxin adjuvants head-to-head. Thus, we can only approximate the effects seen in the great number of published studies, and based on this information we will discuss whether they share the same or different immunomodulating properties. As already been mentioned the holotoxins are very powerful adjuvants that would make a major difference if allowed in human mucosal vaccines. However, the safety aspect precludes their use while the less toxic derivatives of these toxins have been made more attractive from a vaccine point of view. Notwithstanding this, we believe that learning how and why the holotoxins work is the key element to a breakthrough in mucosal vaccine development. As will be further discussed re-targeting of the immunomodulating elements of holotoxin adjuvanticity, as in the CTA1-DD adjuvant, appears to be a particularly promising strategy to the successful development of a mucosal vaccine adjuvant.

Cholera toxin: the prototype for ADP-ribosylating holotoxin adjuvants

CT is perhaps the most studied member of the AB_5 toxins. As mentioned previously, it has been extensively used as an experimental adjuvant owing to its ability to strongly enhance immune responses when administered at mucosal sites (Elson and Ealding, 1984; Hornquist and Lycke, 1993; Yamamoto *et al.*, 1997a,b; Lavelle *et al.*, 2003; Luci *et al.*, 2006; Raghavan *et al.*, 2010; Meza-Sanchez *et al.*, 2011; Olvera-Gomez *et al.*, 2012). The toxin is produced by *Vibrio cholerae*, which causes epidemic and endemic acute diarrheal disease, killing thousands of people every year in developing countries where sanitary conditions are poor (WHO, 2011). The diarrhoea is the result of a massive out-flux of water and electrolytes from epithelial cells in the upper part of the small intestine It is caused by the enzymatically active CTA1-part of the toxin. The molecular mechanisms behind the toxic effects of CT are not fully understood. However, the major intracellular events have been described. Initially the toxin gains entry into the cell through binding with the B-subunit to the GM1-ganglioside receptor present on virtually all nucleated cells (Holmgren *et al.*, 1973; Spangler, 1992; Zhang *et al.*, 1995; Rappuoli *et al.*, 1999). Following receptor-binding, the toxin is endocytosed and transported to the endoplasmic reticulum (ER) via the Golgi apparatus by retrograde vesicular transport (Orlandi *et al.*, 1993; Majoul *et al.*, 1996). The disulfide bond which links the A1 and the A2 subunit is reduced in the ER, separating the subunits, which allows for the unfolding of the A1 subunit, a process that is facilitated by disulfide isomerase (Tsai *et al.*, 2001). To reach its substrate, the α subunit of the stimulatory G protein Gs (Gsα), which is located in the cell membrane, the A1-subunit must be transported into the cytosol, via the ER-associated degradation (ERAD) pathway. The ERAD functions as a protein quality control system, directing misfolded proteins to the proteasome in the cytosol for degradation (Teter and Holmes, 2002). Hence, CT hijacks the ERAD system to allow for CTA1 to pass through the cytosol and access the Gsα-substrate in the cell membrane. While detailed information on how CT is brought into the cytosol is still scarce, earlier studies have implicated the Sec61 and Derlin-1 channels in this process (Schmitz *et al.*, 2000; Bernardi *et al.*, 2008). Importantly, CT is not ubiquitinated in the cytosol, probably due to the paucity of available lysine residues in CT (Rodighiero *et al.*, 2002). Thereby CT avoids protein degradation and it allows CTA1 to be refolded and subsequently enzymatically active. The ADP-ribosylating efficiency is further enhanced by the interaction of CTA1 with ADP-ribosylating factors (ARFs), which cause a conformational change in the CTA1 subunit (Kahn and Gilman, 1984b; O'Neal *et al.*, 2005). The CTA1-subunit catalyses the transfer of an ADP-ribose moiety from nicotinamide adenine dinucleotide (NAD) to Gsα, which activates the G protein subunit by causing it to lose its GTPase activity (Kassis *et al.*, 1982; Kahn and Gilman, 1984a). Then Gsα acts on adenylate cyclase which converts ATP to the second messenger cAMP, dramatically raising intracellular cAMP levels (Iyengar, 1993). To what extent cAMP is responsible for the adjuvant effect or not is a matter of debate at present. Whereas CT and LT clearly affect intracellular cAMP levels it is quite clear that CTA1-DD does not increase intracellular cAMP. Because both holotoxins and CTA1-DD act as adjuvants, but differ in their ability to augment cAMP, it appears reasonable to assume that adjuvanticity is independent of cAMP, while it is heavily dependent on ADP-ribosylation.

Despite that CT has been named the 'golden standard' for mucosal adjuvants its mechanism of action is not fully understood (Elson and Ealding, 1984; Hornquist and Lycke, 1993; Yamamoto *et al.*, 1997a,b; Lavelle *et al.*, 2003; Luci *et al.*, 2006; Raghavan *et al.*, 2010; Meza-Sanchez *et al.*, 2011; Olvera-Gomez *et al.*, 2012). This is partly because CT can involve many different types of cells as it binds GM1-ganglioside receptors and many of the cells at the location of injection or mucosal administration may not directly be involved in the adjuvant function, although they have been activated. However, even if we restrict our interest to cells that belong to the innate and adaptive immune system we will have a problem to answer this question in a simple way. In fact, using LT adjuvant one study attempted to use Affymetrix analysis and in this way identify biomarkers in the lung as their read-out to depict a successful protective immunization (Tritto *et al.*, 2007). Unfortunately, this strategy resulted in a massive amount of data, with little possibility to dissect and distinguish between adjuvant effects and immune responses in general to the vaccine (Tritto *et al.*, 2007). From earlier studies we have learnt that site directed mutagenesis could be a useful tool in deciphering the relative roles of the different subunits, e.g. the contribution of the GM1-ganglioside binding (the B subunit) as compared with the importance of the enzymatic activity (the A1 subunit) in the adjuvant function of CT (Rappuoli *et al.*, 1999). Using mutants of LTB it was observed that lack of GM1-ganglioside binding exerted strongly reduced toxicity, and these mutants were also devoid of most of their adjuvant function (Jobling and Holmes, 1991). Researchers have also used mice lacking GM1-gangliosides to understand the importance of GM1-binding for the immunoenhancing effect of CT (Gustafsson *et al.*, 2013). These mice failed to respond to CT-adjuvant following immunization, as evident from lack of T cell proliferation and antibody responses (Kawamura *et al.*, 2003). Hence, the ability of CTB to bind to target cells is an essential component of the adjuvant function of the holotoxins. On the other hand when CTA1 is formulated as part of the CTA1–DD fusion protein, which cannot bind GM1-ganglioside, it is clear that the binding itself is dispensable for the adjuvant function of CTA1, as the enzymatically active subunit is retargeted to access cells in a different manner, independent of GM1-ganglioside binding. Noteworthy experiments using CTA1-DD also highlight that the ADP-ribosylating activity of the CTA1 moiety is a key element, responsible for the adjuvant function, as mutants unable to ADP-ribosylate lose most of the adjuvant effect (Agren *et al.*, 1997). Therefore, it has become evident that while CTA1-DD retains the full adjuvant function of CT it is safe and non-toxic, because of the lack of CTB-mediated binding to GM1-ganglioside carrying cells. Thus, CTA1-DD has a very much more restricted repertoire of target cells than CT, which explains its low toxicity, albeit it carries the full enzymatic effect of CT (see below) (Agren *et al.*, 1997).

The importance of the ADP-ribosylating effects of the A1-subunit for adjuvanticity was further confirmed in elegant studies by Giuliani *et al.* (1998), who mutated LTA1 in many different ways by single-point mutations and could show that when some enzymatic activity (LTR72) was retained a better adjuvant effect was seen than when mutants lacked enzymatic activity completely (LTK63). However, the LTK63 mutant still retained significant adjuvant activity despite lacking ADP-ribosylation completely. Today many studies have been performed with LTK63, including clinical trials, and it is striking that this molecule exerts strong adjuvant effects even though it lacks ADP-ribosylating function. Undoubtedly, the adjuvant effect is by several magnitudes much stronger than that seen with the LT B subunit alone (Giuliani *et al.*, 1998). Presently, we cannot explain this discrepancy as the main difference between LTB and LTK63 is the AB5-structure and the LTA-molecular elements that are missing from the LTB molecule. As will be discussed later the requirement for cAMP induction may not be critical for the adjuvant effect and one could speculate that also other aspects of the A1-subunit, distinct from the enzymatic activity, are responsible for some of the immunoenhancing functions as seen with LTK63 and other killed mutants, effects that go beyond those that simply can be ascribed to the B-subunit. In fact, we have to predict that LTK63 and other killed mutants are taken up and traffic in the cytosol of the target cell much similar to the holotoxins. Hence, in an antigen-presenting cell (APC), such as a DC, antigens could be subject to an enhanced transport to the ER and Golgi apparatus due to the mutant toxin, leading to an up-regulated ability to prime naïve T cells. More importantly, perhaps, the A1-subunit would still interact with many proteins in the cytosol, such as ARFs (Rappuoli *et al.*, 1999; Gagliardi *et al.*, 2002). Interestingly, a similar mutation in CTA1 as LTK63, i.e. CTK63, lost both the adjuvant effect as well as the ADP-ribosylating function (Fontana *et al.*, 1995; Douce *et al.*, 1997). Presently, we cannot explain why there is a difference between the LTK63 and CTK63, with regard to adjuvant function but it may have to do with conformational changes that are different between LT and CT, which could affect the stability or protein binding abilities of the molecule. Alternatively, CT and LT have different patterns for cell–receptor interactions (discussed below) and it is this distinction rather than the cytosolic fate of the molecules that differs. Arguing in favour of this interpretation is the fact that, multiple other mutants of CT have been generated, including CTE112K and CTS61F, which both have retained adjuvant activity while lacking ADP-ribosylating activity (Yamamoto *et al.*, 1997b, 1998; Watanabe *et al.*, 2002). Also, the CTE29H and CTS106 mutants lack most of the ADP-ribosylating function but have retained adjuvant activity (Fontana *et al.*, 2000; Tebbey *et al.*, 2000).

The immunoenhancing effects of CT are believed to be largely attributed to the effects of the holotoxin on APC. Both B cells and macrophages have exhibited improved ability to function as APCs after exposure to CT (Cong *et al.*, 1997; Sjoblom-Hallen *et al.*, 2010). Notwithstanding this it is likely that the effect on the DC is the most critical for the adjuvant effect *in vivo* (Porgador *et al.*, 1998; Gagliardi *et al.*, 2000; Kawamura *et al.*, 2003; Shreedhar *et al.*, 2003; Anjuere *et al.*, 2004; Lavelle *et al.*, 2004; Grdic *et al.*, 2005; Fahlen-Yrlid *et al.*, 2009). Many studies have reported that CT strongly activates DCs, as demonstrated by their increased expression of a number of co-stimulatory molecules, including CD80, CD86 and CD40, as well as the major histocompatibility complex II (MHC II) following exposure to CT *in vitro* (Williamson *et al.*, 1999; Lavelle *et al.*, 2003). Few studies have, however, reported on the expression of co-stimulatory molecules on DCs after injection of CT *in vivo*. We have repeatedly found that CD86, rather than CD80, CD40 or MHC class II, is up-regulated after immunizations using CT in mice (Grdic *et al.*, 2005). Importantly, the enhanced expression of these molecules on activated DCs allows for an augmented priming of naïve T cells, (Williamson *et al.*, 1999; Lavelle *et al.*, 2003; Grdic *et al.*, 2005) (Fig. 4.1). In addition to increased expression of co-stimulatory molecules, DCs exposed to CT also up-regulate the expression of the chemokine receptors CCR7 and CXCR4 (Gagliardi *et al.*, 2000; Song *et al.*, 2009). This allows antigen-carrying DCs to migrate into lymph nodes and the T cell areas, where they can activate naïve T cells in a cognate fashion. Gene expression profiling have provided some insights as to the impact of CT on APCs. We performed Affymetrix analysis of B cells exposed to CT and found that roughly 100 annotated genes were up-regulated by CT, among those we specifically noted STAT3 (Sjoblom-Hallen *et al.*, 2010). Clearly with such an impact on gene transcription in all APCs, including DCs,

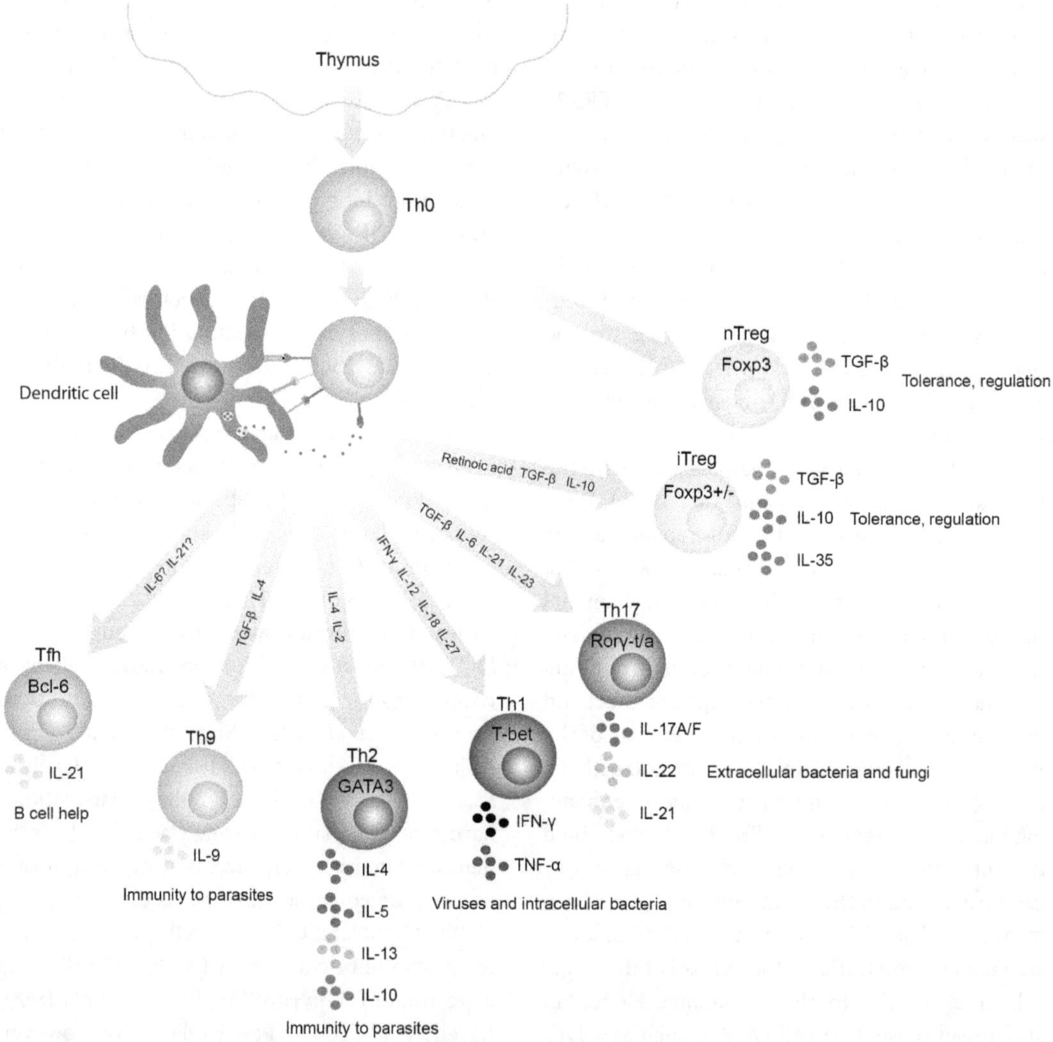

Figure 4.1 An overview of the of T cell differentiation process from naïve Th0 cells to functional subsets of CD4T cells: Naïve undifferentiated T cells (Th0) are derived from the thymus. Upon activation by antigen presenting cells such as dendritic cells, the local cytokine milieu (indicated at the corresponding arrow) drives T cell expansion and differentiation to become either regulatory T cells (Tregs) or effector T cells, termed Th1, Th2, Th17 or Th9 cells, that are specialized to combat different types of infections or cancer development. T cells can also differentiate to become T follicular helper (Tfh) cells which are specialized to provide B cell help and to drive germinal centre reactions. To prevent overt inflammation there are also regulatory subsets inducible Treg (iTreg) or natural Tregs (nTreg). The different subsets all have their unique cytokine secretion patterns, which exert regulatory functions on the immune response.

one could expect to find data to explain the very dramatic immunomodulating effects of CT on these cells. For example, STAT3 is a major regulator of the anti-inflammatory response to IL-10, but it is also involved in IL-6- and IL-22-induced activation, of innate cells, which are both thought to have a mucosal barrier protective effect (Hruz *et al.*, 2010). STAT3 is also critical for Th17 cell differentiation and in this regard CT could promote, rather than prevent inflammation, and thus indirectly augment vaccine efficacy (Hirahara *et al.*, 2010). Nevertheless, additional studies are much needed to dissect the pattern of gene regulation that CT exerts on APCs, so that we may better understand the complexity of this impact.

Complementing the effect on co-stimulatory molecules the CT-targeted cell will release IL-1β, IL-6 and several other pro-inflammatory molecules (Bromander *et al.*, 1991). However, in direct conflict with the documented drive on pro-inflammatory cytokines, it is generally believed that CT primarily promotes Th2 dominated responses. This is because CT-adjuvant has been reported to preferentially augment IL-4, IL-5, IL-6 and IL-10 as well as the IgG1, IgA and IgE antibody subclasses, all factors that are associated with Th2-dominated responses (Yamamoto *et al.*, 1997a; Williamson *et al.*, 1999; Simecka *et al.*, 2000; Yamamoto *et al.*, 2000; Lavelle *et al.*, 2004). The underlying mechanism for the Th2-skewing effect has been attributed primarily to the down-regulation of the Th1-inducing cytokine IL-12. The mechanism behind this effect has been demonstrated by elegant studies by Kelsall and co-workers, who could show that CT acts via blocking IRF-8, a transcription factor that controls IL-12 production (la Sala *et al.*, 2009). However, CT has, in fact, also been found to effectively deplete CD8α$^+$ DCs *in vivo*, leaving little leverage for this DC subset for the adjuvant effect *in vivo* (la Sala *et al.*, 2009). While the effects of CT on IL-12 production appears dramatic, as IL-12 is a key regulator of Th1-differentiation, numerous reports, including our own work, have consistently found strong Th1 and IFN-γ responses after immunizations with CT-adjuvant (Hornquist and Lycke, 1993; Xu-Amano *et al.*, 1994; Belyakov *et al.*, 1998; Jones *et al.*, 2001; Eriksson *et al.*, 2003; Luci *et al.*, 2006; Meza-Sanchez *et al.*, 2011). This conflicting finding is presently under intense investigation, but it appears that CT is not dependent on IL-12 for its adjuvant function (Akhiani *et al.*, 2002).

Recent studies have also described that CT in some ways exert adjuvant function through promoting Th17-differentiation, an effect that depended on IL-6 (Lee *et al.*, 2009; Datta *et al.*, 2010; Hervouet *et al.*, 2010; Meza-Sanchez *et al.*, 2011). One study links adjuvanticity to induction of Th17 cells and this mechanism requires cAMP-dependent secretion of IL-1β and β-calcitonin in DC (Datta *et al.*, 2010). These authors also argued that the mucosal adjuvant effect of CT requires Th17 cells, although in this regard their data are less convincing as IL17–/– mice were used and oral gavage with ovalbumin plus CT-adjuvant stimulated only weak IgA responses in IL-17–/– mice compared to wild-type mice, but no control for the ability of IL-17–/– mice to generate a specific mucosal IgA response to another adjuvant was included in the study. Nevertheless, these findings further add to the complexity when it comes to explaining why CT is such a strong mucosal adjuvant *in vivo*.

Quite in contrast to its augmenting effects on the immune system, CT has been shown to inhibit production of a number of factors, considered to be important in the early stages of an immune response, i.e. several chemokines such as CCL2, CCL3 and CCL4 (Lavelle *et al.*, 2003). Furthermore, CT has been reported to increase production of the regulatory cytokine IL-10, which is normally associated with regulatory T cells, termed Tr1 cells (Lavelle *et al.*, 2003). Thus, despite these potentially inhibitory effects, CT is, undoubtedly, an exceptionally potent adjuvant.

LT: a more complex family of holotoxin adjuvants

LT can be produced by different strains of *Escherichia coli*. The holotoxins can be divided into two main types; the type I toxin is LT-I (previously discussed as LT) (> 80% structural homology to CT), whereas the type II toxins include LT-IIa, b and c (Zhang *et al.*, 1995). The LT-II toxins are structurally similar to each other, although there is a significant homology of around 60% between the A1 subunits of the type I and type II holotoxins. However, as there is no sequence homology between the B subunits of type I and type II toxins they bind very different target cells (Pickett *et al.*, 1987; Pickett *et al.*, 1989). These features also agree well with the functional characteristics of the type II holotoxins, albeit their A1 subunits have a similar function to those of LT-I and CT, the B subunits all have unique and different targets. Thus, holotoxins of both type I and II share the ability to ADP-ribosylate Gsα, resulting in cAMP increases, but the target cells may be strikingly different. Owing to the dissimilar B subunits, the binding properties of the toxins differ significantly. LT-IIa binds a wide range of gangliosides including GD1b, GM1, GT1b, GQ1b, GM2, GD1a, and GM3 with descending order of avidity, whereas LT-IIb binds to GD1a, GM2, GM3, GM1b, GT1b and GD1α (Fukuta *et al.*, 1988; Berenson *et al.*, 2010). The recently discovered LT-IIc toxin has been shown to bind GM1, GM2, GM3, and GD1a (Nawar *et al.*, 2010; Berenson *et al.*, 2012).

Interestingly, all type I and type II toxins have been found to host adjuvant functions. Given that their A1 subunits share a common substrate, the differences are believed to be due to the different binding properties of the B subunits. For example, LT-I generates a mixed Th1/Th2/Th17 response with the production of the cytokines IFN-γ, IL-4, IL-5 and IL-17 as well the antibody isotypes IgG1 and IgG2a upon immunization (Takahashi *et al.*, 1996; Fromantin *et al.*, 2001; Brereton *et al.*, 2011). Similarly to CT, LT-I also induces IL-1 production, this ability was shown to be essential for the production of IFN-γ and IL-17 in mice immunized with LT-I (Brereton *et al.*, 2011). In contrast to CT, LT-I does not induce significant production of IgE antibodies after mucosal administration (Takahashi *et al.*, 1996; Simecka *et al.*, 2000).This latter effect is also not seen with CTA1-DD, which may indicate that CTA1-DD also differ mechanistically in its adjuvant function compared to that employed by the CT holotoxin.

Furthermore, LT-IIa and LT-IIb promote a more Th1-biased response as compared to CT and LT-I, with ample amounts of IFN-γ but with less IL-4 and IL-5 being produced after immunization (Martin *et al.*, 2000; Nawar *et al.*, 2007; Mathias-Santos *et al.*, 2011). The type II toxins also appear to favour IgG2a production over IgG1, which also is indicative of a Th1 skewed response (Nawar *et al.*, 2007). Importantly, all LT toxins induce the production of secretory IgA when administered mucosally (Takahashi *et al.*, 1996; Martin *et al.*, 2000; Nawar *et al.*, 2011). The notion that the different binding specificities of the B subunit is important for adjuvant function is further supported by results from LT-mutants with altered binding properties. A number of different LT-IIa mutants have been described with varying binding activities; those that retained binding to at least one ganglioside, or those with reduced binding affinity but with maintained specificity. These mutants were all capable of enhancing immune responses (Nawar *et al.*, 2007). However, if the binding was completely abolished, adjuvant function was also lost (Nawar *et al.*, 2005). Interestingly the mutants that retained binding, but with altered specificity or affinity, displayed qualitative differences as compared to wild type holotoxins, indicating that the distribution of gangliosides, and, thus, the available target cell population of the different toxins, influenced the outcome of the immune response following immunization (Nawar *et al.*, 2007). The importance of ganglioside-binding was addressed with a mutant of LT-I that lacks GM1-binding, termed LT-G33D, which was completely devoid of toxicity and exerted poor oral adjuvant effects (Guidry *et al.*, 1997). However, when given i.n. or intradermally, LT-G33D had retained significant adjuvant effects, arguing that the route of administration is critical for the adjuvant function (Guidry *et al.*, 1997; de Haan *et al.*, 1998; Zoeteweij *et al.*, 2006). In addition, a recent study showed that the A domain (which includes the A1 and the A2 subunits) as well as the A1 subunit alone could enhance antibody responses when admixed with antigen and delivered i.n., arguing that the LTA subunit may function as an adjuvant independently of the B subunit provided certain routes for administration are used (Norton *et al.*, 2012).

Similarly to CT, a number of mutant LT variants have been constructed with the purpose developing safe and potent mucosal vaccine adjuvants. As aforementioned, LTK63, has retained significant adjuvant function while devoid of toxicity, as confirmed by the ileal loop assay (Giuliani *et al.*, 1998). LTK63 has successfully been used in a large number of studies assessing both parenteral and mucosal immunization protocols (Di Tommaso *et al.*, 1996; Giuliani *et al.*, 1998; Douce *et al.*, 1999; Partidos *et al.*, 1999; Brynjolfsson *et al.*, 2008). Furthermore, immunizations using LTK63 have been shown to stimulate protective immunity in many infectious disease models (Ryan *et al.*, 1999; Bonenfant *et al.*, 2001; Simmons *et al.*, 2001; Tempesta *et al.*, 2007; Brynjolfsson *et al.*, 2008; Romero *et al.*, 2009; Zhou *et al.*, 2009; Brereton *et al.*, 2011). The quality of the immune response stimulated by LTK63 adjuvant is similar to that of the LT holotoxin, generating a mixed Th1/Th2/Th17 response, with the production of IL-4, IFN-γ and IL-17 (Ryan *et al.*, 1999; Brereton *et al.*, 2011) (Fig. 4.1). Similar to CT, LT and LTK63 were found to augment IL-1α and IL-1β, but also IL-23, and it was concluded that all these cytokines were critical for the promoting effect on Th17-development. An alternative approach to LTK63 is the LTR192G mutant adjuvant. In this molecule the mutation is introduced in the proteolytically sensitive region that links the A1 and the A2 subunits, and, thus not directly in the A1-enzyme (Dickinson and Clements, 1995). Unlike CT, in which the protein is cleaved by an enzyme before being secreted by the bacterium, the A1 and A2 subunits are covalently linked in LT (Clements and Finkelstein, 1979; Booth *et al.*, 1984). Therefore LT must undergo proteolytic cleavage before it becomes activated. The LTR192G harbours a mutation at the site at which the protein is normally cleaved by trypsin. This mutation, therefore, prevents the separation of the A1 subunit from the A2 subunit (Dickinson and Clements, 1995). Consequently the LTR192G mutant lacks enzymatic activity, as tested with *in vitro* assays (Grant *et al.*, 1994; Dickinson and Clements, 1995). LTR192G may fail to unfold in the ER, thereby interfering with the cytosolic transport, or it may fail to separate A1 from A2, which could prevent ARF binding, reducing the enzymatic activity (Norton *et al.*, 2011). Noteworthy, though, it has been found that additional proteases may be able to cleave the A1/A2 subunits *in vivo*, thereby the enzyme would become active and toxic (Grant *et al.*, 1994). In fact, this notion has support in that LTR192G was found to be enterotoxic and a delayed cAMP-increase was observed in LTR192G-targeted cells (Giannelli *et al.*, 1997; Cheng *et al.*, 1999). This may, indeed, argue that LTR192G is dependent on the Gsα for its adjuvant effect, which is achieved after delayed ADP-ribosylation. Nevertheless, LTR192G has been reported to be effective and to generate protective immune responses in influenza and rotavirus infection models (Cheng *et al.*, 1999; McNeal *et al.*, 1999, 2007; Tumpey *et al.*, 2001; Lu *et al.*, 2002; McNeal *et al.*, 2002; Bertolotti-Ciarlet *et al.*, 2003). The LTR192G has been tested in clinical trials; such as an oral vaccine against *Helicobacter pylori*, where some cases of enterotoxicity was observed, indicating that the LTR192G retains enzymatic activity (Kotloff *et al.*, 2001). Therefore, to avoid enterotoxigenic side-effects of LTR192G, a double mutant was constructed with an additional mutation in the A2 domain. This additional mutation, termed L211A, targets a putative pepsin site. The dmLT (LT(R192G/L211A) retains significant adjuvant function, but is less toxic compared to the LTR192G mutant (Norton *et al.*, 2011). Used as an adjuvant it has been shown to induce protective immunity in several infectious disease models and is presently in clinical trials (Lu *et al.*, 2010; Summerton *et al.*, 2010; Martinez-Becerra *et al.*, 2012).

Because of the toxicity problem alternative routes have been explored with the holotoxins (Kotloff *et al.*, 2001; Mutsch *et al.*, 2004). One such route is transcutaneous immunization, in which the vaccine formulation is applied via a patch directly to the skin. This approach offers the possibility of taking advantage of the potent adjuvant function of LT, while reducing the risk of side effects that would preclude clinical use. In fact, LT was found to be safe and effective in a number of human trials using the transcutaneous immunization route (Guerena-Burgueno *et al.*, 2002; McKenzie *et al.*, 2007; Frech *et al.*, 2008; Frerichs *et al.*, 2008; Glenn *et al.*, 2009).

Are ADP-ribosylating toxins in general good adjuvants?

We also know other ADP-ribosylating holotoxins that act on different other families of target proteins than Gsα and which exert adjuvant function. The best example of this category is pertussis toxin (PT), which is produced by *Bordetella pertussis*, the causative agent of whooping cough. It shares the AB_5 structure with CT and LT and is transported to the cytosol via retrograde transport in an analogous fashion to the CT- and LT toxins (Tamura *et al.*, 1982; Stein *et al.*, 1994; el Baya *et al.*, 1997). However, in contrast to CT and LT, PT acts primarily on the α-subunit of the inhibitory G protein (Giα), but also the $G_{o\alpha}$ G-proteins, i.e. $G_{i1\alpha}$, $G_{i2\alpha}$, $G_{i3\alpha}$ and $G_{o1\alpha}$ and $G_{o2\alpha}$ respectively (Bokoch *et al.*, 1983; Katada *et al.*, 1983). Because Giα negatively regulates adenyl cyclase activity, and blocking its action with PT binding, causes increased levels of cytosolic cAMP due to a poor control of adenyl cyclase (Milligan and Kostenis, 2006). As of today no specific receptor for PT has been identified, instead it appears to bind a variety of molecules including glycoproteins and glycolipids, suggesting a rather promiscuous target cell profile (Witvliet *et al.*, 1989; Hausman and Burns, 1993). A well-known property of PT is the induction of leucocytosis, or the accumulation of white blood cells in the blood, resulting from the inability of lymphocytes to recirculate to lymph nodes or to the white pulp of the spleen (Morse and Riester, 1967; Munoz *et al.*, 1981; Cyster and Goodnow, 1995). This effect is mediated by the ADP-ribosylation of Giα, which uncouples chemokine signalling and prevents cell movement, affecting also immunocytes (Spangrude *et al.*, 1985; Campbell *et al.*, 1998). However, despite this seemingly counterproductive effect on the immune system, PT functions as a powerful adjuvant and is able to enhance responses to co-administered antigens when delivered by mucosal- as well as parenteral routes (Wilson *et al.*, 1993; Samore and Siber, 1996; Jabbal-Gill *et al.*, 1998; Ryan *et al.*, 1998). In fact, PT has been shown to induce a mixed Th1/Th2 response with the production of the cytokines IFN-γ, IL-4 and IL-5 as well as humoral responses represented by the IgG1, IgG2a, IgE and IgA antibody isotypes (Wilson *et al.*, 1993; Samore and Siber, 1996; Jabbal-Gill *et al.*, 1998; Ryan *et al.*, 1998). An important observation is that PT could reduce the total number of regulatory T cells and impair their function, causing less suppression of immune responses and, thus, in that way enhance T and B cell responses (Cassan *et al.*, 2006; Chen *et al.*, 2006). These effects are believed to be induced by the activation of DCs or other APCs. Indeed, PT has been reported to induce up-regulation of the co-stimulatory molecules CD80 and CD86 on B cells, macrophages and DCs, as well as CD40 and MHC-II on DCs (Ryan *et al.*, 1998; Hou *et al.*, 2003). Recently PT was shown to polyclonally activate $CD8^+$ T cells to up-regulate CD28 expression and produce IFN-γ *in vitro*, however whether this also occurs *in vivo* is currently unclear (Murphey *et al.*, 2011).

A fundamental question is whether the adjuvant effect of PT is dependent on ADP-ribosylation or not. This has not yet been adequately addressed. Although not many attempts have been reported to develop mutant variants of PT, a few non-toxic mutants have been described (Burnette *et al.*, 1988; Pizza *et al.*, 1989). The PT-9K/129G mutant carries two amino acid substitutions which abolishes the ADP-ribosylating enzymatic activity, rendering the molecule non-toxic and unable to induce leukocytosis (Burnette *et al.*, 1988; Pizza *et al.*, 1989). However, as with mutant CT and LT, PT-9K/129G maintained many adjuvant properties, including the induction of T cell proliferation as well as IFN-γ and IL-5 secretion upon immunization (Ryan *et al.*, 1998). Furthermore, it generated specific IgG1, IgG2a and IgA antibody levels comparable to those generated by PT holotoxin (Roberts *et al.*, 1995; Ryan *et al.*, 1998). The PT-9K/129G mutant has also been tested as a component of the acellular pertussis vaccine and in that context found safe and immunogenic its adjuvant properties in the clinic has not directly been assessed (Podda *et al.*, 1990, 1992), but it has been proposed by several researchers that adjuvanticity in PT may not depend on the ADP-ribosylating effect. Ongoing studies in our laboratory would argue against this, though, because we have successfully replaced CTA1 with the ADP-ribosylating S1-moiety from PT, in the fusion protein and i.n. immunizations with S1-DD have revealed very potent enzyme-dependent adjuvant effects. Hence, it appears that also other substrates than Gsα can mediate an adjuvant pathway, which augment immune responses. Whether or not cAMP is a common denominator for these pathways is too early to answer, although work with the CTA1-DD adjuvant would not support such a notion.

Taking toxin adjuvant immunomodulation one step further

Because the holotoxins and the mutants thereof have been found to carry substantial risks for side effects we developed the CTA1-DD adjuvant. This mucosal adjuvant was found to be safe and non-toxic because it does not have the B-subunit and thus cannot bind GM1-ganglioside receptors (Agren *et al.*, 1997). Hence, it has a more limited repertoire of target cells *in vivo*. It was originally designed to target the Ig-receptor on B cells, since protein A and fragments of this protein are well known binders of the Fc-part of immunoglobulins of most isotypes (Ljungberg *et al.*, 1993). Because the ability of CTA1-DD to bind Ig could potentially result in the formation of immune complexes we analysed the involvement of immune complexes in the adjuvant function of CTA1-DD. We found, however, that mice deficient in the Fc-receptors FcγRIIb and FcεR were perfectly susceptible to the adjuvant effect, arguing against immune complexes being important for the adjuvant effect (Agren *et al.*, 2000). Despite this, it has been observed that CTA1-DD/IgG immune complexes formed *ex vivo* could potentiate the adjuvant function by acting directly on mast cells, and in this way augment nasal immune responses (Fang *et al.*, 2010). Recent investigations have clearly demonstrated that DCs are targeted by CTA1-DD and that, even in B cell-deficient mice, these cells fulfil the most important function as APC for T cell priming after CTA1-DD adjuvant administration (Eliasson *et al.*, 2008). Moreover, CTA1-DD adjuvant greatly enhances both $CD4^+$ and $CD8^+$ T cell priming, leading to augmented cytokine production and cytotoxic T lymphocyte (CTL) activity, respectively (Simmons *et al.*, 1999). It generates a mixed Th1/Th2 response, but recent data also support a significant augmenting effect on Th17 cells (McNeal *et al.*, 2007). On humoral immune responses it is its dramatic effects on promoting germinal centre (GC) formation, memory B cell development and specific antibody production that stand out as exceptionally potent (Agren *et al.*, 1997, 2000; Simmons *et al.*, 1999; Bemark *et al.*, 2011) (Fig. 4.2). All these augmenting effects are dependent on the ADP-ribosylating A1 subunit and mutations that disrupt the enzymatic activity have largely caused the loss of adjuvant function. For example, a single point mutation in amino acid sequence position 7 (arginine (R) to lysine (K)) of CTA1, the CTA1R7K-DD mutant, or the E112K mutation, were both unable to augment immune responses (Agren *et al.*, 1999).

Because we found very striking effects on GC formation, we focused efforts to understand the underlying mechanism for this effect. Germinal centres are the sites in lymph nodes or spleen where antigen-activated B cells undergo expansion, selection and differentiation processes to finally develop into plasma cells and memory B cells (Victora and Nussenzweig, 2012). The GC is also a site for cell migration where antigen-activated $CD4^+$ T cells migrate into the GC to perform critical functions as helper T cells, also termed follicular helper cells (T_{FH}) (Crotty, 2011). Activated B cells also migrate to the GC and are attracted to interact with the follicular dendritic cells (FDC) through the release of chemokines and cytokines (Mattsson *et al.*, 2011). We are only beginning to understand the complexity of this system and its central role for regulating humoral immune responses. Therefore, it was especially interesting to find that CTA1-DD directly interacted with FDC after injection (Mattsson *et al.*, 2011). We found that CTA1-DD activated complement via the DD-domain, primarily the alternative pathway, and this way could bind complement fragments, which allowed CTA1-DD to bind complement receptors on the FDC (Mattsson *et al.*, 2011). The binding of CTA1-DD to FDCs via the complement receptor CD21 was found to be critical for the development of GC as mice deficient in complement or CD21 failed to respond to CTA1-DD with fewer GC and significantly reduced antibody responses and impaired B cell memory development as compared to wild-type mice (Mattsson *et al.*, 2011). Given that FDCs are essential elements

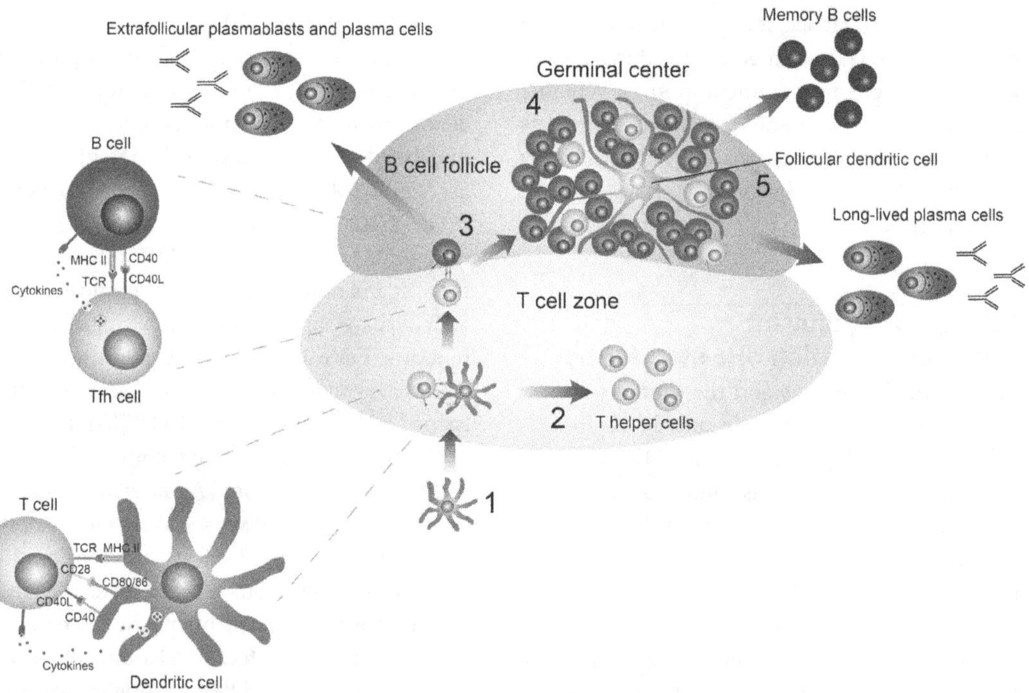

Figure 4.2 Adjuvants derived from bacterial enterotoxins can modulate the immune response at multiple levels. This schematic representation illustrates various stages of the immune response where ADP-ribosylating adjuvants can affect the outcome in a qualitative- or quantitative manner. **1.** Dendritic cells (DCs) present peptides on major histocompatibility complex II (MHC II) to cognate T cells that recognize the peptide-MHC II complex via their T cell receptor (TCR). Upon adjuvant stimulation DCs mature and up-regulate MHC II molecules as well as a number of co-stimulatory molecules, including CD40 and CD80/86. This process is critical for the priming of naive T cells to become effector- or follicular helper cells (Tfh). **2.** The differentiation of T cells into different subsets can be governed by the cytokines secreted by the DCs, which in turn can be influenced by the type of adjuvant used. **3.** T cells activate B cells, which can become extrafollicular plasma cells or enter germinal centre (GC) reactions and expand under the influence of Tfh cells. The cytokines secreted by Tfh cells, Th1 or Th2 type, regulate B cell differentiation and the activated B cells switch antibody class and accumulate somatic hypermutations (SHM) in their Ig-genes, that will increase the function and affinity maturation of the antibody response. In this way adjuvants can skew immune responses to Th1 or Th2 so that B cells make IgG1 or IgG2a antibodies, respectively, rather than IgM antibodies. **4.** Follicular dendritic cells (FDC) are central to GC reactions, they secrete survival signals and provide bound antigen to the proliferating B cells. This generates antibodies of high affinity by SHM, a process that can be enhanced by adjuvant-activation of FDCs. The latter is thought to be a mechanism by which the CTA1-DD adjuvant promotes strong GC reactions. **5.** Ultimately the GC produces memory B cells or long-lived plasma cells.

in GC formation, the site for somatic hypermutation and affinity maturation, this augmenting ability of CTA1-DD is critical for its augmenting effect on antibody responses. Accordingly, CTA1-DD is effective at promoting memory B cell responses as well as stimulating serum antibody levels that remain high for extended periods of time, even in comparison to other established adjuvants, such as alum or monophosphoryl lipid A (MPL) (Bemark *et al.*, 2011). These effects can be explained by a strong enhancing effect on the generation of long lived plasma cells in the bone marrow. These cells emanate from the GC reaction and they are known to be the most important source for serum antibody formation (Slifka and Ahmed, 1998). Another interesting observation after use of CTA1-DD is the enhanced level of neutralizing antibodies. It was found that the ability of anti-chlamydial antibodies to neutralize the infectivity of elementary bodies was significantly augmented compared to antibodies generated with other adjuvants despite that the total specific IgG titres were no different (Cunningham *et al.*, 2009).

The adjuvant has been used in combination with a wide variety of antigens and CTA1-DD adjuvant has been shown to induce protective responses in several different infectious disease models, including influenza, chlamydia, rotavirus, and *Helicobacter pylori* (Belyakov *et al.*, 1998; Akhiani *et al.*, 2006; McNeal *et al.*, 2007; Eliasson *et al.*, 2008; Cunningham *et al.*, 2009). An intriguing aspect of the CTA1-DD adjuvant is the possibility to incorporate peptide epitopes within the fusion protein itself, in addition to the conventional way of simply admixing antigens with the adjuvant. This concept has been proven to be effective in a universal influenza vaccine candidate, exploring the potential of a peptide from influenza virus matrix protein 2 (M2e). The candidate vaccine then possessed three qualities in one fusion protein, CTA1–3M2e-DD, namely cell targeting (DD-element), immunoenhancement (CTA1-element) and antigen delivery (M2e-element) (Eliasson *et al.*, 2008, 2011). It was found to be very effective and complete resistance to a challenge virus infection persisted for more than 1.5 years in the i.n. immunized mice.

Concluding remarks

The overall prospects for a shift of paradigm in mucosal vaccine development now look brighter as we are beginning to unravel the mechanisms responsible for priming of mucosal immune responses. Based on newly acquired knowledge about adjuvant activation of innate immunity at mucosal membranes and a better understanding of the build-up of IgA B-cell responses and long-term memory development we should be expecting many new and effective mucosal vaccines in the future. The bacterial ADP-ribosylating toxins and the less toxic derivatives thereof, play a central role in this development. Through rational targeting strategies toxicity of these molecules can be avoided while retaining strong adjuvant activity. This will allow regulatory authorities to reconsider the use of this family of adjuvants for human vaccines. Identifying correlates of adjuvant effectiveness using different or combinations of mucosal adjuvants will boost the design and formulation of future mucosal vaccines. Although several hurdles still need to be overcome, mucosal vaccinology has already started to explore the new technologies in systems biology and will soon identify biological markers for vaccine adjuvant efficacy. An overriding question is whether there are several ADP-ribosylating targets, such as Gsα and Giα, that lead to an adjuvant effect or whether one common pathway is operational and whether or not this pathway is dependent on cAMP. More basic research will be needed to unravel these elements and to clarify how to effectively explore ADP-ribosylating toxins and their derivatives in future effective mucosal vaccines. We are confident that a next generation of mucosal vaccines soon can be developed and we think that knowledge emanating from the impact of ADP-ribosylating toxins on innate and adaptive immunity will greatly contribute to this development.

References

Agren, L., Sverremark, E., Ekman, L., Schon, K., Lowenadler, B., Fernandez, C., and Lycke, N. (2000). The ADP-ribosylating CTA1-DD adjuvant enhances T cell-dependent and independent responses by direct action on B cells involving anti-apoptotic Bcl-2- and germinal center-promoting effects. J. Immunol. *164*, 6276–6286.

Agren, L.C., Ekman, L., Lowenadler, B., and Lycke, N.Y. (1997). Genetically engineered nontoxic vaccine adjuvant that combines B cell targeting with immunomodulation by cholera toxin A1 subunit. J. Immunol. *158*, 3936–3946.

Agren, L.C., Ekman, L., Lowenadler, B., Nedrud, J.G., and Lycke, N.Y. (1999). Adjuvanticity of the cholera toxin A1-based gene fusion protein, CTA1-DD, is critically dependent on the ADP-ribosyltransferase and Ig-binding activity. J. Immunol. *162*, 2432–2440.

Akhiani, A.A., Pappo, J., Kabok, Z., Schon, K., Gao, W., Franzen, L.E., and Lycke, N. (2002). Protection against *Helicobacter pylori* infection following immunization is IL-12-dependent and mediated by Th1 cells. J. Immunol. *169*, 6977–6984.

Akhiani, A.A., Stensson, A., Schon, K., and Lycke, N. (2006). The nontoxic CTA1-DD adjuvant enhances protective immunity against *Helicobacter pylori* infection following mucosal immunization. Scand. J. Immunol. *63*, 97–105.

Anjuere, F., Luci, C., Lebens, M., Rousseau, D., Hervouet, C., Milon, G., Holmgren, J., Ardavin, C., and Czerkinsky, C. (2004). *In vivo* adjuvant-induced mobilization and maturation of gut dendritic cells after oral administration of cholera toxin. J. Immunol. *173*, 5103–5111.

Anosova, N.G., Chabot, S., Shreedhar, V., Borawski, J.A., Dickinson, B.L., and Neutra, M.R. (2008). Cholera toxin, *E. coli* heat-labile toxin, and non-toxic derivatives induce dendritic cell migration into the follicle-associated epithelium of Peyer's patches. Mucosal Immunol. *1*, 59–67.

Armstrong, M.E., Lavelle, E.C., Loscher, C.E., Lynch, M.A., and Mills, K.H. (2005). Proinflammatory responses in the murine brain after intranasal delivery of cholera toxin: implications for the use of AB toxins as adjuvants in intranasal vaccines. J. Infect. Dis. *192*, 1628–1633.

Baccala, R., Gonzalez-Quintial, R., Lawson, B.R., Stern, M.E., Kono, D.H., Beutler, B., and Theofilopoulos, A.N. (2009). Sensors of the innate immune system: their mode of action. Nat. Rev. Rheumatol. *5*, 448–456.

Belyakov, I.M., Derby, M.A., Ahlers, J.D., Kelsall, B.L., Earl, P., Moss, B., Strober, W., and Berzofsky, J.A. (1998). Mucosal immunization with HIV-1 peptide vaccine induces mucosal and systemic cytotoxic T lymphocytes and protective immunity in mice against intrarectal recombinant HIV-vaccinia challenge. Proc. Natl. Acad. Sci. U.S.A. *95*, 1709–1714.

Bemark, M., Bergqvist, P., Stensson, A., Holmberg, A., Mattsson, J., and Lycke, N.Y. (2011). A unique role of the cholera toxin A1-DD adjuvant for long-term plasma and memory B cell development. J. Immunol. *186*, 1399–1410.

Berenson, C.S., Nawar, H.F., Yohe, H.C., Castle, S.A., Ashline, D.J., Reinhold, V.N., Hajishengallis, G., and Connell, T.D. (2010). Mammalian cell ganglioside-binding specificities of *E. coli* enterotoxins LT-IIb and variant LT-IIb(T13I). Glycobiology *20*, 41–54.

Berenson, C.S., Nawar, H.F., Kruzel, R.L., Mandell, L.M., and Connell, T.D. (2012). Ganglioside-binding specificities of *E. coli* enterotoxin LT-IIc: Importance of long chain fatty acyl ceramide. Glycobiology *23*, 23–31.

Bernardi, K.M., Forster, M.L., Lencer, W.I., and Tsai, B. (2008). Derlin-1 facilitates the retro-translocation of cholera toxin. Mol. Biol. Cell *19*, 877–884.

Bertolotti-Ciarlet, A., Ciarlet, M., Crawford, S.E., Conner, M.E., and Estes, M.K. (2003). Immunogenicity and protective efficacy of rotavirus 2/6-virus-like particles produced by a dual baculovirus expression vector and administered intramuscularly, intranasally, or orally to mice. Vaccine *21*, 3885–3900.

Blazar, B.R., Murphy, W.J., and Abedi, M. (2012). Advances in graft-versus-host disease biology and therapy. Nat. Rev. Immunol. *12*, 443–458.

Bokoch, G.M., Katada, T., Northup, J.K., Hewlett, E.L., and Gilman, A.G. (1983). Identification of the predominant substrate for ADP-ribosylation by islet activating protein. J. Biol. Chem. *258*, 2072–2075.

Bonenfant, C., Dimier-Poisson, I., Velge-Roussel, F., Buzoni-Gatel, D., Del Giudice, G., Rappuoli, R., and Bout, D. (2001). Intranasal immunization with SAG1 and nontoxic mutant heat-labile enterotoxins protects mice against *Toxoplasma gondii*. Infect. Immun. *69*, 1605–1612.

Booth, B.A., Boesman-Finkelstein, M., and Finkelstein, R.A. (1984). *Vibrio cholerae* hemagglutinin/protease nicks cholera enterotoxin. Infect. Immun. *45*, 558–560.

Brereton, C.F., Sutton, C.E., Ross, P.J., Iwakura, Y., Pizza, M., Rappuoli, R., Lavelle, E.C., and Mills, K.H. (2011). *Escherichia coli* heat-labile enterotoxin promotes protective Th17 responses against infection by driving innate IL-1 and IL-23 production. J. Immunol. *186*, 5896–5906.

Bromander, A., Holmgren, J., and Lycke, N. (1991). Cholera toxin stimulates IL-1 production and enhances antigen presentation by macrophages *in vitro*. J. Immunol. *146*, 2908–2914.

Brynjolfsson, S.F., Bjarnarson, S.P., Mori, E., Del Giudice, G., and Jonsdottir, I. (2008). Neonatal immune response and serum bactericidal activity induced by a meningococcal conjugate vaccine is enhanced by LT-K63 and CpG2006. Vaccine *26*, 4557–4562.

Burnette, W.N., Cieplak, W., Mar, V.L., Kaljot, K.T., Sato, H., and Keith, J.M. (1988). Pertussis toxin S1 mutant with reduced enzyme activity and a conserved protective epitope. Science *242*, 72–74.

Campbell, J.J., Hedrick, J., Zlotnik, A., Siani, M.A., Thompson, D.A., and Butcher, E.C. (1998). Chemokines and the arrest of lymphocytes rolling under flow conditions. Science *279*, 381–384.

Cassan, C., Piaggio, E., Zappulla, J.P., Mars, L.T., Couturier, N., Bucciarelli, F., Desbois, S., Bauer, J., Gonzalez-Dunia, D., and Liblau, R.S. (2006). Pertussis toxin reduces the number of splenic Foxp3+ regulatory T cells. J. Immunol. *177*, 1552–1560.

Chen, X., Winkler-Pickett, R.T., Carbonetti, N.H., Ortaldo, J.R., Oppenheim, J.J., and Howard, O.M. (2006). Pertussis toxin as an adjuvant suppresses the number and function of CD4+CD25+ T regulatory cells. Eur. J. Immunol. *36*, 671–680.

Cheng, E., Cardenas-Freytag, L., and Clements, J.D. (1999). The role of cAMP in mucosal adjuvanticity of *Escherichia coli* heat-labile enterotoxin (LT). Vaccine *18*, 38–49.

Clements, J.D., and Finkelstein, R.A. (1979). Isolation and characterization of homogeneous heat-labile enterotoxins with high specific activity from *Escherichia coli* cultures. Infect. Immun. *24*, 760–769.

Cong, Y., Weaver, C.T., and Elson, C.O. (1997). The mucosal adjuvanticity of cholera toxin involves enhancement of costimulatory activity by selective up-regulation of B7.2 expression. J. Immunol. *159*, 5301–5308.

Connell, T.D. (2007). Cholera toxin, LT-I, LT-IIa and LT-IIb: the critical role of ganglioside binding in immunomodulation by type I and type II heat-labile enterotoxins. Expert Rev. Vaccines *6*, 821–834.

Cox, J.C., and Coulter, A.R. (1997). Adjuvants – a classification and review of their modes of action. Vaccine *15*, 248–256.

Crotty, S. (2011). Follicular helper CD4 T cells (TFH). Annu. Rev. Immunol. *29*, 621–663.

Cunningham, K.A., Carey, A.J., Lycke, N., Timms, P., and Beagley, K.W. (2009). CTA1-DD is an effective adjuvant for targeting anti-chlamydial immunity to the murine genital mucosa. J. Reprod. Immunol. *81*, 34–38.

Cyster, J.G., and Goodnow, C.C. (1995). Pertussis toxin inhibits migration of B and T lymphocytes into splenic white pulp cords. J. Exp. Med. *182*, 581–586.

Datta, S.K., Sabet, M., Nguyen, K.P., Valdez, P.A., Gonzalez-Navajas, J.M., Islam, S., Mihajlov, I., Fierer, J., Insel, P.A., Webster, N.J., *et al.* (2010). Mucosal adjuvant activity of cholera toxin requires Th17 cells and protects against inhalation anthrax. Proc. Natl. Acad. Sci. U.S.A. *107*, 10638–10643.

de Haan, L., Verweij, W.R., Feil, I.K., Holtrop, M., Hol, W.G., Agsteribbe, E., and Wilschut, J. (1998). Role of GM1 binding in the mucosal immunogenicity and adjuvant activity of the *Escherichia coli* heat-labile enterotoxin and its B subunit. Immunology *94*, 424–430.

Di Tommaso, A., Saletti, G., Pizza, M., Rappuoli, R., Dougan, G., Abrignani, S., Douce, G., and De Magistris, M.T. (1996). Induction of antigen-specific antibodies in vaginal secretions by using a nontoxic mutant of heat-labile enterotoxin as a mucosal adjuvant. Infect. Immun. *64*, 974–979.

Dickinson, B.L., and Clements, J.D. (1995). Dissociation of *Escherichia coli* heat-labile enterotoxin adjuvanticity from ADP-ribosyltransferase activity. Infect. Immun. *63*, 1617–1623.

Douce, G., Fontana, M., Pizza, M., Rappuoli, R., and Dougan, G. (1997). Intranasal immunogenicity and adjuvanticity of site-directed mutant derivatives of cholera toxin. Infect. Immun. *65*, 2821–2828.

Douce, G., Giannelli, V., Pizza, M., Lewis, D., Everest, P., Rappuoli, R., and Dougan, G. (1999). Genetically detoxified mutants of heat-labile toxin from *Escherichia coli* are able to act as oral adjuvants. Infect. Immun. *67*, 4400–4406.

Douce, G., Giuliani, M.M., Giannelli, V., Pizza, M.G., Rappuoli, R., and Dougan, G. (1998). Mucosal immunogenicity of genetically detoxified derivatives of heat labile toxin from *Escherichia coli*. Vaccine *16*, 1065–1073.

Ebensen, T., and Guzman, C.A. (2008). Immune modulators with defined molecular targets: cornerstone to optimize rational vaccine design. Hum. Vaccin. *4*, 13–22.

el Baya, A., Linnemann, R., von Olleschik-Elbheim, L., Robenek, H., and Schmidt, M.A. (1997). Endocytosis and retrograde transport of pertussis toxin to the Golgi complex as a prerequisite for cellular intoxication. Eur. J. Cell Biol. 73, 40–48.

Eliasson, D.G., El Bakkouri, K., Schon, K., Ramne, A., Festjens, E., Lowenadler, B., Fiers, W., Saelens, X., and Lycke, N. (2008). CTA1-M2e-DD: a novel mucosal adjuvant targeted influenza vaccine. Vaccine *26*, 1243–1252.

Eliasson, D.G., Helgeby, A., Schon, K., Nygren, C., El-Bakkouri, K., Fiers, W., Saelens, X., Lovgren, K.B., Nystrom, I., and Lycke, N.Y. (2011). A novel non-toxic combined CTA1-DD and ISCOMS adjuvant vector for effective mucosal immunization against influenza virus. Vaccine *29*, 3951–3961.

Elson, C.O., and Ealding, W. (1984). Generalized systemic and mucosal immunity in mice after mucosal stimulation with cholera toxin. J. Immunol. *132*, 2736–2741.

Eriksson, A.M., Schon, K.M., and Lycke, N.Y. (2004). The cholera toxin-derived CTA1-DD vaccine adjuvant administered intranasally does not cause inflammation or accumulate in the nervous tissues. J. Immunol. *173*, 3310–3319.

Eriksson, K., Fredriksson, M., Nordstrom, I., and Holmgren, J. (2003). Cholera toxin and its B subunit promote dendritic cell vaccination with different influences on Th1 and Th2 development. Infect. Immun. *71*, 1740–1747.

Fahlen-Yrlid, L., Gustafsson, T., Westlund, J., Holmberg, A., Strombeck, A., Blomquist, M., MacPherson, G.G., Holmgren, J., and Yrlid, U. (2009). CD11c(high) dendritic cells are essential for activation of CD4+ T cells and generation of specific antibodies following mucosal immunization. J. Immunol. *183*, 5032–5041.

Fan, E., Merritt, E.A., Zhang, Z., Pickens, J.C., Roach, C., Ahn, M., and Hol, W.G. (2001). Exploration of the GM1 receptor-binding site of heat-labile enterotoxin and cholera toxin by phenyl-ring-containing galactose derivatives. Acta Crystallogr. D Biol. Crystallogr. *57*, 201–212.

Fan, E., O'Neal, C.J., Mitchell, D.D., Robien, M.A., Zhang, Z., Pickens, J.C., Tan, X.J., Korotkov, K., Roach, C., Krumm, B., *et al.* (2004). Structural biology and structure-based inhibitor design of cholera toxin and heat-labile enterotoxin. Int. J. Med. Microbiol. *294*, 217–223.

Fang, Y., Larsson, L., Mattsson, J., Lycke, N., and Xiang, Z. (2010). Mast cells contribute to the mucosal adjuvant effect of CTA1-DD after IgG-complex formation. J. Immunol. *185*, 2935–2941.

Fontana, M.R., Manetti, R., Giannelli, V., Magagnoli, C., Marchini, A., Olivieri, R., Domenighini, M., Rappuoli, R., and Pizza, M. (1995). Construction of nontoxic derivatives of cholera toxin and characterization of the immunological response against the A subunit. Infect. Immun. *63*, 2356–2360.

Fontana, M.R., Monaci, E., Yanqing, L., Guoming, Q., Duan, G., Rappuoli, R., and Pizza, M. (2000). IEM101, a naturally attenuated *Vibrio cholerae* strain as carrier for genetically detoxified derivatives of cholera toxin. Vaccine *19*, 75–85.

Frech, S.A., Dupont, H.L., Bourgeois, A.L., McKenzie, R., Belkind-Gerson, J., Figueroa, J.F., Okhuysen, P.C., Guerrero, N.H., Martinez-Sandoval, F.G., Melendez-Romero, J.H., *et al.* (2008). Use of a patch containing heat-labile toxin from *Escherichia coli* against travellers' diarrhoea: a phase II, randomised, double-blind, placebo-controlled field trial. Lancet *371*, 2019–2025.

Frerichs, D.M., Ellingsworth, L.R., Frech, S.A., Flyer, D.C., Villar, C.P., Yu, J., and Glenn, G.M. (2008). Controlled, single-step, stratum corneum disruption as a pretreatment for immunization via a patch. Vaccine *26*, 2782–2787.

Freytag, L.C., Clements, J.D., Grdic Eliasson, D., and Lycke, N. (2010). Use of genetically or chemically detoxified mutants of Cholera and *Escherichia coli* heat-labile enterotoxins as mucosal adjuvants. In New Generation Vaccines, Levine, M.M., ed. (informa, New York), pp. 273–283.

Fromantin, C., Jamot, B., Cohen, J., Piroth, L., Pothier, P., and Kohli, E. (2001). Rotavirus 2/6 virus-like particles administered intranasally in mice, with or without the mucosal adjuvants cholera toxin and *Escherichia coli* heat-labile toxin, induce a Th1/Th2-like immune response. J. Virol. *75*, 11010–11016.

Fujihashi, K., Koga, T., van Ginkel, F.W., Hagiwara, Y., and McGhee, J.R. (2002). A dilemma for mucosal vaccination: efficacy versus toxicity using enterotoxin-based adjuvants. Vaccine *20*, 2431–2438.

Fukuta, S., Magnani, J.L., Twiddy, E.M., Holmes, R.K., and Ginsburg, V. (1988). Comparison of the carbohydrate-binding specificities of cholera toxin and *Escherichia coli* heat-labile enterotoxins LTh-I, LT-IIa, and LT-IIb. Infect. Immun. *56*, 1748–1753.

Gagliardi, M.C., Sallusto, F., Marinaro, M., Langenkamp, A., Lanzavecchia, A., and De Magistris, M.T. (2000). Cholera toxin induces maturation of human dendritic cells and licences them for Th2 priming. Eur. J. Immunol. *30*, 2394–2403.

Gagliardi, M.C., Sallusto, F., Marinaro, M., Vendetti, S., Riccomi, A., and De Magistris, M.T. (2002). Effects of the adjuvant cholera toxin on dendritic cells: stimulatory and inhibitory signals that result in the amplification of immune responses. Int. J. Med. Microbiol. *291*, 571–575.

Galli, G., Hancock, K., Hoschler, K., DeVos, J., Praus, M., Bardelli, M., Malzone, C., Castellino, F., Gentile, C., McNally, T., *et al.* (2009). Fast rise of broadly cross-reactive antibodies after boosting long-lived human memory B cells primed by an MF59 adjuvanted prepandemic vaccine. Proc. Natl. Acad. Sci. U.S.A. *106*, 7962–7967.

Giannelli, V., Fontana, M.R., Giuliani, M.M., Guangcai, D., Rappuoli, R., and Pizza, M. (1997). Protease susceptibility and toxicity of heat-labile enterotoxins with a mutation in the active site or in the protease-sensitive loop. Infect. Immun. *65*, 331–334.

Giuliani, M.M., Del Giudice, G., Giannelli, V., Dougan, G., Douce, G., Rappuoli, R., and Pizza, M. (1998). Mucosal adjuvanticity and immunogenicity of LTR72, a novel mutant of *Escherichia coli* heat-labile enterotoxin with partial knockout of ADP-ribosyltransferase activity. J. Exp. Med. *187*, 1123–1132.

Glenn, G.M., Thomas, D.N., Poffenberger, K.L., Flyer, D.C., Ellingsworth, L.R., Andersen, B.H., and Frech, S.A. (2009). Safety and immunogenicity of an influenza vaccine A/H5N1 (A/Vietnam/1194/2004) when coadministered with a heat-labile enterotoxin (LT) adjuvant patch. Vaccine 27 *Suppl 6*, G60–66.

Glueck, R. (2001). Pre-clinical and clinical investigation of the safety of a novel adjuvant for intranasal immunization. Vaccine *20 Suppl 1*, S42–44.

Grant, C.C., Messer, R.J., and Cieplak, Jr, W.. (1994). Role of trypsin-like cleavage at arginine 192 in the enzymatic and cytotonic activities of *Escherichia coli* heat-labile enterotoxin. Infect. Immun. *62*, 4270–4278.

Grdic, D., Ekman, L., Schon, K., Lindgren, K., Mattsson, J., Magnusson, K.E., Ricciardi-Castagnoli, P., and Lycke, N. (2005). Splenic marginal zone dendritic cells mediate the cholera toxin adjuvant effect: dependence on the ADP-ribosyltransferase activity of the holotoxin. J. Immunol. *175*, 5192–5202.

Guerena-Burgueno, F., Hall, E.R., Taylor, D.N., Cassels, F.J., Scott, D.A., Wolf, M.K., Roberts, Z.J., Nesterova, G.V., Alving, C.R., and Glenn, G.M. (2002). Safety and immunogenicity of a prototype enterotoxigenic *Escherichia coli* vaccine administered transcutaneously. Infect. Immun. *70*, 1874–1880.

Guidry, J.J., Cardenas, L., Cheng, E., and Clements, J.D. (1997). Role of receptor binding in toxicity, immunogenicity, and adjuvanticity of *Escherichia coli* heat-labile enterotoxin. Infect. Immun. *65*, 4943–4950.

Gustafsson, T., Hua, Y.-J., Dahlgren, M., Livingston, M., Johansson-Lindbom, B., and Yrlid, U. (2013). Direct interaction between cholera toxin and dendritic cells is required for oral adjuvant activity. Eur. J. Immunol. In press.

Hausman, S.Z., and Burns, D.L. (1993). Binding of pertussis toxin to lipid vesicles containing glycolipids. Infect. Immun. *61*, 335–337.

Hervouet, C., Luci, C., Rol, N., Rousseau, D., Kissenpfennig, A., Malissen, B., Czerkinsky, C., and Anjuere, F. (2010). Langerhans cells prime IL-17-producing T cells and dampen genital cytotoxic responses following mucosal immunization. J. Immunol. *184*, 4842–4851.

Hirahara, K., Ghoreschi, K., Laurence, A., Yang, X.P., Kanno, Y., and O'Shea, J.J. (2010). Signal transduction pathways and transcriptional regulation in Th17 cell differentiation. Cytokine Growth Factor Rev. *21*, 425–434.

Holmgren, J., and Czerkinsky, C. (2005). Mucosal immunity and vaccines. Nat. Med. *11*, S45–53.

Holmgren, J., Lonnroth, I., and Svennerholm, L. (1973). Tissue receptor for cholera exotoxin: postulated structure from studies with GM1 ganglioside and related glycolipids. Infect. Immun. *8*, 208–214.

Hornquist, E., and Lycke, N. (1993). Cholera toxin adjuvant greatly promotes antigen priming of T cells. Eur. J. Immunol. *23*, 2136–2143.

Hou, W., Wu, Y., Sun, S., Shi, M., Sun, Y., Yang, C., Pei, G., Gu, Y., Zhong, C., and Sun, B. (2003). Pertussis toxin enhances Th1 responses by stimulation of dendritic cells. J. Immunol. *170*, 1728–1736.

Hruz, P., Dann, S.M., and Eckmann, L. (2010). STAT3 and its activators in intestinal defense and mucosal homeostasis. Curr. Opin. Gastroenterol. *26*, 109–115.

Iyengar, R. (1993). Molecular and functional diversity of mammalian Gs-stimulated adenylyl cyclases. FASEB J. *7*, 768–775.

Jabbal-Gill, I., Fisher, A.N., Rappuoli, R., Davis, S.S., and Illum, L. (1998). Stimulation of mucosal and systemic antibody responses against *Bordetella pertussis* filamentous haemagglutinin and recombinant pertussis toxin after nasal administration with chitosan in mice. Vaccine *16*, 2039–2046.

Jobling, M.G., and Holmes, R.K. (1991). Analysis of structure and function of the B subunit of cholera toxin by the use of site-directed mutagenesis. Mol. Microbiol. *5*, 1755–1767.

Jones, H.P., Hodge, L.M., Fujihashi, K., Kiyono, H., McGhee, J.R., and Simecka, J.W. (2001). The pulmonary environment promotes Th2 cell responses after nasal-pulmonary immunization with antigen alone, but Th1 responses are induced during instances of intense immune stimulation. J. Immunol. *167*, 4518–4526.

Kahn, R.A., and Gilman, A.G. (1984a). ADP-ribosylation of Gs promotes the dissociation of its alpha and beta subunits. J. Biol. Chem. *259*, 6235–6240.

Kahn, R.A., and Gilman, A.G. (1984b). Purification of a protein cofactor required for ADP-ribosylation of the stimulatory regulatory component of adenylate cyclase by cholera toxin. J. Biol. Chem. *259*, 6228–6234.

Karlsson, K.A., Teneberg, S., Angstrom, J., Kjellberg, A., Hirst, T.R., Berstrom, J., and Miller-Podraza, H. (1996). Unexpected carbohydrate cross-binding by *Escherichia coli* heat-labile enterotoxin. Recognition of human and rabbit target cell glycoconjugates in comparison with cholera toxin. Bioorg. Med. Chem. *4*, 1919–1928.

Kassis, S., Hagmann, J., Fishman, P.H., Chang, P.P., and Moss, J. (1982). Mechanism of action of cholera toxin on intact cells. Generation of A1 peptide and activation of adenylate cyclase. J. Biol. Chem. *257*, 12148–12152.

Katada, T., Tamura, M., and Ui, M. (1983). The A protomer of islet-activating protein, pertussis toxin, as an active peptide catalyzing ADP-ribosylation of a membrane protein. Arch. Biochem. Biophys. *224*, 290–298.

Kawamura, Y.I., Kawashima, R., Shirai, Y., Kato, R., Hamabata, T., Yamamoto, M., Furukawa, K., Fujihashi, K., McGhee, J.R., Hayashi, H., *et al.* (2003). Cholera toxin activates dendritic cells through dependence on GM1-ganglioside which is mediated by NF-kappaB translocation. Eur. J. Immunol. *33*, 3205–3212.

Kotloff, K.L., Sztein, M.B., Wasserman, S.S., Losonsky, G.A., DiLorenzo, S.C., and Walker, R.I. (2001). Safety and immunogenicity of oral inactivated whole-cell *Helicobacter pylori* vaccine with adjuvant among volunteers with or without subclinical infection. Infect. Immun. *69*, 3581–3590.

la Sala, A., He, J., Laricchia-Robbio, L., Gorini, S., Iwasaki, A., Braun, M., Yap, G.S., Sher, A., Ozato, K., and Kelsall, B. (2009). Cholera toxin inhibits IL-12 production and CD8alpha+ dendritic cell differentiation by cAMP-mediated inhibition of IRF8 function. J. Exp. Med. *206*, 1227–1235.

Lavelle, E.C., McNeela, E., Armstrong, M.E., Leavy, O., Higgins, S.C., and Mills, K.H. (2003). Cholera toxin promotes the induction of regulatory T cells specific for bystander antigens by modulating dendritic cell activation. J. Immunol. *171*, 2384–2392.

Lavelle, E.C., Jarnicki, A., McNeela, E., Armstrong, M.E., Higgins, S.C., Leavy, O., and Mills, K.H. (2004). Effects of cholera toxin on innate and adaptive immunity and its application as an immunomodulatory agent. J. Leukoc. Biol. *75*, 756–763.

Lee, J.B., Jang, J.E., Song, M.K., and Chang, J. (2009). Intranasal delivery of cholera toxin induces th17-dominated T-cell response to bystander antigens. PLoS One *4*, e5190.

Levine, M.M., Black, R.E., Clements, M.L., Lanata, C., Sears, S., Honda, T., Young, C.R., and Finkelstein, R.A. (1984). Evaluation in humans of attenuated *Vibrio cholerae* El Tor Ogawa strain Texas Star-SR as a live oral vaccine. Infect. Immun. *43*, 515–522.

Lewis, D.J., Huo, Z., Barnett, S., Kromann, I., Giemza, R., Galiza, E., Woodrow, M., Thierry-Carstensen, B., Andersen, P., Novicki, D., *et al.* (2009). Transient facial nerve paralysis (Bell's palsy) following intranasal delivery of a genetically detoxified mutant of *Escherichia coli* heat labile toxin. PLoS One *4*, e6999.

Liang, S., and Hajishengallis, G. (2010). Heat-labile enterotoxins as adjuvants or anti-inflammatory agents. Immunol. Invest. *39*, 449–467.

Ljungberg, U.K., Jansson, B., Niss, U., Nilsson, R., Sandberg, B.E., and Nilsson, B. (1993). The interaction between different domains of staphylococcal protein A and human polyclonal IgG, IgA, IgM and F(ab')2: separation of affinity from specificity. Mol. Immunol. *30*, 1279–1285.

Llewellyn-Smith, I.J., Minson, J.B., Wright, A.P., and Hodgson, A.J. (1990). Cholera toxin B-gold, a retrograde tracer that can be used in light and electron microscopic immunocytochemical studies. J. Comp. Neurol. *294*, 179–191.

Lu, X., Clements, J.D., and Katz, J.M. (2002). Mutant *Escherichia coli* heat-labile enterotoxin [LT(R192G)] enhances protective humoral and cellular immune responses to orally administered inactivated influenza vaccine. Vaccine *20*, 1019–1029.

Lu, Y.J., Yadav, P., Clements, J.D., Forte, S., Srivastava, A., Thompson, C.M., Seid, R., Look, J., Alderson, M., Tate, A., *et al.* (2010). Options for inactivation, adjuvant, and route of topical administration of a killed, unencapsulated pneumococcal whole-cell vaccine. Clin. Vaccine Immunol. *17*, 1005–1012.

Luci, C., Hervouet, C., Rousseau, D., Holmgren, J., Czerkinsky, C., and Anjuere, F. (2006). Dendritic cell-mediated induction of mucosal cytotoxic responses following intravaginal immunization with the nontoxic B subunit of cholera toxin. J. Immunol. *176*, 2749–2757.

Lycke, N. (2007). Mechanisms of adjuvant action. In Vaccine Adjuvants and Delivery Systems, Singh, M., ed. (Hoboken, New Jersey, USA: John Wiley & Sons, Inc.), pp. 53–79.

Majoul, I.V., Bastiaens, P.I., and Soling, H.D. (1996). Transport of an external Lys-Asp-Glu-Leu (KDEL) protein from the plasma membrane to the endoplasmic reticulum: studies with cholera toxin in Vero cells. J. Cell Biol. *133*, 777–789.

Martin, M., Metzger, D.J., Michalek, S.M., Connell, T.D., and Russell, M.W. (2000). Comparative analysis of the mucosal adjuvanticity of the type II heat-labile enterotoxins LT-IIa and LT-IIb. Infect. Immun. *68*, 281–287.

Martinez-Becerra, F.J., Kissmann, J.M., Diaz-McNair, J., Choudhari, S.P., Quick, A.M., Mellado-Sanchez, G., Clements, J.D., Pasetti, M.F., and Picking, W.L. (2012). Broadly protective Shigella vaccine based on type III secretion apparatus proteins. Infect. Immun. *80*, 1222–1231.

Mathias-Santos, C., Rodrigues, J.F., Sbrogio-Almeida, M.E., Connell, T.D., and Ferreira, L.C. (2011). Distinctive immunomodulatory and inflammatory properties of the *Escherichia coli* type II heat-labile enterotoxin LT-IIa and its B pentamer following intradermal administration. Clin. Vaccine Immunol. *18*, 1243–1251.

Mattsson, J., Yrlid, U., Stensson, A., Schon, K., Karlsson, M.C., Ravetch, J.V., and Lycke, N.Y. (2011). Complement activation and complement receptors on follicular dendritic cells are critical for the function of a targeted adjuvant. J. Immunol. *187*, 3641–3652.

McKenzie, R., Bourgeois, A.L., Frech, S.A., Flyer, D.C., Bloom, A., Kazempour, K., and Glenn, G.M. (2007). Transcutaneous immunization with the heat-labile toxin (LT) of enterotoxigenic *Escherichia coli* (ETEC): protective efficacy in a double-blind, placebo-controlled challenge study. Vaccine *25*, 3684–3691.

McNeal, M.M., Rae, M.N., Bean, J.A., and Ward, R.L. (1999). Antibody-dependent and -independent protection following intranasal immunization of mice with rotavirus particles. J. Virol. *73*, 7565–7573.

McNeal, M.M., VanCott, J.L., Choi, A.H., Basu, M., Flint, J.A., Stone, S.C., Clements, J.D., and Ward, R.L. (2002). CD4 T cells are the only lymphocytes needed to protect mice against rotavirus shedding after intranasal immunization with a chimeric VP6 protein and the adjuvant LT(R192G). J. Virol. *76*, 560–568.

McNeal, M.M., Basu, M., Bean, J.A., Clements, J.D., Lycke, N.Y., Ramne, A., Lowenadler, B., Choi, A.H., and Ward, R.L. (2007). Intrarectal immunization of mice with VP6 and either LT(R192G) or CTA1-DD as adjuvant protects against fecal rotavirus shedding after EDIM challenge. Vaccine *25*, 6224–6231.

Meza-Sanchez, D., Perez-Montesinos, G., Sanchez-Garcia, J., Moreno, J., and Bonifaz, L.C. (2011). Intradermal immunization in the ear with cholera toxin and its non-toxic beta subunit promotes efficient Th1 and Th17 differentiation dependent on migrating DCs. Eur. J. Immunol. *41*, 2894–2904.

Milligan, G., and Kostenis, E. (2006). Heterotrimeric G-proteins: a short history. Br. J. Pharmacol. *147* Suppl 1, S46–55.

Morse, S.I., and Riester, S.K. (1967). Studies on the leukocytosis and lymphocytosis induced by *Bordetella pertussis*. II. The effect of pertussis vaccine on the thoracic duct lymph and lymphocytes of mice. J. Exp. Med. *125*, 619–628.

Munoz, J.J., Arai, H., Bergman, R.K., and Sadowski, P.L. (1981). Biological activities of crystalline pertussigen from *Bordetella pertussis*. Infect. Immun. 33, 820–826.

Murphey, C., Chang, S., Zhang, X., Arulanandam, B., and Forsthuber, T.G. (2011). Induction of polyclonal CD8+ T cell activation and effector function by Pertussis toxin. Cell. Immunol. *267*, 50–55.

Mutsch, M., Zhou, W., Rhodes, P., Bopp, M., Chen, R.T., Linder, T., Spyr, C., and Steffen, R. (2004). Use of the inactivated intranasal influenza vaccine and the risk of Bell's palsy in Switzerland. N. Engl. J. Med. *350*, 896–903.

Nawar, H.F., Arce, S., Russell, M.W., and Connell, T.D. (2005). Mucosal adjuvant properties of mutant LT-IIa and LT-IIb enterotoxins that exhibit altered ganglioside-binding activities. Infect. Immun. *73*, 1330–1342.

Nawar, H.F., Arce, S., Russell, M.W., and Connell, T.D. (2007). Mutants of type II heat-labile enterotoxin LT-IIa with altered ganglioside-binding activities and diminished toxicity are potent mucosal adjuvants. Infect. Immun. *75*, 621–633.

Nawar, H.F., King-Lyons, N.D., Hu, J.C., Pasek, R.C., and Connell, T.D. (2010). LT-IIc, a new member of the type II heat-labile enterotoxin family encoded by an *Escherichia coli* strain obtained from a nonmammalian host. Infect. Immun. *78*, 4705–4713.

Nawar, H.F., Greene, C.J., Lee, C.H., Mandell, L.M., Hajishengallis, G., and Connell, T.D. (2011). LT-IIc, a new member of the type II heat-labile enterotoxin family, exhibits potent immunomodulatory properties that are different from those induced by LT-IIa or LT-IIb. Vaccine *29*, 721–727.

Northrup, R.S., and Fauci, A.S. (1972). Adjuvant effect of cholera enterotoxin on the immune response of the mouse to sheep red blood cells. J. Infect. Dis. *125*, 672–673.

Norton, E.B., Lawson, L.B., Freytag, L.C., and Clements, J.D. (2011). Characterization of a mutant *Escherichia coli* heat-labile toxin, LT(R192G/L211A), as a safe and effective oral adjuvant. Clin. Vaccine Immunol. *18*, 546–551.

Norton, E.B., Lawson, L.B., Mahdi, Z., Freytag, L.C., and Clements, J.D. (2012). The A subunit of *Escherichia coli* heat-labile enterotoxin functions as a mucosal adjuvant and promotes IgG2a, IgA, and Th17 responses to vaccine antigens. Infect. Immun. *80*, 2426–2435.

O'Neal, C.J., Jobling, M.G., Holmes, R.K., and Hol, W.G. (2005). Structural basis for the activation of cholera toxin by human ARF6-GTP. Science *309*, 1093–1096.

Olvera-Gomez, I., Hamilton, S.E., Xiao, Z., Guimaraes, C.P., Ploegh, H.L., Hogquist, K.A., Wang, L., and Jameson, S.C. (2012). Cholera toxin activates nonconventional adjuvant pathways that induce protective CD8 T-cell responses after epicutaneous vaccination. Proc. Natl. Acad. Sci. U.S.A. *109*, 2072–2077.

Orlandi, P.A., Curran, P.K., and Fishman, P.H. (1993). Brefeldin A blocks the response of cultured cells to cholera toxin. Implications for intracellular trafficking in toxin action. J. Biol. Chem. *268*, 12010–12016.

Partidos, C.D., Salani, B.F., Pizza, M., and Rappuoli, R. (1999). Heat-labile enterotoxin of *Escherichia coli* and its site-directed mutant LTK63 enhance the proliferative and cytotoxic T-cell responses to intranasally co-immunized synthetic peptides. Immunol. Lett. *67*, 209–216.

Pickett, C.L., Weinstein, D.L., and Holmes, R.K. (1987). Genetics of type IIa heat-labile enterotoxin of *Escherichia coli*: operon fusions, nucleotide sequence, and hybridization studies. J. Bacteriol. *169*, 5180–5187.

Pickett, C.L., Twiddy, E.M., Coker, C., and Holmes, R.K. (1989). Cloning, nucleotide sequence, and hybridization studies of the type IIb heat-labile enterotoxin gene of *Escherichia coli*. J. Bacteriol. *171*, 4945–4952.

Pizza, M., Covacci, A., Bartoloni, A., Perugini, M., Nencioni, L., De Magistris, M.T., Villa, L., Nucci, D., Manetti, R., Bugnoli, M., *et al.* (1989). Mutants of pertussis toxin suitable for vaccine development. Science *246*, 497–500.

Pizza, M., Giuliani, M.M., Fontana, M.R., Monaci, E., Douce, G., Dougan, G., Mills, K.H., Rappuoli, R., and Del Giudice, G. (2001). Mucosal vaccines: non toxic derivatives of LT and CT as mucosal adjuvants. Vaccine *19*, 2534–2541.

Podda, A., Nencioni, L., De Magistris, M.T., Di Tommaso, A., Bossu, P., Nuti, S., Pileri, P., Peppoloni, S., Bugnoli, M., Ruggiero, P., *et al.* (1990). Metabolic, humoral, and cellular responses in adult volunteers immunized with the genetically inactivated pertussis toxin mutant PT-9K/129G. J. Exp. Med. *172*, 861–868.

Podda, A., De Luca, E.C., Titone, L., Casadei, A.M., Cascio, A., Peppoloni, S., Volpini, G., Marsili, I., Nencioni, L., and Rappuoli, R. (1992). Acellular pertussis vaccine composed of genetically inactivated pertussis toxin: safety and immunogenicity in 12- to 24- and 2- to 4-month-old children. J. Pediatr. *120*, 680–685.

Porgador, A., Staats, H.F., Itoh, Y., and Kelsall, B.L. (1998). Intranasal immunization with cytotoxic T-lymphocyte epitope peptide and mucosal adjuvant cholera toxin: selective augmentation of peptide-presenting dendritic cells in nasal mucosa-associated lymphoid tissue. Infect. Immun. *66*, 5876–5881.

Pulendran, B., and Ahmed, R. (2011). Immunological mechanisms of vaccination. Nat. Immunol. *12*, 509–517.

Raghavan, S., Ostberg, A.K., Flach, C.F., Ekman, A., Blomquist, M., Czerkinsky, C., and Holmgren, J. (2010). Sublingual immunization protects against *Helicobacter pylori* infection and induces T and B cell responses in the stomach. Infect. Immun. *78*, 4251–4260.

Rappuoli, R., Pizza, M., Douce, G., and Dougan, G. (1999). Structure and mucosal adjuvanticity of cholera and *Escherichia coli* heat-labile enterotoxins. Immunol. Today *20*, 493–500.

Roberts, M., Bacon, A., Rappuoli, R., Pizza, M., Cropley, I., Douce, G., Dougan, G., Marinaro, M., McGhee, J., and Chatfield, S. (1995). A mutant pertussis toxin molecule that lacks ADP-ribosyltransferase activity, PT-9K/129G, is an effective mucosal adjuvant for intranasally delivered proteins. Infect. Immun. *63*, 2100–2108.

Rodighiero, C., Tsai, B., Rapoport, T.A., and Lencer, W.I. (2002). Role of ubiquitination in retro-translocation of cholera toxin and escape of cytosolic degradation. EMBO Rep. 3, 1222–1227.

Romero, J.F., Ciabattini, A., Guillaume, P., Frank, G., Ruggiero, P., Pettini, E., Del Giudice, G., Medaglini, D., and Corradin, G. (2009). Intranasal administration of the synthetic polypeptide from the C-terminus of the circumsporozoite protein of Plasmodium berghei with the modified heat-labile toxin of *Escherichia coli* (LTK63) induces a complete protection against malaria challenge. Vaccine *27*, 1266–1271.

Ryan, E.J., McNeela, E., Murphy, G.A., Stewart, H., O'Hagan, D., Pizza, M., Rappuoli, R., and Mills, K.H. (1999). Mutants of *Escherichia coli* heat-labile toxin act as effective mucosal adjuvants for nasal delivery of an acellular pertussis vaccine: differential effects of the nontoxic AB complex and enzyme activity on Th1 and Th2 cells. Infect. Immun. *67*, 6270–6280.

Ryan, M., McCarthy, L., Rappuoli, R., Mahon, B.P., and Mills, K.H. (1998). Pertussis toxin potentiates Th1 and Th2 responses to co-injected antigen: adjuvant action is associated with enhanced regulatory cytokine production and expression of the co-stimulatory molecules B7-1, B7-2 and CD28. Int. Immunol. *10*, 651–662.

Samore, M.H., and Siber, G.R. (1996). Pertussis toxin enhanced IgG1 and IgE responses to primary tetanus immunization are mediated by interleukin-4 and persist during secondary responses to tetanus alone. Vaccine *14*, 290–297.

Schijns, V.E., and Lavelle, E.C. (2011). Trends in vaccine adjuvants. Expert Rev. Vaccines *10*, 539–550.

Schmitz, A., Herrgen, H., Winkeler, A., and Herzog, V. (2000). Cholera toxin is exported from microsomes by the Sec61p complex. J. Cell Biol. *148*, 1203–1212.

Schwartz, R.H. (2012). Historical overview of immunological tolerance. Cold Spring Harb. Perspect. Biol. *4*, a006908.

Shreedhar, V.K., Kelsall, B.L., and Neutra, M.R. (2003). Cholera toxin induces migration of dendritic cells from the subepithelial dome region to T- and B-cell areas of Peyer's patches. Infect. Immun. *71*, 504–509.

Simecka, J.W., Jackson, R.J., Kiyono, H., and McGhee, J.R. (2000). Mucosally induced immunoglobulin E-associated inflammation in the respiratory tract. Infect. Immun. *68*, 672–679.

Simmons, C.P., Mastroeni, P., Fowler, R., Ghaem-maghami, M., Lycke, N., Pizza, M., Rappuoli, R., and Dougan, G. (1999). MHC class I-restricted cytotoxic lymphocyte responses induced by enterotoxin-based mucosal adjuvants. J. Immunol. *163*, 6502–6510.

Simmons, C.P., Hussell, T., Sparer, T., Walzl, G., Openshaw, P., and Dougan, G. (2001). Mucosal delivery of a respiratory syncytial virus CTL peptide with enterotoxin-based adjuvants elicits protective, immunopathogenic, and immunoregulatory antiviral CD8+ T cell responses. J. Immunol. *166*, 1106–1113.

Sjoblom-Hallen, A., Marklund, U., Nerstedt, A., Schon, K., Ekman, L., Bergqvist, P., Lowenadler, B., and Lycke, N.Y. (2010). Gene expression profiling identifies STAT3 as a novel pathway for immunomodulation by cholera toxin adjuvant. Mucosal Immunol. 3, 374–386.

Slifka, M.K., and Ahmed, R. (1998). Long-lived plasma cells: a mechanism for maintaining persistent antibody production. Curr. Opin. Immunol. *10*, 252–258.

Snider, D.P. (1995). The mucosal adjuvant activities of ADP-ribosylating bacterial enterotoxins. Crit. Rev. Immunol. *15*, 317–348.

Song, J.H., Kim, J.I., Kwon, H.J., Shim, D.H., Parajuli, N., Cuburu, N., Czerkinsky, C., and Kweon, M.N. (2009). CCR7-CCL19/CCL21-regulated dendritic cells are responsible for effectiveness of sublingual vaccination. J. Immunol. *182*, 6851–6860.

Spangler, B.D. (1992). Structure and function of cholera toxin and the related *Escherichia coli* heat-labile enterotoxin. Microbiol. Rev. *56*, 622–647.

Spangrude, G.J., Araneo, B.A., and Daynes, R.A. (1985). Site-selective homing of antigen-primed lymphocyte populations can play a crucial role in the efferent limb of cell-mediated immune responses *in vivo*. J. Immunol. *134*, 2900–2907.

Stein, P.E., Boodhoo, A., Armstrong, G.D., Cockle, S.A., Klein, M.H., and Read, R.J. (1994). The crystal structure of pertussis toxin. Structure *2*, 45–57.

Stephenson, I., Zambon, M.C., Rudin, A., Colegate, A., Podda, A., Bugarini, R., Del Giudice, G., Minutello, A., Bonnington, S., Holmgren, J., *et al.* (2006). Phase I evaluation of intranasal trivalent inactivated influenza vaccine with nontoxigenic *Escherichia coli* enterotoxin and novel biovector as mucosal adjuvants, using adult volunteers. J. Virol. *80*, 4962–4970.

Summerton, N.A., Welch, R.W., Bondoc, L., Yang, H.H., Pleune, B., Ramachandran, N., Harris, A.M., Bland, D., Jackson, W.J., Park, S., *et al.* (2010). Toward the development of a stable, freeze-dried formulation of *Helicobacter pylori* killed whole cell vaccine adjuvanted with a novel mutant of *Escherichia coli* heat-labile toxin. Vaccine *28*, 1404–1411.

Sundling, C., Schon, K., Morner, A., Forsell, M.N., Wyatt, R.T., Thorstensson, R., Karlsson Hedestam, G.B., and Lycke, N.Y. (2008). CTA1-DD adjuvant promotes strong immunity against human immunodeficiency virus type 1 envelope glycoproteins following mucosal immunization. J. Gen. Virol. *89*, 2954–2964.

Tagliabue, A., and Rappuoli, R. (2008). Vaccine adjuvants: the dream becomes real. Hum. Vaccin. *4*, 347–349.

Takahashi, I., Marinaro, M., Kiyono, H., Jackson, R.J., Nakagawa, I., Fujihashi, K., Hamada, S., Clements, J.D., Bost, K.L., and McGhee, J.R. (1996). Mechanisms for mucosal immunogenicity and adjuvancy of *Escherichia coli* labile enterotoxin. J. Infect. Dis. *173*, 627–635.

Tamura, M., Nogimori, K., Murai, S., Yajima, M., Ito, K., Katada, T., Ui, M., and Ishii, S. (1982). Subunit structure of islet-activating protein, pertussis toxin, in conformity with the A-B model. Biochemistry *21*, 5516–5522.

Tebbey, P.W., Scheuer, C.A., Peek, J.A., Zhu, D., LaPierre, N.A., Green, B.A., Phillips, E.D., Ibraghimov, A.R., Eldridge, J.H., and Hancock, G.E. (2000). Effective mucosal immunization against respiratory syncytial virus using purified F protein and a genetically detoxified cholera holotoxin, CT-E29H. Vaccine *18*, 2723–2734.

Tempesta, M., Camero, M., Bellacicco, A.L., Tarsitano, E., Lorusso, A., Martella, V., Decaro, N., Del Giudice, G., Cassone, A., Quaranta, A., *et al.* (2007). Caprine herpesvirus 1 vaccine with the LTK63 mutant as a mucosal adjuvant induces strong protection against genital infection in goats. Vaccine *25*, 7927–7930.

Teter, K., and Holmes, R.K. (2002). Inhibition of endoplasmic reticulum-associated degradation in CHO cells resistant to cholera toxin, *Pseudomonas aeruginosa* exotoxin A, and ricin. Infect. Immun. *70*, 6172–6179.

Tritto, E., Muzzi, A., Pesce, I., Monaci, E., Nuti, S., Galli, G., Wack, A., Rappuoli, R., Hussell, T., and De Gregorio, E. (2007). The acquired immune response to the mucosal adjuvant LTK63 imprints the mouse lung with a protective signature. J. Immunol. *179*, 5346–5357.

Tsai, B., Rodighiero, C., Lencer, W.I., and Rapoport, T.A. (2001). Protein disulfide isomerase acts as a redox-dependent chaperone to unfold cholera toxin. Cell *104*, 937–948.

Tumpey, T.M., Renshaw, M., Clements, J.D., and Katz, J.M. (2001). Mucosal delivery of inactivated influenza vaccine induces B-cell-dependent heterosubtypic cross-protection against lethal influenza A H5N1 virus infection. J. Virol. *75*, 5141–5150.

van Ginkel, F.W., Jackson, R.J., Yuki, Y., and McGhee, J.R. (2000). Cutting edge: the mucosal adjuvant cholera toxin redirects vaccine proteins into olfactory tissues. J. Immunol. *165*, 4778–4782.

van Ginkel, F.W., Jackson, R.J., Yoshino, N., Hagiwara, Y., Metzger, D.J., Connell, T.D., Vu, H.L., Martin, M., Fujihashi, K., and McGhee, J.R. (2005). Enterotoxin-based mucosal adjuvants alter antigen trafficking and induce inflammatory responses in the nasal tract. Infect. Immun. *73*, 6892–6902.

Watanabe, I., Hagiwara, Y., Kadowaki, S.E., Yoshikawa, T., Komase, K., Aizawa, C., Kiyono, H., Takeda, Y., McGhee, J.R., Chiba, J., *et al.* (2002). Characterization of protective immune responses induced by nasal influenza vaccine containing mutant cholera toxin as a safe adjuvant (CT112K). Vaccine *20*, 3443–3455.

Victora, G.D., and Nussenzweig, M.C. (2012). Germinal centers. Annu. Rev. Immunol. *30*, 429–457.

WHO (World Health Organization). (2011). Cholera, 2010. Wkly. Epidemiol. Rec. *86*, 325–339.

Williamson, E., Westrich, G.M., and Viney, J.L. (1999). Modulating dendritic cells to optimize mucosal immunization protocols. J. Immunol. *163*, 3668–3675.

Wilson, A.D., Robinson, A., Irons, L., and Stokes, C.R. (1993). Adjuvant action of cholera toxin and pertussis toxin in the induction of IgA antibody response to orally administered antigen. Vaccine *11*, 113–118.

Witvliet, M.H., Burns, D.L., Brennan, M.J., Poolman, J.T., and Manclark, C.R. (1989). Binding of pertussis toxin to eucaryotic cells and glycoproteins. Infect. Immun. *57*, 3324–3330.

Xu-Amano, J., Jackson, R.J., Fujihashi, K., Kiyono, H., Staats, H.F., and McGhee, J.R. (1994). Helper Th1 and Th2 cell responses following mucosal or systemic immunization with cholera toxin. Vaccine *12*, 903–911.

Yamamoto, M., Briles, D.E., Yamamoto, S., Ohmura, M., Kiyono, H., and McGhee, J.R. (1998). A nontoxic adjuvant for mucosal immunity to pneumococcal surface protein A. J. Immunol. *161*, 4115–4121.

Yamamoto, M., Rennert, P., McGhee, J.R., Kweon, M.N., Yamamoto, S., Dohi, T., Otake, S., Bluethmann, H., Fujihashi, K., and Kiyono, H. (2000). Alternate mucosal immune system: organized Peyer's patches are not required for IgA responses in the gastrointestinal tract. J. Immunol. *164*, 5184–5191.

Yamamoto, S., Kiyono, H., Yamamoto, M., Imaoka, K., Fujihashi, K., Van Ginkel, F.W., Noda, M., Takeda, Y., and McGhee, J.R. (1997a). A nontoxic mutant of cholera toxin elicits Th2-type responses for enhanced mucosal immunity. Proc. Natl. Acad. Sci. U.S.A. *94*, 5267–5272.

Yamamoto, S., Takeda, Y., Yamamoto, M., Kurazono, H., Imaoka, K., Fujihashi, K., Noda, M., Kiyono, H., and McGhee, J.R. (1997b). Mutants in the ADP-ribosyltransferase cleft of cholera toxin lack diarrheagenicity but retain adjuvanticity. J. Exp. Med. *185*, 1203–1210.

Zhou, F., Goodsell, A., Uematsu, Y., and Vajdy, M. (2009). Prolonged protection against Intranasal challenge with influenza virus following systemic immunization or combinations of mucosal and systemic immunizations with a heat-labile toxin mutant. Clin. Vaccine Immunol. *16*, 471–478.

Zoeteweij, J.P., Epperson, D.E., Porter, J.D., Zhang, C.X., Frolova, O.Y., Constantinides, A.P., Fuhrmann, S.R., El-Amine, M., Tian, J.H., Ellingsworth, L.R., *et al.* (2006). GM1 binding-deficient exotoxin is a potent noninflammatory broad spectrum intradermal immunoadjuvant. J. Immunol. *177*, 1197–1207.

Mobile Genetic Elements as Carriers for Bacterial Virulence Genes

5

José R. Penadés and J. Ross Fitzgerald

Abstract

The identification of accessory genetic elements (plasmids, bacteriophages, and 'pathogenicity islands') encoding virulence-associated genes has led to enhanced understanding of the evolution of pathogenic bacteria and how they adapt to new host environments. It is evident that mobile genetic elements (MGEs) have had a profound influence on the emergence and spread of pathogenic bacteria. Furthermore, an understanding of the mechanisms of horizontal acquisition of virulence genes may lead to the identification of alternative approaches for preventing the emergence of new pathogenic clones or for controlling existing ones. In this chapter, we provide examples of MGEs, which confer determinants of pathogenicity to selected bacterial species, and summarize current knowledge regarding their mechanisms of transmission.

Core and adaptive genome

The concept that bacterial genomes within a single species vary widely in gene content has been known for many years. Determination of genome size by pulse-field gel electrophoresis in the 1980s and 1990s showed that representatives of the *Escherichia coli* ECOR collection had genomes which varied in size from 4.5 to 5.5 Mbp (Bergthorsson and Ochman, 1998). However, it was only with the advent of the genomic era that the phenomenon could be properly investigated. Not only was the genome size different but a significant portion of each genome included genes not found in other strains.

The shared, 'core' genome accounts for around only 40% of the total gene pool in *E. coli* and it is interrupted by multiple variable regions unique to some strains. These findings have contributed to a change in the concept of bacterial species in recent years (Konstantinidis and Tiedje, 2005a,b) which are more appropriately defined by the 'pan-genome', consisting of the core-genome and an accessory-genome consisting of partially shared and strain-specific genes found after examination of multiple strains of the species. The core genome is the essence of the phylogenetic unit of the species and is thought to be representative at various taxonomic levels (Ochman and Santos, 2005). The accessory (or adaptive) genome, on the other hand, includes key genes required for survival in a specific environment, and are commonly linked to virulence, niche adaptation and resistance, typically reflecting the organisms' predominant lifestyle or habitat.

The core-genome is a good starting point for the identification of essential genes that might be useful as antibiotic targets and genes involved in host–pathogen interactions which being common to all pathogenic strains could be potential vaccine targets (Rappuoli and Nabel, 2001). These shared genes may also represent useful sequences for phylogenetic inference (Daubin *et al.*, 2003). A second consequence of the core-adaptive split is the identification of dispensable and non-dispensable genes. Normally, human pathogens contain relatively large genomes that include many paralogous gene families. The disruption of many of those genes and their deletion in species with a restricted lifestyle shows that they are non-essential (Pushker *et al.*, 2004). For example, a simple model found in nature is given by obligate host-related bacteria (typically symbionts) with a minimalist genome composition: The term 'minimal genome' has been used to describe the set of genes that are required for a self-sustainable cell (Mushegian and Koonin, 1996).

The mobile bacterial genome

In addition to the broadly conserved set of genes that are responsible for the basic cellular processes of growth and multiplication, bacterial species contain a second facultative set that is concerned with adaptations to environmental contingencies (Frost *et al.*, 2005). It has become evident in recent years that the facultative genome includes a large variety of mobile genetic elements (MGEs), and that this 'mobile genome' may exceed 10% of the total for many species and is a key contributor to the plasticity of the bacterial genome. MGEs commonly contribute to antibiotic resistance and pathogenesis. All known classes of bacterial MGEs, including temperate phages, plasmids, transposons and chromosomal islands, have been shown to be involved in pathogenesis. Of particular importance, it is striking that the majority of the bacterial toxins that cause specific toxin-mediated diseases (toxinoses) are encoded by MGEs (Novick, 2003). This includes diseases such as diphtheria, dysentery, toxic shock syndrome, food poisoning, necrotizing pneumonia, scalded skin syndrome, botulism, haemolytic–uraemic syndrome or necrotizing fasciitis. This raises a number of important questions, including what are the evolutionary forces responsible for this association, and what are the selective advantages for the bacterial host and the MGEs respectively.

In this book chapter, we will analyse the role of the MGEs in the dissemination of virulence toxin genes among bacterial pathogens. Additionally, we will analyse the role of the SOS response in the physiology of the MGE-encoded toxinoses.

Mobile genetic elements and their role in virulence

Plasmids

Plasmids are extrachromosomal, autonomously replicating genetic elements found in cells of all kingdoms of life. They have an important impact on bacterial physiology, and can play a role in bacterial virulence. As an example of this, in 1887 Robert Koch published the results of several experiments demonstrating that the causative agent of anthrax was the rod-shaped bacterium *Bacillus anthracis*. Approximately 100 years later it was established that this bacterium harboured two plasmids that were required for its virulence properties (Little and Ivins, 1999).

Plasmids are autonomous replicons in bacterial cells. They display an amazing diversity of characteristics, such as size, model of replication and transmission, host ranges, and the repertoire of genes that they carry. Their genetic constitution reflects their function as gene exchange machines. Thus, they always contain genes required for replication, stability, DNA transfer and establishment in recipient cells. In addition, they carry genes with adaptive functions and others of unknown function. Classically, plasmids are covalently closed, circular double-stranded DNA molecules, but linear double-stranded DNA plasmids have been found in an increasing number of bacterial species (Frost *et al.*, 2005). The role of plasmids in bacterial virulence is linked to the fact that toxins are often encoded by plasmids, which can be easily transferred to a recipient strain by several different mechanisms, including natural transformation, conjugation or transduction. Bacterial transformation is the process by which bacterial cells take up naked DNA molecules. If the foreign DNA has an origin of replication recognized by the host cell DNA polymerases, as the plasmids have, the bacteria will replicate the foreign DNA along with their own. Conjugation is defined as the unidirectional transfer of genetic information between cells by cell-to-cell contact. This latter requirement for contact distinguishes conjugation from transduction and transformation. The term 'unidirectional' refers to the fact that a copy of the plasmid is transferred from one cell, termed the 'donor', to another cell, termed the 'recipient'. One additional significance of the conjugation lies in the fact that many plasmids can also affect the transfer of chromosomal DNA, as exemplified by the high frequency of recombination (Hfr) mode of F plasmids and by the chromosome mobilization ability (Cma) of plasmids in *Streptomyces* species (Frost *et al.*, 2005). Such conjugative elements integrate into the host genome and transfer large sections of the chromosome, along with parts of the conjugative element, into recipient cells. Finally, transduction is defined as the transfer of genetic information between cells through the mediation of a virus (phage) particle. It therefore does not require cell-to-cell contact.

As mentioned previously, the spectrum of diseases caused by most pathogens is due to the acquisition of a variety of specific virulence genes harboured on plasmids. Thus, the virulence in the Gram-negative *Shigella flexneri, Escherichia coli* or *Yersinia pestis,* as well as in the Gram-positive *Clostridium perfringens, Bacillus anthracis, Staphylococcus aureus* or *Enterococcus faecalis* depends in some cases of the plasmid-encoded virulence factors. A detailed explanation of some of these examples will be provided.

Bacteriophages

Phages are the most abundant and the most rapidly replicating life forms on earth and their genetic diversity is enormous (Brussow and Hendrix, 2002; Hendrix, 2003). Commonly used as tools in genetic engineering, they have gained new attention for their potential for use in antibacterial therapy and in nanotechnology (Fischetti, 2001). The genomes of phages can be composed of either single- or double-stranded DNA or RNA and can range in size from a few to several 100 kb. Their characteristic essential genes encode replicases, phage components involved in hijacking the host cell replicative machinery, and proteins that package DNA in a protein coat (capsid). Virulent bacteriophages replicate vigorously and characteristically lyse the host bacteria. Temperate bacteriophages have an alternative, quiescent, non-lytic growth mode called lysogeny. A phage in the lysogenic state is called a prophage. In most known cases of lysogeny, the phage genome integrates into the bacterial chromosome and replicates with it but in a few cases the phage genome replicates autonomously as a circular or linear plasmid. Some prophages alter the phenotype of the host bacterium. These are called converting phages, and the process is known as lysogenic conversion. For example, if the prophage encodes a toxin, then the bacterium will be lysogenically converted for toxin-production (positive lysogenic conversion). Alternatively, when lysogenization results in the loss of a particular phenotype, it is referred to as negative lysogenic conversion. For example, in *S. aureus,* there are at least two examples of negative lysogenic conversion resulting from direct integration of the phage into the gene sequence, which interrupts β-haemolysin and lipase expression, respectively (Coleman *et al.,* 1986, 1989; Ye and Lee, 1989).

The first example that phages alter the phenotype of their bacteria hosts through toxigenic conversion came from the original studies demonstrating that sterile cultures filtrates from 'scarlatina' strains of *Streptococcus pyogenes* caused the toxigenic conversion of non-toxin to toxin producing strains (Cantacuzene and Boncieu, 1926). Since then, a vast body of examples have reported showing the enormous impact that phages have had in altering the host phenotype. Importantly, phages do not only encode toxigenic virulence factors, but also proteins involved in other pathogenic processes. Thus, some temperate phages encode for proteins involved in bacterial resistance to killing by serum or phagocytosis, while others facilitate bacterial persistence or invasiveness. Specific examples of how phages influence bacterial virulence will be described later.

Phages are also involved in the transfer of bacterial genes that they do not directly encode. During the lytic phase, chromosomal or plasmid DNA can be accidentally packaged into phage heads. This DNA can then be transferred to a recipient strain, and incorporated into the bacterial genome. Horizontal transmission of a gene by this process is called generalized transduction, which occurs a low frequency. Finally, recombination between prophages and other mobile elements that reside in the same bacterial host contributes to the well-documented mosaic structure of phages (Hendrix *et al.,* 1999).

Transposable elements

Transposons are segments of DNA that have the capacity to move between locations in the genome by a process known as transposition. Unlike other processes that reorganize DNA, transposition does not require extensive areas of homology between the transposon and its integration site. Transposable elements differ from phages in lacking a virus life cycle and from plasmids in being unable to replicate autonomously. The simplest transposable elements are insertion sequences or IS elements. An IS element is a short DNA sequence (around 1000–1500 bp in length) containing only the genes for those enzymes required for its transposition and bounded at both ends by identical or very similar sequences of nucleotides in reverse orientation knows as inverted repeats. The repeats are typically 15 to 25 bp long and vary in sequence so that each IS has its own characteristic inverted repeat. Between the inverted repeats is a gene that codes for an enzyme called transposase (and sometimes a gene for a resolvase). This enzyme is required for transposition as it recognizes the ends of the IS element. In addition to the genes involved in the transposition process, some transposable elements also contain genes involved in virulence or antibiotic resistance. These elements are called composite transposons consisting of a central region containing the additional genes flanked on both sides by IS elements that are identical or very similar in sequence. Many composite transposon are more simple in organization. They are bounded by short inverted repeats, and the coding region contains both transposition genes and the extra genes. It is believed that composite transposons are formed when two IS elements flanked a central chromosomal region containing one or more genes.

The process of transposition in prokaryotes involves a series of events, including self-replication and recombinational processes. Typically, the original transposon remains at the parental site on the chromosome, while a replicated copy inserts at the 5 to 9 bp target DNA in a process known as replicative transposition which results in duplication of the target site.

Transposable elements contribute to a variety of effects on the host bacterium, including gene mutations (when the element is inserted in the middle of the gene) or rearrangements (deletions) of genetic material (between two of these elements). Additionally, transposons can affect the expression of the flanking genes situated near its insertion site, since some of these elements contain promoters.

Transposons are frequently associated with other MGEs, such as plasmids or chromosomal islands, and in so doing they use the transfer capacities of the MGEs. Transposons are important elements for the spread virulence factors among bacterial species.

Chromosomal and pathogenicity islands

The chromosomal islands are the most recently identified of the MGEs and consequently are somewhat less well defined than the others. They were initially recognized as discrete chromosomal segments that contained virulence genes, lacked essential genes, and represented important genomic differences between closely related organisms that differed in pathogenicity and were accordingly labelled 'pathogenicity islands' (PAIs) (Dobrindt *et al.*, 2000; Hacker and Kaper, 2000). PAIs, however, are a subset of a much broader family of inserted units, the genomic or chromosomal islands (Hacker and Kaper, 2000). The key feature of these elements is, of course, their transferability. However, not all laterally transferred DNA represents a genomic island, and the distinction is not always obvious. It seems important to have a clear view of what constitutes a genomic island as opposed to an accidentally transferred segment of DNA. It is suggested that the primary criterion must be evidence of active mobility.

The minimum requirements for mobility are (1) a site-specific recombination function; (2) a pair of flanking repeats upon which it can act; and (3) a means of inter-cell transfer, which need not be encoded by the island. To be considered a chromosomal island, a genetic unit must either possess one or more of these functions or it must possess vestiges of such functions. Alternatively, it must closely resemble an established genomic island. Well-characterized mobile islands include the staphylococcal PAIs (SaPIs) (Novick and Subedi, 2007; Novick *et al.*, 2010), the symbiosis island of *Rhizobia* (Sullivan and Ronson, 1998) and genomic islands of enteric bacteria known as CONSTINS (conjugative, self-transmissible, integrating elements) (Hochhut and Waldor, 1999), also known as ICE (integrative and conjugative elements) (Schubert *et al.*, 2004) which possess a conjugation system as well as integration-excision capability. Interestingly, the latter have been known for many years as *incJ* plasmids (see Novick, 1969), but their true nature has been appreciated only recently. Many other genomic islands encode functional site-specific recombinases catalysing their excision and re-insertion including the SCCmec islands of staphylococci (Katayama *et al.*, 2000), several islands of *E. coli* (Hochhut *et al.*, 2006), and the island of *Shigella flexneri* (Sakellaris *et al.*, 2004). The PAPI-1 island of *Pseudomonas aeruginosa* is the first of the large pathogenicity islands of Gram-negative bacteria for which inter-strain transfer has been demonstrated (Qiu *et al.*, 2006).

The extent of the role of pathogenicity islands in bacterial virulence has only been realized since the genomic era. In most bacteria, PAIs encode many different functions, which largely depend on the environmental context in which the bacterium lives. The genetic repertoire found within PAIs can be functionally divided into several groups, including adhesins, invasins, iron uptake systems, pore-forming toxins, proteins causing apoptosis, superantigens, secreted lipases, secreted proteases, O antigens, proteins transported by type I, III, IV and V secretion systems, and genes involved in antibiotic resistance. A detailed description of these PAIs can be found in recent reviews (Schmidt and Hensel, 2004; Gal-Mor and Finlay, 2006).

Why MGEs encode exotoxins and other virulence factors?

If a gene confers some benefit to the host bacterium, but an encoding prophage does not otherwise confer any additional advantage to the same bacterium, then over time we would expect evolution to favour deletion of the prophage sequences accompanying a given gene (Lawrence *et al.*, 2001). These deletions result in gene stabilization within a bacterial genome. However, in most cases, the MGE containing the virulence factor is fully functional, indicating that this functionality is beneficial for the bacteria.

From the bacterial perspective, MGE-encoded proteins represent a reservoir of additional (non-core) genes that enable populations to respond and adapt to new environmental conditions. This is the case for the pathogenicity islands of *Staphylococcus aureus* (SaPIs). SaPIs have been implicated in the pathogenesis and evolution of *S. aureus* including adaptation to new hosts (Novick and Subedi, 2007). From bovine *S. aureus* isolates, different SaPIs have been characterized (Fitzgerald *et al.*, 2001; Ubeda *et al.*, 2003). SaPIbov1, was identified in an isolate associated with clinical mastitis carried the toxin genes *tst*, *sel* and *sec*. In contrast, SaPIbov2, identified in an isolate from subclinical mastitis, substituted the toxin genes with a transposon containing *bap*, a gene involved in biofilm formation and bacterial persistence (Cucarella *et al.*, 2004). In addition to a central core region involved in replication and transfer of these elements (Ubeda *et al.*, 2008), both island share the same integrase implying that both islands compete to integrate at the same bacterial att_B site.

As mentioned, SaPIs has been involved not only in pathogenesis but also in host adaptation. It has been previously reported that staphylococci can adapt to ruminants by acquiring SaPIs encoding for animal-specific alleles of the von Willebrand factor-binding protein (vWbp) gene, *vwb*. Expression of these SaPI-encoded proteins confer the bacteria the ability to coagulate the host plasma from the ruminant host (Guinane *et al.*, 2010; Viana *et al.*, 2010), indicating that acquisition of vWbp-encoding SaPIs may be determinative for animal specificity.

In addition to the horizontal gene transfer between different strains of the same species, toxins encoded by MGEs can be transferred between different species. A recent example of inter-species dissemination involved the transfer of vancomycin resistance between *Enterococcus faecalis* and *S. aureus*. A clinical isolate of *S. aureus* with high-level resistance to vancomycin was isolated in June 2002. This isolate harboured a 57.9-kb multiresistance conjugative plasmid within which Tn1546 (*van*A) was integrated. Additional elements on the plasmid encoded resistance to trimethoprim (*dfr*A), β-lactams (*bla*Z), aminoglycosides (*aac*A-*aph*D), and disinfectants (*qac*C). Genetic analyses suggested that the long-anticipated transfer of vancomycin resistance to a methicillin-resistant *S. aureus* occurred *in vivo* by interspecies transfer of Tn1546 from a co-isolate of *E. faecalis* (Weigel *et al.*, 2003).

Additionally, it is important to note that the evolution of MGEs is driven by selective forces that operate to maintain the functionality of these elements, including maintenance of their capabilities to replicate and be transferred. Accordingly, environmental stimuli such as DNA damaging agents that induce the bacterial SOS response compromising the viability of the cells activate the transfer of MGEs. Therefore, the association of virulence factors with 'functional' elements will increase their presence in the bacterial pool genome. Furthermore, virulence genes encoded by PAIs or phages may withstand environmental exposure better than those encoded by bacterial chromosomes. For example, Muniesa *et al.* (1999) found that Stx2-encoding phages persisted in river water longer and were more resistant to chlorination and pasteurization than Stx2-encoding bacteria, suggesting that phage particles serve as a more durable environmental reservoir of the stx_2 genes than bacteria.

Role of SOS induction

The SOS system, first described and thoroughly studied in *Escherichia coli* (Walker, 1984), is a global response designed to guarantee cell survival when massive DNA damage is introduced and the normal DNA replication of the bacterial cell is disturbed. This network is the prototypic cell cycle check-point control and DNA repair system and because of this, a detailed picture of the signal transduction pathway that regulates this response is required. A central part of the SOS response is the de-repression of more than 40 genes under the direct and indirect transcriptional control of the RecA and LexA proteins, which are also member of this regulon (Khil and Camerini-Otero, 2002). The LexA protein is the repressor of the system through its specific biding to the regulatory motifs present in the promoter region of the SOS genes. This regulatory motif, commonly known as the LexA box, has in *E. coli* the CTGTN$_8$ACAG motif as the consensus sequence (Walker, 1984).

The signal transduction pathway leading to an SOS response ensues when RecA protein binds to single stranded DNA (ssDNA), which can be created by processing of DNA damage or stalled replication (Little and Mount, 1982; Walker, 1984; Miller *et al.*, 2004). This binding activates an otherwise dormant co-protease activity of RecA, which facilitates the proteolytic self-cleavage of the LexA repressor (Little, 1991) to trigger the expression of DNA repair genes. Once DNA lesions have been repaired, RecA ceases to be activated and non-cleaved LexA protein returns to its normal levels, repressing again the transcription of the SOS genes. The LexA gene is widespread among bacteria and is present in most phylogenetic groups for which different monophyletic LexA-binding motifs have been described including the GAACN7GTTC motif of Gram-positive bacteria.

The importance of the SOS response in the biology of the MGEs has emerged in recent years. Activated RecA also facilitates the auto-cleavage of phage repressors, which maintain the lysogenic state of temperate bacteriophages (Little, 1993).

Although the frequency of environmental SOS induction by typical DNA-damaging agents, such as radiation and certain chemicals, is probably not very high (though there does not seem to be much literature on this issue), a class of SOS-inducing compounds, the fluoroquinolone antibiotics, is in wide use with serious clinical implications. For example, Shiga toxin is encoded by a prophage and its gene is SOS-induced along with rest of the phage genome (Zhang *et al.*, 2000). Treatment of the Shiga-toxin dependent haemolytic–uraemic syndrome (caused by *E. coli* H57:0157) with fluoroquinolones increases disease severity, sometimes fatally, as well as amplifying the population of phages encoding Shiga toxin (Muniesa *et al.*, 1999; Wong *et al.*, 2000; Zhang *et al.*, 2000). Similarly, we demonstrated that the fluoroquinolone antibiotic ciprofloxacin induced staphylococcal prophages and any co-resident SaPI whose replication is controlled by the resident prophage, which is strongly predicted to promote spread of the SaPI. Although there is no information on whether SOS induction increases the production of any SaPI-encoded superantigen, potentially worsening the associated clinical condition (e.g. TSS), it is highly likely that the increase in gene dosage will have this effect, whether or not transcription of the superantigen gene is under direct SOS induction control.

A third example of clinically adverse SOS induction by antibiotics is that of the integrating conjugative elements (ICEs), mobile elements that are transferred by conjugation and integrate site-specifically into the recipient chromosome. The prototype, SXT, is a ~ 100-kb *V. cholerae* ICE that carries resistance to chloramphenicol, sulphamethoxazole, trimethoprim and streptomycin (Beaber *et al.*, 2004). SXT-related elements were not detected in *V. cholerae* before 1993 but are now present in almost all clinical *V. cholerae* isolates from Asia. ICEs related to SXT are also present in several other bacterial species and encode a variety of antibiotic and heavy metal resistance genes. Beaber and co-workers have shown that an SXT-encoded protein, SetR, represses SXT transfer and that this repression is relieved by SOS induction, stimulating transfer, with the implication that this has contributed to the dissemination of antibiotic resistance in Asian *V. cholerae* in recent years (Beaber *et al.*, 2004). Thus, therapeutic agents can promote the spread of antibiotic resistance genes as well as virulence determinants.

It is suggested, in conclusion, that SOS induction by antibiotics of clinically important processes, such as virulence gene expression and transfer and antibiotic resistance transfer is likely to be of much wider importance than was previously thought.

Bacteriophage-encoding toxins and other virulence factors

Many key virulence factors, including diphtheria toxin, Shiga toxin, cholera toxin, and several staphylococcal toxins, are encoded by genes found in the genomes of lysogenic bacteriophages (Wagner and Waldor, 2002; Brussow *et al.*, 2004; Waldor and Friedman, 2005). In addition, recent studies with Shiga toxin (Stx)-encoding phages in *E. coli* and staphylococcal enterotoxin-encoding phages in *S. aureus* revealed that prophage induction can provide a mechanism to control toxin production. Here we provide a brief description of some of these processes that will not be described in other chapters:

Shiga toxin (Stx)-encoding phages

Shiga toxin-producing *Escherichia coli* (STEC) are considered food- and water-borne pathogens, although person-to-person transmission has also been reported. STEC cause bloody diarrhoea and haemolytic uraemic syndrome and can lead to severe complications. One of the best-known serotypes is O157:H7, which is the most common virulent serogroup in the USA and Canada, while other serotypes are frequently reported in European outbreaks. STEC are normally found in cattle, goat and sheep, which act as the reservoir, releasing STEC in their faeces.

Shiga toxins are the main virulence factors made by STEC, and are homologous to the toxin made by *Shigella dysenteriae* serotype I (O'Brien *et al.*, 1984). Currently two Stx have been described, Stx1 and Stx2 in addition to their variants (Jaeger and Acheson, 2000) which are only found in certain reservoirs (Fraser *et al.*, 2004). Generally, Shiga toxins are characterized by their hexameric conformation comprised of five B subunits, which allow toxin attachment to their enterocyte receptor, Gb3, which is also present in endothelial cells of glomerular capillaries (Hoey *et al.*, 2002), and one A subunit, which is catalytically active and blocks translation of mRNA to protein, leading to cell death (Fraser *et al.*, 2004).

Shiga toxins are encoded in the genome of temperate phages (Schmidt, 2001) considered members of the lambdoid family since they share a common genome arrangement that conserves the relative positions of the genes with similar activities and associated regulatory signals (Campbell, 1994). Although Shiga toxin-encoding bacteriophages are a heterogeneous group, in terms of morphology and their genetic organization (Muniesa *et al.*, 2004; Serra-Moreno *et al.*, 2007), *stx* genes location is conserved, being found next to the lytic genes and downstream of the *Q* anti-terminator (O'Brien *et al.*, 1984). Therefore, Shiga toxin production is linked to the induction or progression of the phage lytic cycle, after activation of the SOS response in the bacterial host. Several studies have showed that treatment of STEC with SOS-inducer molecules, including mitomycin C or quinolone antibiotics, increased Stx expression (Acheson and Donohue-Rolfe, 1989; Zhang *et al.*, 2000) and treatment of human STEC infection with bacteriophage-inducing antibiotics, such as fluoroquinolones, may have significant adverse clinical consequences and enhance the spread of virulence factors *in vivo*. After phage induction, the toxin is released through cell lysis (Wagner *et al.*, 1999). Following the lytic burst, Stx phages act as vectors in the *stx* horizontal transmission to other *Enterobacteriaceae* (Acheson *et al.*, 1998; Schmidt *et al.*, 1999).

In addition to the direct control of toxin expression by the phage, previous studies with STEC O157:H7 strains isolated from a single outbreak showed that isolates harbouring two different Stx prophages produced less toxin than strains from the same clone carrying only one Stx prophage (Muniesa *et al.*, 2003), suggesting that the expression of phage genes may be regulated when other temperate phages are present. In a recent study, Serra-Moreno and Muniesa (2008) demonstrated that the CI repressors of both prophages operating *in trans* could regulate the reduced production of the Stx toxin when a bacterium contains two prophages. The authors hypothesized that although the sequence of the *cI* genes of the phages studied differed, the CI protein conformation was conserved. Consequently, the presence of more than one prophage in the host chromosome could be regarded as a mechanism to allow genetic retention in the cell, by reducing the activation of lytic cycle and hence the pathogenicity of the strains (Serra-Moreno *et al.*, 2008).

Staphylococcal phages

Bacteriophages of *S. aureus* encode many clinically relevant virulence factors, including toxins. For example, a recently emerged community-acquired epidemic strain is responsible for rapidly progressive, fatal diseases including necrotizing pneumonia, severe sepsis and necrotizing fasciitis. In addition to novel methicillin resistance genetic cassettes, these strains harbour a phage encoding Panton–Valentine leukocidin (PVL), which has been implicated in the increased virulence of these clones (Labandeira-Rey *et al.*, 2007). Other known phage-encoded virulence factors include exfoliative toxin type A, staphylokinase, and staphylococcal enterotoxins (Yamaguchi *et al.*, 2000; Baba *et al.*, 2002; Sumby and Waldor, 2003).

Although the presence of virulence factors in different *S. aureus* phages was reported two decades ago by Betley and Mekalanos (1985), who demonstrated that the gene for staphylococcal enterotoxin A (*sea,* formerly *entA*) is carried by related temperate bacteriophages, it is not well known whether the life cycles of the phages influence the expression of the virulence genes and thus *S. aureus* pathogenicity. In a pioneer study, and using as a model ϕSa3ms, a lysogenic bacteriophage encoding the staphylococcal enterotoxins SEA, SEG, and SEK and the fibrinolytic enzyme staphylokinase (Sak), Sumby and Waldor (2003) demonstrated that upon φSa3ms prophage induction, transcription of all four virulence factors were greatly increased. Interestingly, while the increase in *sea* and *sak* transcription was a result of read-through transcription from upstream latent phage promoters and an increase in phage copy number, the majority of the *seg*2 and *sek*2 transcripts were shown to initiate from the upstream phage *cI* promoter and hence were regulated by factors influencing *cI* transcription (Sumby and Waldor, 2003), suggesting that the production of phage-encoded virulence factors in *S. aureus* may be regulated by processes that govern lysogeny.

In addition to this work, several studies of *S. aureus* have described an antibiotic-induced SOS response that affects virulence by modulating phages. For example, Goerke and co-workers analysed the effects of ciprofloxacin and trimethoprim on phage induction and expression of phage-encoded virulence factors. Treatment of lysogens with subinhibitory concentrations of either antibiotic resulted in replication of phages in the bacterial host linked to elevated expression of the phage-encoded virulence genes, chiefly due to the activation of latent phage promoters (Goerke *et al.*, 2006).

The collective conclusion from these and other studies is that phages, together with DNA damaging agents that induce the phages, can promote enhanced horizontal gene transfer of phage-encoded toxins, as well as the expression of these factors, which can alter the course of a clinical infection.

In addition to toxins, bacteriophages are carriers of a vast of virulence factors involved in host adaptation. As a case in point, *S. aureus* strains infecting humans usually carry phages encoding for two new innate immune modulators: staphylococcal complement inhibitor (SCIN) and chemotaxis inhibitory protein of *S. aureus* (CHIPS). SCIN is a C3 convertase inhibitor, blocking the formation of C3b on the surface of the bacterium and the ability of human neutrophils to phagocytose *S. aureus* (Rooijakkers *et al.*, 2007). CHIPS is a bacterial chemokine receptor modulator that specifically binds two chemokine receptors. CHIPS attenuates the response of the C5a receptor (C5aR) as well as the formylated peptide receptor of human neutrophils. This results in inhibition of neutrophil chemotaxis and activation in response to C5a and formylated peptides (Haas *et al.*, 2004). Both SCIN and CHIPS are important virulence factors that protect *S. aureus* from human innate immune defence system.

Interestingly, both *scn* and *chp* are carried by β-haemolysin (*hlb*)-converting bacteriophages, which besides SCIN and CHIPS, also encode for two other toxins that are also involved in opposing the innate immune system. SEA is a superantigen but also has the capability to modulate the function of chemokine receptors like CCR1, CCR2, and CCR5 (Rahimpour *et al.*, 1999). SAK has recently been described as a modulator of different parts of the innate immune system (Jin *et al.*, 2004). Interestingly, all four immune evasion molecules on βC-φs display an extreme human specificity (van Wamel *et al.*, 2006).

Plasmids encoding toxins

Several well-known diseases are caused by toxin-producing clostridia; for example, gas gangrene and necrotic enteritis are caused by *Clostridium perfringens*, diarrhoea and pseudomembranous colitis are caused by *Clostridium difficile*, tetanus disease is caused by *Clostridium tetani*, and food-borne botulism is caused by *Clostridium botulinum*. The last two organisms produce the most powerful neurotoxins known to mankind, the tetanus (TeTX) and the botulinum toxin (BoNT), respectively (Bruggemann, 2005). Whereas the *tetX* gene can only be found on large plasmids in *C. tetani* (see below), different toxinotypes of BoNT exist (A–G), the genes of which are located on the chromosome (*C. botulinum* A, B, E and F; *Clostridium butyricum*), on plasmids (*Clostridium argentinense*) or on bacteriophages (*C. botulinum* C and D) (Raffestin *et al.*, 2004; Bruggemann, 2005). Some strains can produce a mixture of two BoNTs and many type A strains contain cryptic or silent BoNT type B genes. There is strong evidence that the BoNT gene loci are located on (degenerated) mobile genetic elements, which accounts for their presence on chromosomes, plasmids and phages as well as their probable transfer among different clostridial strains in evolutionary history (Dineen *et al.*, 2003).

As mentioned, The TeTx is encoded on a large plasmid and to date has only been found in toxigenic strains of *C. tetani* (Finn *et al.*, 1984). Its genome sequence has been determined for the Massachusetts derivative strain E88. The plasmid pE88 is 74 kb in size with a low G+C content of only 24.5% (Bruggemann *et al.*, 2003; Bruggemann, 2005). It encodes 61 genes, which cover 67% of the plasmid sequence. An additional virulence factor, (a collagenase similar to the k toxin of *C. perfringens)* is encoded on pE88. The only other gene of known function is *tet*R, which encodes a positive regulator of TeTx and is located directly upstream of *tet*X. Additional regulatory genes found on pE88 include three sigma factor-like proteins (CTP05, CTP10 and CTP11) and a two-component system with unknown regulatory function (CTP21 and CTP22). A large portion of genes on pE88 code for transport proteins: five multisubunit ATP-binding cassette (ABC) transporters can be found, some of which show highest homology to peptide transporters responsible for bacteriocin efflux (Bruggemann *et al.*, 2003).

It has been shown that non-toxigenic *C. tetani* strains either completely lack plasmids or contain plasmids of different sizes. Recently, the plasmid of *C. tetani* strain E4222, a non-toxigenic variant, was sequenced (Bruggemann, 2005). It is 47 kb in size with a G+C content of 28.8% (slightly higher than that of pE88) encoding 63 genes, 30% of which show similarity to bacteriophage-related genes but no homology with plasmid pE88 was detected. Finally, it is important to remark that although the role of the plasmids in the *C. tetani* virulence is well established, no further studies have reported the mechanisms involved in the their horizontal transfer nor the environmental stimuli that trigger this process.

In addition to *Clostridium*, plasmids encoded toxins are present in other clinically relevant bacteria. *S. aureus* is capable of producing several plasmid-encoded toxins, including exfoliative toxin B (ETB) or enterotoxins D (SED) or J (SEJ). Warren *et al.*, in a pioneering study, demonstrated that ETB was associated with the presence of plasmid pRW001 (Warren *et al.*, 1974). This was later confirmed by Jackson and Iandolo (1986), who demonstrated that the structural gene for ETB, designated *etb*, was encoded in pRW001. ETB was also encoded in plasmid pETB, a 38.2-kb *S. aureus* plasmid, which in addition encodes for EDIN-C, a novel protein that belongs to the C3 family of ADP-ribosyltransferases modifying Rho GTPases, members of the Ras superfamily of proteins involved in cytoskeletal network regulation within eukaryotic cells (Yamaguchi *et al.*, 2001).

The gene encoding for SED, *entD*, is located on a 27.6-kb penicillinase plasmid designated pIB485 (Bayles and Iandolo, 1989). More recently, Zhang *et al.* (1998) have demonstrated that plasmid pIB485 not only encodes for SED but also for SEJ. The role of this toxin, alone or in conjunction with the SED toxin, to the staphylococcal pathogenesis remains to be determined.

Pathogenicity islands encoding toxins

The term pathogenicity island was coined by Hacker's group to describe two large unstable regions on the chromosome of uropathogenic *E. coli* (Blum *et al.*, 1994). Currently, this term is commonly used to describe regions in the genomes of certain pathogens that are absent is clonally related non-pathogenic strains. PAIs have not only relevance as a repertoire of virulence factors, including toxins, but also changed our way of thinking about the evolution of bacterial pathogenicity. Of special interest are the SaPIs, a family of PAIs present in *S. aureus*, which are the reservoirs of relevant virulence factors that use these mobile elements to be horizontally transferred (Novick *et al.*, 2010).

SaPIs

Staphylococcus aureus pathogenicity islands are a family of related 15–17 kb mobile genetic elements that commonly carry genes for superantigen toxins and other virulence factors. SaPIs were the first pathogenicity islands described for any Gram-positive species and the first pathogenicity islands for which mobility has been demonstrated (Lindsay *et al.*, 1998; Ubeda *et al.*, 2005). They are, at present, probably the best characterized of any of the bacterial pathogenicity islands (Ubeda *et al.*, 2007a,b, 2008). Complete sequences are known for more than 20 SaPIs, which form a highly coherent family with conserved functional and genetic organization. The SaPIs are biologically analogous to coliphage P4 (Lindqvist *et al.*, 1993) and to *Sulfolobus* plasmid pSSVx (Arnold *et al.*, 1999). The key feature of their mobility and spread is the induction by certain phages of their excision, replication and efficient encapsidation into specific small-headed phage-like infectious particles (Lindsay *et al.*, 1998; Ruzin *et al.*, 2001; Ubeda *et al.*, 2005). This sequence of events is referred to as the SaPI excision–replication–packaging (ERP) cycle and is the result of an intimate interaction between phage and SaPI genomes, in which SaPIs divert key phage functions for its own ends. Since the SaPI uses phage proteins for particle formation and packaging (Tallent *et al.*, 2007; Tormo *et al.*, 2008), it initiates its own replication cycle only when an active prophage provides the necessary inducing/derepressing functions (Ubeda *et al.*, 2008; Tormo-Más *et al.*, 2010). SaPIs encode an excisionase (*xis*), a site-specific integrase and proteins that initiate replication from its own specific replication origin (Ubeda *et al.*, 2007a). Other proteins are involved in remodelling the phage capsid to fit its smaller genome, and a terminase small subunit is encoded that recognizes its specific *pac* site, enabling the packaging of SaPI DNA via the phage terminase large subunit and portal protein (Ubeda *et al.*, 2007b). The phage is responsible for lysis. SaPIs are identifiable by several universal features including a directly repeated *att* site core, a specific integrase, a terminase small subunit (but never a large subunit or portal protein), a replication initiator protein and an adjacent replication origin, and a key site of divergent transcription. SaPIs are very widespread among the staphylococci, and using these criteria we have very recently identified them in several other Gram-positive species (unpublished results).

Clinical significance

Staphylococcal superantigens have been found, thus far, only in association with mobile genetic elements, including SaPIs, temperate phages and plasmids. The biological basis for this is unclear, but is consistent with the observation that superantigens in other organisms are also carried by mobile genetic elements. The toxin most frequently encoded by SaPIs is toxic shock syndrome toxin-1 (TSST-1), which has been found, thus far, associated only with SaPIs; and are exclusively responsible for menstrual TSS. Similarly, SEB has been found exclusively in association with SaPIs, which are responsible for a considerable fraction of non-menstrual TSS and food poisoning cases. SEB is also lethal in aerosol form and is listed as a Select Agent by the Department of Homeland Security (USA). Notably, TSST-1 appears to have a role in pathogenicity independently of toxic shock – it contributes importantly to morbidity and mortality in mice, which are not susceptible to TSS. Many of the SaPIs encode two or three superantigen toxins. Therefore, SaPIs may be regarded as major players in staphylococcal pathogenesis, perhaps in addition to the superantigenicity of their toxins. In this context SaPIbov2 contains *bap*, the gene for a ~240 kDa adhesin that is involved in biofilm formation and has a significant role in bovine mastitis (Cucarella *et al.*, 2001, 2002, 2004).

Expression of the toxin genes is increased during SaPI replication owing to the increase in gene dosage (unpublished data). Toxin production could also be specifically induced during SaPI replication if the gene is driven by a replication-induced promoter, as is the case with the phage-coded Shiga toxin (Zhang *et al.*, 2000). SOS induction caused by fluoroquinolones, β-lactams and probably other antibiotics therefore has the unfortunate side effect of inducing toxin synthesis, with adverse clinical consequences (Zhang *et al.*, 2000), as well as the production of phage or SaPI particles (Ubeda *et al.*, 2005; Maiques *et al.*, 2006), which would promote transfer of the toxin genes.

As described below, the vast majority of *S. aureus* strains carry one or more SaPIs, so that the SaPIs and their mobility constitute a significant feature of staphylococcal virulence. The occurrence of similar or identical SaPIs in unrelated strains suggests strongly that transfer occurs under natural circumstances as well as in the laboratory. An improved understanding of SaPI biology, genetics and dissemination will greatly aid in the epidemiological tracking and control of superantigen- and other SaPI-mediated diseases.

Open questions

Only recently the significant link between mobile genetics elements, genome diversity, and bacterial pathogenesis has been fully appreciated. The renewed interest in the biology of the MGEs has been a consequence of the complete genome sequencing of multiple and varied bacterial genomes. The results of comparative bacteria genomics, along with the discovery of novel MGEs, have allowed understanding the role of these elements in the bacterial physiology. Moreover, we are only just beginning to understand the complex relationship between the MGEs' life cycle, the environmental condition that induce their transfer, and their role in the physiology of the recipient strain. Continued investigation in the biology of these elements will reveal additional significant contributions about the role that the MGEs have in bacterial pathogenesis.

Web resources

http://db-mml.sjtu.edu.cn/MobilomeFINDER/
For identification of bacterial strains rich in novel genetic material and for high-throughput genomic island discovery

http://www.ispb.org/
International Society for Plasmid Biology and other Mobile Genetic Elements

http://genomes.urv.es/HGT-DB/
Horizontal gene transfer database

http://phast.wishartlab.com/
For identification of phage or pathogenicity island sequences

References

Acheson, D.W., and Donohue-Rolfe, A. (1989). Cancer-associated hemolytic uremic syndrome: a possible role of mitomycin in relation to Shiga-like toxins. J. Clin. Oncol. 7, 1943.

Acheson, D.W., Reidl, J., Zhang, X., Keusch, G.T., Mekalanos, J.J., and Waldor, M.K. (1998). *In vivo* transduction with shiga toxin 1-encoding phage. Infect. Immun. *66*, 4496–4498.

Arnold, H.P., She, Q., Phan, H., Stedman, K., Prangishvili, D., Holz, I., Kristjansson, J.K., Garrett, R., and Zillig, W. (1999). The genetic element pSSVx of the extremely thermophilic crenarchaeon *Sulfolobus* is a hybrid between a plasmid and a virus. Mol. Microbiol. *34*, 217–226.

Baba, T., Takeuchi, F., Kuroda, M., Yuzawa, H., Aoki, K., Oguchi, A., Nagai, Y., Iwama, N., Asano, K., Naimi, T., *et al.* (2002). Genome and virulence determinants of high virulence community-acquired MRSA. Lancet *359*, 1819–1827.

Beaber, J.W., Hochhut, B., and Waldor, M.K. (2004). SOS response promotes horizontal dissemination of antibiotic resistance genes. Nature *427*, 72–74.

Bergthorsson, U., and Ochman, H. (1998). Distribution of chromosome length variation in natural isolates of *Escherichia coli*. Mol. Biol. Evol. *15*, 6–16.

Betley, M.J., and Mekalanos, J.J. (1985). Staphylococcal enterotoxin A is encoded by phage. Science *229*, 185–187.

Blum, G., Ott, M., Lischewski, A., Ritter, A., Imrich, H., Tschape, H., and Hacker, J. (1994). Excision of large DNA regions termed pathogenicity islands from tRNA-specific loci in the chromosome of an *Escherichia coli* wild-type pathogen. Infect. Immun. *62*, 606–614.

Bruggemann, H. (2005). Genomics of clostridial pathogens: implication of extrachromosomal elements in pathogenicity. Curr. Opin. Microbiol. *8*, 601–605.

Bruggemann, H., Baumer, S., Fricke, W.F., Wiezer, A., Liesegang, H., Decker, I., Herzberg, C., Martinez-Arias, R., Merkl, R., Henne, A., *et al.* (2003). The genome sequence of *Clostridium tetani*, the causative agent of tetanus disease. Proc. Natl. Acad. Sci. U.S.A. *100*, 1316–1321.

Brussow, H., and Hendrix, R.W. (2002). Phage genomics: small is beautiful. Cell *108*, 13–16.

Brussow, H., Canchaya, C., and Hardt, W.D. (2004). Phages and the evolution of bacterial pathogens: from genomic rearrangements to lysogenic conversion. Microbiol. Mol. Biol. Rev. *68*, 560–602.

Campbell, A. (1994). Comparative molecular biology of lambdoid phages. Annu. Rev. Microbiol. *48*, 193–222.

Coleman, D.C., Arbuthnott, J.P., Pomeroy, H.M., and Birkbeck, T.H. (1986). Cloning and expression in *Escherichia coli* and *Staphylococcus aureus* of the beta-lysin determinant from *Staphylococcus aureus*: evidence that bacteriophage conversion of beta-lysin activity is caused by insertional inactivation of the beta-lysin determinant. Microb. Pathog. *1*, 549–564.

Coleman, D.C., Sullivan, D.J., Russell, R.J., Arbuthnott, J.P., Carey, B.F., and Pomeroy, H.M. (1989). *Staphylococcus aureus* bacteriophages mediating the simultaneous lysogenic conversion of beta-lysin, staphylokinase and enterotoxin A: molecular mechanism of triple conversion. J. Gen. Microbiol. *135*, 1679–1697.

Cucarella, C., Solano, C., Valle, J., Amorena, B., Lasa, I., and Penades, J.R. (2001). Bap, a *Staphylococcus aureus* surface protein involved in biofilm formation. J. Bacteriol. *183*, 2888–2896.

Cucarella, C., Tormo, M.A., Knecht, E., Amorena, B., Lasa, I., Foster, T.J., and Penades, J.R. (2002). Expression of the biofilm-associated protein interferes with host protein receptors of *Staphylococcus aureus* and alters the infective process. Infect. Immun. *70*, 3180–3186.

Cucarella, C., Tormo, M.A., Ubeda, C., Trotonda, M.P., Monzon, M., Peris, C., Amorena, B., Lasa, I., and Penades, J.R. (2004). Role of biofilm-associated protein bap in the pathogenesis of bovine *Staphylococcus aureus*. Infect. Immun. *72*, 2177–2185.

Daubin, V., Moran, N.A., and Ochman, H. (2003). Phylogenetics and the cohesion of bacterial genomes. Science *301*, 829–832.

Dineen, S.S., Bradshaw, M., and Johnson, E.A. (2003). Neurotoxin gene clusters in *Clostridium botulinum* type A strains: sequence comparison and evolutionary implications. Curr. Microbiol. *46*, 345–352.

Dobrindt, U., Janke, B., Piechaczek, K., Nagy, G., Ziebuhr, W., Fischer, G., Schierhorn, A., Hecker, M., Blum-Oehler, G., and Hacker, J. (2000). Toxin genes on pathogenicity islands: impact for microbial evolution. Int. J. Med. Microbiol. *290*, 307–311.

Finn, Jr, C.W.., Silver, R.P., Habig, W.H., Hardegree, M.C., Zon, G., and Garon, C.F. (1984). The structural gene for tetanus neurotoxin is on a plasmid. Science *224*, 881–884.

Fischetti, V.A. (2001). Phage antibacterials make a comeback. Nat. Biotechnol. *19*, 734–735.

Fitzgerald, J.R., Monday, S.R., Foster, T.J., Bohach, G.A., Hartigan, P.J., Meaney, W.J., and Smyth, C.J. (2001). Characterization of a putative pathogenicity island from bovine *Staphylococcus aureus* encoding multiple superantigens. J. Bacteriol. *183*, 63–70.

Fraser, M.E., Fujinaga, M., Cherney, M.M., Melton-Celsa, A.R., Twiddy, E.M., O'Brien, A.D., and James, M.N. (2004). Structure of shiga toxin type 2 (Stx2) from *Escherichia coli* O157:H7. J. Biol. Chem. *279*, 27511–27517.

Frost, L.S., Leplae, R., Summers, A.O., and Toussaint, A. (2005). Mobile genetic elements: the agents of open source evolution. Nat. Rev. Microbiol. *3*, 722–732.

Gal-Mor, O., and Finlay, B.B. (2006). Pathogenicity islands: a molecular toolbox for bacterial virulence. Cell. Microbiol. *8*, 1707–1719.

Goerke, C., Koller, J., and Wolz, C. (2006). Ciprofloxacin and trimethoprim cause phage induction and virulence modulation in *Staphylococcus aureus*. Antimicrob. Agents Chemother. *50*, 171–177.

Hacker, J., and Kaper, J.B. (2000). Pathogenicity islands and the evolution of microbes. Annu. Rev. Microbiol. *54*, 641–679.

Hendrix, R.W. (2003). Bacteriophage genomics. Curr. Opin. Microbiol. *6*, 506–511.

Hendrix, R.W., Smith, M.C., Burns, R.N., Ford, M.E., and Hatfull, G.F. (1999). Evolutionary relationships among diverse bacteriophages and prophages: all the world's a phage. Proc. Natl. Acad. Sci. U.S.A. *96*, 2192–2197.

Hochhut, B., and Waldor, M.K. (1999). Site-specific integration of the conjugal *Vibrio cholerae* SXT element into *prfC*. Mol. Microbiol. *32*, 99–110.

Hochhut, B., Wilde, C., Balling, G., Middendorf, B., Dobrindt, U., Brzuszkiewicz, E., Gottschalk, G., Carniel, E., and Hacker, J. (2006). Role of pathogenicity island-associated integrases in the genome plasticity of uropathogenic *Escherichia coli* strain 536. Mol. Microbiol. *61*, 584–595.

Hoey, D.E., Currie, C., Else, R.W., Nutikka, A., Lingwood, C.A., Gally, D.L., and Smith, D.G. (2002). Expression of receptors for verotoxin 1 from *Escherichia coli* O157 on bovine intestinal epithelium. J. Med. Microbiol. *51*, 143–149.

Jaeger, J.L., and Acheson, D.W. (2000). Shiga toxin-producing *Escherichia coli*. Curr. Infect. Dis. Rep. *2*, 61–67.

Katayama, Y., Ito, T., and Hiramatsu, K. (2000). A new class of genetic element, *Staphylococcus* cassette chromosome mec, encodes methicillin resistance in *Staphylococcus aureus*. Antimicrob. Agents Chemother. *44*, 1549–1555.

Khil, P.P., and Camerini-Otero, R.D. (2002). Over 1000 genes are involved in the DNA damage response of *Escherichia coli*. Mol. Microbiol. *44*, 89–105.

Konstantinidis, K.T., and Tiedje, J.M. (2005a). Genomic insights that advance the species definition for prokaryotes. Proc. Natl. Acad. Sci. U.S.A. *102*, 2567–2572.

Konstantinidis, K.T., and Tiedje, J.M. (2005b). Towards a genome-based taxonomy for prokaryotes. J. Bacteriol. *187*, 6258–6264.

Labandeira-Rey, M., Couzon, F., Boisset, S., Brown, E.L., Bes, M., Benito, Y., Barbu, E.M., Vazquez, V., Hook, M., Etienne, J., *et al.* (2007). *Staphylococcus aureus* Panton–Valentine leukocidin causes necrotizing pneumonia. Science *315*, 1130–1133.

Lawrence, J.G., Hendrix, R.W., and Casjens, S. (2001). Where are the pseudogenes in bacterial genomes? Trends Microbiol. *9*, 535–540.

Lindqvist, B.H., Deho, G., and Calendar, R. (1993). Mechanisms of genome propagation and helper exploitation by satellite phage P4. Microbiol. Rev. *57*, 683–702.

Lindsay, J.A., Ruzin, A., Ross, H.F., Kurepina, N., and Novick, R.P. (1998). The gene for toxic shock toxin is carried by a family of mobile pathogenicity islands in *Staphylococcus aureus*. Mol. Microbiol. *29*, 527–543.

Little, J.W. (1991). Mechanism of specific LexA cleavage: autodigestion and the role of RecA coprotease. Biochimie 73, 411–421.

Little, J.W. (1993). LexA cleavage and other self-processing reactions. J. Bacteriol. *175*, 4943–4950.

Little, J.W., and Mount, D.W. (1982). The SOS regulatory system of *Escherichia coli*. Cell 29, 11–22.

Little, S.F., and Ivins, B.E. (1999). Molecular pathogenesis of *Bacillus anthracis* infection. Microbes Infect. *1*, 131–139.

Maiques, E., Ubeda, C., Campoy, S., Salvador, N., Lasa, I., Novick, R.P., Barbe, J., and Penades, J.R. (2006). β-Lactam Antibiotics Induce the SOS Response and Horizontal Transfer of Virulence Factors in *Staphylococcus aureus*. J. Bacteriol. *188*, 2726–2729.

Miller, C., Thomsen, L.E., Gaggero, C., Mosseri, R., Ingmer, H., and Cohen, S.N. (2004). SOS response induction by beta-lactams and bacterial defense against antibiotic lethality. Science *305*, 1629–1631.

Muniesa, M., Lucena, F., and Jofre, J. (1999). Comparative survival of free shiga toxin 2-encoding phages and *Escherichia coli* strains outside the gut. Appl. Environ. Microbiol. *65*, 5615–5618.

Muniesa, M., de Simon, M., Prats, G., Ferrer, D., Panella, H., and Jofre, J. (2003). Shiga toxin 2-converting bacteriophages associated with clonal variability in *Escherichia coli* O157:H7 strains of human origin isolated from a single outbreak. Infect. Immun. *71*, 4554–4562.

Muniesa, M., Blanco, J.E., De Simon, M., Serra-Moreno, R., Blanch, A.R., and Jofre, J. (2004). Diversity of *stx2* converting bacteriophages induced from Shiga-toxin-producing *Escherichia coli* strains isolated from cattle. Microbiology *150*, 2959–2971.

Mushegian, A.R., and Koonin, E.V. (1996). A minimal gene set for cellular life derived by comparison of complete bacterial genomes. Proc. Natl. Acad. Sci. U.S.A. 93, 10268–10273.

Novick, R.P. (1969). Extrachromosomal inheritance in bacteria. Bacteriol. Rev. 33, 210–263.

Novick, R.P. (2003). Mobile genetic elements and bacterial toxinoses: the superantigen-encoding pathogenicity islands of *Staphylococcus aureus*. Plasmid *49*, 93–105.

Novick, R.P., and Subedi, A. (2007). The SaPIs: mobile pathogenicity islands of Staphylococcus. Chem. Immunol. Allergy 93, 42–57.

O'Brien, A.D., Newland, J.W., Miller, S.F., Holmes, R.K., Smith, H.W., and Formal, S.B. (1984). Shiga-like toxin-converting phages from *Escherichia coli* strains that cause hemorrhagic colitis or infantile diarrhea. Science *226*, 694–696.

Ochman, H., and Santos, S.R. (2005). Exploring microbial microevolution with microarrays. Infect. Genet. Evol. *5*, 103–108.

Pushker, R., Mira, A., and Rodriguez-Valera, F. (2004). Comparative genomics of gene-family size in closely related bacteria. Genome Biol. 5, R27.

Qiu, X., Gurkar, A.U., and Lory, S. (2006). Interstrain transfer of the large pathogenicity island (PAPI-1) of *Pseudomonas aeruginosa*. Proc. Natl. Acad. Sci. U.S.A. *103*, 19830–19835.

Raffestin, S., Marvaud, J.C., Cerrato, R., Dupuy, B., and Popoff, M.R. (2004). Organization and regulation of the neurotoxin genes in *Clostridium botulinum* and *Clostridium tetani*. Anaerobe *10*, 93–100.

Rappuoli, R., and Nabel, G. (2001). Vaccines: ideal drugs for the 21st century? Curr. Opin. Investig. Drugs 2, 45–46.

Ruzin, A., Lindsay, J., and Novick, R.P. (2001). Molecular genetics of SaPI1 – a mobile pathogenicity island in *Staphylococcus aureus*. Mol. Microbiol. *41*, 365–377.

Sakellaris, H., Luck, S.N., Al-Hasani, K., Rajakumar, K., Turner, S.A., and Adler, B. (2004). Regulated site-specific recombination of the she pathogenicity island of *Shigella flexneri*. Mol. Microbiol. 52, 1329–1336.

Schmidt, H. (2001). Shiga-toxin-converting bacteriophages. Res. Microbiol. *152*, 687–695.

Schmidt, H., and Hensel, M. (2004). Pathogenicity islands in bacterial pathogenesis. Clin. Microbiol. Rev. *17*, 14–56.

Schmidt, H., Bielaszewska, M., and Karch, H. (1999). Transduction of enteric *Escherichia coli* isolates with a derivative of Shiga toxin 2-encoding bacteriophage phi3538 isolated from *Escherichia coli* O157:H7. Appl. Environ. Microbiol. *65*, 3855–3861.

Schubert, S., Dufke, S., Sorsa, J., and Heesemann, J. (2004). A novel integrative and conjugative element (ICE) of *Escherichia coli*: the putative progenitor of the *Yersinia* high-pathogenicity island. Mol. Microbiol. *51*, 837–848.

Serra-Moreno, R., Jofre, J., and Muniesa, M. (2007). Insertion site occupancy by stx2 bacteriophages depends on the locus availability of the host strain chromosome. J. Bacteriol. *189*, 6645–6654.

Serra-Moreno, R., Jofre, J., and Muniesa, M. (2008). The CI repressors of Shiga toxin-converting prophages are involved in co-infection of *Escherichia coli* strains, which causes a down regulation in the production of Shiga toxin 2. J. Bacteriol. *190*, 4722–4735.

Sullivan, J.T., and Ronson, C.W. (1998). Evolution of rhizobia by acquisition of a 500-kb symbiosis island that integrates into a phe-tRNA gene. Proc. Natl. Acad. Sci. U.S.A. *95*, 5145–5149.

Sumby, P., and Waldor, M.K. (2003). Transcription of the toxin genes present within the staphylococcal phage phiSa3ms is intimately linked with the phage's life cycle. J. Bacteriol. *185*, 6841–6851.

Tallent, S.M., Langston, T.B., Moran, R.G., and Christie, G.E. (2007). Transducing particles of *Staphylococcus aureus* pathogenicity island SaPI1 are comprised of helper phage-encoded proteins. J. Bacteriol. *189*, 7520–7524.

Tormo, M.A., Ferrer, M.D., Maiques, E., Ubeda, C., Selva, L., Lasa, I., Calvete, J.J., Novick, R.P., and Penades, J.R. (2008). *Staphylococcus aureus* pathogenicity island DNA is packaged in particles composed of phage proteins. J. Bacteriol. *190*, 2434–2440.

Tormo-Más, M.A., Mir, I., Shrestha, A., Tallent, S.M., Campoy, S., Lasa, I., Barbé, J., Novick, R.P., Christie, G.E., and Penadés, J.R. (2010). Moonlighting bacteriophage proteins derepress staphylococcal pathogenicity islands. Nature. *465*, 779–82.

Ubeda, C., Tormo, M.A., Cucarella, C., Trotonda, P., Foster, T.J., Lasa, I., and Penades, J.R. (2003). Sip, an integrase protein with excision, circularization and integration activities, defines a new family of mobile *Staphylococcus aureus* pathogenicity islands. Mol. Microbiol. *49*, 193–210.

Ubeda, C., Maiques, E., Knecht, E., Lasa, I., Novick, R.P., and Penades, J.R. (2005). Antibiotic-induced SOS response promotes horizontal dissemination of pathogenicity island-encoded virulence factors in staphylococci. Mol. Microbiol. *56*, 836–844.

Ubeda, C., Barry, P., Penades, J.R., and Novick, R.P. (2007a). A pathogenicity island replicon in *Staphylococcus aureus* replicates as an unstable plasmid. Proc. Natl. Acad. Sci. U.S.A. *104*, 14182–14188.

Ubeda, C., Maiques, E., Tormo, M.A., Campoy, S., Lasa, I., Barbe, J., Novick, R.P., and Penades, J.R. (2007b). SaPI operon I is required for SaPI packaging and is controlled by LexA. Mol. Microbiol. *65*, 41–50.

Ubeda, C., Maiques, E., Barry, P., Matthews, A., Tormo, M.A., Lasa, I., Novick, R.P., and Penades, J.R. (2008). SaPI mutations affecting replication and transfer and enabling autonomous replication in the absence of helper phage. Mol. Microbiol. *67*, 493–503.

Wagner, P.L., and Waldor, M.K. (2002). Bacteriophage control of bacterial virulence. Infect. Immun. *70*, 3985–3993.

Wagner, P.L., Acheson, D.W., and Waldor, M.K. (1999). Isogenic lysogens of diverse shiga toxin 2-encoding bacteriophages produce markedly different amounts of shiga toxin. Infect. Immun. *67*, 6710–6714.

Waldor, M.K., and Friedman, D.I. (2005). Phage regulatory circuits and virulence gene expression. Curr. Opin. Microbiol. *8*, 459–465.

Walker, G.C. (1984). Mutagenesis and inducible responses to deoxyribonucleic acid damage in *Escherichia coli*. Microbiol. Rev. *48*, 60–93.

Weigel, L.M., Clewell, D.B., Gill, S.R., Clark, N.C., McDougal, L.K., Flannagan, S.E., Kolonay, J.F., Shetty, J., Killgore, G.E., and Tenover, F.C. (2003). Genetic analysis of a high-level vancomycin-resistant isolate of *Staphylococcus aureus*. Science *302*, 1569–1571.

Wong, C.S., Jelacic, S., Habeeb, R.L., Watkins, S.L., and Tarr, P.I. (2000). The risk of the hemolytic-uremic syndrome after antibiotic treatment of *Escherichia coli* O157:H7 infections. N. Engl. J. Med. *342*, 1930–1936.

Yamaguchi, T., Hayashi, T., Takami, H., Nakasone, K., Ohnishi, M., Nakayama, K., Yamada, S., Komatsuzawa, H., and Sugai, M. (2000). Phage conversion of exfoliative toxin A production in *Staphylococcus aureus*. Mol. Microbiol. *38*, 694–705.

Ye, Z.H., and Lee, C.Y. (1989). Nucleotide sequence and genetic characterization of staphylococcal bacteriophage L54a *int* and *xis* genes. J. Bacteriol. *171*, 4146–4153.

Zhang, X., McDaniel, A.D., Wolf, L.E., Keusch, G.T., Waldor, M.K., and Acheson, D.W. (2000). Quinolone antibiotics induce Shiga toxin-encoding bacteriophages, toxin production, and death in mice. J. Infect. Dis. *181*, 664–670.

The Staphylococcal Superantigen-like Toxins

6

Ries J. Langley and John D. Fraser

Abstract

Staphylococcus aureus is a serious human pathogen responsible for a wide range of hospital and community acquired infections and deaths. It produces many virulence factors with functions involved first in invasion and then establishment and persistence within the body. Immune evasion is essential for survival of *S. aureus* in the host. The staphylococcal superantigen-like proteins (SSLs), a family of recently discovered proteins expressed by all strains, play key roles in immune evasion by targeting important components of innate immunity. For example SSL7 blocks IgA from binding its Fc receptor on immune cells. IgA is the body's first line of defence against invading pathogens. SSL7 also binds the complement component C5 and prevents the cleavage of this molecule into its active fragments. The active fragment C5a has been shown to be of great importance in clearance of *S. aureus*. The binding of SSL10 to IgG1 prevents effector functions of this important class of antibody. Another subgroup of SSLs bind immune cell receptor proteins in a glycan-dependent manner to inhibit host immune defence against *S. aureus*. It is becoming clear that the SSLs are among the important arsenal of proteins that are responsible for immune evasion in *S. aureus*.

Introduction

Staphylococcus aureus – the pathogen

Staphylococcus aureus is a member of the genus *Staphylococcus* of which there are approximately 40 taxa (Takahashi *et al.*, 1999; Harris *et al.*, 2002), a third of which are human commensals. *S. aureus* is predominantly positive for the gene encoding coagulase while most other staphylococci are coagulase-negative. This gives it the ability to convert fibrinogen to fibrin and thus cause clot formation. It is the only coagulase-positive staphylococcus species that is a common colonizer of humans (Hanselman *et al.*, 2009; Walther *et al.*, 2012) and typically colonizes the anterior nares. Approximately 20% (range 12–30%) of the population are persistent nasal carriers, while around 30% are intermittent carriers (range 16–70%), and approximately 50% (range 16–69%) are non-carriers. The variations in the reported proportions of carrier status reflect differences in culturing techniques, the populations studied, and interpretation of results (Wertheim *et al.*, 2005). Regardless of the actual percentage it is clear that *S. aureus* has a significant presence in human populations. The throat and skin are other major reservoirs for *S. aureus* carriage (Krishna and Miller, 2012) with a recent study revealing that throat carriage is at least as persistent as nasal carriage and often occurs preferentially in this site over the anterior nares (Nilsson and Ripa, 2006).

S. aureus is by far the most virulent of the pathogenic staphylococci, causing the majority of human diseases associated with this genus. These cover the full spectrum of severity, from the superficial skin infections such as boils and carbuncles, impetigo, cellulitis, and folliculitis, to invasive diseases including bacteraemia and sepsis, osteomyelitis, endocarditis, pneumonia, and wound infections. Major toxin-mediated diseases include staphylococcal scalded skin syndrome (SSSS) caused by the epidermolytic toxins A and B (ETA and ETB) while the violent food poisoning and toxic shock syndrome (TSS) associated with *S. aureus* are caused by a family of potent immune modulators named the superantigens (SAgs) (reviewed in Lowy, 1998).

S. aureus is considered to be an opportunistic pathogen with the ability to infect practically any site in the body with infections said to be preceded by colonization. Its success can be attributed to the abundance of virulence factors encoded in its genome. Kuroda *et al.* (2001) identified around 150 known and potential virulence factors in the first published whole genome analysis of *S. aureus*, many of which were found to be carried by mobile genetic elements (MGEs). Types of *S. aureus* MGE include plasmids, transposons (Tn), insertion sequences (IS), bacteriophages, pathogenicity and genomic islands (SaPI or νSa), and staphylococcal cassette chromosomes (SSC) (Malachowa and DeLeo, 2010). MGEs make up the accessory genome and encode proteins such as resistance and virulence factors that are involved in adaptation of the bacteria to different environments. In addition to these adaptation-mediating genes, MGEs possess the machinery required for transfer and integration into new host DNA. The accessory genes generally have a different GC content to the core genome owing to their acquisition from other bacterial species but are usually under the control of global regulators located within the core genome. The pathogenicity/genomic islands (νSa) of *S. aureus* encode a great deal of the toxins or virulence factors (including enterotoxins, exotoxins, leucocidins, and leucotoxins) not found in less pathogenic staphylococci such as *S. epidermidis* (Gill *et al.*, 2005) and explain how *S. aureus* has evolved to be such a successful opportunistic human pathogen. Genes encoding virulence factors tend to be found on phages and pathogenicity/genomic islands while resistance genes are more likely to be transferred by SCC, plasmids, and (Lindsay and Holden, 2006).

In 1961 (shortly after the introduction of methicillin in 1959) the first methicillin-resistant *Staphylococcus aureus* (MRSA) strain was isolated (Barber, 1961). The term MRSA is used to describe strains that are resistant to penicillins and cephalosporins which are β-lactam antibiotics that act by inhibiting bacterial cell wall biosynthesis. The resistance of MRSA is due to the particular penicillin-binding protein (PBP) it produces named PBP2a which has a much lower binding affinity for β-lactam than other PBPs so can participate in cell wall biosynthesis in the presence of otherwise inhibitory concentrations of antibiotic. The *mecA* gene that encodes PBP2a is carried by the staphylococcal cassette chromosome *mec* (SCC*mec*) (Katayama *et al.*, 2000). Since the 1970s MRSA has become the leading cause of nosocomial infection worldwide and while the rates of Hospital Associated MRSA (HA-MRSA) have begun to stabilize, the emergence of Community Associated MRSA (CA-MRSA) in the closing decades of last century has ensured that rates of infection continue to rise (David *et al.*, 2012). The seriousness of antibiotic resistant *S. aureus* is evidently demonstrated by a recent report from the Centers for Disease Control and Prevention that determined MRSA to be the main cause of serious and fatal bacterial infections in the United States of America with an incident rate over twice as high as the next leading bacterial pathogen (Klevens *et al.*, 2007). Of addition concern are the outcomes that misuse of antibiotics can cause. Besides applying selective pressure on the bacteria to develop antibiotic resistance genes, low levels of β-lactam antibiotics induce biofilm formation in *S. aureus* by up-regulating the autolysin AtlA. Cell lysis mediated by AtlA results in the release of extracellular DNA which acts as a biofilm matrix adhesion (Kaplan *et al.*, 2012). There is an association between biofilm formation and the persistence of chronic staphylococcal infections as a consequence of the increased resistance to chemotherapies and host defences that biofilms afford the bacterium (Kiedrowski and Horswill, 2011).

The combined effect of its enormous arsenal of virulence factors and capacity to form biofilms, together with its ever increasing ability to resist antibiotic treatment, has contributed to the rise of *S. aureus* as a serious human pathogen. As antibiotic resistance becomes more common, it is apparent that there is a need for alternative therapies. Fully understanding how *S. aureus* interacts with the host immune system and characterizing the factors it produces to aid in its immune evasion is of utmost importance in designing new strategies to lessening the impact of this pathogen. The wide variety of virulence factors produced by *S. aureus* promote colonization and immune evasion by interfering with the host immune system. There are both secreted and cell-surface expressed factors that predominantly target innate immune defence mechanisms including the complement system, host antimicrobial peptides, and recruitment and activation of innate effector cells (Foster, 2005; Chavakis *et al.*, 2007). Of the recently discovered virulence factors, one family of secreted immune evasion molecules, the SSLs are proving to play a key role in immune evasion by targeting important components of innate immunity. This chapter will describe current knowledge of the SSLs and how they act on the immune system to benefit *S. aureus* survival.

The immune defence against *S. aureus*

Infections caused by *S. aureus* are predominantly preceded by colonization with the bacterium (von Eiff *et al.*, 2001; Wertheim *et al.*, 2004). Whilst it can be considered as commensal on the skin or mucosa its pathogenic potential is realized when these physical barriers are compromised and *S. aureus* comes under challenge from the hosts immune system. The immune defence against *S. aureus* depends primarily on innate mechanisms involving the competent system and the recruitment and activation of neutrophils (Foster, 2005). The complement system comprises of a family of around 30 secreted and membrane bound proteins that assist and complement the actions of antibodies and phagocytes. They achieve this in a number of ways. By binding to antibodies and the microbial cell wall, the process of opsonization, the complement system enhances the phagocytic process. Additionally complement activation results in the generation of (1) anaphylatoxins that mediate chemotaxis and inflammation, and (2) the formation of the terminal complement complex C5b–9, or membrane attack complex (MAC), which forms channels in the microbial cell membrane resulting in cell lysis and death. Like other Gram-positive bacteria, *S. aureus* is considered to be resistant to lysis due to its thick cell wall. So in the context of *S. aureus*, complement functions by attracting and activating phagocytes. The importance of the anaphylatoxin C5a in the defence against *S. aureus* is well documented. It has been shown to be protective in a mouse model of *S. aureus* bloodstream infection (von Köckritz-Blickwede *et al.*, 2010). Deletion of C5αR makes mice more susceptible to *S. aureus* infection (Hopken *et al.*, 1996), while C5 deficiency in humans and mice correlates with the persistence of *S. aureus* infections (Jacobs and Miller, 1972; Cerquetti *et al.*, 1983). Neutrophil localization to the site of infection is crucial for bacterial clearance (Miller and Cho, 2011) and deficiencies in neutrophil function are associated with increased susceptibility to *S. aureus* infection (Rigby and DeLeo, 2012).

Complement has three pathways of activation that converge at a point where the opsonin C3b is generated from the abundant serum protein complement C3 (reviewed in Gasque, 2004; Lambris *et al.*, 2008). Classical complement pathway (CP) activation results from generation of the C1qr^{2}s^2 complex on bacterial surface-bound antibody that subsequently binds and cleaves complement component C4 into C4a and C4b. The larger product, C4b, then binds covalently to the bacterial surface. Activation of the lectin pathway (LP) relies on direct binding of the mannose-binding lectin (MBL) or ficolins to staphylococcal peptidoglycan (PGN) or lipoteichoic acid (LTA), which recruit and activate the MASP proteases that also cleave C4. Rapid recruitment of C2 by the CP- and LP-generated C4b results in C2 cleavage by C1qr^{2}s^2 or MASPs and formation of the C4b2a convertase complex. In the alternative pathway (AP) hydrolysis of C3 in the plasma to C3a and C3b occurs spontaneous and is rapidly inactivated by factor I. Should the C3b covalently bind to the microbial surface it is protected from inactivation so can bind and activated factor B to form the alternative C3bBb convertase complex. C4b2a and C3bBb go on to cleave C3 to C3b which binds to the complex and forms the C5 convertases C4b2a3b and C3bBb3b. Cleavage of C5 by these convertases releases the most potent of the anaphylatoxins, C5a, which activates neutrophils through the C5a receptor.

Activation of complement plays two parts in the clearance of *S. aureus*. The first is C3b coating (opsonization) of the bacteria for recognition by neutrophil complement receptors (CR1, CR2 and CR3) and the second is the release of small soluble anaphylatoxins C3a, C4a, and C5a that activate many cell types including endothelial cells and neutrophils. The anaphylatoxins, in particular C5a acts as a chemoattractant to activate neutrophils through the C5a receptor (C5aR) as an essential first step in phagocytosis. Neutrophils are rapidly recruited to the site of infection by attaching to the wall of blood vessels adjacent to the infection through a process initially mediated by P- and E-selectins that have been unregulated on endothelial cells and platelets (Guyer *et al.*, 1996). Surface expression of P-selectin can be induced by C5a stimulation of endothelial cells (Foreman *et al.*, 1994). Once P-selectin binds the carbohydrate sialyl Lewis X (sLeX) on the P-selectin glycoprotein ligand (PSGL-1) expressed on all leucocytes, including neutrophils (Frenette *et al.*, 2000), it allows for the initial rolling of the neutrophil along the inflamed vessel so that sampling the endothelial cell surface for inflammatory signals can occur. This process culminates in the neutrophil transmigrating through the vessel wall and following the chemotactic gradient towards the site of infection. Additionally, *S. aureus* can be phagocytosed directly by Fc receptor recognition of bound antibody although phagocytosis is significantly amplified and accelerated with the addition of complement.

Direct detection of *S. aureus* by neutrophils can occur through the recognition of pathogen associated molecular patterns (PAMPs) including, PGN, lipoprotein (Lpp), and LTA, by the pattern recognition receptor TLR2 (Fournier and Philpott, 2005; Pietrocola *et al.*, 2011). Signalling through TLR2 activates the transcription factor NF-κB (nuclear factor κB), which controls the expression of genes encoding cytokines, chemokines, and costimulatory molecules that are necessary for the activation of the defence response. It has also been reported that TLR2 is necessary for the efficient oxidative killing of *S. aureus* by murine neutrophils by activating nicotinamide adenine dinucleotide phosphate (NADPH) oxidase (Jann *et al.*, 2011).

Neutrophil killing of Gram-positive bacteria

There are multiple mechanisms attributed to neutrophil intracellular killing such as lysozyme, lactoferrin, defensins, cathepsins, acid hydrolases, elastase, and metal chelation, many of which are stored in azurophilic and specific granules of neutrophils that can fuse with phagocytic vacuoles (and to the plasma membrane in the case of specific granules) to effect their antimicrobial functions (Rigby and DeLeo, 2012). The mobilization and targeting of granules is regulated in part by actin/cytoskeletal rearrangements.

Antimicrobial peptides (AMPs) including defensins, cathelicidins, and peptide fragments such as those derived from lysozyme and lactoferrin, act either by inserting into the cell membrane to form pores or by penetration of the membrane to bind critical intracellular molecules (Brogden, 2005). High concentrations are required and not all anti-microbial peptides are effective against *S. aureus*.

Another mechanism of *S. aureus* killing is by the predominant neutrophil granule protein calprotectin (S100A8) that effectively chelates metals such as Mn^{2+} and Zn^{2+} that are essential for staphylococcal growth. Calprotectin-deficient mice are more susceptible to organ abscess formation and display higher levels of free Mn^{2+} and Zn^{2+} in the infected abscesses than wild-type mice (Corbin *et al.*, 2008).

The main mechanism of *S. aureus* killing by human neutrophils is however believed to be by the production of reactive oxygen species (ROS), such as superoxide, during the respiratory or oxidative burst (reviewed in Rigby and DeLeo, 2012). ROS are produced by enzymatic oxidation within acidic neutrophil granules by the abundant enzymes NADPH oxidase and myeloperoxidase (MPO). NADPH oxidase transfers electrons from NADPH in the cytosol to molecular oxygen to produce superoxide which in itself has limited direct microbicidal capacity but is used to generate secondarily derived ROS such as hypochlorous acid, hydroxyl radical, chloramines, and singlet oxygen that are all effective microbicides. Defects in NADPH oxidase lead to recurrent staphylococcal infection and chronic granulomatous disease (CGD) (Lekstrom-Himes and Gallin, 2000). When MPO is released into the phagosome from azurophils it catalyses a reaction with chloride and hydrogen peroxide to produce hypochlorous acid and secondary products such as chloramines, hydroxyl radical, and singlet oxygen. It has been suggested that entrapment of *S. aureus* in neutrophil extracellular traps (NETs) may not be sufficient to kill the bacteria (Menegazzi *et al.*, 2012), however it is the release of large amounts of NET-bound MPO during NET formation that contributes to *S. aureus* killing via the production of hypochlorous acid from available hydrogen peroxide (Parker *et al.*, 2012). In response to the oxidative burst *S. aureus* has several antioxidant defences such as free manganese, catalase, and superoxide dismutase. *S. aureus* also expresses the anti-oxidant carotenoid pigment staphyloxanthin that gives the organism its classic golden colour. Defectives in biosynthesis of the carotenoid result in decreased *S. aureus* virulence and enhancement of its innate immune clearance (Liu *et al.*, 2005; Liu *et al.*, 2008).

Naturally *S. aureus* wants to prevent destruction by the host immune system so produces a vast array of virulence factors to achieve this. Research within the last decade has revealed that the SSLs are among the important arsenal of proteins that are responsible for immune evasion in *S. aureus*.

The SSLs – an introduction

The SSLs are a family of up to 14 proteins of approximately 22–24 kDa secreted by *S. aureus*. First named the staphylococcal exotoxin-like proteins (SETs) by Williams *et al.* (2000), this family of exoproteins was subsequently remained the SSLs (Lina *et al.*, 2004) to prevent confusion in terminology with the staphylococcal enterotoxin family of superantigens which inevitable extended its membership to include staphylococcal enterotoxin T (SET) in 2008 (Ono *et al.*, 2008). The renaming of these exoproteins prevented the additional confusion that was created when *set* alleles were given separate gene numbers. For example *set6–set15* identified by Kuroda *et al.* (2001) in the strains N315 and Mu50 were in fact the same genes as *set16–set26* reported in the genome analysis of strain MW2 by Baba *et al.* (2002). Bioinformatic analysis of the SSLs revealed the existence of homology to the two most conserved regions of the superantigens (SAgs) consisting of the α4 alpha helix of the β-grasp domain: K-x(2)-[LIVF]-x(4)-[LIVF]-D-x(3)-R-x(2)-L-x(5)-[LIV]-Y (Prosite PS00278) and the opposing face on the OB-domain: Y-GG-[LIV]-T-{I}-{N}-xx-N (Prosite PS00277) (Williams *et al.*, 2000). These consensus SAg amino acid sequences that were used to identify the SSLs are highly conserved because they form the internal interface between the two domains and are responsible for the structural integrity of the protein (Arcus *et al.*, 2002). However there is one key difference in the α4 alpha-helix between these two related protein families. While there is a glutamine (Q) present in all the SAgs except TSST-1, there is an absolutely conserved lysine (K) in the equivalent position in the SSL sequence. Another contrasting feature of the SSLs is that they have been found in every examined strain, unlike the SAgs that are variably carried by *S. aureus*. In fact the presence of so many *ssl* genes within a single genome gives support to the importance of these proteins to *S. aureus* and is highly suggestive of non-redundant functions for the various members of the family. Additional evidence that they are important to the bacteria during infection is the fact that almost all people tested have antibodies against SSLs (Arcus *et al.*, 2002; Fitzgerald *et al.*, 2003; Al-Shangiti *et al.*, 2005; Langley *et al.*, 2010).

Genetics of the *ssls*

Genomic localization

The SSLs are numbered in the order they appear in the genome. The genes encoding SSL1 to SSL10 (*ssl1–ssl10*) are located clustered together on a pathogenicity island initially named SaPIn2 (N315) or SaPIm2 (mu50) (Kuroda *et al.*, 2001) which has subsequently been renamed as the genomic island νSaα (Baba *et al.*, 2002). *ssl11* is located 3.4 kb downstream of this cluster due to insertion of two restriction/modification genes *hsdM* and *hsdS* between *ssl10* and *ssl11*. This event has created considerable allelic sequence diversity at the 5′ end of *ssl11*. The second half of νSaα contains two hypothetical genes of unknown function followed by a tandem cluster of up to 11 putative lipoprotein encoding genes (Fig. 6.1A).

Unlike the majority of genomic and pathogenicity island of *S. aureus*, νSaα has been found in every strain and isolate studied so far, indicating it is both evolutionarily old and important to survival and/or evasion. In fact it has been hypothesized that its acquisition into the ancestral genome has contributed to the subsequent evolution of *S. aureus* as a major human pathogen (Baba *et al.*, 2008; Holt *et al.*, 2011). It is not known if

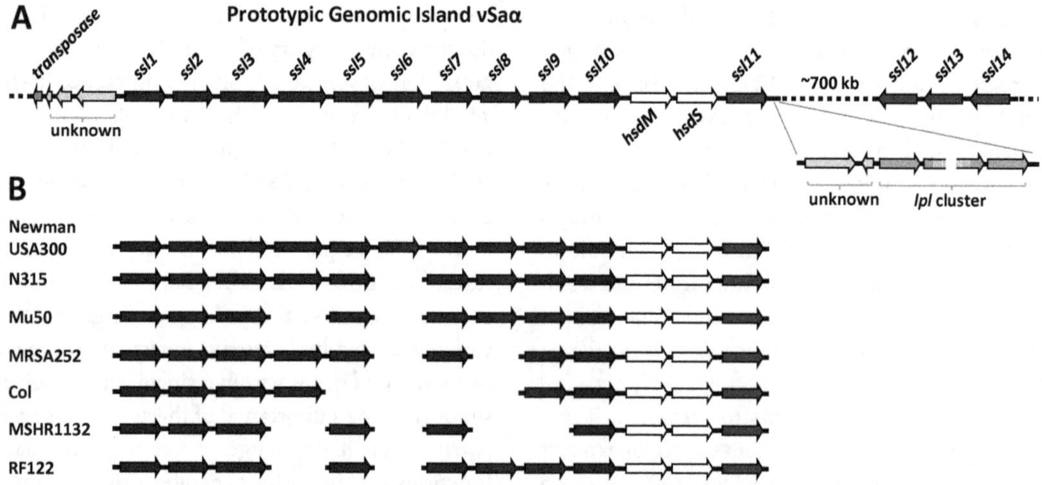

Figure 6.1 The genomic island vSaα. A. The prototypic vSaα genomic island structure possessing the full complement of ssls, preceded by genes of unknown function, a transposases remnant, and followed by the restriction/modification genes *hsdM* and *hsdS*, genes of unknown function, and a tandem cluster of up to 11 *lpl* genes. Included is the relative genome position and orientation of *ssl12*, *ssl13*, and *ssl14*. B. The *ssl* region of vSaα from selected published strains showing the variation *ssl* gene composition.

vSaα is mobile, though if so its ubiquitous presence dictates that it must be selected for and is in some way essential to *S. aureus*. However, the upstream transposase gene associated with vSaα is just a remnant and provides evidence against the mobility of this island (Baba *et al.*, 2002; Malachowa and DeLeo, 2010)

The genes for SSL12, SSL13, and SSL14 are situated in reverse tandem order approximately 700 kb downstream of vSaα in a region of the genome called the second immune evasion cluster (IEC-2) (Jongerius *et al.*, 2007) due to the presence of the many immune evasion molecules (IEMs) located here. IEC-2 also contains the genes for either FLIPr or its homologue FLIPr-like, one of the two SCIN homologues SCIN-B or SCIN-C, Efb, Ecb/Ehp, and α-haemolysin. Additionally, it contains transposases and bacteriophage remnants.

The *ssls* have been identified in every strain of *S. aureus* analysed. Comparison of *ssl* gene clusters from different strains indicates that the order of genes is conserved and that while some strains, such as MW2, carry all 14 genes, others do not carry the full complement. The strain Col for example only has ten *ssl* genes (*ssl1–4*, *ssl9–14*) (Fig. 6.1B). Homology between members of the SSL family ranges from approximately 19% up to 70% and there is also a great degree of variation within alleles. While some *ssls* are highly conserved other alleles display up to 20% difference in their sequences. The more highly conserved *ssls* tend towards the outer members of the cluster with fewer allelic variations seen in *ssl1* and *ssl10* for example than *ssl7*. In addition, *ssls* that are lacking in a strain are more likely to be those towards the centre of the cluster (Fig. 6.1B). For example, *ssl1* and *ssl10* are found in all strains so far analysed, whereas *ssl6* is missing in a number of strains (R.J. Langley, unpublished). Allelic and interstrain copy number variations have most likely occurred through recombination and gene transfer events occurring after an initial gene amplification and subsequent sequence divergence in a distant ancestral strain (Fitzgerald *et al.*, 2003; Tsuru *et al.*, 2006). This *S. aureus* ancestor thus possessed a full complement of *ssl* genes in vSaα.

Expression and regulation

While a good deal of information has been determined on the mechanisms of action of many of the SSLs, less is presently known about the conditions under which *S. aureus* produces SSLs and the levels that they are expressed at during bacterial infection or colonization. Colonization may predispose to invasive infection if host immunity is overcome or breached. SSLs are encoded by all strains of *S. aureus* and have been shown to play important roles in inhibiting host immune processes. They therefore have the potential to aid *S. aureus* in avoiding detection during asymptomatic colonization of the host. Conversely, infection starts with invasion of *S. aureus* into the body, followed by establishment at sites in the body, and finally by persistence due to evasion of the host immune defences. It is therefore also plausible that SSLs are important for immune evasion and survival of *S. aureus* during infection of the host.

Studies have shown the presence of SSL transcripts from *S. aureus* cultured under various *in vitro* conditions many of which involve strains of *S. aureus* possessing mutations in their regulatory systems. *In vitro* analysis revealed SSL11 transcript levels were ~ 16 times higher (Voyich *et al.*, 2009) and ~ 80 times higher (Liang *et al.*, 2006) in wild-type *S. aureus* compared with a mutant deficient in the *S. aureus* exoprotein expression (saeRS) two-component gene regulatory system. The mutant strain showed significantly reduced virulence and survival in murine sepsis, subcutaneous abscess, and kidney infection models. Importantly, it was found that that the SaeRS system negatively affected the master gene regulator agrA. Down-regulation of, or mutations in the accessory gene regulator (Agr) system have been associated with a switch from toxin production to up-regulation of molecules involved in immune evasion. Significantly Agr is often seen to be inactive or mutated *in vivo* and in clinical *S. aureus* isolates (Yarwood *et al.*, 2002; Yarwood and Schlievert, 2003; Goerke and Wolz, 2004; Traber *et al.*, 2008).

Regulation of *ssl* expression appears to be under tight control with increases only seen in particular circumstances most likely involving host immune evasion. A recent study revealed that *ssl* transcript levels were not up-regulated *in vitro* in tryptic soy broth and only marginally so in human serum, however increased levels of expression were observed when *S. aureus* was cultured in whole blood (Malachowa *et al.*, 2011). An exception to this observation was *ssl11* which showed increased gene expression over the base line measurement in all conditions assayed. There is further evidence that the SSLs are produced in higher levels in response to bacterial stress induced by host cell-based mechanisms. SSL gene transcript levels were increased in *S. aureus* that had been phagocytosed by neutrophils (Voyich *et al.*, 2005). Phagocytosis of the organism resulted in increased ROS production and granule–phagosome fusion. It is of particular interest that *ssl* transcripts were preferentially up-regulated in CA strains that were also significantly more resistant to neutrophil killing. When the CA-MRSA strain MW2 was cultured in the presence of neutrophil azurophilic granule proteins it up-regulated *ssl* expression (Palazzolo-Ballance *et al.*, 2008) and an increase in *ssl* promoter activity has been observed when *S. aureus* strain Newman was cultured *in vitro* in the presence of neutrophils (Benson *et al.*, 2012). That SSL expression is increased when the bacteria is subjected to cell membrane damage (Attia *et al.*, 2010) indicates that they are produced to help the bacteria avoid clearance when it finds itself subject to attack from the hosts immune system.

The damage-induced stresses put on the bacterium appear to trigger regulatory systems that control *ssl* transcription. Increases in *sae* transcripts are observed in conditions that promote *ssl* up-regulation (Voyich *et al.*, 2005, 2009; Palazzolo-Ballance *et al.*, 2008; Pantrangi *et al.*, 2010; Benson *et al.*, 2011, 2012). Benson *et al.* (2012) showed recently that co-operation of the transcription regulators SaeR and Rot (repressor of toxins) are required for *ssl* expression to be activated. They established that there is an SaeR binding site in the promoter region of each *ssl* and that both Rot and SaeR interact directly with these promoters. Their proposition is that Rot recruits SaeR to the promoter which subsequently enlists RNA polymerase to initiate transcription. Rot and SaeR have previously only been shown acting in opposition to each other. That they are acting synergistically to activate *ssl* expression describes a unique situation in which this family of immune evasion molecules is produced. Rot expression is negatively regulated by Agr and given that Agr is often inactivated or mutated during infection or in clinical isolates provides for a condition by which Rot can participate synergistically with SaeR, that has been up-regulated in response to neutrophils and their products, to activate *ssl* expression.

A soluble factor in the supernatant of a *Lactobacillus reuteri* strain exhibited a potent inhibitory effect on SSL11 production (Laughton *et al.*, 2006). The authors were analysing the surface proteome of *S. aureus* to look for differentially expressed adhesion molecules given the observation that *L. reuteri* RC-14 inhibited *S. aureus* infection in a surgical-implant model (Gan *et al.*, 2002). The level of SSL11, found associated with the cell wall fraction, was significantly decreased when *S. aureus* was co-cultured with *L. reuteri* RC-14. Under the same conditions a decrease in Agr expression was also observed. It is of great importance to find alternative ways of combatting *S. aureus* and the discovery of compounds that inhibit the expression of immune evasion molecules such as SSLs is an attractive option. It makes sense that these compounds are produced by organisms that can occupy similar niches (in this case the urogenital tract) with *S. aureus* as a mechanism of competitive exclusion.

Molecular biology of the SSLs

Introduction

The *ssl* genes translate to proteins that possess a highly conserved 30 amino acid N-terminal hydrophobic signal peptide which leaves mature secreted exoproteins of approximately 200 amino acids after cleavage. The two exceptions are SSL3 and SSL4 which both have an additional region of variable length and unknown homology at their N-terminus. Mature SSLs (except SSL3 and SSL4) range from ~22 to ~24.5 kDa in size. Isoelectric points range from mildly acidic [e.g. SSL11 Newman (nm) – 6.0] to highly basic (SSL10 nm – 9.8) with the majority of the SSLs falling within the range from 8.5 to 9.5. The SSLs possess a low amino acid sequence identity to the SAgs of less than 20%. Greatest sequence identity is shared with the TSST-1, creating a new subgroup with this previous superantigen outlier (Fig. 6.2). Despite this minimal sequence identity the SSLs share the OB-fold/β-grasp 2-domain structure common to the SAgs. They are however, functionally distinct

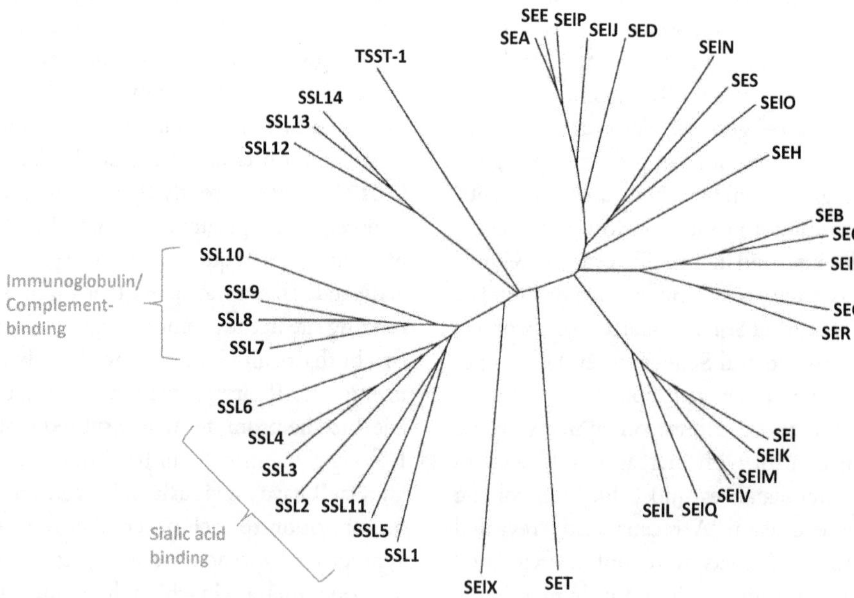

Figure 6.2 Phylogenetic tree of the SSL/SAg super-family of proteins expressed by *S. aureus*. The unrooted dendrogram was constructed from an alignment of the SSLs of strain Newman and all known SAgs of *S. aureus* generated using ClustalW (www.genome.jp/tools/clustalw). SSL subfamilies containing members with known functions are labelled accordingly.

and fail to exhibit the characteristic Vβ-specific stimulation of T-cells caused by the SAgs (Langley *et al.*, 2010). The SSLs do not contain the FL motif of the major histocompatibility complex (MHC) class II α-chain binding SAgs nor the Zn^{2+} co-ordinating residues of the MHC class II β-chain binding SAgs (Fraser and Proft, 2008), and while the SAgs exhibit a diversity of modes to perform the singular role of ligating MHC class II with the T cell receptor (TCR) (Sundberg *et al.*, 2007), the SSLs do not display this commonality in their interactions with the host organism.

Structural aspects of the SSLs

The published crystal structures of four SSLs: SSL4 (Hermans *et al.*, 2012), SSL5 (Arcus *et al.*, 2002), SSL7 (Al-Shangiti *et al.*, 2004; Ramsland *et al.*, 2007), and SSL11 (Chung *et al.*, 2007) reveal that despite very low sequence homology to the bacterial SAgs a high degree of structure conservation exists between the two families. Comparative structural analysis established that this low degree of sequence conservation is limited to residues that preserve the proteins structural stability (Arcus *et al.*, 2002) with no observable homology to key features of the SAg structure that are involved in MHC II or TCR binding. Furthermore residues that are absolutely conserved within the entire SSL family are without exception involved in maintaining structural integrity reflecting the variety of functional interactions displayed by the SSLs. By preserving the minimal number of residues required for maintaining the overall structure the bacteria has given rise to an extended family of proteins based on highly stable domains that have extremely variable surfaces capable of mediated a wide range of interactions.

The compact, globular SAg/SSL structure is composed of two highly stable domains (Fig. 6.3). The N-terminal domain is formed by a mixed β-barrel with Greek-key topology called an oligonucleotide/oligosaccharide-binding (OB)-fold (Murzin, 1993). This five-stranded β-barrel (comprising β-strands β1–β5) is a very common protein domain and can be found in many other secreted bacterial toxins such as the AB5 toxin family, which includes cholera toxin and pertussis toxin, and also in staphylococcal nuclease (Mitchell *et al.*, 2000). The outer concave face of

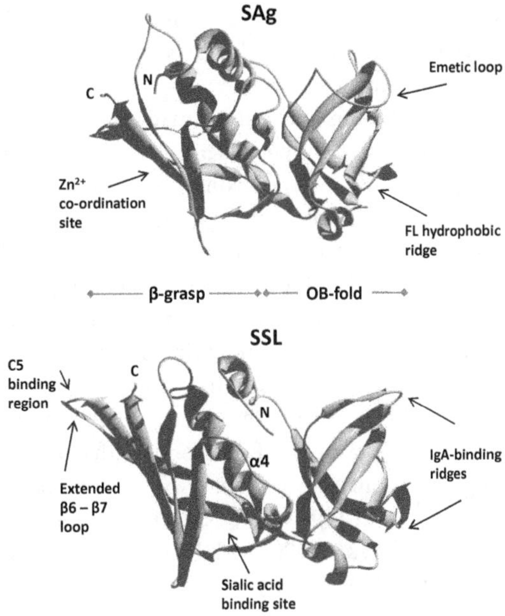

Figure 6.3 The typical SSL/SAg structure. Comparison of the SAg and SSL structures highlighting key features of each. On the left is the β-grasp domain separated from the OB-fold domain on the right by the central α4-alpha helix. Regions of the SAg responsible for MHC class II binding and staphylococcal food poisoning are indicated. Regions involved in C5, IgA, and sialic acid binding are emphasized on the SSL structure. Images were created using DeepView v4.1 from the PDB files 1SXT (SEA) and 1V1O (SSL7) and rendered using POV-Ray v3.6.2.

the OB-fold domain typically forms the binding or active site of the domain (Arcus, 2002). There is a short hydrophobic ridge containing the phenylalanine-leucine (FL) motif in the OB-domain of the MHC class II α-chain-binding SAgs that fits into the hydrophobic pocket formed along the distal α1-helix domain of the MHC class II molecule (Fraser and Proft, 2008). Other SAgs, such as SED and SpeC, use this same region to homodimerize (Sundstrom *et al.*, 1996; Roussel *et al.*, 1997). The disulfide bond-containing loop characteristic to the staphylococcal enterotoxins that protrudes from the top of the concave face has been implicated in the emetic properties of these SAgs (Hovde *et al.*, 1994). SSL7 uses this region to bind to IgA (Ramsland *et al.*, 2007), the implications of which will be discussed later in this chapter. Two loops, forming the upper and lower edges of the concave OB-fold face of SSL7, interact with the Cα2/Cα3 domain junction of the IgA Fc chain.

The C-terminal domain is made up of a five-stranded twisted antiparallel β-sheet (strand order β7–β6–β12–β9–β10) backing on to the central amphipathic α-helix (the 'β-grasp' motif). Also part of this β-grasp domain is the very N-terminal tail and α-helix of the molecule, which packs against the β-sheet. Further coplanar β-strands and an additional helix near the C-terminus are other common features in the SAg/SSL β-grasp domain. This domain shares structural features with the β-grasp motifs present in the immunoglobulin-binding domains of streptococcal protein G and *Peptostreptococcus magnus* protein L (Mitchell *et al.*, 2000) and the immune evasion molecules (IEMs) EAP, CHIPS, and FLIPr of *S. aureus* are all β-grasp motif-containing proteins (Langley *et al.*, 2010). A triad of residues in the C-terminal domain of the MHC class II β-chain-binding SAgs are responsible for forming the tetravalent co-ordination of a zinc molecule together with a conserved histidine in the β-chain α-helix of MHC class II (Fraser and Proft, 2008). Compared to the SAgs, the SSLs commonly display an elongated β6-β7 loop creating a larger face on the C-terminal domain of the SSLs (Fig. 6.3). The SSLs that have been crystallized have done so as homodimers through this extended face although it is not currently known whether this dimerization has any functional significance. In the co-crystal structure of SSL7 with human C5 (Laursen *et al.*, 2010) residues near the end of the elongated face formed by this extended loop can be seen interacting with the MG1 and MG5 domains of C5. The function consequences of this interact will be covered later in this chapter. Co-crystals of SSL4, SSL5, and SSL11 with the ligand sialyl Lewis X (sLeX – Neu5Acα2–3Galβ1–4(Fucα1–3)GlcNAc) reveal a highly conserved binding site for sialylated glycans on the side of their β-grasp domains (Baker *et al.*, 2007; Chung *et al.*, 2007; Hermans *et al.*, 2012). It comprises of a V-shaped depression formed between the β10 strand (opposite edge of the β-sheet to the elongated β6-β7 loop) and a distorted 3_{10}-helix in the following loop that packs against the back side of the OB-fold domain.

Close packing of the two domains and extension of the N-terminus over the top of the C-terminal domain provide stability to the molecule. One side of the central buried α4-helix packs against the β-grasp motif while the other side of the α-helix packs against the inner side of the OB-fold domain. It is this interface between the two domains that holds the protein together and is the only region almost completely conserved among all members of both the SAg and SSL families.

Molecular interactions and functional aspects of the SSLs

The SSLs that possess the conserved binding site for sialylated glycans form a subfamily of SSLs based on homology and function while SSL7 and SSL10 can be defined as a separate subfamily of immunoglobulin-binding SSLs based on their function and close homology (Fig. 6.2). Neither SSL7 nor SSL10 possess identity of residues involved in glycan binding and both have been shown to interact with their corresponding immunoglobulin in a carbohydrate-independent manner (Ramsland *et al.*, 2007; Itoh *et al.*, 2010b).

SSL7

SSL7 (formally SET1) is the prototypic SSL and has been one of the most thoroughly studied. A great deal is now known about its molecular interactions and the consequence of these interactions on host immune recognition of, and immune evasion by *S. aureus*. Two major immune defence proteins, IgA and complement C5, can be bound concurrently by SSL7 (Langley *et al.*, 2005). SSL7 has the ability to bind monomeric serum IgA and dimeric secretory IgA (sIgA), and both C5 as well as activated C5b that has been cleaved from the C5a anaphylatoxin. Langley *et al.* (2005) showed that SSL7 was able to block the binding of IgA with its cellular receptor FcαRI (CD89), prevented serum-mediated bacterial killing, and inhibited complement-mediated haemolysis of human red blood cells.

A combination of structural and mutagenesis analyses of SSL7 and IgA revealed that the OB-fold domain of SSL7 binds to the Fc region of IgA at the Cα2/Cα3 junction (Wines *et al.*, 2006; Ramsland *et al.*, 2007) (Fig. 6.4A). Key residues involved in this interaction are asparagine-38 in

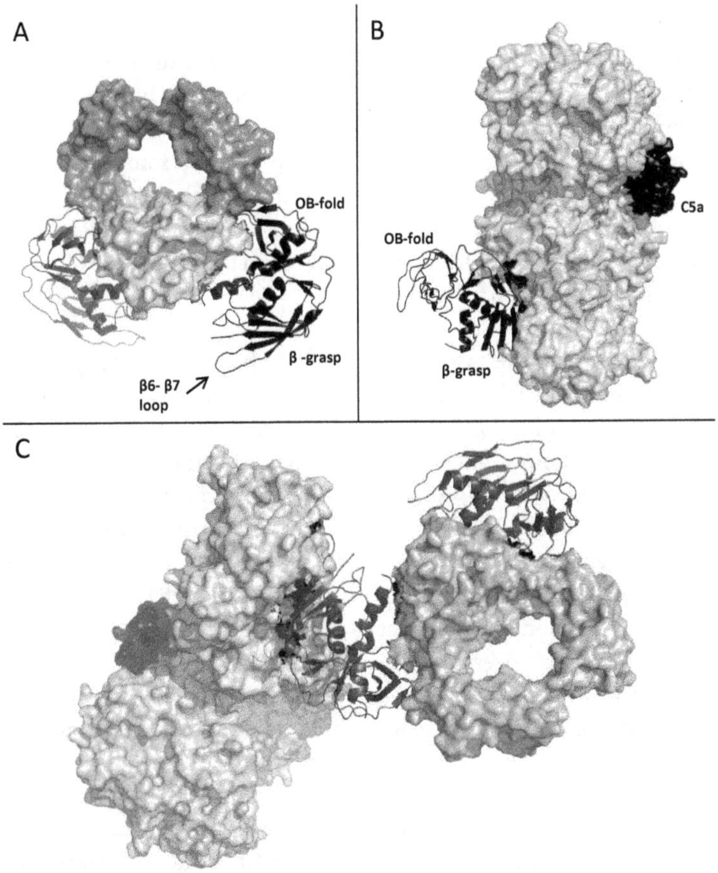

Figure 6.4 Structures showing SSL7 binding to IgA and C5. A. In the crystal structure of SSL7 bound to the IgA Fc region (grey) two SSL7 molecules (black ribbon structures) can be seen binding in the Cα2/Cα3 junction of the Fc region and covering much of the lateral face of Cα3. The Cα2 domain of each Fc chain in shown in dark grey and the Cα3 domain of each chain is shown in light grey. B. The structure on SSL7 (black ribbon) bound to the MG1 and MG5 domains of C5 (grey). The position of the C5a anaphylatoxin on the opposing face of C5 is depicted in black. C. A model of the C5–SSL7-IgA-Fc complex based on the co-crystal structures of SSL7 with IgA-Fc and SSL7 with C5. SSL7 is shown in black as a ribbon structure while C5 and IgA-Fc are shown in grey as surface structures. The C5a domain of C5 is shown in black. Image created using pyMOL from the PDB files 2QEJ (SSL7 complexed with IgA), and 3KLS (SSL7 complexed with C5).

the lower loop and leucine-79 and proline-82 in the upper loop of the OB-fold concave face. Two SSL7 molecules are able to bind the Fc region (one to each heavy chain) and conceal much of the lateral surfaces of the Cα3 domain. This binding overlaps with the recognition site for FcαRI with residues of IgA involved in Fc receptor binding participating in the interaction with SSL7. As such this interaction describes a potential mechanism of immune evasion through the inhibition of IgA-mediated cellular effector functions via FcαRI.

The SSL7-IgA co-crystal structure showed that binding IgA through its OB-fold domain would leave the β-grasp domain of SSL7 available for its interaction with complement C5. It was determined by further co-crystallization and mutagenesis studies, this time between SSL7 and C5 that this indeed was the case (Laursen *et al.*, 2010) (Fig. 6.4B). SSL7 is able to prevent the cleavage of C5 into C5a and C5b. The structure of SSL7 bound to C5 revealed an unexpected mode of action in which SSL7 binds primarily to the MG1 and MG5 domains with minor interactions with the MG2 and MG6 domains of C5 (Laursen *et al.*, 2010). This region of the C5 molecule is on the opposite side to the C5a domain and the C5 convertase cleavage site so rules out a direct inhibition of C5 cleavage. Additionally, no significant conformational changes in C5 were observed upon SSL7 binding excluding the likelihood of SSL7-mediated alteration to the convertase cleavage site. Instead, the binding of SSL7 prevents the binding of the C5 convertase to C5 by a mechanism of steric hindrance that is largely reliant on the simultaneous binding of SSL7 to IgA (Laursen *et al.*, 2010, 2011). A model of the C5–SSL7–IgA complex based on the co-crystal structures of SSL7 with IgA and SSL7 with C5 is possible due to a lack of spatial conflict involving any of the molecules in the complexes (Laursen *et al.*, 2010) (Fig. 6.4C). This model reveals that the position of IgA is such that it overlaps with convertase binding on C5, whereas SSL7 alone has minimal impact on the convertase binding face. SSL7, however, is still in a position to prevent the formation of MAC by binding C5 in the absence of bound IgA. Functional studies on complement inhibition by SSL7 are in complete agreement with the structural data (Laursen *et al.*, 2010). Potent inhibition of C5 cleavage can be achieved by SSL7 in the presence of IgA however an IgA-binding deficient mutant or the β-grasp only mutant of SSL7 exhibit far weaker inhibition. This is reflected in the serum-only bactericidal assay in which *E. coli* is protected from killing in the presence of *wt* SSL7 with a dramatic decrease of protection in the presence of either mutant. In contrast, an assay on MAC function reveals that the IgA binding deficient and β-grasp only mutants show comparable inhibition of red blood cell haemolysis as *wt* SSL7.

The ability of SSL7 to bind IgA and C5 in different species is quite varied. While C5 from many species can be bound the concurrent binding of both molecules is observed mainly in humans and primates (Langley, 2005). This suggests humans as a major host for *S. aureus* because it is only by binding both C5 and IgA that SSL7 can inhibit the potent anti-staphylococcal actions of C5a. Abi-Rached *et al.* (2007) has proposed that selective pressure on IgA by a pathogenic IgA-binding molecule like SSL7 led to changes in the FcαRI binding site of the immunoglobulin so that it was no longer able to bind its receptor. This in turn led to the loss of the FcαRI gene through lack of selective pressure. Examples that support this theory are the mouse and rabbit, neither of which possesses a functional FcαRI, and the IgA of neither of which is bound by SSL7.

SSL10

SSL10 has a restricted specificity for human and primate immunoglobulin of the IgG1 subclass (Itoh *et al.*, 2010b; Patel *et al.*, 2010). The binding of SSL10 competes with the binding of C1q to IgG1 with the consequence of inhibiting activation of the classical pathway of complement (Itoh *et al.*, 2010b). To determine whether the interaction of IgG1 with its FcϒRs was affected by SSL10 binding Patel *et al.* (2010) studied the direct effect of SSL10 competing for binding of IgG1 to monocytes. A dose dependent inhibition of IgG1-FITC binding to monocytes was observed. The ability of neutrophils to phagocytose IgG-opsonized bacteria (*S. aureus* and *Streptococcus pyogenes* were studied) was also compromised in the presence of SSL10, further substantiating the SSL10-mediated inhibition of IgG1-FcϒR binding. Mutation analysis suggests that the binding site for SSL10 is on the outer face of the Cϒ2 domain of IgG1 close to both the FcϒR and C1q binding sites (Patel *et al.*, 2010). The binding affinity of SSL10 for IgG1 was determined to be 220 nM by surface plasmon resonance (SPR) analysis (Itoh *et al.*, 2010b) and 300 nM by isothermal titration calorimetry (Patel *et al.*, 2010). The thermodynamic data fit with a 1:1 stoichiometry for SSL10 binding IgG1 (Patel *et al.*, 2010). This is in contrast to the binding of SSL7 to IgA in which two SSL7 molecules bind one IgA with an up to 300-fold higher affinity of 1.1 nM (Langley *et al.*, 2005; Ramsland *et al.*, 2007). It is possible that the binding of one IgG1 Fc chain by SSL10 may create an allosteric effect that prevents a second SSL10 from binding the IgG1 molecule.

Binding of SSL10 to fibronectin and fibrinogen (Patel *et al.*, 2010), and phosphatidylserine (PS) (Itoh *et al.*, 2012) has been detected. These interactions were localized to the β-grasp domain and the OB-fold respectively. The functional consequence of these interactions is not known although Itoh *et al.* (2012) propose that the binding of SSL10 to PS on apoptotic cells can be used as a diagnostic tool for monitoring apoptosis.

SSL9

SSL9 has been shown to potently inhibit both the classical and lectin mediated complement pathways in a range of species including human. The mechanism of complement inhibition is not yet known, but like SSL7 the activity is located within the β-grasp domain. Another interesting effect of SSL9 is its ability to enhance *S. aureus* growth when cultured under suboptimal conditions such as in Luria broth. Remarkably its influence is maximal at subnanomolar concentrations suggesting that SSL9 may have some regulatory role in addition to its complement inhibiting activity. This growth enhancement induced by SSL9 occurs in low levels of serum (0.01–0.1%) suggesting that SSL9 combines with a serum factor to enhance staphylococcal growth (Langley *et al.*, 2010).

The sialyl lactosamine-binding SSL subfamily

A subfamily of SSLs (Fig. 6.2) is involved in a wide variety of interactions with the host at a cellular and molecular level that in many cases has been shown to be glycan-dependent. SSL2, SSL3, SSL4, SSL5, SSL6, and SSL11 have been shown either experimentally or by homology modelling to possess a binding site for sialylated glycans (Baker *et al.*, 2007; Chung *et al.*, 2007; Hermans *et al.*, 2012). Glycan array screening (www.functionalglycomics.org) has confirmed that the minimal conserved motif recognized by these proteins is the trisaccharide sialyl-lactosamine (sLacNac – NeuAcα2–3Galβ1–4GlcNAc). The discovery that this binding site existed on SSL11 came about when examining the ability of this SSL to block the binding of IgA to its Fc receptor (CD89 or FcαRI). SSL11 was observed to bind CD89 from human leucocytes in pull-down assays and recombinant CD89 produced in Chinese Hamster Ovary (CHO) cells. It however failed to bind an insect cell-expressed recombinant CD89 (Chung *et al.*, 2007) suggesting a requirement for complex carbohydrate not expressed by the insect cells. Neuraminidase treatment of CD89 established that terminal sialic acid was responsible for this recognition. Furthermore, neuraminidase treatment of leucocytes abrogated the ability of SSL11 to bind these cells. Chung *et al.* also determined that SSL11 bound PSGL-1 on the surface of neutrophils in a sialic acid-dependent manner that potently prevented neutrophils from rolling on P-selectin coated surface. Additionally, SSL11 was rapidly internalized into cytoplasmic structures of neutrophils in a glycan and energy dependant manner.

Crystal structures of SSL4, SSL5, and SSL11 in complex with the blood group antigen sialyl Lewis X (sLeX – Neu5Acα2–3Galβ1–4(Fucα1–3) GlcNAc) have been solved that together with mutagenesis experiments have identified key residues in the glycan binding site (Fig. 6.5). One of these residues, threonine (Thr) 171 in SSL4, Thr175 in SSL5, or Thr168 in SSL11, makes two essential hydrogen bonds to the sialic acid portion of sLeX. Mutation to proline in SSL11 results in complete loss of binding to any of the ligands recognized by wild-type SSL and also to a lack of inhibition of neutrophil attachment to P-selectin (Chung *et al.*, 2007). Mutating this residue in the β10 strand or the absolutely conserved arginine located in the 3_{10}-helix on the other side of the glycan binding site results in a loss of cell binding and affinity for sialylated glycans (Baker *et al.*, 2007; Chung *et al.*, 2007; Hu *et al.*, 2011; Hermans *et al.*, 2012)

The majority of the hydrogen bonds that the SSLs make with sLeX are identical and are made to the sialic acid (Neu5Ac) and galactose (Gal) subunits of the tetrasaccharide. Variations in bonding of SSL4, SSL5, and SSL11 to the N-acetylglucosamine (GlcNAc) and fucose (Fuc) units can be observed and were the focus of a comparative analysis of glycan binding by these SSLs published recently (Hermans *et al.*, 2012). In particular more extensive hydrogen bonding from SSL4 to Fuc is possible because the bonding of SSL11 to GlcNAc holds the Fuc further out of the binding site. This explains the differences seen in these SSLs affinities and specificities for sialylated glycans. At low concentrations of around 10nM SSL4 reveals competitive binding to myeloid cells. That this phenomenon is lost at higher concentrations suggests that once the

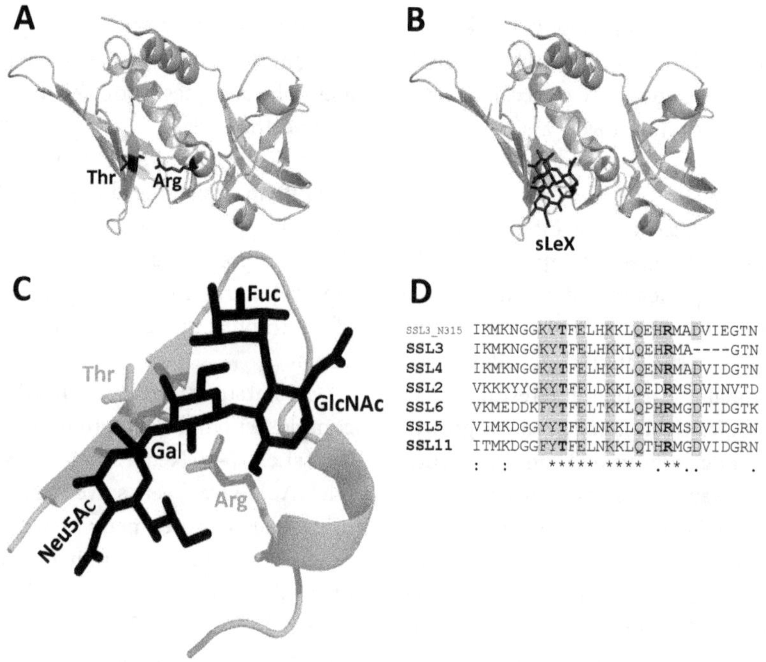

Figure 6.5 The SSL sialic acid binding site. A. Structure of SSL11 showing in black the positions of the conserved threonine (Thr) and arginine (Arg) of the sialic acid binding site that have been mutated to abolish glycan binding. B. Structure of SSL11 showing in black sialyl Lewis X bound in the sialic acid binding site. C. Close up view of the sialic acid binding site (shown in grey) with bound sialyl Lewis X (shown in black). On the left is the β10 strand showing the side chain of the conserved Thr while the side chain of the conserved Arg is shown on the right in the 3_{10}-helix. Images were created using PyMOL from the SSL11 PDB file 2RDG.D. Sequence alignment generated using ClustalW (www.genome.jp/tools/clustalw) from the sialic acid binding SSLs of *S. aureus* strain Newman (nm). SSL3 from N315 was included because SSL3nm has a 4 amino acid deletion in the loop following the sialic acid binding site. Key residues seen in the SSL4, SSL5, and SSL11 co-crystal structures to hydrogen bond with sLeX are highlighted in grey. The conserved Thr and Arg are shown in bold type.

specific receptor is fully bound by SSL4 the exoprotein begins binding off-target to the multitude of sialylated receptors found on the cells surface. Interestingly, SSL4 is unable to compete with SSL11 for cell binding pointing to individual specificities of these proteins for host factors. SSL4 displays higher specificity for sLeX over the core component sLacNac (a subcomponent of sLeX lacking the fucose) and its ability to compete with itself for cell binding suggests a discrete target molecule for this SSL. SSL11 on the other hand shows no difference in binding specificity to sLeX compared to sLacNac and cannot be competed for cell binding by itself or SSL4 indicating more promiscuity in its target selection (Chung *et al.*, 2007; Hermans *et al.*, 2012). This discovery is highly supportive of independent roles for members of the glycan-binding subfamily and offers an explanation as to why *S. aureus* produces up to six sialylated glycan-binding SSLs.

SSL5

SSL5 is by far the most studied of the SSLs. It was the first SSL to be structurally defined (Arcus *et al.*, 2002) and since the initial discovery that it binds PSGL-1 to impede P-selectin mediated neutrophil rolling (Bestebroer *et al.*, 2007) many other interactions have been reported. SSL5 has been shown to activate platelets and promote their adhesion to endothelial cells (De Haas *et al.*, 2009) while inhibiting the adhesion of leukaemia cells to endothelial cells and platelets (Walenkamp *et al.*, 2010). Inhibition of leucocyte activation by chemokines and anaphylatoxins (Bestebroer *et al.*, 2009) as well as the obstruction of matrix metalloproteinase-9 (MMP-9) activation (Itoh *et al.*,

2010a) has also been attributed to SSL5. Much of these observed functions of SSL5 can be credited to the sialic acid binding ability of this molecule.

In addition to the inhibitory effect on neutrophil migration exhibited by the SSL5-dependent blockage of P-selectin mediated rolling, an inhibition of neutrophil response to chemokines and anaphylatoxins has been observed (Bestebroer *et al.*, 2009). Calcium mobilization and actin polymerization of neutrophils in response to chemokines from different chemokine classes as well as the anaphylatoxins C3a and C5a were blocked in the presence of SSL5. This resulted in neutrophils treated with SSL5 exhibiting reduced migration towards the chemokine CXCL8. The sialic acid dependency of these responses was demonstrated with the inhibitory effect of SSL5 being abolished after pre-incubation with soluble sLeX or by pre-treatment of cells with neuraminidase. SSL5 was subsequently determined to bind to all G protein-coupled receptors (GPCRs) but only inhibited the chemokines and anaphylatoxins that require interaction with the N-terminus of their chemoattractant receptor for activation. GPCRs are glycosylated in their N-terminal region so sialic acid dependent binding here by SSL5 blocks activation by chemoattractants that target this region whilst chemoattractants such as FMLP and ATP that bind their GPCR in the transmembrane regions remain unaffected by the binding of SSL5.

It is particularly interesting to note that the effects reported by Bestebroer *et al.* (2009) were not observed for SSL11, which possesses the same glycan binding site as SSL5, suggesting that there are differences in their sialic acid preferences and/or possibly additional binding specificities that augment their glycan binding. One significant difference between the two proteins is that SSL5 is a highly charged cationic protein with a pI of ~9.5 while SSL11 has a pI of ~6.0, giving these two proteins very different surface charges. It is possible that some of the observed activities for SSL5 are a result of additional electrostatic interactions that potentially extend the binding capacity of this SSL. Could the broad spectrum of SSL5's binding capacities a function of its charge? Hu *et al.* (2011) noted that concentrations of SSL5 over 80 nM could not be used in their binding analyses due to the development of inconsistencies in the binding characteristics. They suggest this may be a consequence of the highly cationic nature of SSL5. The other members of the sialic acid dependent binders have pIs in the range of 9.0–9.5 so may be more likely to display similar characteristics to SSL5. Hermans *et al.* (2012) determined however that SSL4 and SSL11 display differences in their ligand specificity due to subtle variations in their respective glycan binding sites. SSL5 was suggested to possess more similar characteristics to SSL4 than to SSL11 in this analysis arguing that glycan specificity rather than additional binding sites is the reason for the differences in activity between SSL5 and SSL11.

Itoh *et al.* (2010a) showed that the binding of MMP-9 by SSL5 resulted in inhibition of gelatin degradation by this endopeptidase which resulted in a reduction in neutrophil migration through reconstituted membranes. The binding of SSL5 to the preform of MMP-9 was proposed to provide a mechanism for immune evasion by interfering with the migration of neutrophils to the site of *S. aureus* infection. It is thought that MMP-9 released from granules of stimulated neutrophils enzymatically degrades components of the extracellular matrix to aid in the infiltration of inflammatory tissues. It is activated by other proteases including elastase. Deficiency of MMP-9 results in an increased severity of *S. aureus*-induced septic arthritis with enhanced persistence of the bacteria in mice (Calendar *et al.*, 2006). Glycan binding dependency was established with SSL5 showing a decreased affinity for desialylated MMP-9. If SSL5 is rapidly internalized into granulocytes in a glycan dependent manner as SSL4 and SSL11 are it could certainly place it in the appropriate environment for interacting with MMP-9 and inhibiting its release and subsequent enzymatic activity.

Two recent articles that report on SSL5-mediated platelet activation and adhesion have added to the expanding repertoire of host modifying functions displayed by this SSL. Both articles demonstrate sialic acid dependent binding to the platelet glycoprotein GPIbα; however, one reveals additional binding of SSL5 to the integrin $\alpha_{IIb}\beta_3$ (GPIIb-IIIa) (De Haas *et al.*, 2009) while the other identified the platelet collagen receptor GPVI as a target for SSL5 binding (Hu *et al.*, 2011). Glycan dependency was not reported for $\alpha_{IIb}\beta_3$ though the binding of SSL5 to GPVI was reliant on its glycan binding site. A consequence of platelet binding by SSL5 was GPIbα-dependent signalling as shown by translocation of the GPIbα-associated adapter protein 14–3–3ζ to the actin cytoskeleton. Platelet activation via GPIbα binding resulted in aggregation and initial platelet adhesion; however, the observed interaction of SSL5 with $\alpha_{IIb}\beta_3$ was proposed to be required for stable platelet adhesion (De Haas *et al.*, 2009). In a follow-up investigation Armstrong *et al.* (2012) implicate SSL5 as having an involvement in the disseminated intravascular coagulation (DIC) associated with *S. aureus* as a consequence of this interaction with platelets. DIC is characterized by thrombosis with simultaneous bleeding and arises when the bacteria interacts and activates platelets systemically resulting in consumption of platelets and coagulation proteins (Fitzgerald *et al.*, 2006). SSL5 introduced intravenously to mice resulted in platelet-rich thrombus formation in the lungs within 15 min. Furthermore, while SSL5 was found to increase the binding of platelets to von Willebrand factor (vWF)-coated plates, it inhibited the binding of ADP-activated platelets to vWF by ~ 50%. The binding of vWF to platelets via its interaction with GPIb-IX-V is important for the adhesion of platelets to damaged vasculature (Berndt *et al.*, 2001). Additionally, platelets become activated following their adhesion to vWF and release platelet-stimulates such as ADP that recruit additional platelets to the developing thrombus (Andrews *et al.*, 2003). Signal transduction from GPIb-IX-V results in activation of integrin $\alpha_{IIb}\beta_3$ which then binds vWF or fibrinogen and mediates platelet aggregation. Because SSL5 binds both GPIbα and integrin $\alpha_{IIb}\beta_3$ it has the potential to interfere with this process and cause increased bleeding. In support of this Armstrong *et al.* (2012) showed that administration of SSL5 to mice 10 min prior to tail cutting resulted in a 4-fold increase in bleeding time. Their investigation revealed that SSL5 can cause both the thrombotic and bleeding complications *in vivo* that are characteristics of DIC. The mortality rate of *S. aureus* sepsis doubles to 40% when DIC is induced. SSL5 was shown to both induce thrombi formation and increase bleeding time in mice and as such may participate in *S. aureus* induced DIC. Armstrong *et al.* (2012) showed that these effects could be inhibited *in vivo* using either a monoclonal antibody specific for SSL5 or the glycan-based therapeutic bimosiamose. Bimosiamose is a synthetic small-molecule pan-selectin antagonist that was developed to inhibit sLeX mediated binding by E-, P- and L-selectins. It has been successfully investigated in different clinical Phase I and Phase IIa trials for the indication of chronic obstructive pulmonary disease (COPD), asthma and psoriasis. Therapeutic targeting of SSL5 by monoclonal antibodies or small molecule inhibitors of its glycan-binding site may prove to be effective in lessening the impact of *S. aureus*-induced DIC.

SSL3 and toll-like receptor 2 (TLR2)

Recently two groups published articles describing the binding of SSL3 to toll-like receptor 2 (TLR2) with the subsequent inhibition of TLR2-induced cell activation (Bardoel *et al.*, 2012; Yokoyama *et al.*, 2012). One group showed SSL3 could compete with the binding of anti-TLR2 to TLR2 on neutrophils and monocytes and confirmed TLR2 binding by ELISA (Bardoel *et al.*, 2012). They followed this up by demonstrating SSL3 inhibition of IL-8 production by cells stimulated with either Pam3Cys (a TLR1/TLR2 agonist) or MALP-2 (a TLR2/TLR6 agonist). The other group identified TLR2 as the protein pulled out from lysates of porcine spleen using SSL3 coupled to Sepharose and showed that SSL3 could inhibit tumour necrosis factor (TNF)-α production in mouse macrophages exposed to heat-killed *S. aureus*, peptidoglycan (PGN), or the synthetic TLR2 ligands Pam_3CSK_4 and MALP-2 (Yokoyama *et al.*, 2012). The Extra N-terminal region of SSL3 was deemed to not play a role in this inhibition. A truncated version of SSL3 corresponding to the prototypically conserved OB-fold – β-grasp SSL structure still retained the ability to prevent an anti-TLR2 mAb binding to monocytes and granulocytes (Bardoel *et al.*, 2012). Bardoel *et al.* (2012) report that desialylation of monocytes and neutrophils resulted in a significant reduction in SSL3's ability to inhibit TLR2. This effect could be replicated by mutating the important conserved arginine in the sialic acid binding site of SSL3. The authors state that while there was an effect on TLR2 inhibition associated with the loss of glycan binding this was only up to 30-fold and suggested that sialic acid binding was involved but not essential for the inhibition of TLR2 function. They state that amino acids in the OB-fold domain are responsible for binding TLR2 with only a secondary contribution from the glycan binding region because a β-grasp only mutant of SSL3 lost its ability to inhibit TLR2 activation. They support this claim with the observations that other SSLs with sialic acid binding capacity do not inhibit TLR2 and that the glycosylation sites on TLR2 are not close to the ligand binding site of the receptor. However Herman *et al.* (2012) showed significant differences in specificity to sialic acid in their comparison of SSL4 with SSL11 that supports the binding of separate host factors by members of the glycan-binding subfamily of SSLs. What's more, it may be necessary to mutate more than one residue in the sialic acid binding site to completely abolish glycan binding. Finally, one side of the glycan binding site packs against the OB-fold so removing the N-terminal domain will very likely have resulted in disruption of the glycan binding. Conversely, Yokoyama *et al.* (2012) state that inhibition of TLR2 by SSL3 does not involve the sialic acid binding site. They mutated conserved residues phenylalanine 297 and glutamic acid 298 in the glycan binding region to isoleucine and aspartic acid – residues in this position that are conserved in the non-glycan binding SSLs – with no effect on TLR2 binding or activation. However, in the structures of SSL4, SSL5 and SSL11 in complex with sLeX the equivalent phenylalanine of SSL4, SSL5 and SSL11 is not seen to make any appreciable contact with sLeX. Therefore, substituting it with isoleucine is not likely to affect glycan binding. Additionally, much of the interaction of the equivalent glutamic acid side chain is with the fucose side chain of sLeX. It has been determined that the fucose does not constitute the core moiety of sLacNac (a subcomponent of sLeX lacking the fucose) recognized by the SSLs. Therefore mutation of the glutamic acid in SSL3 may not result in the desired loss of sialic acid binding required to categorically determine if glycan binding is involved. However, if SSL3 does possess binding properties not wholly reliant on its glycan binding capability this opens up the possibility that other members of the glycan binding SSL subfamily potentially do so too. It may be that like other described SSLs, and their SAg cousins, there are multiple binding sites on each sialic acid binding SSL, perhaps that work synergistically with the glycan binding site or independently to carry out unique functions.

SSL-related *S. aureus* immune evasion molecules

The β-grasp domain of the SSL and the SAg family of toxins can be found in other *S. aureus* virulence factors including the chemotaxis inhibitory protein from *S. aureus* (CHIPS), the formylated peptide receptor like inhibitory protein (FLIPr), the extracellular adhesion protein (EAP), and staphylococcal enterotoxin-like X (SElX).

CHIPS (Chemotaxis inhibitory protein)

Staphylococcus aureus produces a 14-kDa chemotaxis inhibitory protein called CHIPS that is structurally similar to the β-grasp domain of the SSLs (Haas *et al.*, 2005) (Fig. 6.6). CHIPS binds to myeloid cells and inhibits both neutrophil and monocyte chemotaxis. It was found to block C5a and fMLP activation of neutrophils and intracellular calcium mobilization by binding and down-regulating the human C5a receptor and formylated peptide receptors (Veldkamp *et al.*, 2000; de Haas *et al.*, 2004; Postma *et al.*, 2004). The gene encoding CHIPS (*chp*), is carried at the 3′ end of a β-haemolysin converting bacteriophage family, called an immune evasion cluster (IEC), together with SCIN (*scn*), staphylokinase (*sak*), and SEA (*sea*) or SEP (*sep*) (Rooijakkers *et al.*, 2005; van Wamel *et al.*, 2006). Not only does this IEM share structural conservation with the β-grasp domain of the SSLs, it also possesses homology in the glycan-binding region. Furthermore the conserved arginine residue of the glycan-binding site (R84 in CHIPS) was found to negatively affect CHIPS function even though the C5aR-binding residues R44 and K95 are located on the α-helix and the β3 strand respectively, well away from the sialic acid binding site (Haas *et al.*, 2005). Interestingly

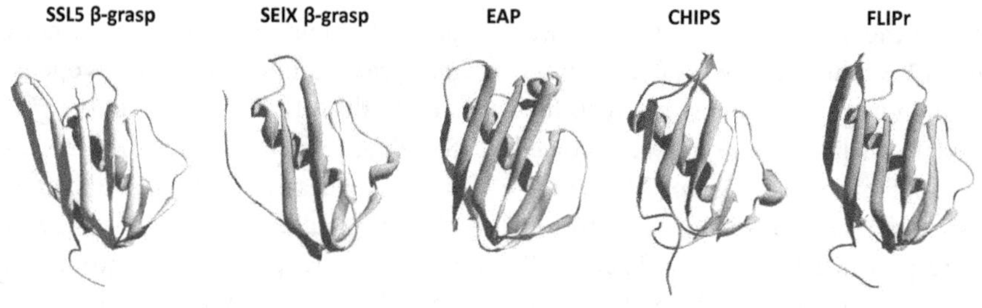

Figure 6.6 Structural comparison of the β-grasp domain-containing immune evasion molecules of *S. aureus*. The β-grasp domain of SSL5 (1M4V) is compared with the β-grasp domain of SEIX (modelled on the structure of SSL7 (PDB 1V1O), the EAP2 domain of EAP (PDB 1YN3), CHIPS (PDB 1XEE), and FLIPr (modelled on SSL11 PDB 2RDG). Images created using DeepView v4.1 and rendered using POV-Ray v3.6.2.

the arrangement of the β-strands in CHIPS differs to that seen in the other β-grasp domain IEMs of *S. aureus*. Strands β1, β2 and β3 of CHIPS correspond to β9, β10 and β12 of SSL5. However, the anti-parallel β4 strand of CHIPS corresponds to the β6 strand of SSL5 which hydrogen bonds in parallel to the C-terminal β12 strand. Despite this rearrangement of the β-sheet CHIPs retains structural conservation of the glycan binding site characterized in the SSLs (Baker *et al.*, 2007) and possesses binding specificity towards sialylated glycans in a glycan array screen (www.functionalglycomics.org).

FLIPr (formylated peptide receptor like inhibitory protein)

FLIPr, a 12.5-kDa exoprotein of *S. aureus* displaying ~ 30% homology to CHIPS, binds primarily to the Formylated peptide receptor-like 1 (FPRL1) to inhibit fMLP-induced calcium mobilization of neutrophils (Prat *et al.*, 2006). The genes encoding FLIPr or its homologue FLIPr-like are carried on IEC-2 together with one of the two SCIN homologues SCIN-B or SCIN-C, Efb, Ecb/Ehp, α-haemolysin, and SSL12, SSL13, and SSL14 (Jongerius *et al.*, 2007). The structure of FLIPr has not been published, however, using the known structures of SSLs or CHIPS, modelling of FLIPr results in a β-grasp topology. The sheet arrangement in FLIPr is in agreement with a model based on the SSL β-grasp as opposed to the β-strand order seen in CHIPS (Fig. 6.6). FLIPr also exhibits conservation to the SSL glycan binding site (authors own observations).

EAP (Extracellular adherence protein)

EAP is a multidomain protein expressed by *S. aureus*, consisting of four to six tandem repeat units, that has extended binding capacity for host extracellular matrix and cell surface proteins including fibrinogen, fibronectin, prothrombin, vitronectin, collagen, osteopontin, bone sialoprotein, and ICAM-1 (Palma *et al.*, 1999; Chavakis *et al.*, 2002). Through its extended interactions with host factors EAP is a virulence factor that contributes to *S. aureus* pathogenesis (Hussain *et al.*, 2002; Harraghy *et al.*, 2003; Haggar *et al.*, 2004; Athanasopoulos *et al.*, 2006). The three-dimensional structures of three EAP domains reveal that the tandem repeat domains adopt a β-grasp fold with homology to the C-terminus of the SSLs and SAgs (Geisbrecht *et al.*, 2005) (Fig. 6.6). EAP does not however possess any superantigenic ability (Haggar *et al.*, 2010).

SEIX (staphylococcal enterotoxin-like X)

The newly described superantigen-like enterotoxin X (SElX) has the typical C-terminal β-grasp domain however the shortened N-terminal domain possesses no sequence homology to the SAg OB-fold domain (Wilson *et al.*, 2011) (Fig. 6.6). Additionally there is no observable homology with any of the conserved MHC class II α- or β-chain binding residues. Taken together this suggests that SElX possesses novel ligand binding residues and may use a novel mechanism of T cell activation. Despite this lack of homology SElX has the ability to stimulate human TCR bearing the Vβ 1, 6, 18, and 21 chains (Wilson *et al.*, 2011).

Unlike the typical bacterial SAgs, SElX is present in the genomes of all strains of *S. aureus* (except CC30). Interestingly SElX shows higher identity to the SSLs than to the SAgs and possesses the conserved lysine of the SSLs in its α4 alpha-helix as opposed to the glutamine present in the SAgs. Moreover it possesses the highly conserved site of the glycan-binding SSLs (authors own observations). These factors suggest that SElX may represent a link between the SSL and SAg families.

Mechanisms of immune evasion mediated by SSLs

When *S. aureus* comes under attack from the host immune system, damage induced stresses placed on the bacteria trigger the up-regulation of many factors that aid in avoidance of destruction. The SSLs are amongst these immune evasion molecules and research has shown they are instrumental in this immune evasion process. Some of the immune evasion mechanisms that SSLs are involved in are pictorially summarized in the cartoon in Fig. 6.7. IgA is the most abundant antibody secreted onto mucosal surfaces where it is the body's first line of defence against invading microorganisms and their toxin products (van Egmond *et al.*, 2001) and *S. aureus* may secrete SSL7 to block its antimicrobial function. Furthermore IgA is the second most predominant isotype found in serum and provides a second line of defence against microbes by initiating FcαRI-mediated phagocytosis (van Egmond *et al.*, 2000). The binding of FcαRI on neutrophils by activated IgA results in phagocytosis of IgA–antigen complexes and stimulation of the chemokine leukotriene B4 that promotes further neutrophil infiltration (van der Steen *et al.*, 2009). SSL7 has shown the ability to block the binding of IgA to its Fc receptor so could potentially prevent

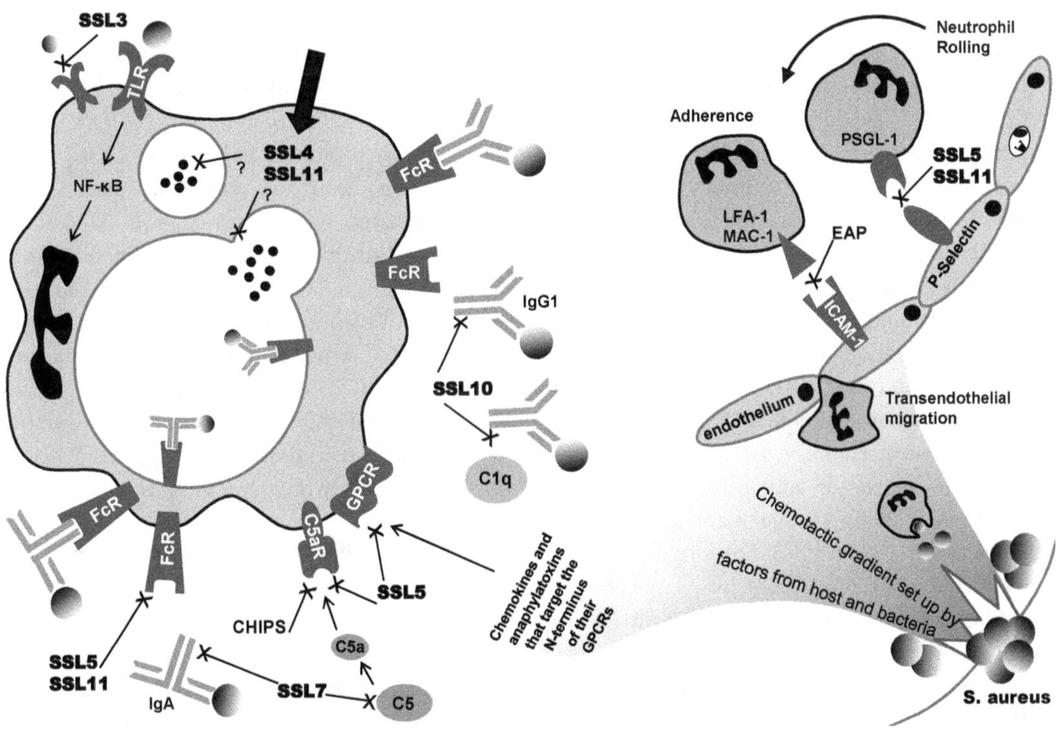

Figure 6.7 Cartoon depicting some of the SSL-mediated immune evasion mechanisms. SSLs and the molecules they interact with to impede processes of neutrophil migration and stimulation, and complement activation, are shown. These include molecules involved in phagocyte migration to the site of infection, antibodies and complement factors involved in recognizing *S. aureus* and signalling to phagocytes, and receptors involved in phagocyte activity.

FcαRI effector functions. A proven functional scenario for binding IgA is that SSL7 utilizes the abundance of this immunoglobulin to build its C5-inhibiting complex in order to prevent the phagocyte-recruiting chemotactic action of C5a. Neutrophils are rapidly recruited to sites of infection and are a key cell-type in the immune defence against *S. aureus*. SSL7 has been shown *in vivo* to inhibit the C5a-mediated influx of neutrophils using a mouse model of peritonitis (Bestebroer *et al.*, 2010). By expressing SSL7 and the structurally related CHIPS the bacterium can prevent both the production of C5a and the recognition of it by its cell-surface receptor.

The recruitment of neutrophils is further hindered by the inhibition of P-selectin mediated vascular rolling by glycan binders such as SSL5 and SSL11 binding to PSGL-1. The structurally related virulence factor EAP then enhances the inhibition of neutrophil extravasation by binding ICAM-1 on endothelial cells to prevent firm neutrophil attachment (Haggar *et al.*, 2004). SSL5 (and possibly other glycan binders) has the added ability to inhibit the activation of MMP-9 thus preventing its capacity to degrade extracellular matrix to aid in neutrophil infiltration of inflammatory tissues. Furthermore it can prevent activation of neutrophils to many chemotactic signals from the site of inflammation by blocking their GPCRs.

SSL10 secreted by *S. aureus* into its immediate environment binds to IgG1. This immunoglobulin is the most abundant of the four human IgG subclasses found in the blood and also has the highest affinity to multiple FcΥRs (Gessner *et al.*, 1998). Furthermore, it is the most effective at complement fixation and antibody-dependent cell-mediated killing (Brüggemann *et al.*, 1987; Bindon *et al.*, 1988). IgG1-opsonized *S. aureus* is protected from engulfment by phagocytes by two SSL10-mediated mechanisms. Firstly SSL10 prevents the recognition of IgG1 by its FcΥRs, and secondly it inhibits activation of the classical pathway of the complement system by blocking the binding of C1q to IgG1.

The antimicrobial mechanisms of neutrophils include ROS, AMPs and other granule proteins that are either released into phagosomes containing *S. aureus* or out of the neutrophil into the site of infection. SSL4 and SSL11 have both been seen to rapidly enter neutrophils and monocyte/macrophages in a sialic acid and an energy-dependent manner (Chung *et al.*, 2007; Hermans *et al.*, 2012) and once internalized they localize to granular structures in neutrophils or cytoskeletal-like structures in macrophages. The consequence of this internalization is not yet known but the fact that they target cells of the innate immune system that are the primary defenders against *S. aureus* suggests this is of some benefit to the organism. Presence of SSLs in the vesicles of neutrophils where many antimicrobial products are stored could potentially be to affect the function of these products. Another possibility is the disruption or dysregulation of bacterial engulfment by these professional phagocytes, or of vesicle fusion, or perhaps on their cellular mobility by the inhibition of actin re-arrangement.

Finally, recognition of staphylococcal PAMPs by TLR2 is important in the immune response to *S. aureus*. Signalling through TLR2 activates innate immune cells to both phagocytose *S. aureus* and to signal the adaptive immune system through the presentation of bacterial antigens (Fournier and Philpott, 2005; Stenzel *et al.*, 2008). It has been shown that mice deficient in TLR2 or its signalling pathway are more susceptible to *S. aureus* infection, with decreased production of the proinflammatory cytokines that ultimately result in a reduction in bacterial clearance (Takeuchi *et al.*, 2000). TLR2 has also been shown to activate ROS in the immune defence against *S. aureus* (Jann *et al.*, 2011). SSL3 inhibits this recognition system and in doing so S. *aureus* is able to promote its persistence within the host by the avoidance of immune detection and clearance.

An SSL-mediated immune evasion mechanism directed at the humoral response?

SSL7 and SSL9 have been reported to be rapidly taken up by monocyte-derived dendritic cells into the same endosomal vesicles as dextran (Al-Shangiti *et al.*, 2004). This internalization was temperature sensitive and suggested of receptor-mediated uptake of the SSLs. Larger SSL9-staining vesicles could be observed that possibly resulted from fusion of many SSL9-containg vesicles. Although specific and non-competitive binding of both these SSLs to monocytes was reported, binding and internalization by macrophages was not seen. Whereas dendritic cells can present antigen to naive T cells, macrophages require cytokine stimulation to present antigen. No discernible changes in dendritic cell activity or morphology were observed upon internalization of these SSLs showing that the SSLs were not affecting some cell-changing function upon internalization. A similar proportion of human volunteers (37%) to those reported to be colonized by *S. aureus* exhibited dose-dependent T-cell responses to dendritic cells loaded with SSL9 (Al-Shangiti *et al.*, 2005). Additionally, the sera of 9 out of 10 volunteers tested positive for IgG specific to SSL7 and SSL9. Antibody class switching indicated by the presence of SSL-specific IgG supported the involvement of SSL-specific T cells.

Al-Shangiti and colleagues speculate that targeting of endocytosis by antigen-presenting dendritic cells may be a means of enhancing immunogenicity to the SSLs. Uptake into the same vesicles as dextran indicates a pathway of antigen trafficking to MHC class II for subsequent presentation to $CD4^+$ helper T cells. The authors suggest that the consequence of T helper cell activation would be to direct an antibody response against the secreted proteins, resulting in a local inflammatory environment that allows for bacterial invasion and avoidance of immune clearance (Al-Shangiti *et al.*, 2005). Whether the immune response to these SSLs is advantageous to *S. aureus* or merely the host defence against the bacterium and its virulence factors remains to be seen.

Novel SSL-associated applications

S. aureus genotyping using SSL profiling

A genotyping method based on variation comparison between alleles of three different SSL genes gave identical or even more discriminatory groupings to those obtained using the standard seven gene multilocus sequence typing (MLST) method (Aguiar-Alves *et al.*, 2006). The SSL-based method was not as discriminatory as pulse field gel electrophoresis (PFGE) on MRSA isolates though it was less expensive and time consuming in comparison to the other two methods.

Anti-inflammatory potential of SSLs

SSL7 has the potential to be used as an anti-inflammatory agent by blocking C5a. The inappropriate or over-production of C5a can be a serious problem and has been associated with inflammation in allergic, infectious, and autoimmune diseases, as well as transplant rejection and tumour progression (reviewed in Klos *et al.*, 2009). Although SSL7 is a microbial protein that would elicit an adaptive immune response, it is largely masked from detection when in complex with IgA and C5 (Laursen *et al.*, 2010). Even so, a more likely SSL7-based anti-C5a therapeutic agent would probably require alteration or molecular mimicking to generate a non-immunogenic reagent. This scenario would also apply to potential therapies based on any of the other SSLs. A C5a inhibitor based on SSL7 would be beneficial in preventing the potent effects of this anaphylatoxin in conditions associated with its dysregulation as well as blocking MAC formation in unwanted situations while leaving other activities of complement such as opsonization unaffected. The humanized monoclonal antibody Eculizumab, the first complement inhibitor approved by the FDA and EMEA (Rother *et al.*, 2007), blocks the cleavage of C5 and is either in clinical use or in trials to treat over thirty complement-related diseases (Schrezenmeier and Höchsmann, 2012). It was reported by Forbes magazine to be the most expensive drug in the world so development of a C5 inhibitor based on SSL7 would provide a more affordable alternative.

Bardoel *et al.* (2012) propose a role for SSL3 in treatment of acute and chronic inflammatory diseases that have a TLR2 involvement. Its ability to block signalling via TLR2 may have a beneficial consequence in preventing the development of some of the inflammatory pathologies associated with TLR2 that can include allergy, sepsis, rheumatoid arthritis, ischaemia–reperfusion injury, systemic lupus erythematosus (SLE) and tumour metastasis (reviewed in O'Neill *et al.*, 2009).

SSL-mediated anti-cancer therapies

The inhibitory effect on cell migration exhibited by various SSLs has the potential to be exploited in anticancer therapies targeting metastasis. Chemokine receptors are involved in the trafficking of leucocytes to sites of infection. They are also influential in the survival, proliferation, and metastasis of tumour cells. CXCR4 is the most widely expressed chemokine receptor in many cancers and has been associated with metastases and a poor prognosis. Walenkamp *et al.* (2009) showed that SSL10 had the ability to obstruct cell binding of a function-blocking antibody against CXCR4. The binding of SSL10 to CXCR4 inhibited CXCL12-induced calcium mobilization and blocked CXCL12-induced migration of a human leukaemia cell line and a human cervical carcinoma cell line. The authors propose that by blocking CXCR4 SSL10 not only gives *S. aureus* an advantage in avoiding clearance by infiltrating neutrophils but has the potential to be used in anti-cancer therapy by interfering with metastasis. Interestingly SSL5 also binds CXCR4 to inhibit the chemotactic response to CXCL12 (Bestebroer *et al.*, 2009) so could possibly be used in a comparable manner to SSL10 in this respect. Another potential use of SSL5 in the treatment of cancer has been highlighted through research conducted by Walenkamp *et al.* (2010). They found that SSL5 blocked the adhesion of a PSGL-1-expressing leukaemia cell line to both endothelial cells and platelets. The interaction with endothelial cells and platelets is important for the metastasis, growth, and survival of cancer cells. SSL5 binding to the sLeX moiety on PSGL-1 was instrumental in preventing the cancer cell from binding to P-selectin expressed on platelets and endothelial cells. Increased and altered glycosylation is common during malignancy and sLeX expression in particular has been linked to increased metastasis and reduced survival of patients (Irimura *et al.*, 1993; Nakamori *et al.*, 1993). SSL5 is not the only SSL that binds to sLeX so this blocking of cancer cell adhesion is likely to be shared by the other sialic acid binding SSLs. The association of sLeX with a poor prognosis would make its targeting using SSLs an attractive proposition.

Concluding remarks

The SSLs are emerging as an important family of immune evasion molecules involved in promoting the survival of *S. aureus* in the hostile environment of the hosts body. The large number of SSLs encoded in a tandem gene cluster represents a toolbox for virulence that *S. aureus* can express from, and adapt, to allow for a wide variety of functionally diverse outcomes depending on its requirements. Phylogenetic analysis of the *ssl* gene region suggests that gene duplication, recombination, and losses have contributed to the diversification of this genomic island. The controlled and synchronized expression of many SSLs with distinct but complementary functions provides broad coverage in inhibiting important aspects of the host's immunity against *S. aureus*. It is unsurprising that the SSLs affect mechanisms of complement activation and phagocyte activity, both having clearly been demonstrated to be essential in protection from staphylococcal infections. SSLs have the potential to be used as tools not only in elucidating mechanisms of immune evasion by *S. aureus* and the corresponding host responses against it, but also as potential anti-staphylococcal targets, therapeutic agents, and as reagents to isolate and purify the host components they interact with. SSL7 for example is a useful tool for rapidly isolating IgA and complement C5, while SSL10 can be used to specifically isolate human IgG1.

While much has been discovered since their discovery just over a decade ago, there are still SSLs that have only been partially characterized, and some that have yet to reveal their purposes. There are certain to be many exciting discoveries yet to be made from continued research of the SSLs and their interactions with the human immune system.

References

Abi-Rached, L., Dorighi, K., Norman, P.J., Yawata, M., and Parham, P. (2007). Episodes of natural selection shaped the interactions of IgA-Fc with FcαRI and bacterial decoy proteins. J. Immunol. *178*, 7943–7954.

Aguiar-Alves, F., Medeiros, F., Fernandes, O., Gudziki Pereira, R.M., Perdreau-Remington, F., and Riley, L.W. (2006). New *Staphylococcus aureus* genotyping method based on exotoxin (set) genes. J. Clin. Microbiol. *44*, 2728–2732.

Al-Shangiti, A.M., Naylor, C.E., Nair, S.P., Briggs, D.C., Henderson, B., and Chain, B.M. (2004). Structural relationships and cellular tropism of staphylococcal superantigen-like proteins. Infect. Immun. *72*, 4261–4270.

Al-Shangiti, A.M., Nair, S.P., and Chain, B.M. (2005). The interaction between staphylococcal superantigen-like proteins and human dendritic cells. Clin. Exp. Immunol. *140*, 461–469.

Andrews, R.K., Gardiner, E.E., Shen, Y., Whisstock, J.C., and Berndt, M.C. (2003). Glycoprotein Ib–IX–V. Int. J. Biochem. Cell Biol. *35*, 1170–1174.

Arcus, V. (2002). OB-fold domains: a snapshot of the evolution of sequence, structure and function. Curr. Opin. Struct. Biol. *12*, 794–801.

Arcus, V.L., Langley, R., Proft, T., Fraser, J.D., and Baker, E.N. (2002). The three-dimensional structure of a superantigen-like protein, SET3, from a pathogenicity island of the *Staphylococcus aureus* genome. J. Biol. Chem. *277*, 32274–32281.

Armstrong, P.C., Hu, H., Rivera, J., Rigby, S., Chen, Y.C., P Howden, B., Gardiner, E., and Peter, K. (2012). Staphylococcal superantigen-like protein 5 induces thrombotic and bleeding complications *in vivo*: Inhibition by an anti-SSL5 antibody and the glycan Bimosiamose. J. Thromb. Haemost. *10*, 2607–2609.

Athanasopoulos, A.N., Economopoulou, M., Orlova, V.V., Sobke, A., Schneider, D., Weber, H., Augustin, H.G., Eming, S.A., Schubert, U., Linn, T., *et al.* (2006). The extracellular adherence protein (Eap) of *Staphylococcus aureus* inhibits wound healing by interfering with host defense and repair mechanisms. Blood *107*, 2720–2727.

Attia, A.S., Benson, M.A., Stauff, D.L., Torres, V.J., and Skaar, E.P. (2010). Membrane Damage Elicits an Immunomodulatory Program in *Staphylococcus aureus*. PLoS Pathog. *6*, e1000802.

Baba, T., Takeuchi, F., Kuroda, M., Yuzawa, H., Aoki, K., Oguchi, A., Nagai, Y., Iwama, N., Asano, K., Naimi, T., *et al.* (2002). Genome and virulence determinants of high virulence community-acquired MRSA. Lancet *359*, 1819–1827.

Baba, T., Bae, T., Schneewind, O., Takeuchi, F., and Hiramatsu, K. (2008). Genome sequence of *Staphylococcus aureus* strain newman and comparative analysis of Staphylococcal genomes: polymorphism and evolution of two major pathogenicity islands. J. Bacteriol. *190*, 300–310.

Baker, H.M., Basu, I., Chung, M.C., Caradoc-Davies, T., Fraser, J.D., and Baker, E.N. (2007). Crystal structures of the Staphylococcal toxin SSL5 in complex with sialyl Lewis X reveal a conserved binding site that shares common features with viral and bacterial sialic acid binding proteins. J. Mol. Biol. *374*, 1298–1308.

Barber, M. (1961). Methicillin-resistant staphylococci. J. Clin. Pathol. *14*, 385–393.

Bardoel, B.W., Vos, R., Bouman, T., Aerts, P.C., Bestebroer, J., Huizinga, E.G., Brondijk, T.H.C., Strijp, J.A.G., and Haas, C.J.C. (2012). Evasion of Toll-like receptor 2 activation by staphylococcal superantigen-like protein 3. J. Mol. Med. *90*, 1109–1120.

Benson, M.A., Lilo, S., Wasserman, G.A., Thoendel, M., Smith, A., Horswill, A.R., Fraser, J., Novick, R.P., Shopsin, B., and Torres, V.J. (2011). *Staphylococcus aureus* regulates the expression and production of the staphylococcal superantigen-like secreted proteins in a Rot-dependent manner. Mol. Microbiol. *81*, 659–675.

Benson, M.A., Lilo, S., Nygaard, T., Voyich, J.M., and Torres, V.J. (2012). Rot and SaeRS cooperate to activate expression of the Staphylococcal superantigen-like exoproteins. J. Bacteriol. *194*, 4355–4365.

Berndt, M.C., Shen, Y., Dopheide, S.M., Gardiner, E.E., and Andrews, R.K. (2001). The vascular biology of the glycoprotein Ib–IX-V complex. Thromb. Haemost. *86*, 178–188.

Bestebroer, J., Poppelier, M.J.J.G., Ulfman, L.H., Lenting, P.J., Denis, C.V., van Kessel, K.P.M., van Strijp, J.A.G., and de Haas, C.J.C. (2007). Staphylococcal superantigen-like 5 binds PSGL-1 and inhibits P-selectin–mediated neutrophil rolling. Blood *109*, 2936–2943.

Bestebroer, J., van Kessel, K.P.M., Azouagh, H., Walenkamp, A.M., Boer, I.G.J., Romijn, R.A., van Strijp, J.A.G., and de Haas, C.J.C. (2009). Staphylococcal SSL5 inhibits leukocyte activation by chemokines and anaphylatoxins. Blood *113*, 328–337.

Bestebroer, J., Aerts, P.C., Rooijakkers, S.H.M., Pandey, M.K., Köhl, J., Van Strijp, J.A.G., and De Haas, C.J.C. (2010). Functional basis for complement evasion by staphylococcal superantigen-like 7. Cell. Microbiol. *12*, 1506–1516.

Bindon, C.I., Hale, G., Brüggemann, M., and Waldmann, H. (1988). Human monoclonal IgG isotypes differ in complement activating function at the level of C4 as well as C1q. J. Exp. Med. *168*, 127–142.

Brogden, K.A. (2005). Antimicrobial peptides: pore formers or metabolic inhibitors in bacteria? Nat. Rev. Microbiol. *3*, 238–250.

Brüggemann, M., Williams, G.T., Bindon, C.I., Clark, M.R., Walker, M.R., Jefferis, R., Waldmann, H., and Neuberger, M.S. (1987). Comparison of the effector functions of human immunoglobulins using a matched set of chimeric antibodies. J. Exp. Med. *166*, 1351–1361.

Calander, A.-M., Starckx, S., Opdenakker, G., Bergin, P., Quiding-Järbrink, M., and Tarkowski, A. (2006). Matrix metalloproteinase-9 (gelatinase B) deficiency leads to increased severity of *Staphylococcus aureus*-triggered septic arthritis. Microbes Infect. *8*, 1434–1439.

Cerquetti, M.C., Sordelli, D.O., Ortegon, R.A., and Bellanti, J.A. (1983). Impaired lung defenses against *Staphylococcus aureus* in mice with hereditary deficiency of the fifth component of complement. Infect. Immun. *41*, 1071–1076.

Chavakis, T., Hussain, M., Kanse, S.M., Peters, G., Bretzel, R.G., Flock, J.-I., Herrmann, M., and Preissner, K.T. (2002). *Staphylococcus aureus* extracellular adherence protein serves as anti-inflammatory factor by inhibiting the recruitment of host leukocytes. Nat. Med. *8*, 687–693.

Chavakis, T., Preissner, K.T., and Herrmann, M. (2007). The anti-inflammatory activities of *Staphylococcus aureus*. Trends Immunol. *28*, 408–418.

Chung, M.C., Wines, B.D., Baker, H., Langley, R.J., Baker, E.N., and Fraser, J.D. (2007). The crystal structure of staphylococcal superantigen-like protein 11 in complex with sialyl Lewis X reveals the mechanism for cell binding and immune inhibition. Mol. Microbiol. *66*, 1342–1355.

Corbin, B.D., Seeley, E.H., Raab, A., Feldmann, J., Miller, M.R., Torres, V.J., Anderson, K.L., Dattilo, B.M., Dunman, P.M., Gerads, R., *et al.* (2008). Metal chelation and inhibition of bacterial growth in tissue abscesses. Science *319*, 962–965.

David, M.Z.M.D.P., Medvedev, S.P., Hohmann, S.F.P., Ewigman, B.M.D., and Daum, R.S.M.D. (2012). Increasing burden of methicillin-resistant *Staphylococcus aureus* hospitalizations at US academic medical centers, 2003–2008. Infect. Control Hosp. Epidemiol. *33*, 782–789.

de Haas, C.J., Veldkamp, K.E., Peschel, A., Weerkamp, F., Van Wamel, W.J., Heezius, E.C., Poppelier, M.J., Van Kessel, K.P., and van Strijp, J.A. (2004). Chemotaxis inhibitory protein of *Staphylococcus aureus*, a bacterial antiinflammatory agent. J. Exp. Med. *199*, 687–695.

De Haas, C.J.C., Weeterings, C., Vughs, M.M., De Groot, P.G., Van Strijp, J.A., and Lisman, T. (2009). Staphylococcal superantigen-like 5 activates platelets and supports platelet adhesion under flow conditions, which involves glycoprotein Ibα and αIIbβ3. J. Thromb. Haemost. *7*, 1867–1874.

Fitzgerald, J.R., Reid, S.D., Ruotsalainen, E., Tripp, T.J., Liu, M., Cole, R., Kuusela, P., Schlievert, P.M., Järvinen, A., and Musser, J.M. (2003). Genome diversification in *Staphylococcus aureus*: molecular evolution of a highly variable chromosomal region encoding the Staphylococcal exotoxin-like family of proteins. Infect. Immun. *71*, 2827–2838.

Fitzgerald, J.R., Foster, T.J., and Cox, D. (2006). The interaction of bacterial pathogens with platelets. Nat. Rev. Micro. *4*, 445–457.

Foreman, K.E., Vaporciyan, A.A., Bonish, B.K., Jones, M.L., Johnson, K.J., Glovsky, M.M., Eddy, S.M., and Ward, P.A. (1994). C5a-induced expression of P-selectin in endothelial cells. J. Clin. Invest. *94*, 1147–1155.

Foster, T.J. (2005). Immune evasion by staphylococci. Nat. Rev. Microbiol. *3*, 948–958.

Fournier, B., and Philpott, D.J. (2005). Recognition of *Staphylococcus aureus* by the innate immune system. Clin. Microbiol. Rev. *18*, 521–540.

Fraser, J.D., and Proft, T. (2008). The bacterial superantigen and superantigen-like proteins. Immunol. Rev. *225*, 226–243.

Frenette, P.S., Denis, C.V., Weiss, L., Jurk, K., Subbarao, S., Kehrel, B., Hartwig, J.H., Vestweber, D., and Wagner, D.D. (2000). P-Selectin glycoprotein ligand 1 (PSGL-1) is expressed on platelets and can mediate platelet–endothelial interactions *in vivo*. J. Exp. Med. *191*, 1413–1422.

Gan, B.S., Kim, J., Reid, G., Cadieux, P., and Howard, J.C. (2002). Lactobacillus fermentum RC-14 inhibits *Staphylococcus aureus* infection of surgical implants in rats. J. Infect. Dis. *185*, 1369–1372.

Gasque, P. (2004). Complement: a unique innate immune sensor for danger signals. Mol. Immunol. *41*, 1089–1098.

Geisbrecht, B.V., Hamaoka, B.Y., Perman, B., Zemla, A., and Leahy, D.J. (2005). The crystal structures of EAP domains from *Staphylococcus aureus* reveal an unexpected homology to bacterial superantigens. J. Biol. Chem. *280*, 17243–17250.

Gessner, J.E., Heiken, H., Tamm, A., and Schmidt, R.E. (1998). The IgG Fc receptor family. Ann. Hematol. *76*, 231–248.

Gill, S.R., Fouts, D.E., Archer, G.L., Mongodin, E.F., Deboy, R.T., Ravel, J., Paulsen, I.T., Kolonay, J.F., Brinkac, L., Beanan, M., *et al.* (2005). Insights on evolution of virulence and resistance from the complete genome analysis of an early methicillin-resistant *Staphylococcus aureus* strain and a biofilm-producing methicillin-resistant Staphylococcus epidermidis strain. J. Bacteriol. *187*, 2426–2438.

Goerke, C., and Wolz, C. (2004). Regulatory and genomic plasticity of *Staphylococcus aureus* during persistent colonization and infection. Int. J. Med. Microbiol. *294*, 195–202.

Guyer, D.A., Moore, K.L., Lynam, E.B., Schammel, C.M., Rogelj, S., McEver, R.P., and Sklar, L.A. (1996). P-selectin glycoprotein ligand-1 (PSGL-1) is a ligand for L-selectin in neutrophil aggregation. Blood *88*, 2415–2421.

Haas, P.-J., de Haas, C.J.C., Poppelier, M.J.J.C., van Kessel, K.P.M., van Strijp, J.A.G., Dijkstra, K., Scheek, R.M., Fan, H., Kruijtzer, J.A.W., Liskamp, R.M.J., *et al.* (2005). The structure of the C5a receptor-blocking domain of chemotaxis inhibitory protein of *Staphylococcus aureus* is related to a group of immune evasive molecules. J. Mol. Biol. *353*, 859–872.

Haggar, A., Ehrnfelt, C., Holgersson, J., and Flock, J.-I. (2004). The extracellular adherence protein from *Staphylococcus aureus* inhibits neutrophil binding to endothelial cells. Infect. Immun. *72*, 6164–6167.

Haggar, A., Flock, J.I., and Norrby-Teglund, A. (2010). Extracellular adherence protein (Eap) from *Staphylococcus aureus* does not function as a superantigen. Clin. Microbiol. Infect. *16*, 1155–1158.

Hanselman, B.A., Kruth, S.A., Rousseau, J., and Weese, J.S. (2009). Coagulase positive staphylococcal colonization of humans and their household pets. Can. Vet. J. *50*, 954–958.

Harraghy, N., Hussain, M., Haggar, A., Chavakis, T., Sinha, B., Herrmann, M., and Flock, J.-I. (2003). The adhesive and immunomodulating properties of the multifunctional *Staphylococcus aureus* protein Eap. Microbiology *149*, 2701–2707.

Harris, L.G., Foster, S.J., and Richards, R.G. (2002). An introduction to *Staphylococcus aureus*, and techniques for identifying and quantifying S. aureus adhesins in relation to adhesion to biomaterials: review. Eur. Cell Mater. *4*, 39–60.

Hermans, S.J., Baker, H.M., Sequeira, R.P., Langley, R.J., Baker, E.N., and Fraser, J.D. (2012). Structural and functional properties of Staphylococcal superantigen-like protein 4. Infect. Immun. *80*, 4004–4013.

Holt, D.C., Holden, M.T.G., Tong, S.Y.C., Castillo-Ramirez, S., Clarke, L., Quail, M.A., Currie, B.J., Parkhill, J., Bentley, S.D., Feil, E.J., *et al.* (2011). A very early-branching *Staphylococcus aureus* lineage lacking the carotenoid pigment staphyloxanthin. Genome Biol. Evol. 3, 881–895.

Hopken, U.E., Lu, B., Gerard, N.P., and Gerard, C. (1996). The C5a chemoattractant receptor mediates mucosal defence to infection. Nature *383*, 86–89.

Hovde, C.J., Marr, J.C., Hoffmann, M.L., Hackett, S.P., Chi, Y.-i., Crum, K.K., Stevens, D.L., Stauffacher, C.V., and Bohach, G.A. (1994). Investigation of the role of the disulphide bond in the activity and structure of staphylococcal enterotoxin C1. Mol. Microbiol. *13*, 897–909.

Hu, H., Armstrong, P.C.J., Khalil, E., Chen, Y.-C., Straub, A., Li, M., Soosairajah, J., Hagemeyer, C.E., Bassler, N., Huang, D., *et al.* (2011). GPVI and GPIbα mediate staphylococcal superantigen-like protein 5 (SSL5) induced platelet activation and direct toward glycans as potential inhibitors. PLoS One *6*, e19190.

Hussain, M., Haggar, A., Heilmann, C., Peters, G., Flock, J.-I., and Herrmann, M. (2002). Insertional inactivation of eap in *Staphylococcus aureus* strain newman confers reduced staphylococcal binding to fibroblasts. Infect. Immun. *70*, 2933–2940.

Irimura, T., Nakamori, S., Matsushita, Y., Taniuchi, Y., Todoroki, N., Tsuji, T., Izumi, Y., Kawamura, Y., Hoff, S.D., Cleary, K.R., *et al.* (1993). Colorectal cancer metastasis determined by carbohydrate-mediated cell adhesion: role of sialyl-LeX antigens. Semin. Cancer Biol. *4*, 319–324.

Itoh, S., Hamada, E., Kamoshida, G., Takeshita, K., Oku, T., and Tsuji, T. (2010a). Staphylococcal superantigen-like protein 5 inhibits matrix metalloproteinase 9 from human neutrophils. Infect. Immun. *78*, 3298–3305.

Itoh, S., Hamada, E., Kamoshida, G., Yokoyama, R., Takii, T., Onozaki, K., and Tsuji, T. (2010b). Staphylococcal superantigen-like protein 10 (SSL10) binds to human immunoglobulin G (IgG) and inhibits complement activation via the classical pathway. Mol. Immunol. *47*, 932–938.

Itoh, S., Yokoyama, R., Murase, C., Takii, T., Tsuji, T., and Onozaki, K. (2012). Staphylococcal superantigen-like protein 10 binds to phosphatidylserine and apoptotic cells. Microbiol. Immunol. *56*, 363–371.

Jacobs, J.C., and Miller, M.E. (1972). Fatal familial Leiner's disease: a deficiency of the opsonic activity of serum complement. Pediatrics *49*, 225–232.

Jann, N.J., Schmaler, M., Ferracin, F., and Landmann, R. (2011). TLR2 enhances NADPH oxidase activity and killing of *Staphylococcus aureus* by PMN. Immunol. Lett. *135*, 17–23.

Jongerius, I., Köhl, J., Pandey, M.K., Ruyken, M., van Kessel, K.P.M., van Strijp, J.A.G., and Rooijakkers, S.H.M. (2007). Staphylococcal complement evasion by various convertase-blocking molecules. J. Exp. Med. *204*, 2461–2471.

Kaplan, J.B., Izano, E.A., Gopal, P., Karwacki, M.T., Kim, S., Bose, J.L., Bayles, K.W., and Horswill, A.R. (2012). Low levels of beta-lactam antibiotics induce extracellular DNA release and biofilm formation in *Staphylococcus aureus*. MBio. 3, 00198–00112.

Katayama, Y., Ito, T., and Hiramatsu, K. (2000). A new class of genetic element, staphylococcus cassette chromosome mec, encodes methicillin resistance in *Staphylococcus aureus*. Antimicrob. Agents Chemother. *44*, 1549–1555.

Kiedrowski, M.R., and Horswill, A.R. (2011). New approaches for treating staphylococcal biofilm infections. Ann. N. Y. Acad. Sci. *1241*, 104–121.

Klevens, R.M., Morrison, M.A., Nadle, J., Petit, S., Gershman, K., Ray, S., Harrison, L.H., Lynfield, R., Dumyati, G., Townes, J.M., *et al.* (2007). Invasive methicillin-resistant *Staphylococcus aureus* infections in the United States. Jama *298*, 1763–1771.

Klos, A., Tenner, A.J., Johswich, K.-O., Ager, R.R., Reis, E.S., and Köhl, J. (2009). The role of the anaphylatoxins in health and disease. Mol. Immunol. *46*, 2753–2766.

Krishna, S., and Miller, L.S. (2012). Host–pathogen interactions between the skin and *Staphylococcus aureus*. Curr. Opin. Microbiol. *15*, 28–35.

Kuroda, M., Ohta, T., Uchiyama, I., Baba, T., Yuzawa, H., Kobayashi, I., Cui, L., Oguchi, A., Aoki, K., Nagai, Y., *et al.* (2001). Whole genome sequencing of meticillin-resistant *Staphylococcus aureus*. Lancet *357*, 1225–1240.

Lambris, J.D., Ricklin, D., and Geisbrecht, B.V. (2008). Complement evasion by human pathogens. Nat. Rev. Micro. *6*, 132–142.

Langley, R., Wines, B., Willoughby, N., Basu, I., Proft, T., and Fraser, J.D. (2005). The Staphylococcal superantigen-like protein 7 binds IgA and complement C5 and inhibits IgA-FcαRI binding and serum killing of bacteria. J. Immunol. *174*, 2926–2933.

Langley, R., Patel, D., Jackson, N., Clow, F., and Fraser, J.D. (2010). Staphylococcal superantigen super-domains in immune evasion. Crit. Rev. Immunol. *30*, 149–165.

Laughton, J.M., Devillard, E., Heinrichs, D.E., Reid, G., and McCormick, J.K. (2006). Inhibition of expression of a staphylococcal superantigen-like protein by a soluble factor from *Lactobacillus reuteri*. Microbiology *152*, 1155–1167.

Laursen, N.S., Gordon, N., Hermans, S., Lorenz, N., Jackson, N., Wines, B., Spillner, E., Christensen, J.B., Jensen, M., Fredslund, F., *et al.* (2010). Structural basis for inhibition of complement C5 by the SSL7 protein from *Staphylococcus aureus*. Proc. Natl. Acad. Sci. U.S.A. *107*, 3681–3686.

Laursen, N.S., Andersen, K.R., Braren, I., Spillner, E., Sottrup-Jensen, L., and Andersen, G.R. (2011). Substrate recognition by complement convertases revealed in the C5-cobra venom factor complex. EMBO J. *30*, 606–616.

Lekstrom-Himes, J.A., and Gallin, J.I. (2000). Immunodeficiency diseases caused by defects in phagocytes. N. Engl. J. Med. 343, 1703–1714.

Liang, X., Yu, C., Sun, J., Liu, H., Landwehr, C., Holmes, D., and Ji, Y. (2006). Inactivation of a two-component signal transduction system, SaeRS, eliminates adherence and attenuates virulence of *Staphylococcus aureus*. Infect. Immun. *74*, 4655–4665.

Lina, G., Bohach, G.A., Nair, S.P., Hiramatsu, K., Jouvin-Marche, E., and Mariuzza, R. (2004). Standard nomenclature for the superantigens expressed by *Staphylococcus*. J. Infect. Dis. *189*, 2334–2336.

Lindsay, J., and Holden, M.G. (2006). Understanding the rise of the superbug: investigation of the evolution and genomic variation of *Staphylococcus aureus*. Funct. Integr. Genomics *6*, 186–201.

Liu, C.-I., Liu, G.Y., Song, Y., Yin, F., Hensler, M.E., Jeng, W.-Y., Nizet, V., Wang, A.H.-J., and Oldfield, E. (2008). A cholesterol biosynthesis inhibitor blocks *Staphylococcus aureus* virulence. Science *319*, 1391–1394.

Liu, G.Y., Essex, A., Buchanan, J.T., Datta, V., Hoffman, H.M., Bastian, J.F., Fierer, J., and Nizet, V. (2005). *Staphylococcus aureus* golden pigment impairs neutrophil killing and promotes virulence through its antioxidant activity. J. Exp. Med. *202*, 209–215.

Lowy, F.D. (1998). *Staphylococcus aureus* infections. N. Engl. J. Med. *339*, 520–532.

Malachowa, N., and DeLeo, F. (2010). Mobile genetic elements of *Staphylococcus aureus*. Cell. Mol. Life Sci. *67*, 3057–3071.

Malachowa, N., Whitney, A.R., Kobayashi, S.D., Sturdevant, D.E., Kennedy, A.D., Braughton, K.R., Shabb, D.W., Diep, B.A., Chambers, H.F., Otto, M., *et al.* (2011). Global changes in *Staphylococcus aureus* gene expression in human blood. PLoS One *6*, e18617.

Menegazzi, R., Decleva, E., and Dri, P. (2012). Killing by neutrophil extracellular traps: fact or folklore? Blood *119*, 1214–1216.

Miller, L.S., and Cho, J.S. (2011). Immunity against *Staphylococcus aureus* cutaneous infections. Nat. Rev. Immunol. *11*, 505–518.

Mitchell, D.T., Levitt, D.G., Schlievert, P.M., and Ohlendorf, D.H. (2000). Structural Evidence for the Evolution of Pyrogenic Toxin Superantigens. J. Mol. Evol. *51*, 520–531.

Murzin, A.G. (1993). OB(oligonucleotide/oligosaccharide binding)-fold: common structural and functional solution for non-homologous sequences. EMBO J. *12*, 861–867.

Nakamori, S., Kameyama, M., Imaoka, S., Furukawa, H., Ishikawa, O., Sasaki, Y., Kabuto, T., Iwanaga, T., Matsushita, Y., and Irimura, T. (1993). Increased expression of sialyl Lewisx antigen correlates with poor survival in patients with colorectal carcinoma: clinicopathological and immunohistochemical study. Cancer Res. *53*, 3632–3637.

Nilsson, P., and Ripa, T. (2006). *Staphylococcus aureus* throat colonization is more frequent than colonization in the anterior nares. J. Clin. Microbiol. *44*, 3334–3339.

O'Neill, L.A.J., Bryant, C.E., and Doyle, S.L. (2009). Therapeutic targeting of Toll-like receptors for infectious and inflammatory diseases and cancer. Pharmacol. Rev. *61*, 177–197.

Ono, H.K., Omoe, K., Imanishi, K.i., Iwakabe, Y., Hu, D.-L., Kato, H., Saito, N., Nakane, A., Uchiyama, T., and Shinagawa, K. (2008). Identification and characterization of two novel staphylococcal enterotoxins, types S and T. Infect. Immun. *76*, 4999–5005.

Palazzolo-Ballance, A.M., Reniere, M.L., Braughton, K.R., Sturdevant, D.E., Otto, M., Kreiswirth, B.N., Skaar, E.P., and DeLeo, F.R. (2008). Neutrophil microbicides induce a pathogen survival response in community-associated methicillin-resistant *Staphylococcus aureus*. J. Immunol. *180*, 500–509.

Palma, M., Haggar, A., and Flock, J.-I. (1999). Adherence of *Staphylococcus aureus* is enhanced by an endogenous secreted protein with broad binding activity. J. Bacteriol. *181*, 2840–2845.

Pantrangi, M., Singh, V.K., Wolz, C., and Shukla, S.K. (2010). Staphylococcal superantigen-like genes, ssl5 and ssl8, are positively regulated by Sae and negatively by Agr in the Newman strain. FEMS Microbiol. Lett. *308*, 175–184.

Parker, H., Albrett, A.M., Kettle, A.J., and Winterbourn, C.C. (2012). Myeloperoxidase associated with neutrophil extracellular traps is active and mediates bacterial killing in the presence of hydrogen peroxide. J. Leukoc. Biol. *91*, 369–376.

Patel, D., Wines, B.D., Langley, R.J., and Fraser, J.D. (2010). Specificity of staphylococcal superantigen-like protein 10 toward the human IgG1 Fc domain. J. Immunol. *184*, 6283–6292.

Pietrocola, G., Arciola, C.R., Rindi, S., Di Poto, A., Missineo, A., Montanaro, L., and Speziale, P. (2011). Toll-like receptors (TLRs) in innate immune defense against *Staphylococcus aureus*. Int. J. Artif. Organs *34*, 799–810.

Postma, B., Poppelier, M.J., van Galen, J.C., Prossnitz, E.R., van Strijp, J.A.G., de Haas, C.J.C., and van Kessel, K.P.M. (2004). Chemotaxis inhibitory protein of *Staphylococcus aureus* binds specifically to the C5a and formylated peptide receptor. J. Immunol. *172*, 6994–7001.

Prat, C., Bestebroer, J., de Haas, C.J.C., van Strijp, J.A.G., and van Kessel, K.P.M. (2006). A new staphylococcal anti-inflammatory protein that antagonizes the formyl peptide receptor-like 1. J. Immunol. *177*, 8017–8026.

Ramsland, P.A., Willoughby, N., Trist, H.M., Farrugia, W., Hogarth, P.M., Fraser, J.D., and Wines, B.D. (2007). Structural basis for evasion of IgA immunity by *Staphylococcus aureus* revealed in the complex of SSL7 with Fc of human IgA1. Proc. Natl. Acad. Sci. U.S.A. *104*, 15051–15056.

Rigby, K.M., and DeLeo, F.R. (2012). Neutrophils in innate host defense against *Staphylococcus aureus* infections. Semin. Immunopathol. *34*, 237–259.

Rooijakkers, S.H.M., Ruyken, M., Roos, A., Daha, M.R., Presanis, J.S., Sim, R.B., van Wamel, W.J.B., van Kessel, K.P.M., and van Strijp, J.A.G. (2005). Immune evasion by a staphylococcal complement inhibitor that acts on C3 convertases. Nat. Immunol. *6*, 920–927.

Rother, R.P., Rollins, S.A., Mojcik, C.F., Brodsky, R.A., and Bell, L. (2007). Discovery and development of the complement inhibitor eculizumab for the treatment of paroxysmal nocturnal hemoglobinuria. Nat. Biotech. *25*, 1256–1264.

Roussel, A., Anderson, B.F., Baker, H.M., Fraser, J.D., and Baker, E.N. (1997). Crystal structure of the streptococcal superantigen SPE-C: dimerization and zinc binding suggest a novel mode of interaction with MHC class II molecules. Nat. Struct. Mol. Biol. *4*, 635–643.

Schrezenmeier, H., and Höchsmann, B. (2012). Drugs that inhibit complement. Transfus. Apher. Sci. *46*, 87–92.

Stenzel, W., Soltek, S., Sanchez-Ruiz, M., Akira, S., Miletic, H., Schlüter, D., and Deckert, M. (2008). Both TLR2 and TLR4 Are Required for the effective immune response in *Staphylococcus aureus*-induced experimental murine brain abscess. Am. J. Pathol. *172*, 132–145.

Sundberg, E.J., Deng, L., and Mariuzza, R.A. (2007). TCR recognition of peptide/MHC class II complexes and superantigens. Semin. Immunol. *19*, 262–271.

Sundstrom, M., Abrahmsen, L., Antonsson, P., Mehindate, K., Mourad, W., and Dohlsten, M. (1996). The crystal structure of staphylococcal enterotoxin type D reveals Zn^{2+}-mediated homodimerization. EMBO J. *15*, 6832–6840.

Takahashi, T., Satoh, I., and Kikuchi, N. (1999). Phylogenetic relationships of 38 taxa of the genus Staphylococcus based on 16S rRNA gene sequence analysis. Int. J. Syst. Bacteriol. *49*, 725–728.

Takeuchi, O., Hoshino, K., and Akira, S. (2000). Cutting edge: TLR2-deficient and MyD88-deficient mice are highly susceptible to *Staphylococcus aureus* infection. J. Immunol. *165*, 5392–5396.

Traber, K.E., Lee, E., Benson, S., Corrigan, R., Cantera, M., Shopsin, B., and Novick, R.P. (2008). agr function in clinical *Staphylococcus aureus* isolates. Microbiology *154*, 2265–2274.

Tsuru, T., Kawai, M., Mizutani-Ui, Y., Uchiyama, I., and Kobayashi, I. (2006). Evolution of paralogous genes: reconstruction of genome rearrangements through comparison of multiple genomes within *Staphylococcus aureus*. Mol. Biol. Evol. *23*, 1269–1285.

van der Steen, L., Tuk, C.W., Bakema, J.E., Kooij, G., Reijerkerk, A., Vidarsson, G., Bouma, G., Kraal, G., de Vries, H.E., Beelen, R.H.J., *et al.* (2009). Immunoglobulin A: FcαRI interactions induce neutrophil migration through release of leukotriene B4. Gastroenterology *137*, 2018–2029.e2013.

van Egmond, M., Damen, C.A., van Spriel, A.B., Vidarsson, G., van Garderen, E., and van de Winkel, J.G.J. (2001). IgA and the IgA Fc receptor. Trends Immunol. *22*, 205–211.

van Egmond, M., van Garderen, E., van Spriel, A.B., Damen, C.A., van Amersfoort, E.S., van Zandbergen, G., van Hattum, J., Kuiper, J., and van de Winkel, J.G.J. (2000). FcαRI-positive liver Kupffer cells: reappraisal of the function of immunoglobulin A in immunity. Nat. Med. *6*, 680–685.

van Wamel, W.J.B., Rooijakkers, S.H.M., Ruyken, M., van Kessel, K.P.M., and van Strijp, J.A.G. (2006). The innate immune modulators Staphylococcal complement inhibitor and chemotaxis inhibitory protein of *Staphylococcus aureus* are located on β-hemolysin-converting bacteriophages. J. Bacteriol. *188*, 1310–1315.

Veldkamp, K.E., Heezius, H.C.J.M., Verhoef, J., van Strijp, J.A.G., and van Kessel, K.P.M. (2000). Modulation of neutrophil chemokine receptors by *Staphylococcus aureus* supernate. Infect. Immun. *68*, 5908–5913.

von Eiff, C., Becker, K., Machka, K., Stammer, H., and Peters, G. (2001). Nasal carriage as a source of *Staphylococcus aureus* bacteremia. Study Group. N. Engl. J. Med. *344*, 11–16.

von Köckritz-Blickwede, M., Konrad, S., Foster, S., Gessner, J.E., and Medina, E. (2010). Protective role of complement C5a in an experimental model of *Staphylococcus aureus* bacteremia. J. Innate Immun. *2*, 87–92.

Voyich, J.M., Braughton, K.R., Sturdevant, D.E., Whitney, A.R., Saïd-Salim, B., Porcella, S.F., Long, R.D., Dorward, D.W., Gardner, D.J., Kreiswirth, B.N., *et al.* (2005). Insights into mechanisms used by *Staphylococcus aureus* to avoid destruction by human neutrophils. J. Immunol. *175*, 3907–3919.

Voyich, J.M., Vuong, C., DeWald, M., Nygaard, T.K., Kocianova, S., Griffith, S., Jones, J., Iverson, C., Sturdevant, D.E., Braughton, K.R., *et al.* (2009). The SaeR/S gene regulatory system is essential for innate immune evasion by *Staphylococcus aureus*. J. Infect. Dis. *199*, 1698–1706.

Walenkamp, A.M., Boer, I.G., Bestebroer, J., Rozeveld, D., Timmer-Bosscha, H., Hemrika, W., van Strijp, J.A., and de Haas, C.J. (2009). Staphylococcal superantigen-like 10 inhibits CXCL12-induced human tumor cell migration. Neoplasia *11*, 333–344.

Walenkamp, A.M.E., Bestebroer, J., Boer, I.G.J., Kruizinga, R., Verheul, H.M., van Strijp, J.A.G., and de Haas, C.J.C. (2010). Staphylococcal SSL5 binding to human leukemia cells inhibits cell adhesion to endothelial cells and platelets. Cell. Oncol. *32*, 1–10.

Walther, B., Hermes, J., Cuny, C., Wieler, L.H., Vincze, S., Abou Elnaga, Y., Stamm, I., Kopp, P.A., Kohn, B., Witte, W., *et al.* (2012). Sharing more than friendship – nasal colonization with coagulase-positive staphylococci (CPS) and co-habitation aspects of dogs and their owners. PLoS One 7, 18.

Wertheim, H.F., Melles, D.C., Vos, M.C., van Leeuwen, W., van Belkum, A., Verbrugh, H.A., and Nouwen, J.L. (2005). The role of nasal carriage in *Staphylococcus aureus* infections. Lancet Infect. Dis. *5*, 751–762.

Wertheim, H.F., Vos, M.C., Ott, A., van Belkum, A., Voss, A., Kluytmans, J.A., van Keulen, P.H., Vandenbroucke-Grauls, C.M., Meester, M.H., and Verbrugh, H.A. (2004). Risk and outcome of nosocomial *Staphylococcus aureus* bacteraemia in nasal carriers versus non-carriers. Lancet *364*, 703–705.

Williams, R.J., Ward, J.M., Henderson, B., Poole, S., O'Hara, B.P., Wilson, M., and Nair, S.P. (2000). Identification of a novel gene cluster encoding staphylococcal exotoxin-like proteins: characterization of the prototypic gene and its protein product, SET1. Infect. Immun. *68*, 4407–4415.

Wilson, G.J., Seo, K.S., Cartwright, R.A., Connelley, T., Chuang-Smith, O.N., Merriman, J.A., Guinane, C.M., Park, J.Y., Bohach, G.A., Schlievert, P.M., *et al.* (2011). A novel core genome-encoded superantigen contributes to lethality of community-associated MRSA necrotizing pneumonia. PLoS Pathog. *7*, e1002271.

Wines, B.D., Willoughby, N., Fraser, J.D., and Hogarth, P.M. (2006). A competitive mechanism for Staphylococcal toxin SSL7 Inhibiting the leukocyte IgA receptor, FcαRI, is revealed by SSL7 binding at the Cα2/Cα3 interface of IgA. J. Biol. Chem. *281*, 1389–1393.

Yarwood, J.M., McCormick, J.K., Paustian, M.L., Kapur, V., and Schlievert, P.M. (2002). Repression of the *Staphylococcus aureus* accessory gene regulator in serum and *in vivo*. J. Bacteriol. *184*, 1095–1101.

Yarwood, J.M., and Schlievert, P.M. (2003). Quorum sensing in Staphylococcus infections. J. Clin. Invest. *112*, 1620–1625.

Yokoyama, R., Itoh, S., Kamoshida, G., Takii, T., Fujii, S., Tsuji, T., and Onozaki, K. (2012). Staphylococcal superantigen-like protein 3 binds to the Toll-like receptor 2 extracellular domain and inhibits cytokine production induced by *Staphylococcus aureus*, cell wall component, or lipopeptides in murine macrophages. Infect. Immun. *80*, 2816–2825.

Botulinum Neurotoxins as Therapeutics

7

Sheng Chen

Abstract

Botulinum neurotoxins (BoNTs) cause flaccid paralysis by interfering with vesicle fusion and neurotransmitter release in the neuronal cells. Due to their high efficacy, prolonged activity and satisfactory safety profile, BoNTs are now the most widely used therapeutic proteins. BoNT/A was approved by the US FDA to treat strabismus, blepharospasm, hemifacial spasm, cervical dystonia, glabellar facial lines, axillary hyperhidrosis and chronic migraine and for cosmetic use. The efficacy of BoNT/A in treating dystonia and other neuronal disorders, coupled with the satisfactory safety profile, has prompted its empirical use in a variety of indications. Currently available BoNT therapies have certain limitations such as the neuronal specific indications and immunoresistance issues resulting from periodic injections. Recent studies on the structure–function characterization of HCs and LCs of Botulinum Neurotoxins have advanced our knowledge on the mechanisms of BoNT receptor binding, internalization and substrate recognition. Advanced understanding in these areas has opened up new opportunities to engineer recombinant proteins to treat diseases that are not amenable to therapy with native neurotoxins, or to give better outcome than with the native neurotoxins. In conclusion, the future of BoNTs in medical applications is bright, yet more research is needed to improve their medical uses.

Introduction

Botulinum neurotoxins (BoNTs) cause flaccid paralysis by interfering with vesicle fusion and neurotransmitter release in the neuronal cells (Schiavo *et al.*, 1994; Cherington, 1998; Dembek *et al.*, 2007). BoNTs are 150 kDa di-chain proteins with typical **A-B** structure–function properties, where the **B** (binding) domain binds to surface components on the mammalian cell and translocates the **A** (active) domain to an intracellular location (Montecucco and Schiavo, 1994; Poulain and Humeau, 2003). BoNTs are organized into three functional domains: an N-terminal catalytic domain (light chain, LC), an internal translocation domain (heavy chain, HC_T), and a C-terminal receptor binding domain (heavy chain, HC_R)(Davletov *et al.*, 2005). Mammalian neuronal exocytosis is driven by the formation of protein complexes between the vesicle SNARE, VAMP2, and the plasma membrane SNAREs, SNAP25 and syntaxin 1a (Brunger, 2005). Thus, BoNTs inhibit exocytosis by the cleavage of one of the three SNARE proteins. Due to their high efficacy, tolerance, longevity and satisfactory safety profile, BoNTs are now the most widely used therapeutic proteins. They have been approved by the US FDA to treat various neuronal disorders and are applied in many other empirical/off-label indications. This chapter describes the current usage of BoNTs and efforts to further extend and optimize their values in therapeutic interventions.

Mechanistic basis of BoNTs as therapeutics

Whether by direct absorption through the GI tract or injection into muscles, BoNTs undergo multiple defined processes to achieve their intoxication functions. These processes include neuron-specific receptor binding and internalization, light chain translocation to the cytosol of neuronal cells and specific neuronal substrate cleavage and inhibition of exocytosis (Pellizzari *et al.*, 1999; Popoff *et al.*, 2001) (Fig. 7.1).

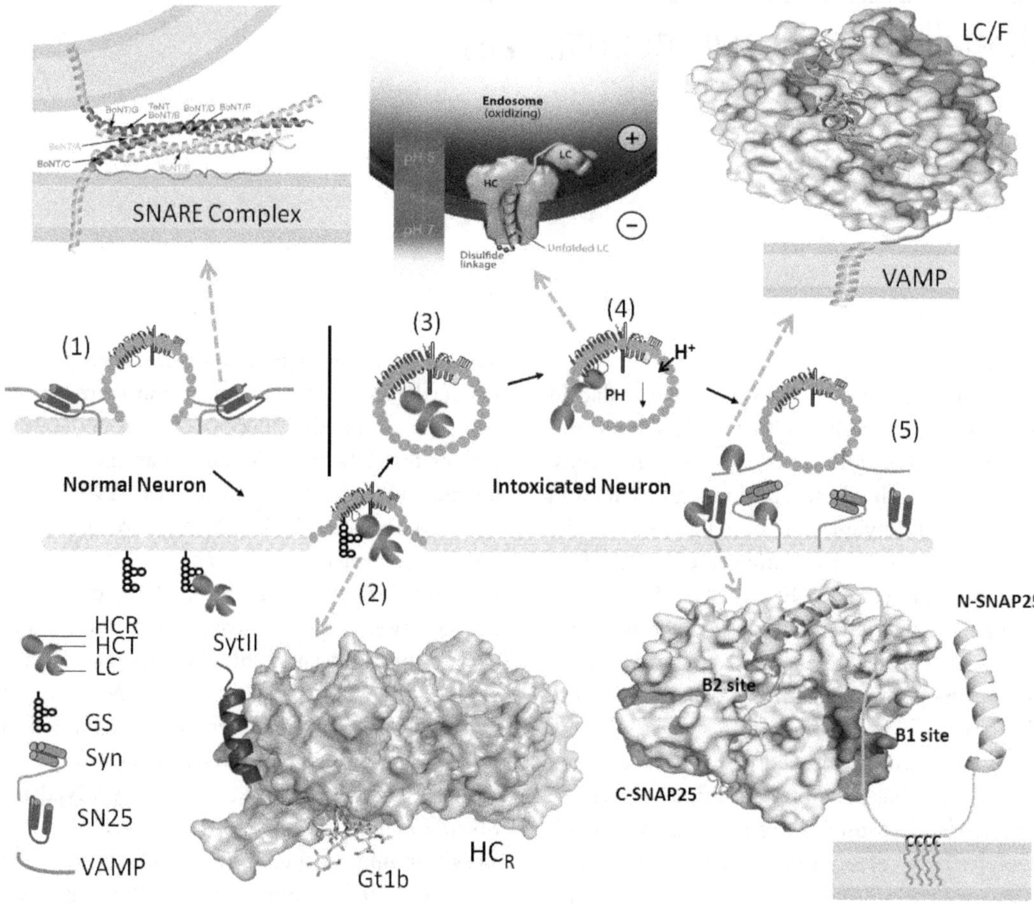

Figure 7.1 Multiple-step process of neuron intoxication by BoNTs. (1) in normal neurons, neurotransmitter release is mediated by vesicle exocytosis facilitated by SNARE complex; (2) the HC_R of BoNT binds gangliosides (GS) on the plasma membrane. Fusion of synaptic vesicles to the plasma membrane exposes loops of synaptic vesicles acting as BoNT protein receptor. The HC_R binds to both GS and protein receptor simultaneously; (3) complexes of synaptic vesicle proteins are endocytosed to be recycled, which internalize BoNT into neuronal cells; (4) the acidic environment triggers insertion of the HC_T domain, which facilitates translocation of a partially unfolded LC through a channel made by the HC_T; (5) after translocation into the cytosol of neurons, BoNT LC cleave SNARE proteins to inhibit exocytosis. Cleavage of VAMP2 and SNAP25 by LC/F and LC/A, respectively, is indicated. GS, gangliosides; Syn, syntaxin 1a; SN25, SNAP25; HC_T, translocation domain (heavy chain); HC_R, receptor binding domain (heavy chain); LC, light chain.

Receptor binding and internalization

The first step of BoNT's interaction with neuronal cells after its transport to the neuromuscular junction (NMJ) area involves specific binding to the surface receptors, a step for which the HC_R domain of BoNT is responsible. Abundant complex gangliosides have been shown to function as receptors for BoNTs and Tetanus Neurotoxin (TeNT), which are structurally related (Pickett and Perrow, 2011). The recent identification of synaptic vesicles as BoNT protein receptors has put forward the double-receptor theory of BoNTs. This theory suggests that the HC_R domain of BoNT initially binds to the abundant complex gangliosides to facilitate the anchoring of BoNT to the plasma membrane surface of neurons (Binz and Rummel, 2009; Brunger and Rummel, 2009). This low affinity binding enables BoNTs to move along the cell surface while retaining their binding properties. During the process of vesicle exocytosis and the exposure of the luminal domain of synaptic proteins to the outer membrane, the

HC_R domain will bind to the specific protein receptors. Simultaneous interaction with gangliosides and protein receptors may create a high affinity binding force, a force that may be critical for the subsequent endocytosis process of BoNTs (Fig. 7.1) (Binz and Rummel, 2009; Brunger and Rummel, 2009; Pickett and Perrow, 2011).

Complex gangliosides are glycosphingolipids with a large oligosaccharide head group and a double-tailed hydrophobic moiety found predominantly in the outer leaflet of the membrane of neuronal cells (Chen *et al.*, 2009). The binding of gangliosides to the HC_T of tetanus neurotoxin was first demonstrated by Halpern and Loftus (1993). Since then numerous reports have shown that BoNT can bind directly to gangliosides. Prior to incubation of BoNT with excess gangliosides such as GD2, GD1b, GT1b and GQ1b reduced BoNT toxicity both *in vivo* and *in vitro* (Marxen *et al.*, 1989, 1991). Pre-incubation of BoNTs with monoclonal antibody to GD1b reduced their toxicity on rat neuron (Kozaki *et al.*, 1998). Knockout mice that were unable to express complex gangliosides displayed lower sensitivity to BoNT/A, B, C, E and G (Bullens *et al.*, 2002; Kitamura *et al.*, 2005; Tsukamoto *et al.*, 2005; Rummel *et al.*, 2007). Diminishing gangliosides by neuraminidase treatment and inhibiting ganglioside biosynthesis using fumonisin inhibited the binding and entry of BoNT/A, B, E and G into cultured neurons (Bullens *et al.*, 2002; Kitamura *et al.*, 2005; Tsukamoto *et al.*, 2005; Rummel *et al.*, 2007). Hence, complex gangliosides such as GD2, GD1b and GT1b play an important role in the specific binding of BoNT to neurons. The phenomenon that stimulation of neurons accelerated the uptake of BoNT/A prompts the hypothesis that luminal segments of the internal synaptic proteins may act as a second receptor for BoNT in addition to gangliosides, since stimulation of neurons causes increased exo- and endocytosis and exposure of the luminal segments of vesicle proteins. Recent studies have identified the protein receptors for BoNT/A, B, D, E, G, F and D-SA, a chimera of C and D. Initial studies have shown that synaptotagmin (syn) II binds to BoNT/B (Nishiki *et al.*, 1994; Dong *et al.*, 2003). The physiological role of syn-I as receptor of BoNT/B was confirmed through the observation that the entry process of BoNT involved synaptic vesicle exocytosis and that the luminal domain of SytI and SytII could mediate the binding and entry of BoNT/B to cultured neuroendocrine cell and hippocampal neurons (Dong *et al.*, 2007). Similar approaches demonstrated that SytI and SytII also mediate the binding and entry of BoNT/G and BoNT/D-SA (Dong *et al.*, 2007; Peng *et al.*, 2012). The 12-transmembrane domain synaptic vesicle glycoprotein 2 (SV2) was identified as protein receptor for BoNT/A, D, E and F (Dong *et al.*, 2006, 2008; Fu *et al.*, 2009; Peng *et al.*, 2011). It was also shown that BoNT/A and BoNT/B are associated with detergent resistant synaptic vesicle protein complex consisting of SV2, SytI, synaptophysin, synaptobrevin 2, and the vacuolar proton pump vATPase (Baldwin and Barbieri, 2007), suggesting that multiple proteins may be involved in the binding of BoNT to the neuronal cell surface and that one of the vesicle proteins has higher affinity to specific BoNT.

The mode of interaction between HCs of BoNT and their receptors, gangliosides and proteins was addressed by some recent structural and biochemical studies. The neurotoxin–ganglioside interaction was first studied in TeNT. A ganglioside binding pocket (GBP) formed by Ser^{1287}, Trp^{1289}, and His^{1293} and a sialic acid binding pocket comprising Arg^{1226} were identified in TeNT HC_R, which can bind to two different molecules of gangliosides at the plasma membrane using these two pockets (Rummel *et al.*, 2003; Chen *et al.*, 2008a, 2009). A similar GBP was found in BoNT/A, B, E, F and G characterized with a conservative sequence/residues (H ... SXWY) (Rummel *et al.*, 2004, 2007; Fu *et al.*, 2009; Benson *et al.*, 2011). In BoNT/A, the GBP is formed by several residues including His^{1253}, Ser^{1264}, Trp^{1266} and Tyr^{1267} located on several antiparallel β-sheets (Rummel *et al.*, 2004). The co-crystal structure of BoNT/A with GT1b shows that galactose sugar ring 4 of the ganglioside and the indole ring of Trp form a hydrophobic ring stacking interaction that is further stabilized by contacts with Glu^{1203}, His^{1253}, and Ser^{1264} (Fig. 7.1) (Stenmark *et al.*, 2008). BoNT/C, D and D-SA (a chimera of C and D) also show overall main chain organization at GBP while lacking the conservative residues that are in other BoNTs (Karalewitz *et al.*, 2010; Kroken *et al.*, 2011). Current structural and biochemical studies have identified a new ganglioside binding loop (GBL), which is responsible for ganglioside binding. The Trp in the GBL contributes to the ganglioside binding in BoNT/C, D and D-SA (Karalewitz *et al.*, 2010; Kroken *et al.*, 2011).

The mode of BoNT HC binding to the respective protein receptors was demonstrated by the recent co-crystal structure analysis of BoNT/B bound to Syt-II (Chai *et al.*, 2006; Jin *et al.*, 2006). The complex structure revealed that the originally unstructured N-terminus of Syt-II (residues 44–60) formed an α-helical confirmation and bound to a sialic acid pocket adjacent to the GBP at the distal tip of HC of BoNT/B (Fig. 7.1) (Chai *et al.*, 2006; Jin *et al.*, 2006). The Sty-II α-helix interacted with two adjacent pockets on BoNT/B through hydrophobic interactions as well as the salt-bridge between Syt-II $E^{57}K$ and BoNT/B K^{1192} (Rummel *et al.*, 2007). Comparative characterization of binding of Syt-I and Syt-II, originally characterized as protein receptor for BoNT/B and G, led to the identification of two residues of Sty-II, Phe^{55} and Ile^{58} that were critical to the binding to BoNT/B (Rummel *et al.*, 2007). This study also pointed out that BoNT/G and BoNT/B showed a similar binding mode to Sty-II, except for some subtle differences concerning interaction with side chain of Glu^{57} of Sty-II. Mutations of residues in the pocket of BoNT/B and G reduced the binding to Syt-I and Syt-II and the neurotoxicity in the mouse phrenic nerve toxicity assay (Rummel *et al.*, 2007).

Translocation of LC into the cytosol of neuronal cells

After specific receptor binding and internalization triggered by the vesicle recycling pathway, the BoNTs are internalized into the vesicle of neuronal cells. The exact mechanisms of BoNT trafficking inside the vesicle is not known yet, but it is clear that the light chain (LC) of BoNT is translocated into the cytosol, where it cleaves its substrate and causes the inhibition of exocytosis of neurotransmitter-carrying vesicles. To date, how LC of BoNT translocates through the translocation domain is not clear. However, current research has begun to advance our understanding of this process. Using a single-molecular translocation assay, Fisher *et al.* addressed the real-time translocation of LC/A through the HC channel (Fig. 7.1) (Fischer and Montal, 2007b). Using lipid bilayers and membrane patches of neuroblastoma cells with controlled pH and redox gradients at different phases to mimic the membrane of endosomes and channel conductance as indicator of translocation of LC/A, a sequence of progressive LC/A translocation process was proposed: the charged surface of the BoNT/A translocation domain inserts into the lipid bilayers and forms the channel; the belt region of BoNT/A may act as a chaperon for LC unfolding and drag the unfolded LC through the HC channel; the process is evidenced by the transiently occluded conductance indicating the transit process followed by the unoccluded transductance suggesting the completion of the translocation; the dissociation of LC from HC by breaking the disulfide bond is triggered by the neutral pH and reducing environment at the cytosol, which promotes the release of LC from the HC (Montal, 2009). Numerous unanswered questions remain regarding the detailed process of the translocation, such as the function of the belt, the unfolding and refolding process, and the fate of the HC channel. More research will be needed to understand the physiology of this most important process in most AB toxins (Brunger *et al.*, 2007; Fischer and Montal, 2007a).

Intracellular trafficking and longevity of BoNT LCs

After being translocated into the cytosol of neuronal cells, the net outcome of the presence of LCs of BoNT is clear: it is the cleavage of the SNARE proteins and inhibition of the exocytosis of neurotransmitter-carrying vesicles. Yet how LCs travel around and eventually find their substrate in the cytosol is not clear. Could the LCs be just floating in the cytosol and randomly find their substrate and cleave it? If the answer is yes, the efficiency of LCs to cleave their substrate may not be very high and this is not consistent with the high potency of the toxins. A few current researches in this area also support that LCs may develop an efficient way to find their substrates, although data are still limited. It was shown that LC/A displayed membrane localization in neuronal cells, while LC/E did not show a plasma membrane association as strong as LC/A (Fernandez-Salas *et al.*, 2004; Tsai *et al.*, 2010). Another study showed that LC/A membrane localization is due to the high affinity binding to its substrate SNAP25, which is membrane-localized. The high-affinity binding of LC/A to SNAP25 is through two discontinuous interactions sites, one through the interaction of the LC/A N-terminus to the (80–110) region of SNAP25 and the other through the binding of SNAP25(141–206) to the substrate binding cleft of LC/A (Fig. 7.1) (Chen and Barbieri, 2011). The biological role of LC/A binding to SNAP25 on the plasma membrane might be an enhancement of the SNAP25 cleavage efficiency. These findings help to explain the potency of BoNT/A by utilizing the substrate, SNAP-25, as an intracellular receptor to increase the efficiency of substrate cleavage. Since biologically functional SNAP25 is always associated with syntaxin or forms a SNARE complex (SNAP25, Syntaxin and VAMP) on the plasma membrane, the binding of the LC/A N-terminus to the available region of SNAP25 (80–110) in the SNARE complex will help the LC/A to compete for the subsequent substrate binding and cleavage. It was also hypothesized that membrane localization of LCs may contribute to another important feature of BoNT, its longevity, since LC/A with the longest activity in cells shows plasma membrane localization, while LC/E with the shortest activity shows less plasma membrane localization. Another study showed that LC membrane localization does not contribute to the longevity of LCs and that instead LC longevity in cells is due to their different protein degradation pathways (Tsai *et al.*, 2010). LC/E is shown to be associated with TRAF2, a RING finger protein implicated in ubiquitylation that promotes fast degradation of recombinant LC/E in neuronal cells (Tsai *et al.*, 2010). In addition, the targeting of LC/A to similar ubiquitylation through a chimeric substrate, TRAF2-SNAP25, dramatically reduces its duration in a cellular model for toxin persistence. Differential susceptibility of the catalytic LCs to ubiquitin-dependent proteolysis therefore might explain the differential persistence of BoNT serotypes (Tsai *et al.*, 2010).

Mode of substrate cleavage and inhibition of exocytosis

There are seven serotypes of BoNTs (termed A–G) that cleave specific residues on one of the three SNARE proteins: serotypes B, D, F, and G cleave VAMP-2, serotypes A and E cleave SNAP25, and serotype C cleaves SNAP25 and syntaxin 1a (Montecucco and Schiavo, 1994). Each LC cleaves one of the SNARE proteins except for LC/C, which cleaves both SNAP25 and syntaxin 1a. The crystal structures of all LCs of BoNT have been resolved and have shown a very similar structural confirmation with a Zinc ion coordinated in the active site of the LCs (Agarwal *et al.*, 2005a,b; Arndt *et al.*, 2005, 2006; Jin *et al.*, 2007; Silvaggi *et al.*, 2007). Unlike other metalloprotease such as thermolysin, biochemical characterization of LC substrate indicated that BoNT LCs required extended substrate for efficient cleavage, while short peptide containing the scissile bond was not the optimal substrate (Binz *et al.*, 1994; Vaidyanathan *et al.*, 1999). The major residue that affects the substrate cleavage is the P1' site of the substrate and mutation of the residue will dramatically decrease LC substrate hydrolysis (Binz *et al.*, 1994; Vaidyanathan *et al.*, 1999). Furthermore, saturation mutagenesis of substrate VAMP-2 or SNAP25 has identified a region around the scissile bond that contributes to the substrate specific recognition. The 10-amino acid peptide derived from SNAP25 or VAMP2 that contains the scissile bond in the middle can be specifically cleaved by LCs although the efficiency of peptide hydrolysis is much lower than for the full length substrate, suggesting that this region was enough for substrate recognition and it was designated as AS (active site) region (Chen and Barbieri, 2006; Sikorra *et al.*, 2006). In addition to the AS region, a region distal to the AS region, designated as B (binding) region, contributes to substrate affinity and binding (Chen and Barbieri, 2006; Sikorra *et al.*, 2006). BoNT LCs require both AS and B regions of the substrate for efficient substrate hydrolysis; the AS region contributes to LC substrate specificity and the B region contributes to substrate binding and affinity.

Decoding the complex structures of LC/A-SNAP25 and LC/F-VAMP2 has depicted the mode of LC recognition of their substrates,

SNAP25 or VAMP2 (Fig. 7.1) (Breidenbach and Brunger, 2004; Agarwal *et al.*, 2009). The recognition of SNAP25 by LC/A is mediated by exosites: two exosites distal to the active site of LC/A facilitated the SNAP25 binding and active site substrate recognition Another biochemical study resolved the step-by-step binding and recognition of SNAP25 by LC/A(Chen *et al.*, 2007). The mechanism of LC/A recognition and cleavage of SNAP25 involves sequential steps representing SNAP25 recognition and active site organization. Initial interactions involve discontinuous surfaces between residues within the belt region of LC/A and the B region residues of SNAP25. The Velcro-like binding of SNAP25 to LC/A aligns the P5 residue Asp^{193} to form a salt bridge with Arg^{177}, an S5 pocket residue at the periphery of one side of the active site. Although the exact order of each step of recognition of SNAP25 by BoNT/A at the active site is not clear, the initial binding could subsequently orient SNAP25 for the formation of a salt bridge between the P4′-residue of SNAP25(Lys^{201}) and the S4′-residue of LC/A(Asp^{257}). These interactions broaden the LC/A active site cavity and dock the P1′-residue (Arg^{198}), via electrostatic and hydrophobic interactions within the S1′-pocket. The fine tuning of the alignment of Arg^{198} into the S1′-pocket resulting in the precise alignment of the scissile bond is facilitated by the binding of the P3 residue, SNAP25-Ala^{195}, into the hydrophobic S3 pocket of LC/A. The proper docking of the P1′-P1 sites into the AS site initiates substrate cleavage. After cleavage, the P4′-residue dissociates from the S4′-residue of LC/A, which converts the AS into a smaller conformation, facilitating dissociation of the P1′-residue from the AS (Chen *et al.*, 2007).

Clinical applications of botulinum neurotoxins

BoNT intoxication is reversible and muscles will function again upon clearance of BoNT from the affected neuronal cells. In addition, local application of BoNT largely limits BoNT toxicity in the applied area and does not spread to the central neuron or does it only really slowly; the action of BoNT can last for up to 6 months, thus frequent applications are not needed. These features of BoNT have turned it from a deadly agent into novel therapies for a range of neuromuscular conditions (Atassi and Oshima, 1999; Mahant *et al.*, 2000; Glogau, 2002; Bell and Williams, 2003; Thant and Tan, 2003; Ascher and Rossi, 2004; Atassi, 2004; Benedetto, 2004; Cheng *et al.*, 2006; Mahajan and Brubaker, 2007). BoNT/A was approved by the US FDA to treat strabismus, blepharospasm, and hemifacial spasm as early as 1989 and then for treatment of cervical dystonia,, glabellar facial lines, axillary hyperhidrosis, chronic migraine and cosmetic use. The efficacy of BoNT/A in treating dystonia and other disorders related to involuntary skeletal muscle activity, coupled with the satisfactory safety profile, has prompted its empirical/off-label use in a variety of ophthalmological, gastrointestinal, urological, orthopaedic, dermatological, secretory, and painful disorders. There are three types of BoNT/A, namely BOTOX (onabotulinumtoxinA, Allergan, approved by FDA in 1989), Dysport (abobotulinumtoxinA, Medicis, approved by FDA in 2009) and Xeomin (incobotulinumtoxinA, Merz, approved by FDA in 2010). On 11 December 2000, a Botulinum Neurotoxin serotype B product (MYOBLOC™) was approved by the FDA in the United States as a treatment for patients with cervical dystonia to reduce the severity of abnormal head position and neck pain associated with cervical dystonia. MYOBLOC is the US trade name for Solstice Neurosciences' Botulinum Neurotoxin serotype B product. This product also received marketing authorization from the *Committee for Proprietary Medicinal Products* of the European Union and is available there under the name Neurobloc. The duration of effect is approximately 12–16 weeks, a little shorter than that of BoNT/A (Brashear, 2001; Figgitt and Noble, 2002).

FDA approved clinical use of BoNTs

Alan Scott and Edward Schantz were the first scientists to apply BoNT into therapeutic use. In 1973, Scott used BoNT/A in a monkey experiment and in 1980 he used the toxin to treat strabismus in humans for the first time. Since then the applications of BoNT in different kinds of neuronal disorders have been widely explored. The US FDA has approved the use of BoNTs in several indications that have been proven to be efficient and safe.

Strabismus

Strabismus is a condition in which the eyes are not properly aligned with each other which results in lack of coordination between the extraocular muscles. It can affect proper binocular vision and in depth perception in severe cases. In 1980, Alan Scott first reported the use of BoNT/A in strabismus patients and concluded that BoNT/A injections were effective in these patients without systemic complications and could be used as a practical adjunct or alternative to surgical correction of strabismus (Elston, 1985). Subsequently, several case controlled studies showed that BoNT/A was effective and safe for different types of strabismus and side-effects were very rare and minor (Scott, 1980; Jampolsky, 1986). In 1989, the US Food and Drug Administration (FDA) approved the use of BoNT/A (BOTOX) to treat strabismus. Further research also indicated its effectiveness in most strabismus patients. However, due to the requirement of repeated injections, BoNT/A may not be considered as an alternative to surgery (Kowal *et al.*, 2007; Krzizok, 2007; Rowe and Noonan, 2009, 2012). It is therefore more suitable for temporary use or in cases where surgery procedure is undesirable. Long-term treatment with BoNT/A has also proven to be safe and efficient (Rowe and Noonan, 2009; Hancox *et al.*, 2012). The major complications of BoNT/A treatment for strabismus are ptosis and acquired vertical deviations, while vision-threatening complications are rare (Krzizok, 2007; Rowe and Noonan, 2009, 2012; Hancox *et al.*, 2012).

Blepharospasm

Blepharospasm is a focal dystonia characterized by an involuntary spasm of periocular muscles resulting in forceful eye closure. The condition varies from frequent blinking to persistent closure of eyelids. The history of using BoNT/A to treat blepharospasm can be traced back to 1984 and it was deemed as a very effective treatment (Frueh *et al.*, 1984; Scott *et al.*, 1985; Shorr *et al.*, 1985). In 1989, the US FDA officially approved the use of BoNT/A for blepharospasm treatment. Further studies suggested that over 90% of patients who received BoNT/A treatment alone showed sustained improvement on their disease onset and the safety profile of BoNT/A treatment on blepharospasm patients is excellent (Chang *et al.*, 1999; Drummond and Hinz, 2001). It was also suggested that there is no need to conduct placebo-controlled trials even though information on the efficiency of randomized and controlled studies is lacking so far. Further trials should focus on the optimization of the treatment, such as on dosages and treatment interval, different delivering techniques and different formulas of BoNTs to achieve better outcome (Costa *et al.*, 2005b). Long-term follow-up studies have confirmed the high efficacy and good safety profile of BoNT/A for treatment of blepharospasm (Drummond and Hinz, 2001; Cillino *et al.*, 2010). There might be a trend towards a decreasing duration of relief from symptoms in patients with blepharospasm over the long-term, but the treatment remains effective in relieving symptoms and signs (Gill and Kraft, 2010). Surgical or oral medication could be used in patients who do not respond to BoNT/A treatment.

Hemifacial spasm

Hemifacial spasm is a neurological disorder characterized by unilateral, periodic, tonic contractions of facial muscles. This condition is thought to be caused by mechanical compression at the root-exit zone of the facial nerve. Although it is a benign condition, it can cause significant cosmetic and functional disability. It is a chronic disease and spontaneous recovery is very rare. Reports on using BoNT/A as treatment for hemifacial spasm dates back to 1985 and preliminary results showed that BoNT/A was effective in treating hemifacial spasm (Mauriello, 1985; Elston, 1986; Tolosa *et al.*, 1988). In 1989, the US FDA approved the use of BoNT/A to treat hemifacial spasm. A large open, case–controlled study showed that 76–100% of the patients benefited from BoNT/A treatment (Duzynski and Slawek, 1998; Gouider-Khouja *et al.*, 1999; Costa *et al.*, 2005d). A mega-analysis that reviewed most of the reported studies suggested that no new large placebo-controlled trials were needed, even although only one placebo-controlled trial has been reported to date (Costa *et al.*, 2005d). Future trials should explore technical factors such as the optimum treatment intervals, different injection techniques, dosages, BoNT types and formulations. Long-term treatment has been proven effective and no significant side-effects were observed (Cillino *et al.*, 2010; Gill and Kraft, 2010).

Cervical dystonia

Cervical dystonia (CD) is characterized by involuntary spasms of neck muscles resulting in abnormal head and neck movements and postures, which are often associated with pain. CD is the most common form of focal dystonia presented to movement disorders clinics. The first placebo-controlled trial of using BoNT/A in CD treatment was conducted in 1987 and showed promising results regarding the improvement of disease onset (Jankovic and Orman, 1987). Subsequent case studies and placebo-controlled studies indicated that over 70% of the patients who received BoNT/A treatment showed significant improvement in their symptoms and pain relief and the beneficial effect usually lasted for about 12 weeks (Tsui and Calne, 1988; Jankovic and Schwartz, 1990; Poewe and Wissel, 1993; Edwards *et al.*, 1995; Bhaumik and Behari, 1999). On 21 December 2000, BoNT/A received FDA's approval for treatment of cervical dystonia. In addition to Botox, there are two other forms of BoNT/A products, Dysport and Xeomin, and BoNT/B(MYOBLOC™ or NeuroBloc) that are also approved by FDA to treat cervical dystonia. There are no significant differences in term of their efficiency in cervical dystonia treatment, although BoNT/B shows generally higher immunogenicity in patients. The greater benefit of BoNT/B treatment is for BoNT/A-resistant patients (Comella *et al.*, 2005; Costa *et al.*, 2005a,c; Factor *et al.*, 2005; Truong *et al.*, 2005). Dysphagia is potentially the most serious side-effect but its incidence and severity could be decreased by injecting lower doses. Some major drawbacks at present are a lack of prospective data to establish the optimal dosage and volume of injection guidelines to maintain good efficacy at a reduced risk of side-effects, and the need to continue indefinitely with repeated injections approximately every 3 months (Edwards *et al.*, 1995; Bhaumik and Behari, 1999; Truong *et al.*, 2005). Future studies should explore technical factors such as the optimum treatment intervals and use of image or electromyographic guidance in administration. Other issues include service delivery, quality of life, long-term efficacy and safety, and the relative indications for BoNT/A, BoNT/B and other treatments such as deep brain stimulation. The long-term use of BoNT/A is confirmed to be safe, effective, and well-tolerated in patients with CD (Maia *et al.*, 2010; Truong *et al.*, 2010; Camargo *et al.*, 2011).

Cosmetic use

The successful uses of BoNT/A for facial hemispasm, strabismus, and blepharospasm promoted the trial use of BoNT/A in cosmetics for reducing glabellar frown lines and facial rejuvenation. The first trial of BoNT/A to relax muscles of facial expression was reported in 1996, and it turned out to be a convenient, effective, and well-tolerated treatment for facial wrinkles (Garcia and Fulton, 1996). Subsequent case studies and placebo-controlled trials proved that BoNT/A is highly effective as an adjuvant therapy for facial rejuvenation. When injected into hyperactive corrugator superciliaris and/or procerus muscles of the face that predominantly control frowning, Botox produces a transient (3 to 6 months), dose-dependent and localized muscle weakness, resulting in temporary improvement in glabellar frown lines ('brow furrows') (Carter and Seiff, 1997; 1999; Carruthers, 2002; Jaspers *et al.*, 2011; Larkin, 2002). In 2000, Canada approved the use of Botox for focal muscle spasticity and cosmetic treatment of wrinkles at the brow line. On 15 April 2002, the US FDA announced the approval of the use of Botox for cosmetic uses. Recent BoNT/A indications for cosmetics applications include glabellar frown lines, horizontal forehead lines, crow's feet, bunny lines, perioral lines, mental crease and dimpled chin, mouth frown, platysmal bands and horizontal neck lines. Cosmetic to temporarily improve the appearance of moderate-to-severe frown lines between the eyebrows (glabellar lines) became very popular and in that year there were approximately 1.1–1.6 million patients using cosmetic BoNT/A (Wollina and Konrad, 2005). Headache and temporary blepharoptosis may occur as adverse events, but they are quite rare and the incidence tends to decrease with repeated treatment sessions (Wollina and Konrad, 2005). Long-term use of BoNT/A as cosmetic procedure is safe and effective and the formation of BoNT/A antibodies in patients is really rare and has only been reported in a few cases so far (Dressler *et al.*, 2010).

Axillary hyperhidrosis

Axillary hyperhidrosis is characterized by excessive sweating in the armpit of a person. It is also known as underarm sweating. The main reason of axillary hyperhidrosis is the overstimulation of the sympathetic nervous system. BoNT/A was tried in 1986 by Bushara *et al.* (1996) to treat axillary hyperhidrosis and showed promising result. In 1998, Heckmann *et al.* successfully used BoNT/A to treat axillary hyperhidrosis (Heckmann *et al.*, 1998). Case studies and placebo-controlled trials showed that BoNT/A was safe, well-tolerated and effective in treating regular and severe axillary hyperhidrosis. Over 95% of the patients have satisfactory improvement of the symptoms (Odderson, 1998; Goldman, 2000; Heckmann *et al.*, 2001; Whatling and Collin, 2001; Maillard *et al.*, 2003; Naumann *et al.*, 2003; Absar and Onwudike, 2008). In 2001, the United Kingdom and Canada approved the use of Botox for axillary hyperhidrosis (excessive sweating). In July 2004, the FDA approved the use of Botox to treat severe axillary hyperhidrosis that cannot be managed by topical agents, such as prescription of antiperspirants. In addition to axillary hyperhidrosis, BoNT/A is also clinically used to treat palmar hyperhidrosis. One of the most troublesome disadvantages associated with this therapy is pain at the injection sites. It was suggested that the reconstitution of botulinum toxin A in a solution of lidocaine could be an easy alternative procedure to reduce the discomfort associated with those injections (Vadoud-Seyedi and Simonart, 2007; Benohanian, 2009). Long-term use of BoNT/A for over a few years is still safe and effective and no neutralizing antibody and retrograde effect of central neuron have been observed (Naumann *et al.*, 2003).

Chronic migraine

Migraines are typically characterized with recurrent severe headache associated with autonomic symptoms. The severity of pain, duration of headache, and frequency of attacks are variable. Migraines are divided into episodic and chronic based on their attack frequency. In episodic migraine, attacks occur for less than 15 days per month, whereas in chronic migraine a patient has at least 15 headache days each month, and at least eight of those are migraine headaches (Solomon, 2007). Current therapies such as β-adrenergic blockers, antidepressants, calcium channel blockers, and anticonvulsants are of limited benefit and can be associated with potentially serious side effects (Silberstein, 2008). Therefore, alternative preventative therapies that are effective and well tolerated, with limited systemic effects are highly demanded. The use of BoNT/A as a potential migraine therapy could be traced back as early as 1991 by Binder *et al* (Binder *et al.*, 2000; Gobel *et al.*, 2001; Silberstein, 2002; Binder and Blitzer, 2003). Most randomized studies showed that BoNT/A was a safe and effective therapy for both acute and prophylactic treatment of migraine headaches, yet placebo-controlled studies have not proven the effectiveness of BoNT/A treatment on migraine (Binder *et al.*, 2000; Gobel *et al.*, 2001; Silberstein, 2002; Binder and Blitzer, 2003; Evers *et al.*, 2004; Aurora *et al.*, 2007; Relja *et al.*, 2007; Saper *et al.*, 2007). A recent review critically analysed available data on BoNT/A treatment of migraine made the following summary (Robertson and Garza, 2012). Several placebo-controlled studies did not convincingly show the effective prevention of episodic migraine by BoNT/A (Evers *et al.*, 2004; Aurora *et al.*, 2007; Relja *et al.*, 2007; Saper *et al.*, 2007; Vo *et al.*, 2007). In chronic migraine prophylaxis, however, available randomized, double-blinded, placebo-controlled trials suggested that BoNT/A is effective (Aurora *et al.*, 2010, 2011; Diener *et al.*, 2010). The therapeutic gain over placebo was modest, even if statistically significant. In addition, the high cost of the toxin and the complicated treatment procedure raise question about using BoNT/A as primary therapy for chronic migraine. Still, the excellent tolerability of BoNT/A makes it an extremely attractive alternative for patients who fail to tolerate, and therefore discontinue traditional oral prophylactics. Although the mechanisms through which BoNT/A may exert its benefit remain unknown, BoNT/A is a welcome addition to the available treatment options for chronic migraine, which is often a disabling and difficult-to-manage condition. In 2010, the US FDA approved the use of BoNT/A to treat chronic migraine.

Off-label use of BoNTs

In addition to FDA-approved indications, BoNT/A has been empirically used in a variety of urological, ophthalmological, gastrointestinal, orthopaedic, dermatological, secretory, and pain disorders. Indications were grouped into different categories and discussed as follows.

Lower urinary tract disorders

The application of BoNT/A in the treatment of lower urinary tract disorders has expanded in the past few years. The indications include detrusor sphincter dyssynergia, neurogenic and idiopathic detrusor overactivity, painful bladder syndrome and prostatic obstruction. To date, the mechanisms of the actions of BoNT/A in these indications are not clear, but it is apparently more than the blockage of the efferent neural pathway, thereby decreasing involuntary contractions of the detrusor smooth muscle. Recent studies have shown that BoNT/A's analgesic properties and the inhibitory effects on the release of ATP, substance P and growth factors may contribute to its efficacy in the treatment of painful bladder syndrome and relief from symptom of urgency (Chancellor *et al.*, 2008). The application of BoNT/A for lower urinary tract disorder was first reported in the treatment of detrusor sphincter dyssynergia (DSD) in 1988 and in spinal cord injury patients, the first double-blinded study of its kind in 1990 (Dykstra *et al.*, 1988; Dykstra and Sidi, 1990). Subsequently, it was used in DSD alone or together with related disorders such as hypocontractility (Petit *et al.*, 1998; Phelan *et al.*, 2001; Kuo, 2003; Gallien *et al.*, 2005; Smith *et al.*, 2005). Over 80% efficacy was found in most trials with significant improvement of symptoms. To date, one of the most widespread urological applications of BoNT/A is the treatment of detrusor overactivity (DO), which is characterized as a major pathology underlying urge urinary incontinence and urgency-frequency syndromes. Several multicentre randomized controlled trials involving more than 600 patients showed notable improvement of incontinence in most of the neurogenic detrusor overactivity (NDO) patients (Reitz *et al.*, 2004; Kuo, 2006; Game *et al.*, 2008; Herschorn *et al.*, 2011; Tincello *et al.*, 2012). Another major benefit of BoNT/A treatment is the reduction in urinary infections such as pyelonephritis, orchitis, and prostatitis commonly observed in NDO patients (Game *et al.*, 2008; Giannantoni *et al.*, 2009). The average duration of effect was about a year and adverse side effects are rare. The treatment was subsequently applied to treat paediatric NDO or idiopathic detrusor overactivity (IDO) and efficacy was observed to be similar as NDO (Patel *et al.*, 2006; Sahai *et al.*, 2007; Sahai *et al.*, 2009). BoNT/A was also reported in the successful treatment of painful bladder syndrome/interstitial cystitis and prostatic obstruction (Giannantoni *et al.*, 2006, 2008, 2010; Gottsch *et al.*, 2011). The application in these indications still needs much work both scientifically and clinically with long follow-up period. In conclusion, BoNT/A treatment of lower urinary tract disorders has remarkable efficacy and minimal side effects and thus will be a key future treatment option; meanwhile, more controlled trials are needed to test the feasibility of various treatment options.

Gastrointestinal tract disorders

The first trial of BoNT/A for treatment of gastrointestinal (GI) tract disorders was reported in a piglet model for lower oesophageal sphincter (LES) (Pasricha *et al.*, 1993). Over the past 15 years, BoNT has been used for a large number of GI tract disorders. The most common GI tract indications with the best available data are achalasia, anal fissures and gastroparesis. In other GI tract disorders, such as oesophageal spasm, sphincter of Oddi dysfunction and anismus as well as obesity, only preliminary data are available. The use of BoNT/A in the treatment of achalasia has been well-studied and it was suggested that this therapy was effective for the majority of patients (Annese *et al.*, 1996, 1998; Sood *et al.*, 1996; Vaezi *et al.*, 1999; Zaninotto *et al.*, 2004). In vigorous achalasia, especially in the elderly, BoNT/A may be the preferred agent. The long-term efficiency of BoNT/A treatment in achalasia is inferior to other traditional treatments such as surgery and repeated injections are often needed (Gordon and Eaker, 1997; Annese *et al.*, 1998; Wehrmann *et al.*, 1999). Safety appears to be excellent, although there still remains concerns over fibrosis around the LES leading to more difficult surgical myotomy. BoNT/A treatment of anal fissure has also been well studied; it is very effective and long-term sequelae have rarely been seen. Short-term side effects include incontinence of flatus or faeces, which usually resolves quickly, as well as other mild symptoms (Jost and Schimrigk, 1995; Minguez *et al.*, 2002; Stankovic and Mirkovic, 2004; Brisinda *et al.*, 2007; Husberg *et al.*, 2009). For gastroparesis, the data remain encouraging despite the fact that further controlled trials are needed (Lacy *et al.*, 2002; Friedenberg *et al.*, 2004; Tcherniak *et al.*, 2006; Arts *et al.*, 2007). In conclusion, for GI tract disorders, the concern remains that the effects of BoNT/A are relatively short-lived and definitive treatment is delayed. Additional trials are necessary to assess its efficacy duration of action, and for comparison with other therapeutic agents.

Spasticity

Spasticity is one of the components of upper motor neuron syndrome characterized by a velocity-dependent increase in muscle tone associated with exaggerated deep tendon reflexes. Spasticity may result from various aetiologies including stroke, spinal cord injury, multiple sclerosis, traumatic brain injury, cerebral palsy and neurodegenerative diseases such as Parkinson's disease. Spasticity is characterized by upper motor neuron dysfunction and if severe, can lead to considerable motion restriction and eventually serious disability. Available therapeutic interventions for spasticity are often of limited benefit. In as early as 1989, BoNT/A was used to treat patients with severe spasticity due to stroke-related hemiplegia. It produced both subjective and objective improvement and the toxin treatment was well-tolerated and no significant side effects were noted (Das and Park, 1989a,b). In the last decade, many open-label and several double-blinded placebo-controlled studies have demonstrated the effectiveness of intramuscular BoNT/A injections for the management of spasticity caused by multiple sclerosis, brain/spinal cord injury, cerebral palsy, or stroke (Bhakta *et al.*, 2000; Hyman *et al.*, 2000; Wissel and Entner, 2001; Feve, 2003; Yelnik *et al.*, 2007; Rosales and Chua-Yap, 2008; Alba-vera-Hernandez *et al.*, 2009). BoNT/A can also be beneficial in the treatment of spasticity, or a mixed condition with spasticity and rigidity, in many neurodegenerative disorders (Grazko *et al.*, 1995; Sastre-Garriga *et al.*, 2001). However, evidence of botulinum toxin injections associated with improved function and improved quality of life is not as compelling (Moore *et al.*, 2008; McCrory *et al.*, 2009; Dubinsky, 2010; Mohammadi *et al.*, 2010). There are a number of challenges with BoN/A therapy, including uncertainty over its role in improving motor dysfunction following stroke, the determination of which subsets of patients may benefit, the cost of treatment, and the identification of meaningful outcome measures. In conclusion, BoNT/A can be considered a first-line treatment for focal or multifocal spasticity. It should be used at the early stage to prevent soft tissue shortening from occurring as a result of the combined effect of spasticity and limb immobility. BoNT/A showed good safety profile and the side effects are minimal and rare in spasticity patients.

Spasmodic dysphonia

Spasmodic dysphonia (SD) is a focal laryngeal dystonia characterized by involuntary action-induced spasm of the muscles that control vocal fold motion. The most common presentation of SD is the adductor form (adductor spasmodic dysphonia or ADSD) and less common is the abductor form of the disorder (abductor spasmodic dysphonia or ABSD). Currently available pharmaceutical agents have the shortcomings of partial efficacy, unwanted adverse effects and drug interactions. BoNT/A has been used to treat ADSD in as early as 1987 in two patients with severe spasmodic dysphonia and significant improvement was seen in both patients without any complications (Miller *et al.*, 1987). Over the past decades, more than 100 published studies on the use of BoNT/A to treat spasmodic dysphonia have been reported. Although randomized double-blind placebo-controlled trials are still limited, there is overwhelming evidence for the effectiveness of BoNT/A in treating ADSD (Blitzer *et al.*, 1998; Watts *et al.*, 2008; Blitzer, 2010). BoNT/A has been established as an effective treatment for ADSD based on one convincing randomized double–blinded cross-over, placebo-controlled study and three double-blind, placebo-controlled/randomized group cohort studies (Watts *et al.*, 2006, 2008). In conclusion, BoNT/A treatment of ADSD is safe and highly tolerated. Side effects are minimal and rare. Most of the randomized and designed studies showed that BoNT/A is more effective in treatment of ADSD as compared to ABSD and other types of dysphonia (Ludlow *et al.*, 1991; Watts *et al.*, 2004; Watts *et al.*, 2006), while recent studies using bilateral or asymmetric dose posterior cricoarytenoid muscle BoNT/A injections showed the efficacy of treatment of ABSD. Simultaneous injection to bilateral posterior cricoarytenoid muscle with botulinum toxin is safe even for the highest dosage reported, 7.5 U. Complications with this approach are consistent with those previously reported using other methods (Stong *et al.*, 2005; Woodson *et al.*, 2006; Klein *et al.*, 2008).

Sialorrhoea

Sialorrhoea or excessive salivation and drooling are common disabling manifestations in different neurological disorders. Sialorrhoea is most commonly associated with infant cerebral palsy, Parkinson's disease and amyotrophic lateral sclerosis (Harris and Purdy, 1987; Hyson *et al.*, 2002). BoNT/A can reduce excessive or uncontrolled salivation via autonomic denervation rather than muscle denervation. In several controlled trials more significant improvements in subjective and objective measures of drooling were found in the treatment group as compared with the corresponding placebo groups (Lipp *et al.*, 2003; Mancini *et al.*, 2003; Lagalla *et al.*, 2006; Alrefai *et al.*, 2009; Wu *et al.*, 2011). A similar efficiency was found in other uncontrolled studies. BoNT/B was found to be effective in the treatment of excessive salivation (Ondo *et al.*, 2004; Jackson *et al.*, 2009; Lagalla *et al.*, 2009; Chinnapongse *et al.*, 2012). Adverse effects such as dysphagia, xerostomia and chewing difficulties were found in some patients, even though they were mild and transient.In conclusion, available evidence indicates that BoNT/A is a safe and effective treatment for sialorrhoea. The effective therapeutic dosages and ideal form of application remain to be established, and require further controlled clinical trials involving larger sample sizes. More research effort is needed to determine the ideal dosages and injection location, as well as to improve the technique of BoNT/A injections.

Temporomandibular disorder

Temporomandibular disorders (TMDs) are a set of musculoskeletal dysfunctions within the masticatory system with multiple aetiologies. They can be divided into two groups – those related to the muscles themselves and those related to the temporomandibular joint (TMJ). TMD may be associated with headache, periauricular pain, neck pain, decreased jaw excursion, locking episodes, and noisy joint movement. BoNT/A has been used to treat TMD since 1999 (Freund *et al.*, 1999). BoNT/A injections produced statistically significant improvement in pain, function, mouth opening, and tenderness. BoNT/A was also reported as analgesic treatment for TMJ and treatment for habitual dislocation and disk disfigurement of TMJ (Freund *et al.*, 2000; Chang, 2005; Karacalar *et al.*, 2005; Denglehem *et al.*, 2012). In addition, injection of BoNT/A to masticatory muscles was found to be efficient in controlling chronic facial pain resulting from hyperactivity of the masticatory muscles and chronic tension headaches (Freund and Schwartz, 2002; Schwartz and Freund, 2002; von Lindern *et al.*, 2003). Due to the complex nature of TMDs and proximity of affected muscles to facial nerves, correct injection technique and appropriate dosing guidelines are very important for successful results. Further studies are needed to optimize the current procedure of BoNT/A treatment.

Chronic musculoskeletal pain

Chronic musculoskeletal pain is due to musculoskeletal disorders in various locations in the human body. BoNT/A has been successfully used in the treatment of spasmodic torticollis, limb dystonia, and spasticity. Investigators have thus become interested in its potential use in treating many chronic pain conditions. There is strong evidence supporting the use of BoNT/A as a treatment to pelvic pain, plantar fasciitis, temporomandibular joint dysfunction associated facial pain, chronic lower back pain (LBP), carpal tunnel syndrome, joint pain, and in complex regional pain syndrome and selected neuropathic pain syndromes (Jabbari, 2008; Zhang *et al.*, 2011). The weight of evidence is also in favour of BoNT type A and type B in piriformis syndrome. There is conflicting evidence regarding the use of BoNT/A in the treatment of whiplash, myofascial pain, and myogenous jaw pain (Jabbari, 2008). A number of open-label and retrospective studies have been done to assess the usefulness of BoNT/A in treating myofascial pain. The preliminary results were encouraging as Botox was shown to be effective in relieving pain. However, these findings have not been confirmed by randomized controlled trials (RCTs) (Lang, 2002; Zhang *et al.*, 2011). It does appear that BoNT/A is useful to certain patients, especially those patients who have not responded favourably to first-line treatments, providing a window of opportunity, and its duration of action may exceed that of conventional treatments. This seems a promising treatment that must be further evaluated.

Other indications

BoNT/A was also tried in several other less common but effective treatments such as vaginismus, wound healing, and diabetic neuropathy. To date, only a few reports regarding to the use of BoNT/A in treatment of patients with vaginismus have been published. Almost all patients showed positive response to BoNT/A treatment (Bertolasi *et al.*, 2009; Fageeh, 2011). BoNT/A is effective, safe and highly tolerated in severe cases of vaginismus, where conventional therapy is not effective (Park and Paraiso, 2009). In some patients, BoNT/A injections are also curative for vaginismus (Ghazizadeh and Nikzad, 2004).

Wounds of the face, especially those lying perpendicular to the lines of Langer, are known to heal poorly with conspicuous scarring. A double-blind randomized study showed the improvement of wound healing after haemorrhoidectomy (Patti *et al.*, 2005). BoNT/A was also effective on forehead wound healing and ugly scars of the face (Gassner *et al.*, 2006; Wilson, 2006). BoNT/A was thought to decrease the expression of substance P, calcitonin gene-related peptide, transforming growth factor beta-1 and alpha smooth muscle actin A in wound healing in a dose-dependent manner with no effect on the healing time (Wang *et al.*, 2009). Some research are testing the effect of BoNT/A on urethral wound healing and collagen deposition in hypertrophic scars using animal models (Sahinkanat *et al.*, 2009; Xiao and Qu, 2012).

Diabetic neuropathy is a common complication in diabetes and includes diverse sensory symptoms such as pain and dysaesthesias in the feet, gustatory hyperhidrosis, oropharyngeal dysphagia, diabetic gastroparesis and more. BoNT/A treatment for different types of diabetic neuropathy has been reported (Argoff, 2002; Restivo *et al.*, 2006; Hummel *et al.*, 2008; Yuan *et al.*, 2009; Bach-Rojecky *et al.*, 2010; Lund *et al.*, 2011). Although the data are still very limited, all case studies supported BoNT/A as an effective and safe treatment. Further randomized and case studies will be needed to gain more insights in the application of BoNT/A to treat diabetic neuropathy.

Limitations of currently available BoNT therapies and novel product development

Currently available BoNT therapies have certain limitations. Firstly, clinical use of BoNTs is limited to conditions that affect neuromuscular activity (Glogau, 2002; Cheng *et al.*, 2006) due to the neuronal tropism of BoNT. For example, BoNT/A only internalizes into neuronal cells and cleaves SNAP25: it is unable to enter non-neuronal cells and cleave the non-neuronal isoform SNAP23 (Sadoul *et al.*, 1997; Vaidyanathan *et al.*, 1999). Non-neuronal SNARE isoforms are involved in divergent cellular processes that include fusion reactions in cell growth, membrane repair, cytokines and synaptic transmission (Jahn and Scheller, 2006). For example, SNAP23 complexes with non-neuronal VAMP and syntaxin isoforms mediate non-neuronal vesicle exocytic processes, including the secretion of airway mucus, antibodies, insulin, gastric acids, and ions (Sadoul *et al.*, 1997; Chen *et al.*, 2000; Hickson *et al.*, 2000;

Martin-Martin *et al.*, 2000; Paumet *et al.*, 2000; Castle *et al.*, 2002; Pagan *et al.*, 2003; Reales *et al.*, 2005; Foster *et al.*, 2006; Saxena *et al.*, 2006; Rogers, 2007; Davis and Dickey, 2008). Targeting SNAP23 with a modified BoNT may reduce the secretion processes of hypersecretion syndromes such as asthma and gastro-oesophageal reflux. SNAP23-specific BoNT/A derivatives can also be targeted for other therapeutic applications that include diabetes, inflammatory and immune disorders that include a hypersecretory component.

The therapeutic benefits of BoNT for treatment of conditions associated with involuntary muscle spasm and contractions, for cosmetic use, and other applications are transient and repeated injections are necessary. In some patients, BoNT could elicit neutralizing antibodies against the corresponding toxin, thus reducing the beneficial effects or rendering the patient completely unresponsive to further treatment (Goschel *et al.*, 1997; Atassi and Oshima, 1999; Dolimbek *et al.*, 2002; Jankovic *et al.*, 2003; Atassi, 2004; Atassi and Dolimbek, 2004; Jankovic, 2004a,b, 2006). Factors that can influence the immune response to the toxin include the dosage used, duration of treatment, frequency of immunization, and quality of the toxin. The appearance of neutralizing antibodies in patients might be controlled by manipulation of MHC presentation pathway (Atassi, 2004). The exact percentage of patients who may develop immunoresistance to BoNT treatment is unknown, but it is commonly believed that there are fewer patients who develop blocking antibodies when treated with BoNT/A than with BoNT/B (Atassi, 2004; Jankovic, 2006). This is probably due to the use of lower doses of BoNT/A complex (11.8ng/treatment for dystonia patient using Botox, Allergan Inc.) than BoNT/B complex (25–100 ng/treatment for dystonia patient using Myobloc, Solstice Neurosciences Inc.) (Brashear *et al.*, 1999; Aoki, 2002). The development of blocking antibodies is also more common in patients who receive treatment of cervical dystonia or spasticity, which requires larger doses and periodic administration of toxin, while it is less common in patients who are treated for laryngeal dystonia, blepharospasm or cosmetic use, all of which require smaller doses for treatments (Atassi, 2004; Comella, 2008; Swope and Barbano, 2008). To date, there is no effective solution for immunoresistance of BoNT therapies. Once a patient develops immunoresistance to one toxin, the benefit of using another toxin instead will most likely be limited and short-lived, because the patient is very likely to also become immunoresistant to the second toxin. BoNT/B has been shown to be efficient in patients who have developed immunoresistance to BoNT/A (Truong *et al.*, 1997; Brashear *et al.*, 1999; Dressler *et al.*, 2003), but most of them will become resistant to BoNT/B treatment after a few injections (Brashear *et al.*, 1999; Aoki, 2002). In a study that involved 100 cervical dystonia patients, a third of the patients who were negative for BoNT/B antibodies became positive for BoNT/B antibodies during a 42-month follow-up, which included about five treatment visits per patient. Thus, the high antigenicity of BoNT/B limits its long-term efficacy (Jankovic *et al.*, 2006).

Retargeting BoNT to extend therapeutic interventions

Recent studies on the structure–function characterization of HCs and LCs of botulinum neurotoxins and tetanus neurotoxins advanced our knowledge on the mechanisms of BoNT receptor binding, internalization and substrate recognition. Understanding the structure–function relationship of BoNTs has opened up new opportunities to engineer recombinant proteins to treat diseases that are not amenable to therapy with native neurotoxins, or to give better outcome than with the native neurotoxins. Attempts have been made to engineer novel BoNT derivatives, and one of them is the development of therapeutic proteins that possess the endopeptidase activity of BoNT, with different cellular specificity but without the inherent toxicity of neurotoxins. By replacing the HC_R of BoNT with a new peptide or protein-targeting domain, the resulting chimeric protein can be retargeted to a new cell type defined by the new binding domain. This includes the NGF-LHC_T/A and wheat germ agglutinin-LHC_T/A for both neuronal and non-neuronal cells. A therapeutically relevant application is to target ECL (*Erythrina cristagalli* lectin)–LHC_T/A conjugate to nociceptive afferents and airway epithelium cells (Chaddock *et al.*, 2000a,b, 2004;

Duggan *et al.*, 2002; Foster, 2005; Foster *et al.*, 2006). The selective cleavage of neuronal-specific SNARE proteins, SNAP25, VAMP-2 and syntaxin 1a by the catalytic domain of BoNT may limit the development of novel therapies for these non-neuronal systems.

The retargeting of the catalytic activity of BoNTs to non-neuronal SNARE isoforms has been investigated (Chen and Barbieri, 2009). SNAP23 is a non-neuronal isoform of SNAP25 and it mediates the process of different secretion events in non-neuronal systems such as airway mucus, gastric acid and antibody secretion. Excessive secretion of these substances is associated with different diseases such as asthma, chronic obstructive pulmonary disease, gastric efflux, diabetes and inflammatory and immune disorders. Targeting SNAP23 with a novel BoNT derivative may reduce the secretion processes of hypersecretion syndromes. A mutated BoNT/E light chain, LC/E(K^{224}D), was engineered and showed extended substrate specificity to cleave SNAP23 and the natural substrate, SNAP25, but not SNAP29 or SNAP47. Upon direct protein delivery into cultured human epithelial cells, LC/E(K^{224}D) cleaved endogenous SNAP23, which inhibited secretion of mucin and IL-8 (Chen and Barbieri, 2009) (Fig. 7.2). These studies show the feasibility of genetically modified LCs to target a non-neuronal SNARE protein, which will extend therapeutic potential of BoNT for treatment of human hypersecretion diseases.

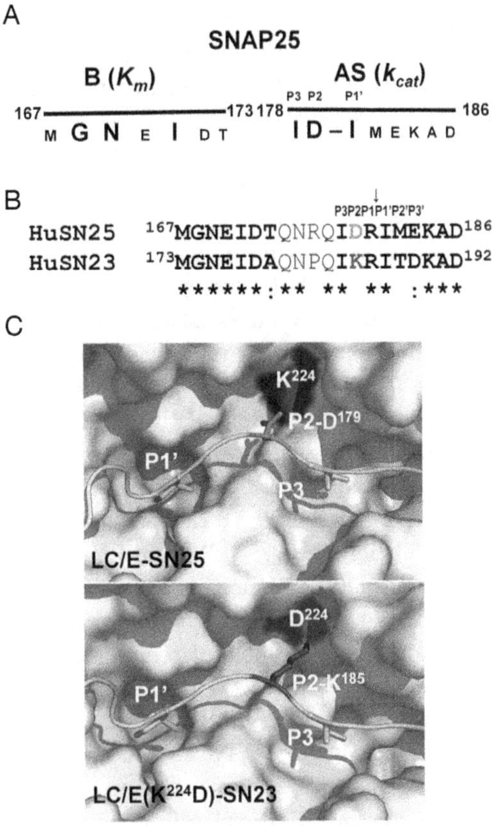

Figure 7.2 K^{185} of human SNAP23 contributes to substrate recognition by BoNT/E. (A) Substrate recognition by LC/E. Two subsites in SNAP25 contribute to substrate binding 'B' (K_m) and catalysis 'AS' (k_{cat}), where the P3, P2, and P1′ residues contribute to recognition by LC/E. (B) Sequence alignment of human SNAP25 (SN25) and human SNAP23 (SN23). (C) (*Upper*) modelled complex structure of LC/E-SNAP25 predicts the recognition of P site residues of SNAP25 by LC/E. (*Lower*) modelled complex structure of LC/E(K^{224}D)-SNAP23 predicts the recognition of P site residues of SNAP23 by LC/E(K^{224}D).

Combating BoNT immunoresistance issues

One way to overcome the immunoresistance problem in BoNTs is to engineer more active BoNTs, which will reduce the amounts of protein required for therapy and may decrease the development of immune response to the therapy. Rummel *et al.* have modified a ganglioside binding motif of the HC domain of BoNT/B that enhances the binding and toxicity to up to 3-fold relative to the wild type toxin (Foster, 2009). The engineering of a mutated BoNT with modified, and in particular enhanced potency, can be a solution to the immunoresistance issue in BoNT/B therapy, since increasing the potency of BoNT/B will reduce the amount of protein required for treatment, hence reducing or eliminating the development of immunoresistance. However, the engineering of BoNT through modification(s) of its receptor binding sites may affect the selectivity of the binding event and protection by current vaccine derived from the HCs of BoNTs. In addition, the modification of binding site(s) may not successfully increase the potency enough to prevent the development of immunoresistance. Modification of LC to alter its activity may well be a better way to achieve this goal.

A comparative study was conducted for LC/B and LC/T to provide proof of principle to engineer LC of CNTs with elevated activities. LC/B and LC/T cleave VAMP2, at the same scissile bond, but differ in catalytic activity, substrate requirement, and sensitivity to inhibitors (Foran *et al.*, 1994; Chen *et al.*, 2008b). Alanine-scanning mutagenesis and kinetic analysis identified three regions within VAMP2 that were recognized by LC/B and LC/T: residues adjacent to the site of scissile bond cleavage (cleavage region) and residues located within the N-terminal region and the C-terminal region relative to the cleavage region (Chen *et al.*, 2008b). Mutations at the P7, P4, P2, and P1′ residues of VAMP2 had the greatest inhibition of LC/B cleavage (>32-fold), while mutations at P7, P4, P1′ and P2′ residues of VAMP2 had the greatest inhibition of LC/T cleavage (>64-fold) (Chen *et al.*, 2008b). Another study addressed the molecular mechanisms of LC/B and LC/T substrate recognition and specificity. Major P sites of VAMP2 (P7, P6, P3, P2, P1, P1′ and P2′) and (P7, P6, P4, P2, P1, P1′ and P2′), contributed to their substrate recognition and catalysis by LC/B and LC/T, respectively. Upon understanding LC/B and LC/T substrate recognition, it was found that the S1 pocket mutation LC/T ($K^{168}E$) increased the rate of native VAMP2 cleavage approaching the rate of LC/B, which explains the molecular basis for the lower k_{cat} that LC/T possesses for VAMP2 cleavage relative to LC/B (Fig. 7.3). In addition, $R^{188}M$, a S4 pocket mutation, increased LC/T substrate hydrolysis by ~5-fold (Chen *et al.*, 2012). For LC/A, the screening of residues around the active site of LC/A identified residue Lys^{165} and the mutation $K^{165}L$ resulted in a 4-fold increase in substrate hydrolysis. These results suggest the possibility to achieve BoNT with higher activity through LC engineering. This analysis explains the molecular basis underlining the VAMP2 recognition and cleavage by LC/B and LC/T and provides insight into the possibility of extending the pharmacological utility of these neurological reagents.

The other way to counter the immunoresistance issue of BoNT therapy is to block the epitopes on the BoNTs that are involved in neutralizing antibody production. Studies with regard to the mapping of neutralization antibodies produced through BoNT/A or BoNT/B therapies have identified an array of human antibodies that are produced by patients who have developed resistance to BoNT/A or BoNT/B treatments. Detailed mapping of these antibodies on the HC and LC of BoNT/A and BoNT/B have identified the regions that are responsible for the development of these neutralizing antibodies (Dolimbek *et al.*, 2007; Atassi *et al.*, 2008, 2012). Attempts have been made to reduce the levels of the Ab response against immunodominant antigenic sites on the heavy chain of BoNT/A in order to diminish immunoresistance caused by neutralizing Abs. Four peptides representing four antigenic regions on BoNT/A were conjugated to monomethoxypolyethylene glycol (mPEG) at the N-terminus. Tolerization with a given mPEG-peptide reduced the Ab levels against the correlated region and

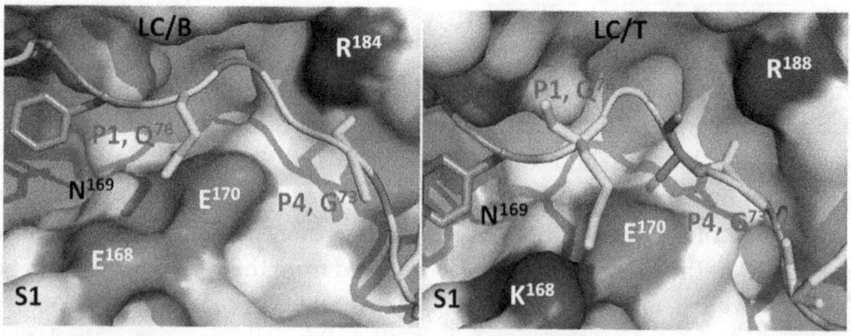

Figure 7.3 Optimization of LC/T substrate recognition pockets increases its catalytic activities. The S1 pocket of LC/B and LC/T formed by residues $E^{168}N^{169}E^{170}$ and $K^{168}N^{169}E^{170}$, respectively, recognizes P1, Q^{76} of VAMP2. Mutation LC/T $K^{168}E$ increases k_{cat} by ~8-fold. The S4 pocket of LC/B and LC/T involves the residues R^{184} and R^{188}. The mutation LC/T $R^{188}M$ increases k_{cat} by ~5-fold.

the antisera became less protective than antisera of non-tolerized controls that were immunized only with inactive BoNT/A. One of the peptides showed effective blocking of the production of neutralizing antibodies against BoNT/A in mice after being traced to up to 50 months (Dolimbek *et al.*, 2011). This result suggests that the tolerization procedure might be potentially useful for clinical applications to immunoresistant patients.

BoNT-related therapies under investigation

In addition to research to counter the current limitations of BoNT therapy, several other avenues that will extend the current use of BoNT/A are also actively investigated; they include the anti-nociceptive application of BoNTs and BoNT as neuronal drug delivery system.

Anti-nociceptive application of BoNTs

Chronic pain is a huge human health problem that always lacks effective long-term treatment. Chronic pain is due to the sensitization of peripheral stimulation and release of neuropeptides and inflammatory mediators such as substance P, calcitonin gene-related peptide (CGRP), and nerve growth factor. This amplifies the impulses transferred to the soma in both the dorsal root and trigeminal ganglia, resulting in sensitization of the central nervous system (Dolly and O'Connell, 2012). The ability of BoNT/A to alleviate pain was observed during the clinical use of BoNT/A. Patients given intramuscular injection of BoNT/A for treating dystonia experienced relief of pain before the onset of muscle relaxation (Brin *et al.*, 1987). The ability of the toxin to ease pain symptoms was evidenced in BoNT/A treatment for glabellar lines, in which an associated improvement in the severity and frequency of migraine episodes was reported (Binder *et al.*, 1998). BoNT/A has been used to treat chronic painful musculoskeletal conditions such as temporomandibular joint dysfunction, cervicothoracic pain, lower back pain and myofascial pain (Jeynes and Gauci, 2008; Zhang *et al.*, 2011). Data from several studies also suggest a role for BoNT in alleviating inflammatory pain, reducing neuropathic pain and in treating chronic headache conditions such as chronic migraine, tension-type headache, cervicogenic headache and cluster headache (Borodic *et al.*, 2001; Argoff, 2002; Dodick, 2003; Dodick *et al.*, 2004; Relja and Telarovic, 2004; Aguggia, 2008; Diener *et al.*, 2010; Lipton *et al.*, 2011). Following several randomized, placebo-controlled clinical trials, the encouraging results allow BoNT/A product from Allergan Inc. to be granted approval by the FDA for treatment of chronic migraine in the UK and USA.

BoNT/A has been shown to be a feeble inhibitor of the release of the pain mediator CGRP from sensory neurons evoked by activating transient receptor potential vanilloid receptor type 1 (TRPV1) with capsaicin (Meng *et al.*, 2009). However, the neuromuscular paralysis induced by BoNT/A can be reversed, at least transiently, by a [Ca^{2+}] ionophore, a sustained elevation of intra-neuronal Ca^{2+} by this TRPV1 agonist might attenuate the toxin's inhibition of CGRP exocytosis elicited by K^+ depolarization (Meng *et al.*, 2009; Dolly and O'Connell, 2012). Therefore, BoNT/E is considered a better CGRP inhibitor since it offers faster action and pseudo-irreversibly abolishes release of neurotransmitters. Limitations on BoNT/E application in pain control are mostly due to its lack of an acceptor for sensory neuron and short duration of action. Application of BoNT/E to cultured sensory neurons from rat trigeminal ganglia failed to cleave SNAP-25 or decrease stimulated CGRP release (Dolly *et al.*, 2009; Dolly and O'Connell, 2012). To overcome this limitation, a chimera of BoNT/A and E, which includes the HC domain of BoNT/A and LC domain of BoNT/E, was constructed. This chimera can be taken up by trigeminal neurons and block CGRP release triggered by all stimuli tested, including capsaicin. However, the short half-life of BoNT/E limits its application in chronic pain treatment (Meng *et al.*, 2009). The long-lasting action of BoNT is a unique feature of these types of proteases, but the exact mechanism underlying such property is not clear. It was suggested that the LC of BoNT/E is ubiquitylated and rapidly degraded inside cells. In contrast, BoNT/A LC is considerably more stable. TRAF2, a RING finger protein implicated in ubiquitylation, was found to be selectively associated with BoNT/E LC, promoting its proteasomal degradation. Based on these data, a chimeric SNAP25-based ubiquitin ligase, which contains BoNT/A resistant SNAP25 and RING finger domains of the X-linked mammalian inhibitor of apoptosis (XIAP) or the homologous to the E6-AP C-terminal (HECT) domain of E6-AP, was constructed and targeted BoNT/A LC for degradation, reducing its duration in a cellular model for toxin persistence. In this regard, it is possible to engineer a LC of BoNT/E with longer half-life by blocking its interaction with ubiquitin ligases (Tsai *et al.*, 2010). Another study unveiled that the prolonged neuromuscular paralysis could be attributed to a di-leucine motif near the C-terminus of its LC (Wang *et al.*, 2011). Based on this finding, a chimeric BoNT was constructed by fusing the LC domain of BoNT/E to the N-terminus of an inactive holotoxin of BoNT/A. The resultant BoNT/E-A chimera exhibited highly specific neurotoxicity in mice and cleaved SNAP-25 at the LC/E scissile bond in central neurons and motor nerve terminals, and potently blocked synaptic transmission in a phrenic nerve diaphragm. Most importantly, it caused prolonged weakness of murine muscles *in vivo*, demonstrating that the normal transient activity of BoNT/E can be extended by several folds to approach a duration typical of BoNT/A (Wang *et al.*, 2011). Although there is still a long way before a clinically useful product appears, the protein engineering strategy has facilitated advancement towards obtaining a potent, long-acting and versatile inhibitor of exocytosis, which displays unique properties advantageous for treating chronic pain.

BoNT as neuronal drug delivery system

Another promising area of BoNT-related therapy is the application of BoNT as a drug delivery system. BoNTs' tropism for neurons and their independent abilities to bind, internalize and translocate into the cytosol of neuronal cells make them an attractive vehicle for delivering drug molecules to neurons. In particular, BoNT offers an opportunity to deliver drugs into peripheral neurons such as cholinergic terminals, whereas TeNT has the ability to deliver them to the central neurons. Tetanus neurotoxin (TeNT) in drug delivery has been studied since 1980 (Bizzini *et al.*, 1980). Beaude *et al.* (1990) showed that a TeNT B-IIb fragment conjugated to glucose oxidase can be taken up by axon terminals and convey retrogradely to spinal-cord motor neurons (Beaude *et al.*, 1990). In addition, TeNT has been shown to deliver different cargoes such as human Cu/Zn superoxide dismutase, neurotrophic factor (GDNF) and DNA (Box *et al.*, 2003; Benn *et al.*, 2005; Larsen *et al.*, 2006). It is accepted that the TeNT neuron-binding fragment can be an efficient drug delivery vehicle, yet its clinical application value is questionable considering that the general population have mostly received vaccination with tetanus toxoid.

Unlike TeNT, there is less concern on pre-existing immunity against BoNT/A, as botulism is an extremely rare disease and only occupational workers are vaccinated against it. Therapeutic doses used for neuromuscular disorder treatments are extremely low, thereby avoiding systemic immune responses. There are two types of BoNT-based drug delivery systems, namely inactive holotoxin and heavy chain (B domain) -based delivery systems. BoNT/D has been tested for its ability to carry different cargo proteins into neurons by fusing them to active holotoxin of BoNT/D (Bade *et al.*, 2004). Different cargo proteins were found to be delivered with different efficiency into neurons by BoNT/D. Protein unfolding is required for translocation into neuron and cargo proteins that are not flexible enough in their conformation are not transported. A recent study has shown that the HC of BoNT/A can be used as an efficient delivery system (Ho *et al.*, 2011). In addition, an inactive BoNT/A labelled with alexa-488 can specifically bind to and become internalized into neuronal cells, suggesting a role of inactivated BoNT/A as neuronal drug delivery system. A GFP protein conjugated to the HC of BoNT/A through a 40-amino acid linker can be delivered into cultured neuronal cells and neuronal cells in mouse motor nerve endings and is functionally active. In addition, a generic HC-based drug delivery system consisting of a targeting molecule, cy3-labelled HC/A, linked by a disulfide bond to a drug stimulant, Oregon green 488-labelled 10-kDa dextran was developed (Goodnough *et al.*, 2002; Zhang *et al.*, 2009). A PDPH linker is bound to one of four possible cysteine sulfhydryl groups on HC/A. The dextran is conjugated to the HC/A by a C–N bond in one of the glucose residues. This consists of the basic structure of the drug delivery system and multiple drug molecules could be attached to a dextran carrier for neuronal delivery. The efficiency of this drug delivery system has been tested in mice spinal cord neuron culture. Sixteen hours after the binding of the Oregon green 488 labelled dextran conjugated to HC/A-based drug delivery system, 40% of the drug carriers were separated from the HC/A and diffused into the cytosol of the neuronal cells. The separation of the drug from the delivery system is dependent on cell maturation (Zhang *et al.*, 2009).

BoNT-based drug delivery systems can be applied to deliver therapeutic cargoes to treat neuronal diseases and BoNT intoxications. BoNT/A is thought to be an attractive option to ensure direct delivery to the cytosol of intoxicated nerve cells, while the use of BoNT-based delivery system has its intrinsic limitations. With regards to the delivery of therapeutic cargoes to treat neuronal diseases, the limitation for BoNT-based delivery system is its inability in retrograde trafficking to the central neurons. Although compelling evidence shows that BoNTs have the ability to undergo intra-axonal retrograde transport and neuronal transcytosis, the mechanisms underlying the retrograde transport for BoNTs remain undefined (Alexiades-Armenakas, 2008; Antonucci *et al.*, 2008). Further studies are needed to delineate the conditions that trigger BoNT retrograde transport. Another specific application of BoNT-based delivery is for antitoxins and antibodies to inhibit BoNT LC activity inside neurons to counteract the damaging effects of BoNT or TeNT poisoning. Current studies have shown that the protein receptor for BoNT/A is the luminal domain of a synaptic vesicle protein SV2 (Dong *et al.*, 2006). The exposure of the BoNT/A receptor on the surface of neurons is dependent on the active recycling of neurotransmitter-carrying vesicles (Dong *et al.*, 2006). Therefore, in intoxicated neurons the presentation and recycling of such a receptor will be impaired, which would prevent the effective delivery of therapeutic cargoes to the intoxicated neurons. Inhibition of BoNT/A uptake by neurons previously treated with BoNT/B has indeed been reported. However, similar to HC of TeNT, the HC of BoNT/C did not use luminal domains of synaptic vesicle proteins as protein receptor and its internalization was not dependent on the active recycling of synaptic vesicles (Nishikawa *et al.*, 2004; Peng *et al.*, 2012). Therefore it can be a potential delivery system for BoNT therapies. Further studies are needed to test the potential of HC of BoNT/C being used as a drug delivery system for BoNT antitoxins.

Conclusion

BoNTs, in particular BoNT/A and B, have been successfully used to treat a large number of neuronal disorders through exploitation of their potential in interfering with a wide spectrum of physiological functions ranging from reduction of muscular contractions to pain alleviations. Their unique characteristics and pharmacological properties have made BoNTs a versatile treatment option for a growing number of indications. The future of BoNTs in medical applications is bright, yet more research is needed to improve the medical uses of BoNTs. These include but are not limited to the optimization of current therapies including choices of dosage, injection sites, injection techniques, injection interval, development of different types of BoNT/A products with different pharmacological properties, exploration of new indications for BoNT, standardization of off-label uses of BoNT and designation of more randomized, placebo-controlled trials, investigation of mechanisms of anti-nociceptive activity of BoNT and its application to analgesia uses, development of novel BoNT-based products to control non-neuronal hypersecretion diseases, development of BoNT products with lower immunogenicity, and exploration of other BoNT serotypes as therapeutic agents.

Acknowledgements

The research was supported through RGC and Hong Kong PolyU Competitive Research Grants, B-Q25N, A-PK05 and G-YJ15 and National Institutes of Health, NIAID, Grant U54 AI057153 (Great Lakes Regional Centre of Excellence).

References

(1999). Cosmetic use of botulinum toxin. Med Lett Drugs Ther *41*, 63–64.

(2001). Neurotoxins licensed as treatment for neck neuromuscular disorder. Am. J. Health Syst. Pharm. *58*, 200, 202.

Absar, M.S., and Onwudike, M. (2008). Efficacy of botulinum toxin type A in the treatment of focal axillary hyperhidrosis. Dermatol. Surg. *34*, 751–755.

Agarwal, R., Binz, T., and Swaminathan, S. (2005a). Analysis of active site residues of botulinum neurotoxin E by mutational, functional, and structural studies: Glu335Gln is an apoenzyme. Biochemistry *44*, 8291–8302.

Agarwal, R., Binz, T., and Swaminathan, S. (2005b). Structural analysis of botulinum neurotoxin serotype F light chain: implications on substrate binding and inhibitor design. Biochemistry *44*, 11758–11765.

Agarwal, R., Schmidt, J.J., Stafford, R.G., and Swaminathan, S. (2009). Mode of VAMP substrate recognition and inhibition of *Clostridium botulinum* neurotoxin F. Nat. Struct. Mol. Biol. *16*, 789–794.

Aguggia, M. (2008). Treating headaches with botulinum toxin. Neurol. Sci. *29 Suppl 1*, S137–139.

Albavera-Hernandez, C., Rodriguez, J.M., and Idrovo, A.J. (2009). Safety of botulinum toxin type A among children with spasticity secondary to cerebral palsy: a systematic review of randomized clinical trials. Clin. Rehabil. *23*, 394–407.

Alexiades-Armenakas, M. (2008). Retrograde transport and transcytosis of botulinum toxin serotypes to the brain: analysis of potential neurotoxicity. Journal of drugs in dermatology: J. Drugs. Dermatol. *7*, 1006–1007.

Alrefai, A.H., Aburahma, S.K., and Khader, Y.S. (2009). Treatment of sialorrhea in children with cerebral palsy: a double-blind placebo controlled trial. Clin. Neurol. Neurosurg. *111*, 79–82.

Annese, V., Basciani, M., Perri, F., Lombardi, G., Frusciante, V., Simone, P., Andriulli, A., and Vantrappen, G. (1996). Controlled trial of botulinum toxin injection versus placebo and pneumatic dilation in achalasia. Gastroenterology *111*, 1418–1424.

Annese, V., D'Onofrio, V., and Andriulli, A. (1998). Botulinum toxin in long-term therapy for achalasia. Ann. Intern. Med. *128*, 696.

Antonucci, F., Rossi, C., Gianfranceschi, L., Rossetto, O., and Caleo, M. (2008). Long-distance retrograde effects of botulinum neurotoxin A. J. Neurosci. *28*, 3689–3696.

Aoki, K.R. (2002). Immunologic and other properties of therapeutic botulinum toxin serotypes. In: Scientific and Therapeutic Aspects of Botulinum Toxin, Brin, MF., Hallett, M., and Jankovic, J., eds. (Lippincott Willams & Wilkins, Philadephia), pp.103–113.

Argoff, C.E. (2002). A focused review on the use of botulinum toxins for neuropathic pain. Clin. J. Pain *18*, S177–181.

Arndt, J.W., Yu, W., Bi, F., and Stevens, R.C. (2005). Crystal structure of botulinum neurotoxin type G light chain: serotype divergence in substrate recognition. Biochemistry *44*, 9574–9580.

Arndt, J.W., Chai, Q., Christian, T., and Stevens, R.C. (2006). Structure of botulinum neurotoxin type D light chain at 1.65 A resolution: repercussions for VAMP-2 substrate specificity. Biochemistry *45*, 3255–3262.

Arts, J., Holvoet, L., Caenepeel, P., Bisschops, R., Sifrim, D., Verbeke, K., Janssens, J., and Tack, J. (2007). Clinical trial: a randomized-controlled crossover study of intrapyloric injection of botulinum toxin in gastroparesis. Aliment. Pharmacol. Ther. *26*, 1251–1258.

Ascher, B., and Rossi, B. (2004). Botulinum toxin and wrinkles: few side effects and effective combining procedures with other treatments. Ann. Chir. Plast. Esthet. *49*, 537–552.

Atassi, M.Z. (2004). Basic immunological aspects of botulinum toxin therapy. Mov. Disord. *19 Suppl 8*, S68–84.

Atassi, M.Z., and Oshima, M. (1999). Structure, activity, and immune (T and B cell) recognition of botulinum neurotoxins. Crit. Rev. Immunol. *19*, 219–260.

Atassi, M.Z., and Dolimbek, B.Z. (2004). Mapping of the antibody-binding regions on the HN-domain (residues 449–859) of botulinum neurotoxin A with antitoxin antibodies from four host species. Full profile of the continuous antigenic regions of the H-chain of botulinum neurotoxin A. Protein J. *23*, 39–52.

Atassi, M.Z., Dolimbek, B.Z., Jankovic, J., Steward, L.E., and Aoki, K.R. (2008). Molecular recognition of botulinum neurotoxin B heavy chain by human antibodies from cervical dystonia patients that develop immunoresistance to toxin treatment. Mol. Immunol. *45*, 3878–3888.

Atassi, M.Z., Jankovic, J., Steward, L.E., Aoki, K.R., and Dolimbek, B.Z. (2012). Molecular immune recognition of botulinum neurotoxin B. The light chain regions that bind human blocking antibodies from toxin-treated cervical dystonia patients. Antigenic structure of the entire BoNT/B molecule. Immunobiology *217*, 17–27.

Aurora, S.K., Gawel, M., Brandes, J.L., Pokta, S., and Vandenburgh, A.M. (2007). Botulinum toxin type a prophylactic treatment of episodic migraine: a randomized, double-blind, placebo-controlled exploratory study. Headache *47*, 486–499.

Aurora, S.K., Dodick, D.W., Turkel, C.C., DeGryse, R.E., Silberstein, S.D., Lipton, R.B., Diener, H.C., and Brin, M.F. (2010). OnabotulinumtoxinA for treatment of chronic migraine: results from the double-blind, randomized, placebo-controlled phase of the PREEMPT 1 trial. Cephalalgia *30*, 793–803.

Aurora, S.K., Winner, P., Freeman, M.C., Spierings, E.L., Heiring, J.O., DeGryse, R.E., VanDenburgh, A.M., Nolan, M.E., and Turkel, C.C. (2011). OnabotulinumtoxinA for treatment of chronic migraine: pooled analyses of the 56-Week PREEMPT clinical program. Headache *51*, 1358–1373.

Bach-Rojecky, L., Salkovic-Petrisic, M., and Lackovic, Z. (2010). Botulinum toxin type A reduces pain supersensitivity in experimental diabetic neuropathy: bilateral effect after unilateral injection. Eur. J. Pharmacol. *633*, 10–14.

Bade, S., Rummel, A., Reisinger, C., Karnath, T., Ahnert-Hilger, G., Bigalke, H., and Binz, T. (2004). Botulinum neurotoxin type D enables cytosolic delivery of enzymatically active cargo proteins to neurones via unfolded translocation intermediates. J. Neurochem. *91*, 1461–1472.

Baldwin, M.R., and Barbieri, J.T. (2007). Association of botulinum neurotoxin serotypes a and B with synaptic vesicle protein complexes. Biochemistry *46*, 3200–3210.

Beaude, P., Delacour, A., Bizzini, B., Domuado, D., and Remy, M.H. (1990). Retrograde axonal transport of an exogenous enzyme covalently linked to B-IIb fragment of tetanus toxin. Biochem. J. *271*, 87–91.

Bell, K.R., and Williams, F. (2003). Use of botulinum toxin type A and type B for spasticity in upper and lower limbs. Phys. Med. Rehabil. Clin. N. Am. *14*, 821–835.

Benedetto, A.V. (2004). Commentary: Botulinum toxin in clinical medicine: part II. Clin. Dermatol. *22*, 1–2.

Benn, S.C., Ay, I., Bastia, E., Chian, R.J., Celia, S.A., Pepinsky, R.B., Fishman, P.S., Brown, Jr, R.H.., and Francis, J.W. (2005). Tetanus toxin fragment C fusion facilitates protein delivery to CNS neurons from cerebrospinal fluid in mice. J. Neurochem. *95*, 1118–1131.

Benohanian, A. (2009). What stands in the way of treating palmar hyperhidrosis as effectively as axillary hyperhidrosis with botulinum toxin type A. Dermatol. Online J. *15*, 12.

Benson, M.A., Fu, Z., Kim, J.J., and Baldwin, M.R. (2011). Unique ganglioside recognition strategies for clostridial neurotoxins. J. Biol. Chem. *286*, 34015–34022.

Bertolasi, L., Frasson, E., Cappelletti, J.Y., Vicentini, S., Bordignon, M., and Graziottin, A. (2009). Botulinum neurotoxin type A injections for vaginismus secondary to vulvar vestibulitis syndrome. Obstet. Gynecol. *114*, 1008–1016.

Bhakta, B.B., Cozens, J.A., Chamberlain, M.A., and Bamford, J.M. (2000). Impact of botulinum toxin type A on disability and carer burden due to arm spasticity after stroke: a randomised double blind placebo controlled trial. J. Neurol. Neurosurg. Psychiatry *69*, 217–221.

Bhaumik, S., and Behari, M. (1999). Botulinum toxin A – injection for cervical dystonia. J. Assoc. Physicians India 47, 267–270.

Binder, W.J., and Blitzer, A. (2003). Treatment of migraine headache with botulinum toxin type A. Facial Plast Surg Clin North Am *11*, 465–475.

Binder, W.J., Blitzer, A., and Brin, M.F. (1998). Treatment of hyperfunctional lines of the face with botulinum toxin A. Dermatol. Surg. *24*, 1198–1205.

Binder, W.J., Brin, M.F., Blitzer, A., Schoenrock, L.D., and Pogoda, J.M. (2000). Botulinum toxin type A (BOTOX) for treatment of migraine headaches: an open-label study. Otolaryngol. Head Neck Surg. *123*, 669–676.

Binz, T., Blasi, J., Yamasaki, S., Baumeister, A., Link, E., Sudhof, T.C., Jahn, R., and Niemann, H. (1994). Proteolysis of SNAP-25 by types E and A botulinal neurotoxins. J. Biol. Chem. *269*, 1617–1620.

Binz, T., and Rummel, A. (2009). Cell entry strategy of clostridial neurotoxins. J. Neurochem. *109*, 1584–1595.

Bizzini, B., Grob, P., Glicksman, M.A., and Akert, K. (1980). Use of the B-IIb tetanus toxin derived fragment as a specific neuropharmacological transport agent. Brain Res. *193*, 221–227.

Blitzer, A. (2010). Spasmodic dysphonia and botulinum toxin: experience from the largest treatment series. Eur. J. Neurol. *17 Suppl 1*, 28–30.

Blitzer, A., Brin, M.F., and Stewart, C.F. (1998). Botulinum toxin management of spasmodic dysphonia (laryngeal dystonia): a 12-year experience in more than 900 patients. Laryngoscope *108*, 1435–1441.

Borodic, G.E., Acquadro, M., and Johnson, E.A. (2001). Botulinum toxin therapy for pain and inflammatory disorders: mechanisms and therapeutic effects. Expert Opin. Investig. Drugs *10*, 1531–1544.

Box, M., Parks, D.A., Knight, A., Hale, C., Fishman, P.S., and Fairweather, N.F. (2003). A multi-domain protein system based on the HC fragment of tetanus toxin for targeting DNA to neuronal cells. J. Drug Target. *11*, 333–343.

Brashear, A. (2001). Botulinum toxin type B: a new injectable treatment for cervical dystonia. Expert Opin. Investig. Drugs *10*, 2191–2199.

Brashear, A., Lew, M.F., Dykstra, D.D., Comella, C.L., Factor, S.A., Rodnitzky, R.L., Trosch, R., Singer, C., Brin, M.F., Murray, J.J., *et al.* (1999). Safety and efficacy of NeuroBloc (botulinum toxin type B) in type A-responsive cervical dystonia. Neurology *53*, 1439–1446.

Breidenbach, M.A., and Brunger, A.T. (2004). Substrate recognition strategy for botulinum neurotoxin serotype A. Nature *432*, 925–929.

Brin, M.F., Fahn, S., Moskowitz, C., Friedman, A., Shale, H.M., Greene, P.E., Blitzer, A., List, T., Lange, D., Lovelace, R.E., *et al.* (1987). Localized injections of botulinum toxin for the treatment of focal dystonia and hemifacial spasm. Mov. Disord. 2, 237–254.

Brisinda, G., Cadeddu, F., Mazzeo, P., and Maria, G. (2007). Botulinum toxin A for the treatment of chronic anal fissure. Expert Rev. Gastroenterol. Hepatol. *1*, 219–228.

Brunger, A.T. (2005). Structure and function of SNARE and SNARE-interacting proteins. Q Rev. Biophys. *38*, 1–47.

Brunger, A.T., and Rummel, A. (2009). Receptor and substrate interactions of clostridial neurotoxins. Toxicon *54*, 550–560.

Brunger, A.T., Breidenbach, M.A., Jin, R., Fischer, A., Santos, J.S., and Montal, M. (2007). Botulinum neurotoxin heavy chain belt as an intramolecular chaperone for the light chain. PLoS Pathog. *3*, 1191–1194.

Bullens, R.W., O'Hanlon, G.M., Wagner, E., Molenaar, P.C., Furukawa, K., Plomp, J.J., and Willison, H.J. (2002). Complex gangliosides at the neuromuscular junction are membrane receptors for autoantibodies and botulinum neurotoxin but redundant for normal synaptic function. J. Neurosci. *22*, 6876–6884.

Bushara, K.O., Park, D.M., Jones, J.C., and Schutta, H.S. (1996). Botulinum toxin – a possible new treatment for axillary hyperhidrosis. Clin. Exp. Dermatol. *21*, 276–278.

Camargo, C.H., Teive, H.A., Becker, N., Munhoz, R.P., and Werneck, L.C. (2011). Botulinum toxin type A and cervical dystonia: a seven-year follow-up. Arq. Neuropsiquiatr. *69*, 745–750.

Carruthers, A. (2002). Botulinum toxin type A: history and current cosmetic use in the upper face. Dis. Mon. *48*, 299–322.

Carter, S.R., and Seiff, S.R. (1997). Cosmetic botulinum toxin injections. Int. Ophthalmol. Clin. *37*, 69–79.

Castle, J.D., Guo, Z., and Liu, L. (2002). Function of the t-SNARE SNAP-23 and secretory carrier membrane proteins (SCAMPs) in exocytosis in mast cells. Mol. Immunol. *38*, 1337–1340.

Chaddock, J.A., Purkiss, J.R., Duggan, M.J., Quinn, C.P., Shone, C.C., and Foster, K.A. (2000a). A conjugate composed of nerve growth factor coupled to a non-toxic derivative of *Clostridium botulinum* neurotoxin type A can inhibit neurotransmitter release *in vitro*. Growth Factors *18*, 147–155.

Chaddock, J.A., Purkiss, J.R., Friis, L.M., Broadbridge, J.D., Duggan, M.J., Fooks, S.J., Shone, C.C., Quinn, C.P., and Foster, K.A. (2000b). Inhibition of vesicular secretion in both neuronal and nonneuronal cells by a retargeted endopeptidase derivative of *Clostridium botulinum* neurotoxin type A. Infect. Immun. *68*, 2587–2593.

Chaddock, J.A., Purkiss, J.R., Alexander, F.C., Doward, S., Fooks, S.J., Friis, L.M., Hall, Y.H., Kirby, E.R., Leeds, N., Moulsdale, H.J., *et al.* (2004). Retargeted clostridial endopeptidases: inhibition of nociceptive neurotransmitter release *in vitro*, and antinociceptive activity in *in vivo* models of pain. Mov. Disord. *19 Suppl 8*, S42–47.

Chai, Q., Arndt, J.W., Dong, M., Tepp, W.H., Johnson, E.A., Chapman, E.R., and Stevens, R.C. (2006). Structural basis of cell surface receptor recognition by botulinum neurotoxin B. Nature *444*, 1096–1100.

Chancellor, M.B., Fowler, C.J., Apostolidis, A., de Groat, W.C., Smith, C.P., Somogyi, G.T., and Aoki, K.R. (2008). Drug Insight: biological effects of botulinum toxin A in the lower urinary tract. Nat. Clin. Pract. Urol. *5*, 319–328.

Chang, H. (2005). Botulism toxin: use in disorders of the temporomandibular joint. Dent. Today *24*, 48, 50–41; quiz 51.

Chang, L.B., Tsai, C.P., Liao, K.K., Kao, K.P., Yuan, C.L., Yen, D.J., and Lin, K.P. (1999). Use of botulinum toxin A in the treatment of hemifacial spasm and blepharospasm. Zhonghua yi xue za zhi (Chinese Medical Journal); Free China ed *62*, 1–5.

Chen, C., Baldwin, M.R., and Barbieri, J.T. (2008a). Molecular basis for tetanus toxin coreceptor interactions. Biochemistry *47*, 7179–7186.

Chen, C., Fu, Z., Kim, J.J., Barbieri, J.T., and Baldwin, M.R. (2009). Gangliosides as high affinity receptors for tetanus neurotoxin. J. Biol. Chem. *284*, 26569–26577.

Chen, D., Bernstein, A.M., Lemons, P.P., and Whiteheart, S.W. (2000). Molecular mechanisms of platelet exocytosis: role of SNAP-23 and syntaxin 2 in dense core granule release. Blood *95*, 921–929.

Chen, S., and Barbieri, J.T. (2006). Unique substrate recognition by botulinum neurotoxins serotypes A and E. J. Biol. Chem. *281*, 10906–10911.

Chen, S., and Barbieri, J.T. (2009). Engineering botulinum neurotoxin to extend therapeutic intervention. Proc. Natl. Acad. Sci. U.S.A. *106*, 9180–9184.

Chen, S., and Barbieri, J.T. (2011). Association of botulinum neurotoxin serotype A light chain with plasma membrane-bound SNAP-25. J. Biol. Chem. *286*, 15067–15072.

Chen, S., Kim, J.J., and Barbieri, J.T. (2007). Mechanism of substrate recognition by botulinum neurotoxin serotype A. J. Biol. Chem. *282*, 9621–9627.

Chen, S., Hall, C., and Barbieri, J.T. (2008b). Substrate recognition of VAMP-2 by botulinum neurotoxin B and tetanus neurotoxin. J. Biol. Chem. *283*, 21153–21159.

Chen, S., Karalewitz, A.P., and Barbieri, J.T. (2012). Insights into the different catalytic activities of clostridium neurotoxins. Biochemistry *51*, 3941–3947.

Cheng, C.M., Chen, J.S., and Patel, R.P. (2006). Unlabeled uses of botulinum toxins: a review, part 2. Am. J. Health Syst. Pharm. *63*, 225–232.

Cherington, M. (1998). Clinical spectrum of botulism. Muscle Nerve *21*, 701–710.

Chinnapongse, R., Gullo, K., Nemeth, P., Zhang, Y., and Griggs, L. (2012). Safety and efficacy of botulinum toxin type B for treatment of sialorrhea in Parkinson's disease: a prospective double-blind trial. Mov. Disord. *27*, 219–226.

Cillino, S., Raimondi, G., Gueprattte, N., Damiani, S., Cillino, M., Di Pace, F., and Casuccio, A. (2010). Long-term efficacy of botulinum toxin A for treatment of blepharospasm, hemifacial spasm, and spastic entropion: a multicentre study using two drug-dose escalation indexes. Eye (Lond) *24*, 600–607.

Comella, C.L. (2008). The treatment of cervical dystonia with botulinum toxins. J. Neural Transm. *115*, 579–583.

Comella, C.L., Jankovic, J., Shannon, K.M., Tsui, J., Swenson, M., Leurgans, S., and Fan, W. (2005). Comparison of botulinum toxin serotypes A and B for the treatment of cervical dystonia. Neurology *65*, 1423–1429.

Costa, J., Borges, A., Espirito-Santo, C., Ferreira, J., Coelho, M., Moore, P., and Sampaio, C. (2005a). Botulinum toxin type A versus botulinum toxin type B for cervical dystonia. Cochrane Database Syst. Rev., CD004314.

Costa, J., Espirito-Santo, C., Borges, A., Ferreira, J.J., Coelho, M., Moore, P., and Sampaio, C. (2005b).

Botulinum toxin type A therapy for blepharospasm. Cochrane Database Syst. Rev., CD004900.

Costa, J., Espirito-Santo, C., Borges, A., Ferreira, J.J., Coelho, M., Moore, P., and Sampaio, C. (2005c). Botulinum toxin type A therapy for cervical dystonia. Cochrane Database Syst. Rev., CD003633.

Costa, J., Espirito-Santo, C., Borges, A., Ferreira, J.J., Coelho, M., Moore, P., and Sampaio, C. (2005d). Botulinum toxin type A therapy for hemifacial spasm. Cochrane Database Syst. Rev., CD004899.

Das, T.K., and Park, D.M. (1989a). Botulinum toxin in treating spasticity. Br. J. Clin. Pract. *43*, 401–403.

Das, T.K., and Park, D.M. (1989b). Effect of treatment with botulinum toxin on spasticity. Postgrad. Med. J. *65*, 208–210.

Davis, C.W., and Dickey, B.F. (2008). Regulated airway goblet cell mucin secretion. Annu. Rev. Physiol. *70*, 487–512.

Davletov, B., Bajohrs, M., and Binz, T. (2005). Beyond BOTOX: advantages and limitations of individual botulinum neurotoxins. Trends Neurosci. *28*, 446–452.

Dembek, Z.F., Smith, L.A., and Rusnak, J.M. (2007). Botulism: cause, effects, diagnosis, clinical and laboratory identification, and treatment modalities. Disaster Med. Public Health Prep. *1*, 122–134.

Denglehem, C., Maes, J.M., Raoul, G., and Ferri, J. (2012). Botulinum toxin A: analgesic treatment for temporomandibular joint disorders. Rev. Stomatol. Chir. Maxillofac. *113*, 27–31.

Diener, H.C., Dodick, D.W., Aurora, S.K., Turkel, C.C., DeGryse, R.E., Lipton, R.B., Silberstein, S.D., and Brin, M.F. (2010). OnabotulinumtoxinA for treatment of chronic migraine: results from the double-blind, randomized, placebo-controlled phase of the PREEMPT 2 trial. Cephalalgia *30*, 804–814.

Dodick, D., Blumenfeld, A., and Silberstein, S.D. (2004). Botulinum neurotoxin for the treatment of migraine and other primary headache disorders. Clin. Dermatol. *22*, 76–81.

Dodick, D.W. (2003). Botulinum neurotoxin for the treatment of migraine and other primary headache disorders: from bench to bedside. Headache *43* Suppl 1, S25–33.

Dolimbek, B.Z., Jankovic, J., and Atassi, M.Z. (2002). Cross reaction of tetanus and botulinum neurotoxins A and B and the boosting effect of botulinum neurotoxins A and B on a primary anti-tetanus antibody response. Immunol. Invest. *31*, 247–262.

Dolimbek, B.Z., Aoki, K.R., Steward, L.E., Jankovic, J., and Atassi, M.Z. (2007). Mapping of the regions on the heavy chain of botulinum neurotoxin A (BoNT/A) recognized by antibodies of cervical dystonia patients with immunoresistance to BoNT/A. Mol. Immunol. *44*, 1029–1041.

Dolimbek, B.Z., Aoki, K.R., and Atassi, M.Z. (2011). Reduction of antibody response against botulinum neurotoxin A by synthetic monomethoxypolyethylene glycol–peptide conjugates. Immunol. Lett. *137*, 46–52.

Dolly, J.O., Lawrence, G.W., Meng, J., Wang, J., and Ovsepian, S.V. (2009). Neuro-exocytosis: botulinum toxins as inhibitory probes and versatile therapeutics. Curr. Opin. Pharmacol. *9*, 326–335.

Dolly, J.O., and O'Connell, M.A. (2012). Neurotherapeutics to inhibit exocytosis from sensory neurons for the control of chronic pain. Curr. Opin. Pharmacol. *12*, 100–108.

Dong, M., Liu, H., Tepp, W.H., Johnson, E.A., Janz, R., and Chapman, E.R. (2008). Glycosylated SV2A and SV2B mediate the entry of botulinum neurotoxin E into neurons. Mol. Biol. Cell *19*, 5226–5237.

Dong, M., Richards, D.A., Goodnough, M.C., Tepp, W.H., Johnson, E.A., and Chapman, E.R. (2003). Synaptotagmins I and II mediate entry of botulinum neurotoxin B into cells. J. Cell Biol. *162*, 1293–1303.

Dong, M., Yeh, F., Tepp, W.H., Dean, C., Johnson, E.A., Janz, R., and Chapman, E.R. (2006). SV2 is the protein receptor for botulinum neurotoxin A. Science *312*, 592–596.

Dong, M., Tepp, W.H., Liu, H., Johnson, E.A., and Chapman, E.R. (2007). Mechanism of botulinum neurotoxin B and G entry into hippocampal neurons. J. Cell Biol. *179*, 1511–1522.

Dressler, D., Bigalke, H., and Benecke, R. (2003). Botulinum toxin type B in antibody-induced botulinum toxin type A therapy failure. J. Neurol. *250*, 967–969.

Dressler, D., Wohlfahrt, K., Meyer-Rogge, E., Wiest, L., and Bigalke, H. (2010). Antibody-induced failure of botulinum toxin a therapy in cosmetic indications. Dermatol. Surg. *36* Suppl 4, 2182–2187.

Drummond, G.T., and Hinz, B.J. (2001). Botulinum toxin for blepharospasm and hemifacial spasm: stability of duration of effect and dosage over time. Can. J. Ophthalmol. *36*, 398–403.

Dubinsky, R.M. (2010). Letter to the editor: Re: Safety of botulinum toxin type A among children with spasticity secondary to cerebral palsy: a systematic review of randomized clinical trials. Clin. Rehabil. *24*, 766.

Duggan, M.J., Quinn, C.P., Chaddock, J.A., Purkiss, J.R., Alexander, F.C., Doward, S., Fooks, S.J., Friis, L.M., Hall, Y.H., Kirby, E.R., *et al.* (2002). Inhibition of release of neurotransmitters from rat dorsal root ganglia by a novel conjugate of a *Clostridium botulinum* toxin A endopeptidase fragment and Erythrina cristagalli lectin. J. Biol. Chem. *277*, 34846–34852.

Dughera, L., Battaglia, E., Maggio, D., Cassolino, P., Mioli, P.R., Morelli, A., Emanuelli, G., and Bassotti, G. (2005). Botulinum toxin treatment of oesophageal achalasia in the old old and oldest old: a 1-year follow-up study. Drugs Aging *22*, 779–783.

Duzynski, W., and Slawek, J. (1998). Treatment of hemifacial spasm with botulinum A toxin. Neurol. Neurochir. Pol. *32 Suppl 1*, 61–69.

Dykstra, D.D., Sidi, A.A., Scott, A.B., Pagel, J.M., and Goldish, G.D. (1988). Effects of botulinum A toxin on detrusor-sphincter dyssynergia in spinal cord injury patients. J. Urol. *139*, 919–922.

Dykstra, D.D., and Sidi, A.A. (1990). Treatment of detrusor-sphincter dyssynergia with botulinum A toxin: a double-blind study. Arch. Phys. Med. Rehabil. *71*, 24–26.

Edwards, L.L., Normand, M.M., and Wszolek, Z.K. (1995). Cervical dystonia: a review the role of botulinum toxin. Nebr. Med. J. *80*, 109–115.

Elston, J.S. (1985). The use of botulinum toxin A in the treatment of strabismus. Trans. Ophthalmol. Soc. UK *104 (Pt 2)*, 208–210.

Elston, J.S. (1986). Botulinum toxin treatment of hemifacial spasm. J. Neurol. Neurosurg. Psychiatry *49*, 827–829.

Evers, S., Vollmer-Haase, J., Schwaag, S., Rahmann, A., Husstedt, I.W., and Frese, A. (2004). Botulinum toxin A in the prophylactic treatment of migraine – a randomized, double-blind, placebo-controlled study. Cephalalgia *24*, 838–843.

Factor, S.A., Molho, E.S., Evans, S., and Feustel, P.J. (2005). Efficacy and safety of repeated doses of botulinum toxin type B in type A resistant and responsive cervical dystonia. Mov. Disord. *20*, 1152–1160.

Fageeh, W.M. (2011). Different treatment modalities for refractory vaginismus in western Saudi Arabia. J. Sex. Med. *8*, 1735–1739.

Fernandez-Salas, E., Steward, L.E., Ho, H., Garay, P.E., Sun, S.W., Gilmore, M.A., Ordas, J.V., Wang, J., Francis, J., and Aoki, K.R. (2004). Plasma membrane localization signals in the light chain of botulinum neurotoxin. Proc. Natl. Acad. Sci. U.S.A. *101*, 3208–3213.

Feve, A. (2003). Spasticity and botulinum toxin in 2003. An update. Neurochirurgie *49*, 265–270.

Figgitt, D.P., and Noble, S. (2002). Botulinum toxin B: a review of its therapeutic potential in the management of cervical dystonia. Drugs *62*, 705–722.

Fischer, A., and Montal, M. (2007a). Crucial role of the disulfide bridge between botulinum neurotoxin light and heavy chains in protease translocation across membranes. J. Biol. Chem. *282*, 29604–29611.

Fischer, A., and Montal, M. (2007b). Single molecule detection of intermediates during botulinum neurotoxin translocation across membranes. Proc. Natl. Acad. Sci. U.S.A. *104*, 10447–10452.

Foran, P., Shone, C.C., and Dolly, J.O. (1994). Differences in the protease activities of tetanus and botulinum B toxins revealed by the cleavage of vesicle-associated membrane protein and various sized fragments. Biochemistry 33, 15365–15374.

Foster, K.A. (2005). A new wrinkle on pain relief: re-engineering clostridial neurotoxins for analgesics. Drug Discov. Today *10*, 563–569.

Foster, K.A. (2009). Engineered toxins: new therapeutics. Toxicon *54*, 587–592.

Foster, K.A., Adams, E.J., Durose, L., Cruttwell, C.J., Marks, E., Shone, C.C., Chaddock, J.A., Cox, C.L., Heaton, C., Sutton, J.M., *et al.* (2006). Re-engineering the target specificity of Clostridial neurotoxins – a route to novel therapeutics. Neurotox. Res. *9*, 101–107.

Freund, B., Schwartz, M., and Symington, J.M. (1999). The use of botulinum toxin for the treatment of temporomandibular disorders: preliminary findings. J. Oral Maxillofac. Surg. *57*, 916–920; discussion 920–911.

Freund, B., Schwartz, M., and Symington, J.M. (2000). Botulinum toxin: new treatment for temporomandibular disorders. J. Oral Maxillofac. Surg. *38*, 466–471.

Freund, B.J., and Schwartz, M. (2002). Relief of tension-type headache symptoms in subjects with temporomandibular disorders treated with botulinum toxin-A. Headache *42*, 1033–1037.

Friedenberg, F., Gollamudi, S., and Parkman, H.P. (2004). The use of botulinum toxin for the treatment of gastrointestinal motility disorders. Dig. Dis. Sci. *49*, 165–175.

Frueh, B.R., Felt, D.P., Wojno, T.H., and Musch, D.C. (1984). Treatment of blepharospasm with botulinum toxin. A preliminary report. Arch. Ophthalmol. *102*, 1464–1468.

Fu, Z., Chen, C., Barbieri, J.T., Kim, J.J., and Baldwin, M.R. (2009). Glycosylated SV2 and gangliosides as dual receptors for botulinum neurotoxin serotype F. Biochemistry *48*, 5631–5641.

Gallien, P., Reymann, J.M., Amarenco, G., Nicolas, B., de Seze, M., and Bellissant, E. (2005). Placebo controlled, randomised, double blind study of the effects of botulinum A toxin on detrusor sphincter dyssynergia in multiple sclerosis patients. J. Neurol. Neurosurg. Psychiatry *76*, 1670–1676.

Game, X., Castel-Lacanal, E., Bentaleb, Y., Thiry-Escudie, I., De Boissezon, X., Malavaud, B., Marque, P., and Rischmann, P. (2008). Botulinum toxin A detrusor injections in patients with neurogenic detrusor overactivity significantly decrease the incidence of symptomatic urinary tract infections. Eur. Urol. *53*, 613–618.

Garcia, A., and Fulton, Jr, J.E.. (1996). Cosmetic denervation of the muscles of facial expression with botulinum toxin. A dose–response study. Dermatol. Surg. *22*, 39–43.

Gassner, H.G., Brissett, A.E., Otley, C.C., Boahene, D.K., Boggust, A.J., Weaver, A.L., and Sherris, D.A. (2006). Botulinum toxin to improve facial wound healing: a prospective, blinded, placebo-controlled study. Mayo. Clin. Proc. *81*, 1023–1028.

Ghazizadeh, S., and Nikzad, M. (2004). Botulinum toxin in the treatment of refractory vaginismus. Obstet. Gynecol. *104*, 922–925.

Giannantoni, A., Costantini, E., Di Stasi, S.M., Tascini, M.C., Bini, V., and Porena, M. (2006). Botulinum A toxin intravesical injections in the treatment of painful bladder syndrome: a pilot study. Eur. Urol. *49*, 704–709.

Giannantoni, A., Porena, M., Costantini, E., Zucchi, A., Mearini, L., and Mearini, E. (2008). Botulinum A toxin intravesical injection in patients with painful bladder syndrome: 1-year followup. J. Urol. *179*, 1031–1034.

Giannantoni, A., Mearini, E., Del Zingaro, M., and Porena, M. (2009). Six-year follow-up of botulinum toxin A intradetrusorial injections in patients with refractory neurogenic detrusor overactivity: clinical and urodynamic results. Eur. Urol. *55*, 705–711.

Giannantoni, A., Mearini, E., Del Zingaro, M., Proietti, S., and Porena, M. (2010). Two-year efficacy and safety of botulinum a toxin intravesical injections in patients affected by refractory painful bladder syndrome. Curr. Drug Deliv. *7*, 1–4.

Gill, H.S., and Kraft, S.P. (2010). Long-term efficacy of botulinum a toxin for blepharospasm and hemifacial spasm. Can. J. Neurol. Sci. *37*, 631–636.

Glogau, R.G. (2002). Review of the use of botulinum toxin for hyperhidrosis and cosmetic purposes. Clin. J. Pain. *18*, S191–197.

Gobel, H., Heinze, A., Heinze-Kuhn, K., and Jost, W.H. (2001). Evidence-based medicine: botulinum toxin A in migraine and tension-type headache. J. Neurol. *248* Suppl *1*, 34–38.

Goldman, A. (2000). Treatment of axillary and palmar hyperhidrosis with botulinum toxin. Aesthetic Plast. Surg. *24*, 280–282.

Goodnough, M.C., Oyler, G., Fishman, P.S., Johnson, E.A., Neale, E.A., Keller, J.E., Tepp, W.H., Clark, M., Hartz, S., and Adler, M. (2002). Development of a delivery vehicle for intracellular transport of botulinum neurotoxin antagonists. FEBS Lett. *513*, 163–168.

Gordon, J.M., and Eaker, E.Y. (1997). Prospective study of esophageal botulinum toxin injection in high-risk achalasia patients. Am. J. Gastroenterol. *92*, 1812–1817.

Goschel, H., Wohlfarth, K., Frevert, J., Dengler, R., and Bigalke, H. (1997). Botulinum A toxin therapy: neutralizing and nonneutralizing antibodies – therapeutic consequences. Exp. Neurol. *147*, 96–102.

Gottsch, H.P., Miller, J.L., Yang, C.C., and Berger, R.E. (2011). A pilot study of botulinum toxin for interstitial cystitis/painful bladder syndrome. Neurourol. Urodyn. *30*, 93–96.

Gouider-Khouja, N., Turki, I., Ben Hamida, M., and Hentati, F. (1999). Hemifacial spasm and its treatment with botulinum toxin. Tunis Med. *77*, 41–44.

Grazko, M.A., Polo, K.B., and Jabbari, B. (1995). Botulinum toxin A for spasticity, muscle spasms, and rigidity. Neurology *45*, 712–717.

Halpern, J.L., and Loftus, A. (1993). Characterization of the receptor-binding domain of tetanus toxin. J. Biol. Chem. *268*, 11188–11192.

Hancox, J., Sharma, S., Mackenzie, K., and Adams, G. (2012). The effect on quality of life of long-term botulinum toxin A injections to maintain ocular alignment in adult patients with strabismus. Br. J. Ophthalmol. *96*, 838–840.

Harris, S.R., and Purdy, A.H. (1987). Drooling and its management in cerebral palsy. Dev. Med. Child Neurol. *29*, 807–811.

Heckmann, M., Breit, S., Ceballos-Baumann, A., Schaller, M., and Plewig, G. (1998). Axillary hyperhidrosis: successful treatment with botulinum toxin A. Hautarzt *49*, 101–103.

Heckmann, M., Ceballos-Baumann, A.O., and Plewig, G. (2001). Botulinum toxin A for axillary hyperhidrosis (excessive sweating). N. Engl. J. Med. *344*, 488–493.

Herschorn, S., Gajewski, J., Ethans, K., Corcos, J., Carlson, K., Bailly, G., Bard, R., Valiquette, L., Baverstock, R., Carr, L., *et al.* (2011). Efficacy of botulinum toxin A injection for neurogenic detrusor overactivity and urinary incontinence: a randomized, double-blind trial. J. Urol. *185*, 2229–2235.

Hickson, G.R., Chamberlain, L.H., Maier, V.H., and Gould, G.W. (2000). Quantification of SNARE protein levels in 3T3-L1 adipocytes: implications for insulin-stimulated glucose transport. Biochem. Biophys. Res. Commun. *270*, 841–845.

Ho, M., Chang, L.H., Pires-Alves, M., Thyagarajan, B., Bloom, J.E., Gu, Z., Aberle, K.K., Teymorian, S.A., Bannai, Y., Johnson, S.C., *et al.* (2011). Recombinant botulinum neurotoxin A heavy chain-based delivery vehicles for neuronal cell targeting. Protein Eng. Des. Sel. *24*, 247–253.

Hummel, M., Geigenberger, G., Brand, J., Ziegler, A.G., and Fuchtenbusch, M. (2008). New therapeutic approaches to diabetic gastroparesis. Med. Klin. (Munich) *103*, 514–518.

Husberg, B., Malmborg, P., and Strigard, K. (2009). Treatment with botulinum toxin in children with chronic anal fissure. Eur. J. Pediatr. Surg. *19*, 290–292.

Hyman, N., Barnes, M., Bhakta, B., Cozens, A., Bakheit, M., Kreczy-Kleedorfer, B., Poewe, W., Wissel, J., Bain, P., Glickman, S., *et al.* (2000). Botulinum toxin (Dysport) treatment of hip adductor spasticity in multiple sclerosis: a prospective, randomised, double blind, placebo controlled, dose ranging study. J. Neurol. Neurosurg. Psychiatry *68*, 707–712.

Hyson, H.C., Johnson, A.M., and Jog, M.S. (2002). Sublingual atropine for sialorrhea secondary to parkinsonism: a pilot study. Mov. Disord. *17*, 1318–1320.

Jabbari, B. (2008). Evidence based medicine in the use of botulinum toxin for back pain. J. Neural Transm. *115*, 637–640.

Jackson, C.E., Gronseth, G., Rosenfeld, J., Barohn, R.J., Dubinsky, R., Simpson, C.B., McVey, A., Kittrell, P.P., King, R., and Herbelin, L. (2009). Randomized double-blind study of botulinum toxin type B for sialorrhea in ALS patients. Muscle Nerve *39*, 137–143.

Jahn, R., and Scheller, R.H. (2006). SNAREs – engines for membrane fusion. Nat. Rev. Mol. Cell Biol. *7*, 631–643.

Jampolsky, A. (1986). Botulinum toxin injections in strabismus. Trans. New Orleans Acad. Ophthalmol. *34*, 526–536.

Jankovic, J. (2004a). Botulinum toxin in clinical practice. J. Neurol. Neurosurg. Psychiatry *75*, 951–957.

Jankovic, J. (2004b). Treatment of cervical dystonia with botulinum toxin. Mov. Disord. *19 Suppl 8*, S109–115.

Jankovic, J. (2006). Botulinum toxin therapy for cervical dystonia. Neurotox. Res. *9*, 145–148.

Jankovic, J., and Orman, J. (1987). Botulinum A toxin for cranial-cervical dystonia: a double-blind, placebo-controlled study. Neurology *37*, 616–623.

Jankovic, J., and Schwartz, K. (1990). Botulinum toxin injections for cervical dystonia. Neurology *40*, 277–280.

Jankovic, J., Vuong, K.D., and Ahsan, J. (2003). Comparison of efficacy and immunogenicity of original versus current botulinum toxin in cervical dystonia. Neurology *60*, 1186–1188.

Jankovic, J., Hunter, C., Dolimbek, B.Z., Dolimbek, G.S., Adler, C.H., Brashear, A., Comella, C.L., Gordon, M., Riley, D.E., Sethi, K., *et al.* (2006). Clinico-immunologic aspects of botulinum toxin type B treatment of cervical dystonia. Neurology *67*, 2233–2235.

Jaspers, G.W., Pijpe, J., and Jansma, J. (2011). The use of botulinum toxin type A in cosmetic facial procedures. Int. J. Oral Maxillofac. Surg. *40*, 127–133.

Jeynes, L.C., and Gauci, C.A. (2008). Evidence for the use of botulinum toxin in the chronic pain setting – a review of the literature. Pain Pract. *8*, 269–276.

Jin, R., Rummel, A., Binz, T., and Brunger, A.T. (2006). Botulinum neurotoxin B recognizes its protein receptor with high affinity and specificity. Nature *444*, 1092–1095.

Jin, R., Sikorra, S., Stegmann, C.M., Pich, A., Binz, T., and Brunger, A.T. (2007). Structural and biochemical studies of botulinum neurotoxin serotype C1 light chain protease: implications for dual substrate specificity. Biochemistry *46*, 10685–10693.

Jost, W.H., and Schimrigk, K. (1995). Botulinum toxin in therapy of anal fissure. Lancet *345*, 188–189.

Karacalar, A., Yilmaz, N., Bilgici, A., Bas, B., and Akan, H. (2005). Botulinum toxin for the treatment of temporomandibular joint disk disfigurement: clinical experience. J. Craniofac. Surg. *16*, 476–481.

Karalewitz, A.P., Kroken, A.R., Fu, Z., Baldwin, M.R., Kim, J.J., and Barbieri, J.T. (2010). Identification of a unique ganglioside binding loop within botulinum neurotoxins C and D-SA. Biochemistry *49*, 8117–8126.

Kitamura, M., Igimi, S., and Furukawa, K. (2005). Different response of the knockout mice lacking b-series gangliosides against botulinum and tetanus toxins. Biochim. Biophys. Acta *1741*, 1–3.

Klein, A.M., Stong, B.C., Wise, J., DelGaudio, J.M., Hapner, E.R., and Johns, M.M., 3rd (2008). Vocal outcome measures after bilateral posterior cricoarytenoid muscle botulinum toxin injections for abductor spasmodic dysphonia. Otolaryngol. Head Neck Surg. *139*, 421–423.

Kowal, L., Wong, E., and Yahalom, C. (2007). Botulinum toxin in the treatment of strabismus. A review of its use and effects. Disabil. Rehabil. *29*, 1823–1831.

Kozaki, S., Kamata, Y., Watarai, S., Nishiki, T., and Mochida, S. (1998). Ganglioside GT1b as a complementary receptor component for *Clostridium botulinum* neurotoxins. Microb. Pathog. 25, 91–99.

Kroken, A.R., Karalewitz, A.P., Fu, Z., Baldwin, M.R., Kim, J.J., and Barbieri, J.T. (2011). Unique ganglioside binding by botulinum neurotoxins C and D-SA. FEBS J. 278, 4486–4496.

Krzizok, T. (2007). Botulinum toxin injections for the treatment of strabismus. Which indications are still useful today? Ophthalmologe *104*, 759–762.

Kuo, H.C. (2003). Botulinum A toxin urethral injection for the treatment of lower urinary tract dysfunction. J. Urol. *170*, 1908–1912.

Kuo, H.C. (2006). Therapeutic effects of suburothelial injection of botulinum a toxin for neurogenic detrusor overactivity due to chronic cerebrovascular accident and spinal cord lesions. Urology *67*, 232–236.

Lacy, B.E., Zayat, E.N., Crowell, M.D., and Schuster, M.M. (2002). Botulinum toxin for the treatment of gastroparesis: a preliminary report. Am. J. Gastroenterol. *97*, 1548–1552.

Lagalla, G., Millevolte, M., Capecci, M., Provinciali, L., and Ceravolo, M.G. (2006). Botulinum toxin type A for drooling in Parkinson's disease: a double-blind, randomized, placebo-controlled study. Mov. Disord. *21*, 704–707.

Lagalla, G., Millevolte, M., Capecci, M., Provinciali, L., and Ceravolo, M.G. (2009). Long-lasting benefits of botulinum toxin type B in Parkinson's disease-related drooling. J. Neurol. *256*, 563–567.

Lang, A.M. (2002). Botulinum toxin therapy for myofascial pain disorders. Curr. Pain Headache Rep. *6*, 355–360.

Larkin, M. (2002). Promotion of cosmetic botulinum toxin A frowned upon. Lancet *360*, 929.

Larsen, K.E., Benn, S.C., Ay, I., Chian, R.J., Celia, S.A., Remington, M.P., Bejarano, M., Liu, M., Ross, J., Carmillo, P., *et al.* (2006). A glial cell line-derived neurotrophic factor (GDNF):tetanus toxin fragment C protein conjugate improves delivery of GDNF to spinal cord motor neurons in mice. Brain Res. *1120*, 1–12.

Lipp, A., Trottenberg, T., Schink, T., Kupsch, A., and Arnold, G. (2003). A randomized trial of botulinum toxin A for treatment of drooling. Neurology *61*, 1279–1281.

Lipton, R.B., Varon, S.F., Grosberg, B., McAllister, P.J., Freitag, F., Aurora, S.K., Dodick, D.W., Silberstein, S.D., Diener, H.C., DeGryse, R.E., *et al.* (2011). OnabotulinumtoxinA improves quality of life and reduces impact of chronic migraine. Neurology *77*, 1465–1472.

Ludlow, C.L., Naunton, R.F., Terada, S., and Anderson, B.J. (1991). Successful treatment of selected cases of abductor spasmodic dysphonia using botulinum toxin injection. Otolaryngol. Head Neck Surg. *104*, 849–855.

Lund, N.S., Maxel, T., and Rungby, J. (2011). Successful treatment of diabetic gustatory hyperhidrosis with topical glycopyrrolate. Ugeskrift Laeger *173*, 2200–2201.

Mahajan, S.T., and Brubaker, L. (2007). Botulinum toxin: from life-threatening disease to novel medical therapy. Am. J. Obstet. Gynecol. *196*, 7–15.

Mahant, N., Clouston, P.D., and Lorentz, I.T. (2000). The current use of botulinum toxin. J. Clin. Neurosci. *7*, 389–394.

Maia, F.M., Kanashiro, A.K., Chien, H.F., Goncalves, L.R., and Barbosa, E.R. (2010). Clinical changes of cervical dystonia pattern in long-term botulinum toxin treated patients. Parkinsonism Relat. Disord. *16*, 8–11.

Maillard, H., Briand, N., Bara, C., and Celerier, P. (2003). Efficacy of botulinum toxin A in the treatment of axillary and palmar hyperhidrosis: 10 cases. Ann. Dermatol. Venereol. *130*, 511–513.

Mancini, F., Zangaglia, R., Cristina, S., Sommaruga, M.G., Martignoni, E., Nappi, G., and Pacchetti, C. (2003). Double-blind, placebo-controlled study to evaluate the efficacy and safety of botulinum toxin type A in the treatment of drooling in parkinsonism. Mov. Disord. *18*, 685–688.

Martin-Martin, B., Nabokina, S.M., Blasi, J., Lazo, P.A., and Mollinedo, F. (2000). Involvement of SNAP-23 and syntaxin 6 in human neutrophil exocytosis. Blood *96*, 2574–2583.

Marxen, P., Fuhrmann, U., and Bigalke, H. (1989). Gangliosides mediate inhibitory effects of tetanus and botulinum A neurotoxins on exocytosis in chromaffin cells. Toxicon *27*, 849–859.

Marxen, P., Erdmann, G., and Bigalke, H. (1991). The translocation of botulinum A neurotoxin by chromaffin cells is promoted in low ionic strength solution and is insensitive to trypsin. Toxicon *29*, 181–189.

Mauriello, Jr, J.A.. (1985). Blepharospasm, Meige syndrome, and hemifacial spasm: treatment with botulinum toxin. Neurology *35*, 1499–1500.

McCrory, P., Turner-Stokes, L., Baguley, I.J., De Graaff, S., Katrak, P., Sandanam, J., Davies, L., Munns, M., and Hughes, A. (2009). Botulinum toxin A for treatment of upper limb spasticity following stroke: a multi-centre randomized placebo-controlled study of the effects on quality of life and other person-centred outcomes. J. Rehabil. Med. *41*, 536–544.

Meng, J., Ovsepian, S.V., Wang, J., Pickering, M., Sasse, A., Aoki, K.R., Lawrence, G.W., and Dolly, J.O. (2009). Activation of TRPV1 mediates calcitonin gene-related peptide release, which excites trigeminal sensory neurons and is attenuated by a retargeted botulinum toxin with anti-nociceptive potential. J. Neurosci. *29*, 4981–4992.

Miller, R.H., Woodson, G.E., and Jankovic, J. (1987). Botulinum toxin injection of the vocal fold for spasmodic dysphonia. A preliminary report. Arch. Otolaryngol. Head Neck Surg. *113*, 603–605.

Minguez, M., Herreros, B., Espi, A., Garcia-Granero, E., Sanchiz, V., Mora, F., Lledo, S., and Benages, A. (2002). Long-term follow-up (42 months) of chronic anal fissure after healing with botulinum toxin. Gastroenterology *123*, 112–117.

Mohammadi, B., Balouch, S.A., Dengler, R., and Kollewe, K. (2010). Long-term treatment of spasticity with botulinum toxin type A: an analysis of 1221 treatments in 137 patients. Neurol. Res. *32*, 309–313.

Montal, M. (2009). Translocation of botulinum neurotoxin light chain protease by the heavy chain protein-conducting channel. Toxicon *54*, 565–569.

Montecucco, C., and Schiavo, G. (1994). Mechanism of action of tetanus and botulinum neurotoxins. Mol. Microbiol. *13*, 1–8.

Moore, A.P., Ade-Hall, R.A., Smith, C.T., Rosenbloom, L., Walsh, H.P., Mohamed, K., and Williamson, P.R. (2008). Two-year placebo-controlled trial of botulinum toxin A for leg spasticity in cerebral palsy. Neurology *71*, 122–128.

Naumann, M., Lowe, N.J., Kumar, C.R., and Hamm, H. (2003). Botulinum toxin type a is a safe and effective treatment for axillary hyperhidrosis over 16 months: a prospective study. Arch. Dermatol. *139*, 731–736.

Nishikawa, A., Uotsu, N., Arimitsu, H., Lee, J.C., Miura, Y., Fujinaga, Y., Nakada, H., Watanabe, T., Ohyama, T., Sakano, Y., *et al.* (2004). The receptor and transporter for internalization of F type C progenitor toxin into HT-29 cells. Biochem. Biophys. Res. Comm. *319*, 327–333.

Nishiki, T., Kamata, Y., Nemoto, Y., Omori, A., Ito, T., Takahashi, M., and Kozaki, S. (1994). Identification of protein receptor for *Clostridium botulinum* type B neurotoxin in rat brain synaptosomes. J. Biol. Chem. *269*, 10498–10503.

Odderson, I.R. (1998). Axillary hyperhidrosis: treatment with botulinum toxin A. Arch. Phys. Med. Rehabil. *79*, 350–352.

Ondo, W.G., Hunter, C., and Moore, W. (2004). A double-blind placebo-controlled trial of botulinum toxin B for sialorrhea in Parkinson's disease. Neurology *62*, 37–40.

Pagan, J.K., Wylie, F.G., Joseph, S., Widberg, C., Bryant, N.J., James, D.E., and Stow, J.L. (2003). The t-SNARE syntaxin 4 is regulated during macrophage activation to function in membrane traffic and cytokine secretion. Curr. Biol. *13*, 156–160.

Park, A.J., and Paraiso, M.F. (2009). Successful use of botulinum toxin type a in the treatment of refractory postoperative dyspareunia. Obstet. Gynecol. *114*, 484–487.

Pasricha, P.J., Ravich, W.J., and Kalloo, A.N. (1993). Effects of intrasphincteric botulinum toxin on the lower esophageal sphincter in piglets. Gastroenterology *105*, 1045–1049.

Patel, A.K., Patterson, J.M., and Chapple, C.R. (2006). The emerging role of intravesical botulinum toxin therapy in idiopathic detrusor overactivity. Int. J. Clin. Pract. Suppl., 27–32.

Patti, R., Almasio, P.L., Muggeo, V.M., Buscemi, S., Arcara, M., Matranga, S., and Di Vita, G. (2005). Improvement of wound healing after hemorrhoidectomy: a double-blind, randomized study of botulinum toxin injection. Dis. Colon Rectum. *48*, 2173–2179.

Paumet, F., Le Mao, J., Martin, S., Galli, T., David, B., Blank, U., and Roa, M. (2000). Soluble NSF attachment protein receptors (SNAREs) in RBL-2H3 mast cells: functional role of syntaxin 4 in exocytosis and identification of a vesicle-associated membrane protein 8-containing secretory compartment. J. Immunol. *164*, 5850–5857.

Pellizzari, R., Rossetto, O., Schiavo, G., and Montecucco, C. (1999). Tetanus and botulinum neurotoxins: mechanism of action and therapeutic uses. Philos. Trans. R. Soc. Lond. B Biol. Sci. *354*, 259–268.

Peng, L., Tepp, W.H., Johnson, E.A., and Dong, M. (2011). Botulinum neurotoxin D uses synaptic vesicle protein SV2 and gangliosides as receptors. PLoS Pathog. *7*, e1002008.

Peng, L., Berntsson, R.P., Tepp, W.H., Pitkin, R.M., Johnson, E.A., Stenmark, P., and Dong, M. (2012). Botulinum neurotoxin D-C uses synaptotagmin I/II as receptors and human synaptotagmin II is not an effective receptor for type B, D-C, and G toxins. J. Cell Sci. *25*, 3233–3242.

Petit, H., Wiart, L., Gaujard, E., Le Breton, F., Ferriere, J.M., Lagueny, A., Joseph, P.A., and Barat, M. (1998). Botulinum A toxin treatment for detrusor-sphincter dyssynergia in spinal cord disease. Spinal Cord *36*, 91–94.

Phelan, M.W., Franks, M., Somogyi, G.T., Yokoyama, T., Fraser, M.O., Lavelle, J.P., Yoshimura, N., and Chancellor, M.B. (2001). Botulinum toxin urethral sphincter injection to restore bladder emptying in men and women with voiding dysfunction. J. Urol. *165*, 1107–1110.

Pickett, A., and Perrow, K. (2011). Towards new uses of botulinum toxin as a novel therapeutic tool. Toxins (Basel) *3*, 63–81.

Poewe, W., and Wissel, J. (1993). Use of botulinum toxin in the treatment of cervical dystonia. Baillieres Clin. Neurol. *2*, 179–185.

Popoff, M.R., Marvaud, J.C., and Raffestin, S. (2001). Mechanism of action and therapeutic uses of botulinum and tetanus neurotoxins. Ann. Pharm. Fr. *59*, 176–190.

Poulain, B., and Humeau, Y. (2003). Mode of action of botulinum neurotoxin: pathological, cellular and molecular aspect. Ann. Readapt. Med. Phys. *46*, 265–275.

Reales, E., Mora-Lopez, F., Rivas, V., Garcia-Poley, A., Brieva, J.A., and Campos-Caro, A. (2005). Identification of soluble N-ethylmaleimide-sensitive factor attachment protein receptor exocytotic machinery in human plasma cells: SNAP-23 is essential for antibody secretion. J. Immunol. *175*, 6686–6693.

Reitz, A., Stohrer, M., Kramer, G., Del Popolo, G., Chartier-Kastler, E., Pannek, J., Burgdorfer, H., Gocking, K., Madersbacher, H., Schumacher, S., *et al.* (2004). European experience of 200 cases treated with botulinum-A toxin injections into the detrusor muscle for urinary incontinence due to neurogenic detrusor overactivity. Eur. Urol. *45*, 510–515.

Relja, M., and Telarovic, S. (2004). Botulinum toxin in tension-type headache. J. Neurol. *251 Suppl 1*, I12–14.

Relja, M., Poole, A.C., Schoenen, J., Pascual, J., Lei, X., and Thompson, C. (2007). A multicentre, double-blind, randomized, placebo-controlled, parallel group study of multiple treatments of botulinum toxin type A (BoNTA) for the prophylaxis of episodic migraine headaches. Cephalalgia *27*, 492–503.

Restivo, D.A., Marchese-Ragona, R., Lauria, G., Squatrito, S., Gullo, D., and Vigneri, R. (2006). Botulinum toxin treatment for oropharyngeal dysphagia associated with diabetic neuropathy. Diabetes Care *29*, 2650–2653.

Robertson, C.E., and Garza, I. (2012). Critical analysis of the use of onabotulinumtoxinA (botulinum toxin type A) in migraine. Neuropsychiatr. Dis. Treat. *8*, 35–48.

Rogers, D.F. (2007). Physiology of airway mucus secretion and pathophysiology of hypersecretion. Respir. Care *52*, 1134–1146; discussion 1146–1139.

Rosales, R.L., and Chua-Yap, A.S. (2008). Evidence-based systematic review on the efficacy and safety of botulinum toxin-A therapy in post-stroke spasticity. J. Neural Transm. *115*, 617–623.

Rowe, F.J., and Noonan, C.P. (2009). Botulinum toxin for the treatment of strabismus. Cochrane Database Syst. Rev., CD006499.

Rowe, F.J., and Noonan, C.P. (2012). Botulinum toxin for the treatment of strabismus. Cochrane Database Syst. Rev. *2*, CD006499.

Rummel, A., Bade, S., Alves, J., Bigalke, H., and Binz, T. (2003). Two carbohydrate binding sites in the H(CC)-domain of tetanus neurotoxin are required for toxicity. J. Mol. Biol. *326*, 835–847.

Rummel, A., Mahrhold, S., Bigalke, H., and Binz, T. (2004). The HCC-domain of botulinum neurotoxins A and B exhibits a singular ganglioside binding site displaying serotype specific carbohydrate interaction. Mol. Microbiol. *51*, 631–643.

Rummel, A., Eichner, T., Weil, T., Karnath, T., Gutcaits, A., Mahrhold, S., Sandhoff, K., Proia, R.L., Acharya, K.R., Bigalke, H., *et al.* (2007). Identification of the protein receptor binding site of botulinum neurotoxins B and G proves the double-receptor concept. Proc. Nat. Acad. Sci. U.S.A. *104*, 359–364.

Sadoul, K., Berger, A., Niemann, H., Weller, U., Roche, P.A., Klip, A., Trimble, W.S., Regazzi, R., Catsicas, S., and Halban, P.A. (1997). SNAP-23 is not cleaved by botulinum neurotoxin E and can replace SNAP-25 in the process of insulin secretion. J. Biol. Chem. *272*, 33023–33027.

Sahai, A., Khan, M.S., and Dasgupta, P. (2007). Efficacy of botulinum toxin-A for treating idiopathic detrusor overactivity: results from a single center, randomized, double-blind, placebo controlled trial. J. Urol. *177*, 2231–2236.

Sahai, A., Dowson, C., Khan, M.S., and Dasgupta, P. (2009). Improvement in quality of life after botulinum toxin-A injections for idiopathic detrusor overactivity: results from a randomized double-blind placebo-controlled trial. BJU Int. *103*, 1509–1515.

Sahinkanat, T., Ozkan, K.U., Ciralik, H., Ozturk, S., and Resim, S. (2009). Botulinum toxin-A to improve urethral wound healing: an experimental study in a rat model. Urology 73, 405–409.

Saper, J.R., Mathew, N.T., Loder, E.W., DeGryse, R., and VanDenburgh, A.M. (2007). A double-blind, randomized, placebo-controlled comparison of botulinum toxin type a injection sites and doses in the prevention of episodic migraine. Pain Med. *8*, 478–485.

Sastre-Garriga, J., Tintore, M., Montabalan, X., Bago, J., and Ferrer, I. (2001). Response to botulinum toxin in a case of rigid spine syndrome. J. Neurol. Neurosurg. Psychiatry *71*, 564–565.

Saxena, S.K., George, C.M., Pinskiy, V., and McConnell, B. (2006). Epithelial sodium channel is regulated by SNAP-23/syntaxin 1A interplay. Biochem. Biophys. Res. Commun. *343*, 1279–1285.

Schiavo, G., Rossetto, O., Benfenati, F., Poulain, B., and Montecucco, C. (1994). Tetanus and botulinum neurotoxins are zinc proteases specific for components of the neuroexocytosis apparatus. Ann. N. Y. Acad. Sci. *710*, 65–75.

Schwartz, M., and Freund, B. (2002). Treatment of temporomandibular disorders with botulinum toxin. Clin. J. Pain *18*, S198–203.

Scott, A.B. (1980). Botulinum toxin injection into extraocular muscles as an alternative to strabismus surgery. Ophthalmology *87*, 1044–1049.

Scott, A.B., Kennedy, R.A., and Stubbs, H.A. (1985). Botulinum A toxin injection as a treatment for blepharospasm. Arch. Ophthalmol. *103*, 347–350.

Shorr, N., Seiff, S.R., and Kopelman, J. (1985). The use of botulinum toxin in blepharospasm. Am. J. Ophthalmol. *99*, 542–546.

Sikorra, S., Henke, T., Swaminathan, S., Galli, T., and Binz, T. (2006). Identification of the amino acid residues rendering TI-VAMP insensitive toward botulinum neurotoxin B. J. Mol. Biol. *357*, 574–582.

Silberstein, S.D. (2002). Clinical results of botulinum toxin type a treatment of migraine headache. Aesthet. Surg. J. *22*, 91–93.

Silberstein, S.D. (2008). Treatment recommendations for migraine. Nat. Clin. Pract. Neurol. *4*, 482–489.

Silvaggi, N.R., Boldt, G.E., Hixon, M.S., Kennedy, J.P., Tzipori, S., Janda, K.D., and Allen, K.N. (2007). Structures of *Clostridium botulinum* Neurotoxin Serotype A Light Chain complexed with small-molecule inhibitors highlight active-site flexibility. Chem. Biol. *14*, 533–542.

Smith, C.P., Nishiguchi, J., O'Leary, M., Yoshimura, N., and Chancellor, M.B. (2005). Single-institution experience in 110 patients with botulinum toxin A injection into bladder or urethra. Urology *65*, 37–41.

Solomon, S. (2007). New appendix criteria open for a broader concept of chronic migraine. Cephalalgia *27*, 469; author reply 469–470.

Sood, A., Singh, G., and Kharay, A.S. (1996). Role of botulinum toxin in treatment of achalasia cardia. Indian J. Gastroenterol. *15*, 97–98.

Stankovic, N., and Mirkovic, D. (2004). [Botulinum toxin A in the treatment of anal fissure]. Vojnosanit Pregl *61*, 531–535.

Stenmark, P., Dupuy, J., Imamura, A., Kiso, M., and Stevens, R.C. (2008). Crystal structure of botulinum neurotoxin type A in complex with the cell surface co-receptor GT1b-insight into the toxin–neuron interaction. PLoS Pathog. *4*, e1000129.

Stong, B.C., DelGaudio, J.M., Hapner, E.R., and Johns, M.M., 3rd (2005). Safety of simultaneous bilateral botulinum toxin injections for abductor spasmodic dysphonia. Arch. Otolaryngol. Head Neck Surg. *131*, 793–795.

Swope, D., and Barbano, R. (2008). Treatment recommendations and practical applications of botulinum toxin treatment of cervical dystonia. Neurol. Clin. *26 Suppl 1*, 54–65.

Tcherniak, A., Kashtan, D.H., and Melzer, E. (2006). Successful treatment of gastroparesis following total esophagectomy using botulinum toxin. Endoscopy *38*, 196.

Thant, Z.S., and Tan, E.K. (2003). Emerging therapeutic applications of botulinum toxin. Med. Sci. Monit. *9*, RA40–48.

Tincello, D.G., Kenyon, S., Abrams, K.R., Mayne, C., Toozs-Hobson, P., Taylor, D., and Slack, M. (2012). Botulinum toxin A versus placebo for refractory detrusor overactivity in women: a randomised blinded placebo-controlled trial of 240 women (the RELAX study). Eur. Urol. *62*, 507–514.

Tolosa, E., Marti, M.J., and Kulisevsky, J. (1988). Botulinum toxin injection therapy for hemifacial spasm. Adv. Neurol. *49*, 479–491.

Truong, D., Duane, D.D., Jankovic, J., Singer, C., Seeberger, L.C., Comella, C.L., Lew, M.F., Rodnitzky, R.L., Danisi, F.O., Sutton, J.P., *et al.* (2005). Efficacy and safety of botulinum type A toxin (Dysport) in cervical dystonia: results of the first US randomized, double-blind, placebo-controlled study. Mov. Disord. *20*, 783–791.

Truong, D., Brodsky, M., Lew, M., Brashear, A., Jankovic, J., Molho, E., Orlova, O., and Timerbaeva, S. (2010). Long-term efficacy and safety of botulinum toxin type A (Dysport) in cervical dystonia. Parkinsonism Relat. Disord. *16*, 316–323.

Truong, D.D., Cullis, P.A., O'Brien, C.F., Koller, M., Villegas, T.P., and Wallace, J.D. (1997). BotB (botulinum toxin type B): evaluation of safety and tolerability in botulinum toxin type A-resistant cervical dystonia patients (preliminary study). Mov. Disord. *12*, 772–775.

Tsai, Y.C., Maditz, R., Kuo, C.L., Fishman, P.S., Shoemaker, C.B., Oyler, G.A., and Weissman, A.M. (2010). Targeting botulinum neurotoxin persistence by the ubiquitin-proteasome system. Proc. Natl. Acad. Sci. U.S.A. *107*, 16554–16559.

Tsui, J.K., and Calne, D.B. (1988). Botulinum toxin in cervical dystonia. Adv. Neurol. *49*, 473–478.

Tsukamoto, K., Kohda, T., Mukamoto, M., Takeuchi, K., Ihara, H., Saito, M., and Kozaki, S. (2005). Binding of *Clostridium botulinum* type C and D neurotoxins to ganglioside and phospholipid. Novel insights into the receptor for clostridial neurotoxins. J. Biol. Chem. *280*, 35164–35171.

Vadoud-Seyedi, J., and Simonart, T. (2007). Treatment of axillary hyperhidrosis with botulinum toxin type A reconstituted in lidocaine or in normal saline: a randomized, side-by-side, double-blind study. Br. J. Dermatol. *156*, 986–989.

Vaezi, M.F., Richter, J.E., Wilcox, C.M., Schroeder, P.L., Birgisson, S., Slaughter, R.L., Koehler, R.E., and Baker, M.E. (1999). Botulinum toxin versus pneumatic dilatation in the treatment of achalasia: a randomised trial. Gut *44*, 231–239.

Vaidyanathan, V.V., Yoshino, K., Jahnz, M., Dorries, C., Bade, S., Nauenburg, S., Niemann, H., and Binz, T. (1999). Proteolysis of SNAP-25 isoforms by botulinum neurotoxin types A, C, and E: domains and amino acid residues controlling the formation of enzyme–substrate complexes and cleavage. J. Neurochem. 72, 327–337.

Vo, A.H., Satori, R., Jabbari, B., Green, J., Killgore, W.D., Labutta, R., and Campbell, W.W. (2007). Botulinum toxin type-a in the prevention of migraine: a double-blind controlled trial. Aviat. Space Environ. Med. *78*, B113–118.

von Lindern, J.J., Niederhagen, B., Berge, S., and Appel, T. (2003). Type A botulinum toxin in the treatment of chronic facial pain associated with masticatory hyperactivity. J. Oral Maxillofac. Surg. *61*, 774–778.

Wang, J., Zurawski, T.H., Meng, J., Lawrence, G., Olango, W.M., Finn, D.P., Wheeler, L., and Dolly, J.O. (2011). A dileucine in the protease of botulinum toxin A underlies its long-lived neuroparalysis: transfer of longevity to a novel potential therapeutic. J. Biol. Chem. *286*, 6375–6385.

Wang, L., Tai, N.Z., and Fan, Z.H. (2009). Effect of botulinum toxin type A on the expression of substance P, calcitonin gene-related peptide, transforming growth factor beta-1 and alpha smooth muscle actin A in wound healing in rats. Zhonghua Zheng Xing Wai Ke Za Zhi *25*, 50–53.

Watts, C., Nye, C., and Whurr, R. (2006). Botulinum toxin for treating spasmodic dysphonia (laryngeal dystonia): a systematic Cochrane review. Clin. Rehabil. *20*, 112–122.

Watts, C.C., Whurr, R., and Nye, C. (2004). Botulinum toxin injections for the treatment of spasmodic dysphonia. Cochrane Database Syst. Rev., CD004327.

Watts, C.R., Truong, D.D., and Nye, C. (2008). Evidence for the effectiveness of botulinum toxin for spasmodic dysphonia from high-quality research designs. J. Neural Transm. *115*, 625–630.

Wehrmann, T., Kokabpick, H., Jacobi, V., Seifert, H., Lembcke, B., and Caspary, W.F. (1999). Long-term results of endoscopic injection of botulinum toxin in elderly achalasic patients with tortuous megaesophagus or epiphrenic diverticulum. Endoscopy *31*, 352–358.

Whatling, P.J., and Collin, J. (2001). Botulinum toxin injection is an effective treatment for axillary hyperhidrosis. Br. J. Surg. *88*, 814–815.

Wilson, A.M. (2006). Use of botulinum toxin type A to prevent widening of facial scars. Plast. Reconstr. Surg. *117*, 1758–1766; discussion 1767–1758.

Wissel, J., and Entner, T. (2001). Botulinum toxin treatment of hip adductor spasticity in multiple sclerosis. Wien. Klin. Wochenschr. *113 Suppl 4*, 20–24.

Wollina, U., and Konrad, H. (2005). Managing adverse events associated with botulinum toxin type A: a focus on cosmetic procedures. Am. J. Clin. Dermatol. *6*, 141–150.

Woodson, G., Hochstetler, H., and Murry, T. (2006). Botulinum toxin therapy for abductor spasmodic dysphonia. J. Voice *20*, 137–143.

Wu, K.P., Ke, J.Y., Chen, C.Y., Chen, C.L., Chou, M.Y., and Pei, Y.C. (2011). Botulinum toxin type A on oral health in treating sialorrhea in children with cerebral palsy: a randomized, double-blind, placebo-controlled study. J. Child. Neurol. *26*, 838–843.

Xiao, Z., and Qu, G. (2012). Effects of botulinum toxin type a on collagen deposition in hypertrophic scars. Molecules *17*, 2169–2177.

Yelnik, A.P., Colle, F.M., Bonan, I.V., and Vicaut, E. (2007). Treatment of shoulder pain in spastic hemiplegia by reducing spasticity of the subscapular muscle: a randomised, double blind, placebo controlled study of botulinum toxin A. J. Neurol. Neurosurg. Psychiatry *78*, 845–848.

Yuan, R.Y., Sheu, J.J., Yu, J.M., Chen, W.T., Tseng, I.J., Chang, H.H., and Hu, C.J. (2009). Botulinum toxin for diabetic neuropathic pain: a randomized double-blind crossover trial. Neurology 72, 1473–1478.

Zaninotto, G., Annese, V., Costantini, M., Del Genio, A., Costantino, M., Epifani, M., Gatto, G., D'Onofrio, V., Benini, L., Contini, S., *et al.* (2004). Randomized controlled trial of botulinum toxin versus laparoscopic heller myotomy for esophageal achalasia. Ann. Surg. *239*, 364–370.

Zhang, P., Ray, R., Singh, B.R., Li, D., Adler, M., and Ray, P. (2009). An efficient drug delivery vehicle for botulism countermeasure. BMC Pharmacol. *9*, 12.

Zhang, T., Adatia, A., Zarin, W., Moitri, M., Vijenthira, A., Chu, R., Thabane, L., and Kean, W. (2011). The efficacy of botulinum toxin type A in managing chronic musculoskeletal pain: a systematic review and meta analysis. Inflammopharmacology *19*, 21–34.

Microbial Toxins as Tools in Cell Biology

8

Julie Claudinon, Gustaf E. Rydell and Winfried Römer

Abstract

Microbial toxins are important virulence factors of many bacteria and still a significant threat to human health. Over the years, many toxins have attracted remarkable attention not only from microbiologists, but also in particular from the field of cell biology, where they have become valuable tools to manipulate and investigate fundamental cellular and physiological processes.

In this review, we highlight the use of microbial toxins by life scientists for permeabilizing cell membranes, targeting cell surface receptors, elucidating intracellular trafficking pathways and signalling mechanisms, and for specifically inactivating DNA and protein functions, amongst others. The use of microbial toxins as important cell biology tools for a multitude of applications benefits from many of the characteristics that they have naturally acquired through interactions with their hosts during co-evolution. Microbial toxins have emerged from being the patient's 'foe' to becoming a highly useful scientist's 'friend'.

Discovery of the first bacterial toxin

More than a century ago, in 1888, Emile Roux and Alexandre Yersin, both scientists at Institut Pasteur (Paris, France), identified the first bacterial protein toxin – diphtheria toxin. They demonstrated the thermolabile and protease-sensitive diphtheria 'poison' in sterile filtrates from cultures of the *Corynebacterium diphtheriae*. For the first time, the mechanism of pathogenicity of a microorganism for humans could be explained in terms of a soluble toxic substance (Roux and Yersin, 1888). Since then, microbial toxins were recognized as highly specific and sophisticated virulence factors for a variety of pathogenic bacteria. Today, hundreds of such bacterial toxins are described. Continuously shaped by co-evolution between host and pathogen microbial toxins are the most powerful human poisons known since they retain high activity at very high dilutions. In recent years and during past wars, bacterial toxins gained sad notoriety as bioterrorism weapons to destroy human lives, partly owing to the high potency of certain toxins and the ease of producing them by relatively unskilled persons in primitively equipped laboratories (Mobley, 1995; Atlas, 1998).

There is conclusive evidence for the pathogenic role of bacterial toxins (e.g. diphtheria toxin, tetanus toxin) and good evidence for the pathological involvement of toxins in bacterial disease (e.g. Shiga toxin, pertussis toxin, anthrax toxin). But why do certain bacteria produce such potent toxins? Bacterial toxins likely have more subtle biological functions in the producers' physiology since many of them have evolved before the appearance of eukaryotic multicellular organisms that are targets of many toxins. The production of a toxin may play a role in adapting a bacterium to a particular niche, but it is not essential to its viability (Mobley, 1995; Nichols *et al.*, 2001).

Our understanding of the mode of action of these toxins has incredibly advanced during the last decades. Many of them share the common feature of being highly specialized enzymes, capable of entering eukaryotic cells and causing the irreversible modification of key components of the host cellular machinery, often resulting in morphological changes, cellular damage or cell death. Some bacterial toxins inhibit protein synthesis, like diphtheria toxin (Dtx), Shiga toxin (Stx) and *Pseudomonas aeruginosa* exotoxin A (PEx). Others, such as cholera toxin (Ctx), the *Escherichia coli* heat-labile enterotoxin (LTx), pertussis toxin (Ptx) and anthrax toxin, affect cell signalling. No absolute correlation exists between the uptake routes and the poisonous actions of the toxins.

Because of their highly specific action in cells, bacterial protein toxins are more and more appreciated and increasingly used as selective and efficient tools for dissecting and modulating the molecular mechanisms of various physiological processes. Furthermore, some are used clinically for treatment of human diseases. As long ago as 1897, Paul Ehrlich proposed that exposure to detoxified proteins that retained antigenicity could result in immunity on subsequent exposure to the active toxins. Moreover, he put forward the idea that retargeting these protein toxins, molecules that he named 'magic bullets', specifically to tumour cells might cure cancer (reprinted in Himmelweit, 1956).

Structure of bacterial toxins

The determination of the structure of toxin components and complete holotoxins from various bacteria by X-ray crystallography over the past two decades can be considered as one of the major breakthroughs for the full understanding of the mode of action of bacterial toxins (Fig. 8.1). Various microbial toxins that act inside host cells represent AB-toxins since they are composed of two components – an A-subunit and a B-subunit (Alouf, 2000a).

The B-subunit, either monomeric or oligomeric, is responsible for binding to host cell surface receptors. It can also play a role in the translocation of the A-subunit to the cytosol. Isolated B-subunits are non-toxic and still bind to target cells; they even block the binding of native holotoxin.

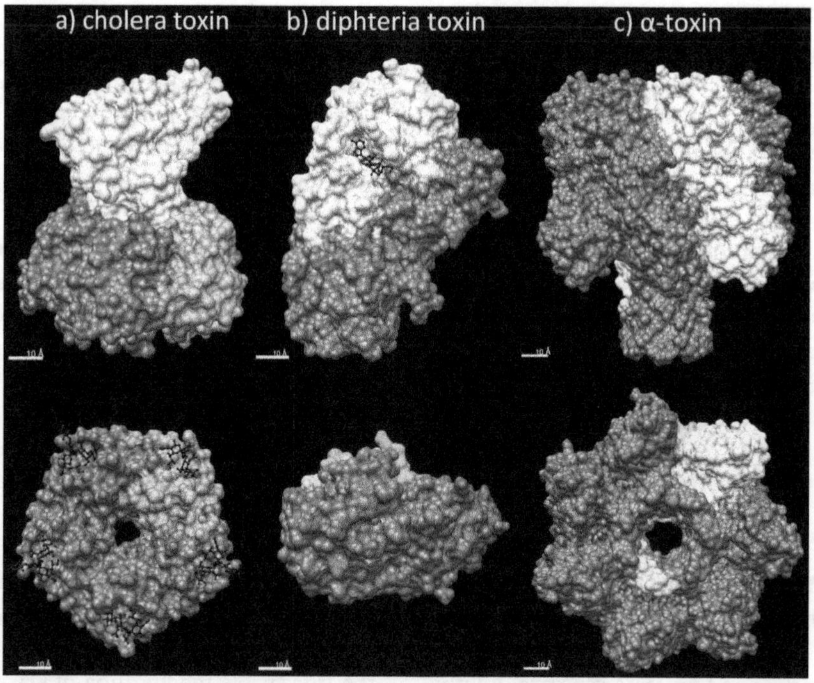

Figure 8.1 Structures of bacterial toxins. Representative x-ray crystallography structures of AB-, AB_5- and pore-forming toxins, shown from the side (top panels) and from below (bottom panels). Cholera toxin is an AB_5-toxin shown with the A-subunit in light grey and the pentameric B-subunit in different nuances of dark grey (a). The lower panel shows the B-subunit with one glycan moiety of the receptor glycosphingolipid, GM1 (balls and sticks) bound to each monomer. Diphtheria toxin is an AB-toxin shown with the A-subunit in light grey, the B-subunit in dark grey and a bound nucleotide in balls and sticks (b). The α-toxin from *Staphylococcus aureus* is a pore-forming toxin shown with the seven monomers coloured in grey scale (c). One monomer is highlighted in light grey. The coordinates were obtained from PDB ID 1XTC (Zhang *et al.*, 1995), 2CHB (Merritt *et al.*, 1997), 1MDT (Bennett and Eisenberg, 1994) and 7AHL (Song *et al.*, 1996). The images were constructed using the chimera package (Pettersen *et al.*, 2004).

The A-subunit has catalytic activity and modifies a specific cellular target upon entry into the cytosol, which affects the physiology of the host cell, or in the worst case, leads to cell death. Intracellular targets, for instance, are actin, ribosomes, small GTP-binding proteins and heterotrimeric G-proteins. In its isolated form, the A-subunit is still enzymatically active, but is not able to bind and enter the target cell.

Bacterial toxins can be synthesized and arranged in a variety of ways:

1 The A- and B-subunits are synthesized and secreted as two separate subunits that meet and interact at the host cell surface (anthrax toxin).
2 The A- and B-subunits are synthesized separately, but associate by non–covalent bonds during secretion (e.g. Ctx, Ptx).
3 A toxin can also be synthesized as a single polypeptide, divided into A- and B-fragments that can be separated by proteolytic cleavage (e.g. Dtx).

Commonly, AB-toxins are synthesized in an inactive form that is activated by proteolytic processing. Some toxins are cleaved by the producing organism (e.g. Ctx) and some by the host cell protease furin (e.g. Stx, Dtx, PEx) (Gordon and Leppla, 1994).

The α-toxin from *Staphylococcus aureus* that is depicted in Fig. 8.1c represents another type of microbial toxin in comparison with AB-toxins. It is one of the prototypes of pore-forming toxins. Membrane spanning pores are formed by oligomerization of monomers (see below).

Microbial toxins as cell biology tools

Owing to length restrictions, we present a rather subjective selection of microbial toxins that are recognized as useful tools for life scientists to explore cellular processes. The following text is divided into sections highlighting the use of microbial toxins for permeabilizing cell membranes, targeting cell surface receptors, investigating endocytosis pathways, characterizing SNARE-mediated membrane fusion and studying cell signalling (Fig. 8.2). Their properties to inactivate DNA, specific proteins, or complete transport or signalling pathways, is discussed throughout the sections whenever appropriate.

Microbial toxins as tools to permeabilize membranes

To gain access to the cytosol while keeping the cell alive is still a major challenge for experimental cell biologists. Pore-forming toxins have been used over the years as tools to permeabilize cell

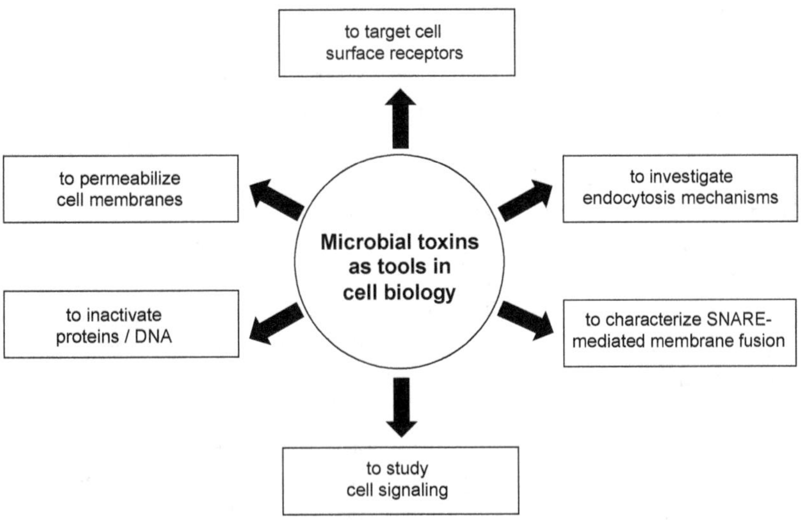

Figure 8.2 Schematic overview of the use of microbial toxins as tools in cell biology.

membranes. As the name suggests, pore-forming toxins insert a membrane-spanning pore into a host cell membrane. Hereby, the selective efflux and influx of ions across the membrane is disrupted. By using this tool, researchers should always keep in mind that pore-forming toxins activate a wide spectrum of cellular events as a result of the permeabilization process e.g. activation of G-proteins or production of cytokines (Dragneva *et al.*, 2001; Kayal and Charbit, 2006).

In the following section, we briefly present pore-forming toxins, their mode of action and highlight their use as research tool for permeabilizing membranes. We refer to Chapter 3 for a much more detailed description.

Pore-forming toxins are secreted as water-soluble toxins and bind to cell surface receptors of target cells. They can be broadly classified into two groups – those, which interact with the cell membrane through α-helices, and those, which utilize mostly β-sheet conformations, such as membrane-spanning β-barrels (Lesieur *et al.*, 1997).

β-Sheet structures

Several microbial toxins interact with cell membranes through β-sheet structures. Examples of this group are α-toxin of *Staphylococcus aureus* and cholesterol-binding toxins from various bacteria (for reviews see Parker, 1997; Heuck *et al.*, 2001).

In general, water-soluble toxin monomers oligomerize into a ring-like amphipathic structure that forms a pore in cell membranes. These pores can be composed of just a few monomers (leading to small pores), or they may be formed from up to 80 monomers (creating large pores that allow the exchange of proteins).

The *S. aureus* α-toxin (see structure in Fig. 8.1c) creates small hydrophilic pores of a size of approximately 1.5 nm in diameter that allow the passage of small ions as well as nucleotides (Bhakdi *et al.*, 1996). It is widely used to permeabilize cell membranes, however, high toxin concentrations in the range of 100 nM are usually necessary (Bhakdi and Tranum-Jensen, 1991).

The α-toxin is synthesized as a 319 aa precursor molecule and gets activated by cleavage of the N-terminal signal sequence of 26 aa. The secreted mature toxin (also called protomer) is a hydrophilic molecule. Seven protomers assemble on the plasma membrane to form a mushroom-shaped pore complex that comprises three domains: The stem domain is a transmembrane channel that is formed by a series of conformational changes of the heptameric pre–pore complex while the cap and rim domains of the heptamer are located at the surface of the plasma membrane (Schmitt *et al.*, 1999).

The cholesterol-binding toxins (also known as cholesterol-dependent cytolysins) are structurally and functionally related and have been identified in five genera of bacterium: *Streptococcus, Listeria, Clostridium, Bacillus* and *Arcanobacterium* (Alouf, 2000b). The presence of cholesterol in the cell membrane is thought to be a prerequisite for binding of these toxins. Streptolysin-O (SLO) from *Streptococcus pyogenes,* a prototype of the cholesterol-binding pore-forming toxins, represents one of the most used research tools for permeabilizing cell membranes (Walev *et al.*, 2001). Proteins of the SLO family oligomerize to form transmembrane pores of around 30 nm in external diameter, containing 20–80 monomers per pore, which appear as spectacular, large, circularized structures by electron microscopy (Sekiya *et al.*, 1993; Bhakdi *et al.*, 1996). Under appropriate experimental conditions pore formation is reversible, i.e. after a short exposure of cells to SLO, SLO-induced membrane lesions can be repaired by incubating cells in a toxin-free buffer solution containing 1–2 mM calcium (Walev *et al.*, 2001).

α-Helical structures

The group of pore-forming α-helical structures comprises amongst others the AB-toxins. They are used as tools for the understanding of the translocation mechanism of bacterial toxins. The mechanism of toxin penetration of cellular membranes by formation of pores is best-studied and understood for diphtheria toxin. The B-subunit of Dtx consists of a C-terminal receptor-binding domain and an N-terminal α-helical translocation (T-) domain, while the A-subunit is the catalytic moiety. A low endosomal pH is needed for toxin translocation from the lumen of endosomes to the cytosol. When Dtx was bound to the plasma membrane and subsequently exposed to acidic pH (thereby mimicking the conditions in the endosome), the translocation from the A-subunit to the cytosol was induced (Sandvig and Olsnes, 1980).

The unfolding of the A- and B-subunits of Dtx at an endosomal pH around five results in the exposure of hydrophobic regions (T-domain), which have an increased tendency to interact with membrane lipids, and thus, to insert into the vesicle membrane (Blewitt *et al.*, 1985; Montecucco *et al.*, 1985). The T-domain may function as a transmembrane chaperone for the unfolded A-subunit in the translocation process (Ren *et al.*, 1999). The channel-forming properties of the T-domain in lipid bilayers have been proven to be useful for elucidating the mechanism of toxin translocation (Oh *et al.*, 1999).

Microbial toxins as tools to target cell surface receptors

Many toxins bind to their cellular receptors with high specificity and may consequently be used to trace the molecules with which they interact. In particular carbohydrate-binding toxins have been used for this purpose. For glycosphingolipids, toxin and pathogen binding is among the best-characterized functions. The glycosphingolipid receptor function was first described not for an endogenous protein, but for cholera toxin (Ctx) (Holmgren *et al.*, 1973). To identify novel receptors for proteins, microbial toxins may be used in inhibition experiments, to block specific cellular glycans.

Ctx binds to the glycan GM1 (Galβ3GalNacβ4(Neu5Acα3)Galβ4Glcβ) (Holmgren *et al.*, 1973). Another well-characterized microbial toxin is Stx which binds to Gb3 (Galα4Galβ4Glcβ) (Lindberg *et al.*, 1987). Stx and Ctx are both very specific for their receptor carbohydrate structures. Being absent from glycoproteins in humans, the expression of both of these carbohydrate structures is assumed to be limited to glycosphingolipids (Yang *et al.*, 1994). Some Stx (Stx2e) bind to Gb4, which has an additional terminal β3-linked N-acetylgalactosamine residue (GalNAcβ3Galα4Galβ4Glcβ) (Johannes and Romer, 2010). Other toxins however have broader binding specificities. The pertussis toxin, for instance, binds to several sialylated and non-sialylated glycans found on both glycoproteins and glycosphingolipids (Millen *et al.*, 2010).

Receptor specificity is important for pathogenicity, since it determines host susceptibility, tissue tropism and the intracellular trafficking of a toxin. CtxB (cholera toxin B-subunit) binds to almost all cell lines because of the ubiquitous expression of the GM1 glycosphingolipid. It is in this context noteworthy that the expression of GM1 is cell-cycle dependent (Majoul *et al.*, 2002). The heat-labile toxin (LTx), which is related to Ctx, also uses the glycolipid GM1 as a receptor, whereas LTx a and LTx b bind to the GD1b (Galβ3GalNacβ4(Neu5Acα8Neu5Acα3) Galβ4Glβ) and GD1a (Neu5Acα3Galβ3 GalNacβ4(Neu5Acα3)Galβ4Glβ) glycosphingolipids, respectively (Beddoe *et al.*, 2010).

Studies of the subtilase cytotoxin (SubAB) have demonstrated how intake of certain food types makes humans susceptible for a microbial toxin (Byres *et al.*, 2008). The SubAB toxin shows an unusual binding specificity for glycans terminated with α3-linked N-glycolylneuraminic acid (Neu5Gc). The sialic acid Neu5Gc is not synthesized by humans, but has been shown to be incorporated into cellular glycoconjugates following dietary uptake. Interestingly, dietary products rich in Neu5Gc are also the most common source of SubAB producing bacteria. In addition, human cell lines have been demonstrated to metabolically incorporate Neu5Gc when grown in medium containing calf serum (Tangvoranuntakul *et al.*, 2003).

Despite a low level of sequence identity, the B-subunits of the Ctx, LTx a, LTx b, Stx, and SubAB toxin families share a common 3D-fold (Beddoe *et al.*, 2010). The monovalent interaction between the toxins and their receptor glycans is weak. However, when binding to several copies of the receptor glycosphingolipid the binding strength of these pentavalent molecules increases. Such an increase in binding avidity caused by multivalency is characteristic for protein–carbohydrate interactions (Mammen *et al.*, 1998). Clusters of glycosphingolipids are found in the cholesterol- and glycosphingolipid-rich microdomains of the plasma membrane often referred to as lipid rafts. These have been subject to intensive research because of their role in fundamental cellular processes such as membrane sorting, signalling and cholesterol homeostasis (Simons and Gerl, 2010). However, the raft concept has been subject to controversies (Munro, 2003). Early evidence of the existence of rafts was based on indirect methods such as detergent extraction, while the direct studies gave conflicting results e.g. concerning the size of the rafts. According to more recent work, lipid rafts (or membrane rafts) can be defined as 'dynamic, nanoscale, sterol–sphingolipid-enriched, ordered assemblies of proteins and lipids, in which the metastable raft resting state can be stimulated to coalesce into larger, more stable raft domains by specific lipid–lipid, protein–lipid and protein–protein oligomerizing interactions' (Simons and Gerl, 2010). Because of the localization of GM1, CtxB has commonly been used as a marker for lipid rafts in biochemical assays, electron as well as fluorescence microscopy (Munro, 2003). However, when using CtxB as a lipid raft marker on living cells before fixation it is important to be aware of the ability of the toxin to cluster GM1 molecules by multivalent binding. Thus, the toxin induces it own domains in analogy with the recent definition of lipid rafts. Furthermore, CtxB should not be considered a marker for all lipid rafts, since all markers are not expected to be enriched in each domain. Instead, only a limited subset of lipids and proteins, not necessarily including GM1, is expected to be enriched in each raft (Simons and Gerl, 2010).

Many AB-toxins use proteins as receptors. For instance, the receptor for PEx is the low-density lipoprotein receptor-related protein (LRP) (Kounnas *et al.*, 1992). LRP is known to also mediate the endocytosis of several endogenous ligands (Herz and Strickland, 2001).

Anthrax toxin uses the proteins TEM8 and CMG2 as receptors (Bradley *et al.*, 2001; Scobie *et al.*, 2003). These proteins are also denoted ANTXR1 and ANTXR2, for anthrax toxin receptor 1 and 2. For this toxin, the A- and B-subunits associate on the cell surface after receptor binding of the B-subunit, the protective antigen (PA) (Abrami *et al.*, 2005; Young and Collier, 2007). Following binding to either one of the receptors, PA is cleaved by cellular proteases into an N-terminal 20 kDa fragment (PA20) and a C-terminal 63 kDa fragment (PA63), which remains bound to the receptor (Inoue *et al.*, 2011). PA63, unlike PA83 can oligomerize to form a heptameric ring structure, referred to as pre-pore, which is capable of binding up to three A-subunits (Young and Collier, 2007). Two types of A-subunits exist, the oedema factor (EF) and the lethal factor (LF). The toxin may either consist of one type of A-subunit or a mixture of both types. After endocytosis, the pre-pore undergoes additional conformational changes in order to convert into a pore.

Microbial toxins as tools to investigate endocytosis mechanisms

Like Trojan horses, toxins have evolved sophisticated mechanisms to exploit cellular transport machineries for their own benefit. Exploring these mechanisms may not only aid the development of antibacterial drugs; it may also help understanding of cellular transport in general.

Regarding the use of toxins for tracing intracellular transport, we will describe how toxins provide us with information about several transport mechanisms on their long route to their targets in the cytosol. Instead of simply penetrating the plasma membrane, these toxins undergo endocytosis followed by transport to a specific cellular location (Fig. 8.3). Once reaching this location, the toxins pass the lipid bilayer in order to enter the cytosol.

Endocytosis is a selective process important not only for the uptake of nutrients, but also for other physiological processes (Conner and Schmid, 2003; Gonnord *et al.*, 2012). These include maintenance of cellular homeostasis, cell polarity and signalling. Thus, while transporting essential factors into the cell, most of the known endocytic processes can also be hi-jacked by bacterial toxins and/or pathogens.

The transport route of many toxins has been studied in great detail. In several cases, toxins have become essential for the characterization of the endocytic mechanisms and transport processes, which they exploit. For instance, studies of microbial toxins provided early suggestions of clathrin-independent endocytosis mechanisms (Montesano *et al.*, 1982). Similarly, Shiga toxin (Stx) was used to demonstrate for the first time

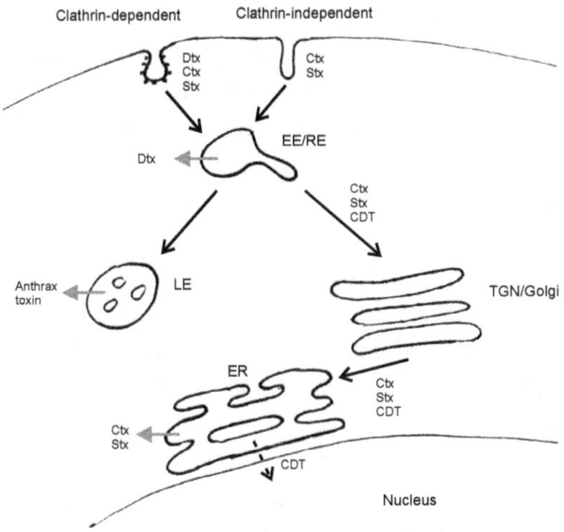

Figure 8.3 Endocytosis and cellular transport of bacterial toxins in cells. The toxins are endocytosed by clathrin-dependent or -independent mechanisms and are transported to the early/recycling endosome. Some toxins enter the cytosol from the early or late endosome. Other toxins are transported by the retrograde route through the TGN/Golgi apparatus to the endoplasmic reticulum before retranslocation to the cytosol. Some toxins are transported to the nucleus. Dtx, diphtheria toxin; Ctx, cholera toxin; Stx, Shiga toxin; CDT, cytolethal distending toxins.

that a molecule can be transported from the cell surface all the way to the endoplasmic reticulum (ER), revealing an unknown cellular pathway (Sandvig *et al.*, 1992). The analysis of the Stx transport pathway has subsequently been of great importance for the characterization of retrograde trafficking (Johannes and Popoff, 2008).

The toxins described in this section belong to the AB-family of toxins (see section 'Structures of bacterial toxins'). The B-subunit of these toxins is responsible for the transport of the toxin, and may therefore be expressed without the A-subunit to be used as a non-toxic cellular tracer. Being bacterial products, it is not surprising that the B-subunits can be recombinantly expressed in large quantities, making them inexpensive cellular probes. In contrast to endogenous proteins, transported in an anterograde manner during their synthesis in the cell, the toxin B-subunits are exclusively transported in a retrograde manner in most cell types. This facilitates studies of the transport pathways of toxins.

When entering the target cell, most bacterial toxins do not cross the lipid bilayer directly from the plasma membrane. Instead, the toxins exploit various endocytic processes for transport to a specific cellular compartment before crossing the membrane to enter the cytosol. A large number of parallel endocytic pathways exists. These are distinguished by the molecular machineries involved (Doherty and McMahon, 2009). Bacterial toxins have been demonstrated to use most of these pathways and have been important for their characterization.

The two major endocytic pathways are distinguishable as clathrin-dependent and clathrin-independent. Out of these, the clathrin-dependent process is the best characterized (McMahon and Boucrot, 2011). This constitutive process begins with the formation of a clathrin coated pit or cavity. Clathrin, as a cytosolic protein, is recruited to the forming pit by the Adaptor protein 2 (AP-2). As an initial step, AP-2, together with other adaptor proteins, recruits the cargo molecules, which will be subject to endocytosis. Clathrin forms a lattice, or a coat, which stabilizes the invagination. The clathrin-coated pit is subsequently scissioned into a clathrin-coated vesicle by a process involving the mechanochemical enzyme dynamin. Relatively few toxins seem to exclusively use clathrin-dependent endocytosis. One example is Dtx. In the absence of functional clathrin-coated pits cells are not sensitive to this toxin (Madshus *et al.*, 1991; Simpson *et al.*, 1998).

Another coat protein found on the plasma membrane is caveolin, which forms invaginations called caveolae. The role of caveolae in endocytic processes has been extensively studied. Yet the contribution of caveolae to endocytosis is not clear, for several reasons (Doherty and McMahon, 2009). Fluorescence recovery after photobleaching (FRAP) studies of GFP-tagged caveolin has demonstrated that caveolae are normally immobile structures on the plasma membrane (Thomsen *et al.*, 2002). Studies using EM techniques distinguishing between surface-connected and non-surface-connected membranes have suggested that many of the structures previously regarded as internalized caveolae are in fact surface-connected and thus not scissioned from the plasma membrane (Parton *et al.*, 2002; Sandvig *et al.*, 2008).

Several endocytic processes appear to be operational in the absence of coat proteins (Sens *et al.*, 2008; Eierhoff *et al.*, 2012). These clathrin- and caveolin-independent processes are frequently exploited by bacterial toxins (Doherty and McMahon, 2009). A mechanism for coat-independent uptake has recently been described using StxB (Römer *et al.*, 2007). Stx can exploit several endocytic pathways in order to enter the cell. In early studies, Stx has been shown to utilize clathrin-dependent endocytosis (Sandvig *et al.*, 1989). However, clathrin is not required for the early steps of toxin cell entry, since Stx is still taken up efficiently when clathrin-dependent endocytosis is blocked (Nichols *et al.*, 2001; Lauvrak *et al.*, 2004; Saint-Pol *et al.*, 2004). The pentameric B-subunit of the toxin can bind up to 15 Gb3 molecules (Johannes and Romer, 2010). The toxin-driven clustering of Gb3 on the plasma membrane results in the formation of lipid-nanodomains, which spontaneously invaginate to form tubular structures (Römer *et al.*, 2007). The process has been reconstituted using giant unilamellar vesicles (GUV) as a model membrane system containing 1,2-dioleoylphosphatidyl choline, cholesterol and Gb3 (Römer *et al.*, 2007). According to this study, no cytosolic coats or other cellular proteins are required for the formation of tubular membrane invaginations. Instead, it could be concluded that the binding of the toxin induces an asymmetric stress on the plasma membrane because of the arrangement of the binding sites on the toxin. In support of this, antibodies against Gb3 failed to induce tubules. Furthermore, tubules were formed on GUVs containing unsaturated Gb3 species, which have voluminous tails, but not on GUVs with less space filling saturated Gb3 species. Line-tension probably also contributes to the process by favouring a reduction of the contact area between the induced domain and the surrounding membrane. In support of this, the tubule formation was inhibited when the GUVs were kept under high membrane tension (Römer *et al.*, 2007).

The cytosolic protein machinery is not required for the formation of tubules, but has proven necessary for their processing into endocytic vesicles. Even though tubular plasma membrane invaginations were observed on cells when dynamin was inactivated by the small molecule dynasore, dynamin was not absolutely required for the scission of StxB-induced tubules *in vitro* (Römer *et al.*, 2010). Experiments on cells and model membranes showed that StxB-induced tubules were stable at 37°C, but underwent scission upon a shift to a lower temperature, suggesting domain formation to be of importance. This was further supported by the observation that temperature-induced scission did not occur when cellular cholesterol levels were reduced. At physiological temperature, actin was demonstrated to polymerize on StxB-induced membrane tubules, inducing membrane reorganization followed by scission. These observations suggest that the StxB tubules are close to a lipid demixing point and that domain formation may induce a line tension-driven tubule constriction, which is independent of dynamin pinchase activity (Römer *et al.*, 2010). Many of the clathrin- and caveolin-independent endocytosis pathways are dependent of cholesterol, but not of dynamin (Doherty and McMahon, 2009). This indicates that the mechanism identified for Stx may be of general interest (Eierhoff *et al.*, 2012). In addition, CtxB has been shown to induce tubules upon binding to receptor glycosphingolipids (Ewers *et al.*, 2010). Another study has demonstrated that CtxB induces an orientationally textured phase on lipid monolayers, suggesting changes in lipid tilt to be important for the invagination process (Watkins *et al.*, 2011).

The uptake of Ctx shares many characteristics with that of Stx. Even though both GM1 and CtxB have been reported to be enriched in caveolae, several studies have shown that CtxB can enter cells in the absence of caveolae (Orlandi and Fishman, 1998; Nichols *et al.*, 2001; Torgersen *et al.*, 2001). The internalization of CtxB by caveolae seems to be a minor mechanism for uptake in certain cell types (Doherty and McMahon, 2009). In cells devoid of caveolae, the uptake of Ctx is dependent on lipid domains, which are sensitive to cholesterol depletion (Wolf *et al.*, 1998), a typical characteristic of clathrin- and caveolae-independent endocytosis mechanisms. However, at least in some cell types CtxB can also be internalized by clathrin-dependent endocytosis (Nichols *et al.*,

2001; Torgersen *et al.*, 2001). An EM characterization of early clathrin-independent CtxB carriers in mouse primary embryonic fibroblasts (MEFs) has shown that these are tubular- and ring-shaped structures similar to the clathrin-independent carriers (CLICs) identified in the GEEC (GPI-AP enriched early endosomal compartment) pathway (Kirkham *et al.*, 2005). The CLIC/GEEC pathway is involved in the uptake of GPI anchored proteins and fluid phase markers (Sabharanjak *et al.*, 2002). These, in turn, were also identified in the CtxB carriers (Kirkham *et al.*, 2005). Some of the CtxB carriers contained caveolin1. However, the CtxB uptake mechanism operated also in caveolin1-null MEFs, and the morphology of the CtxB carriers was ultrastructurally identical in these cells, suggesting that caveolin1 was not important for their function. Whether Stx is also transported by CLICs has to our knowledge not yet been determined.

The uptake mechanism of the anthrax toxin is a special case, since it is dependent on both clathrin and cholesterol (Abrami *et al.*, 2005). The oligomerization of PA63 induces membrane domain formation and the pre-pore has been reported to be associated with specific lipid microdomains (Abrami *et al.*, 2003). Cholesterol depletion was shown to inhibit intracellular accumulation of PA63. However, at the same time the internalization of the PA63/EF/LF complex is clathrin dependent. This was shown by overexpression of a dominant negative mutant of Eps15, a protein of importance for the initiation of clathrin-coated pit formation. When clathrin-dependent endocytosis is inhibited, anthrax toxin may to some extent be internalized by other pathways (Abrami *et al.*, 2003; Boll *et al.*, 2004).

Studying intracellular transport routes

As described in the previous section, several toxins, such as Ctx and Stx, may enter cells by both clathrin-dependent and clathrin-independent endocytosis. After internalization most toxins are found in the early endosome. From there, two main pathways are used to enter the cytosol. The toxins enter the cytosol either directly from the early or late endosomes, or by using the so-called retrograde route from the endosome via the Golgi apparatus to the ER followed by translocation from this destination (Sandvig and van Deurs, 2005; Lamaze and Johannes, 2006). The retrograde route is important also for retrieval and localization of several endogenous proteins, e.g. Golgi-resident proteins such as TGN46, mannose 6-phosphate receptors (MPRs) and furin (Johannes and Popoff, 2008).

The toxins following the retrograde route travel a long intracellular pathway and may consequently be used as tools in studies of several different transport mechanisms. As will be described in this section the transport route of several toxins is very well characterized. Toxins may also be used as tools in co-localization studies, to investigate whether the transport route of a certain molecule overlaps with the path used by a toxin. Furthermore, toxins may be used as tracers for characterization of transport effects resulting from manipulation of cellular factors expected to be involved in transport processes.

Translocation from endosomes

A number of bacterial toxins use the endocytic pathway to reach a low pH environment, which activates the translocation of their catalytic subunits across the membrane into the cytosol. For the diphteria toxin, the low pH triggers a striking conformational change in the toxin structure, exposing hydrophobic areas, which are subsequently inserted into the lipid bilayer to form a pore (see 'Microbial toxins as tools to permeabilize membranes, α-helical structures'). A number of other toxins, including anthrax toxin, *Clostridium difficile* toxin B and *Pasteurella multocida* toxin have been demonstrated to use similar mechanisms for translocation (Sandvig and van Deurs, 2002).

The membrane insertion of the anthrax toxin does not occur randomly into the membrane of the endosomes, but preferentially into that of intraluminal vesicles (Abrami *et al.*, 2004). Consequently, the EF and/or LF subunits are transferred into the lumen of intraluminal vesicles and not into the cytoplasm. The release of the EF and/or LF subunits probably occurs when the intraluminal vesicles undergo back fusion with the limiting membrane of multivesicular endosomes. This event occurs during the late stages of the endocytic pathway, providing an option for delivery of the EF and LF in the perinuclear region of the cell that may be important for the physiological outcome (Abrami *et al.*, 2005). The diphtheria toxin, in contrast, does not show any preference for intraluminal vesicles, but is concentrated on the limiting membrane of the endosome. Thereby it is translocated directly into the cytosol starting from the early endosome.

Retrograde transport to TGN/Golgi apparatus

A large number of bacterial toxins use the retrograde route to be transported to the ER. These include Stx, Ctx and PEx (Lamaze and Johannes, 2006). Other toxins, such as the cytolethal distending toxins, use this pathway for transport all the way to the nucleus (Guerra *et al.*, 2011).

Already in 1992, Stx was reported to be transported to the ER in a retrograde manner through the secretory pathway (Sandvig *et al.*, 1992). Later, the transport routes of the toxin have been studied intensively using not only morphological tools, but also reconstitution approaches, and quantitative methods to measure the arrival of engineered StxB variants carrying sulfation and glycosylation motifs to the Golgi and ER (Mallard and Johannes, 2003). Subsequently, it has become clear that the retrograde trafficking route linking the early endocytic pathway and the TGN is used also by endogenous proteins involved in processes such as signalling, glucose transport and morphogen trafficking (Johannes and Popoff, 2008). The majority of the factors identified as important for the retrograde transport of StxB have later been shown to be involved in the transport of endogenous proteins (Johannes and Popoff, 2008). In many aspects, the StxB pathway differs from the late endosome-to-TGN pathway followed by the mannose 6-phosphate receptor (Johannes and Popoff, 2008).

In cells that are sensitive to Stx intoxication, it has been demonstrated that not much Stx is recycled back to the plasma membrane after endocytosis, and that Stx is absent from the compartments of the late endocytic pathway (Mallard *et al.*, 1998; Schapiro *et al.*, 1998). Instead, the toxin is transported from the early/recycling endosomes to the TGN and the Golgi apparatus. The mechanism by which Stx escapes from the endocytic pathway involves clathrin and retromer.

As mentioned earlier, StxB may be internalized from the plasma membrane in clathrin-coated vesicles. In addition clathrin has been reported to be involved in StxB transport at the level of the early/recycling endosome. The depletion of clathrin activity by various means has been shown to inhibit StxB from exiting early/recycling endosomes (Lauvrak *et al.*, 2004; Saint-Pol *et al.*, 2004). The adaptor protein epsinR was found to be required for the process (Saint-Pol *et al.*, 2004). No effect was observed when interfering with the AP1 function, even though this clathrin adaptor has been associated with retrograde transport in several studies.

Retromer is also a coat protein. It is composed of a curvature recognition subunit and a cargo recognition subunit (Bonifacino and Hurley, 2008). In cells depleted of retromer function, StxB was found in transferrin-free tubular structures connected to transferrin receptor positive endosomes, suggesting retromer to act after clathrin in the Stx pathway (Popoff *et al.*, 2007, 2009). A model has been proposed for the process (Johannes and Popoff, 2008; Johannes and Wunder, 2011). According to this model, clathrin drives curvature changes on membranes in the early endosome. In addition to clathrin adaptors such as epsinR, retromer may be involved in the curvature generation. Cargo subsequently uses clathrin adaptors together with the cargo recognition subunit of retromer to localize to the sites of transport intermediate formation. The model suggests the retrograde tubules to be scissioned into transport carriers in a process that probably involves retromer.

Two SNARE complexes involved in the fusion of the early/recycling endosome-derived StxB transport carriers with the TGN have been characterized. One TGN-localized tSNARE complex is composed of the heavy chain tSNARE syntaxin 16, the light chain tSNAREs Vti1a and syntaxin 6. Two vSNAREs VAMP3/cellubrevin and VAMP4 that form independent complexes with the tSNAREs were identified (Mallard *et al.*, 2002). In another study a SNARE complex composed of syntaxin 5, GS15, GS28, and Ykt6 was found to be involved in early/recycling endosome to TGN transport of StxB (Birkeli *et al.*, 2003). These studies suggest that several molecular machineries may co-exist for the same transport intermediates at the early/recycling endosome–TGN interface or, alternatively, that parallel pathways exist.

The Rab proteins constitute a family of small GTPases that have been shown to regulate a large number of intracellular trafficking steps (Stenmark, 2009). By controlling membrane identity, specific Rab proteins ensure that transport vesicles are delivered to their correct destinations. The trans-Golgi/TGN localized Rab6a' has been shown to be important for StxB transport. Upon expression of dominant negative Rab6a' mutants, StxB accumulates in the early/recycling endosome (Mallard *et al.*, 2002).

Retrograde trafficking of StxB is also dependent on GPP130, a protein of unknown function (Natarajan and Linstedt, 2004; Mukhopadhyay and Linstedt, 2012), the GRIP domain protein Golgin-97 (Lu *et al.*, 2004) and its effector ARL1 (Lu *et al.*, 2004; Nishimoto-Morita *et al.*, 2009), tGolgin-1 (Yoshino *et al.*, 2005), the conserved oligomeric Golgi (COG)-complex (Zolov and Lupashin, 2005), GARP (Golgi-associated retrograde protein) complex (Perez-Victoria *et al.*, 2008), Rab6-binding TATA element modulatory factor (TMF) (Yamane *et al.*, 2007) and dynamin (Lauvrak *et al.*, 2004). So far, the data available make it hard to determine whether the protein toxins that use the retrograde route depend on the same molecular machineries.

TGN/Golgi apparatus to ER transport

After the TGN, protein toxins following the retrograde route are transported through the Golgi apparatus to the ER and the retro-translocation machinery. The necessity of a passage through the Golgi apparatus was suggested by experiments demonstrating that the treatment with brefeldin A protects cells from intoxication with Ctx (Nambiar *et al.*, 1993), Stx (Donta *et al.*, 1995) and PEx (Yoshida *et al.*, 1991). Brefeldin A is a fungal metabolite, which redistributes Golgi markers into the ER and blocks anterograde as well as retrograde transport. A recent study further supports the necessity of Golgi passage for Stx (McKenzie *et al.*, 2009). The study showed that when Stx was prevented from entering the Golgi apparatus either by treatment with aluminium fluoride, a temperature block or by removing the Golgi apparatus with subcellular surgery, the toxin was also inhibited from reaching the ER. The transport of Stx from the TGN to the ER is not so well characterized, but has been demonstrated to be coat protein complex I (COPI)-independent (Girod *et al.*, 1999).

The KDEL receptor-mediated transport in COPI-coated vesicles is a well-characterized pathway between the Golgi apparatus and ER. Soluble ER-resident proteins often contain a KDEL sequence recognized by the KDEL receptor. PEx contains a KDEL-like sequence and uses the KDEL pathway (Jackson *et al.*, 1999). Also the Ctx carries a KDEL-motif, which is found on the A-subunit. However, since the B-subunit of Ctx can travel independently to the ER, the KDEL-motif is not necessary for ER transport (Fujinaga *et al.*, 2003). Stx does not carry a KDEL signal and interference with KDEL-receptor activity did not prevent Stx from reaching the ER (Girod *et al.*, 1999; Jackson *et al.*, 1999). Furthermore, addition of a KDEL sequence to StxB had no effect on the retrograde transport of the toxin to the ER (Johannes *et al.*, 1997). However, the addition of the KDEL sequence induced an increase in the retention of StxB in the ER. This observation suggests that the small effect observed when mutating KDEL motifs on microbial toxins is caused rather by an increase in the escape from the ER than by a decrease in the retrograde transport to the ER (Lencer *et al.*, 1995; Fujinaga *et al.*, 2003). The Golgi to ER transport of Stx has been suggested to be dependent on a number of factors, including Rab6', actin, cdc42, microtubules and myosin motors (Valderrama *et al.*, 2001; Luna *et al.*, 2002; Duran *et al.*, 2003; Del Nery *et al.*, 2006).

The target of the SubAB toxin is the binding immunoglobulin protein (BiP), which is a chaperone with essential functions for protein folding. The localization of this chaperone in the ER lumen suggests that the SubAB does not cross the ER membrane to enter the cytosol (Beddoe *et al.*, 2010). Since SubAB toxin is specific for BiP, it may be used as a tool to deplete this protein in several cell lines.

Translocation from ER

For many of the bacterial toxins exploiting the retrograde route, the ER is the final destination. From this location the subunit with the enzymatic activity crosses the membrane to enter the cytosol. The ER is also the location of the ERAD translocation machinery. This complex is used by the cell to transport misfolded proteins to the cytosol for ubiquitylation and proteasomal degradation.

The translocation process of Ctx has been studied in detail and is dependent on the ERAD translocation machinery (Teter *et al.*, 2002). The CtxB subunit directs the toxin to the ERAD complex, whereafter a reduction reaction releases the A1-subunit from the rest of the toxin. After beeing unfolded by the protein disulfide isomerase (PDI), the A1-subunit passes the membrane through a channel formed either by the Sec61 or the Derlin1–Hrd1 complex (Inoue *et al.*, 2011). In the cytoplasm the A1-subunit is rapidly refolded, causing cellular toxicity. Co-immunoprecipitation experiments have suggested that Stx also uses the ERAD pathway for translocation, by demonstrating that Stx bind to the host chaperones HEDJ, BiP and GRP94 (Yu and Haslam, 2005). Furthermore, this study showed that HEDJ associated Stx interact with Sec61. The amount of lysine residues on the A-subunits of the Ctx, Stx, PEx is very low, something which is presumably important in order for the toxins to avoid ubiquitinylation and degradation (Hazes and Read, 1997).

Transport to the nucleus

Some bacterial toxins do not have their targets in the cytoplasm, but continue their journey to the nucleus. The cytolethal distending toxins (CDTs) have a deoxyribonuclease I-like activity and cleave DNA during replication (Guerra *et al.*, 2011). This DNA damage activates host cellular checkpoint mechanisms, resulting in cell-cycle arrest. The CDTs have been described to take the retrograde route from the plasma membrane, via the Golgi, to the ER. Whether the toxin is transported directly to the nucleus from the ER or via a cytoplasmic intermediate is unclear. However, translocation from the ER does not seem to require the ERAD pathway (Guerra *et al.*, 2005).

Microbial toxins as tools to characterize SNARE-mediated membrane fusion

AB-toxins belonging to the clostridial neurotoxin family have been crucial for the characterization of membrane fusion during neurotransmitter release, which involves SNARE proteins (Schiavo and van der Goot, 2001). The tetanus and botulinum toxins belong to this group and bind with high specificity to the unmyelinated areas of motor nerve terminals (Brunger and Rummel, 2009). Whereas the botulinium toxin remains at the synapse, the tetanus toxin is transported to the cell body and then transcytosed to inhibitory interneurons.

The catalytic A-subunit of clostridial neurotoxins is amongst the most selective proteases known and cleaves specific synaptic members of the SNARE family. The tetanus toxin and botulinium toxin B, D, F and G cleave VAMP2, whereas botulinium toxin A, C and E cleave SNAP-25. Botulinium toxin C has an additional target: syntaxin 1 (Proux-Gillardeaux and Galli, 2008). These toxins can however not only be used for studies of synaptic SNARE proteins. The tetanus toxin has activity also towards VAMP3, which has a high sequence similarity to VAMP2. The non-neuronal cells that normally express VAMP3 are not susceptible to the tetanus toxin. However, the A-subunit may be artificially introduced into such cells using pore-forming toxins or transfection and then be used as a tool to inhibit VAMP3 activity (Proux-Gillardeaux *et al.*, 2005).

Microbial toxins as tools to study cell signalling

Many bacterial toxins possess a highly specific enzymatic activity interfering with signal transduction processes (Fig. 8.4). Thus, they can be used as efficient and specific tools for studying molecular mechanisms controlling signal transduction (Schiavo and van der Goot, 2001).

Here, we will review how some bacterial toxins, by interfering with cell signalling, have already allowed to define the molecular mechanisms of some signal transduction pathways, and how others could be used to modulate and study

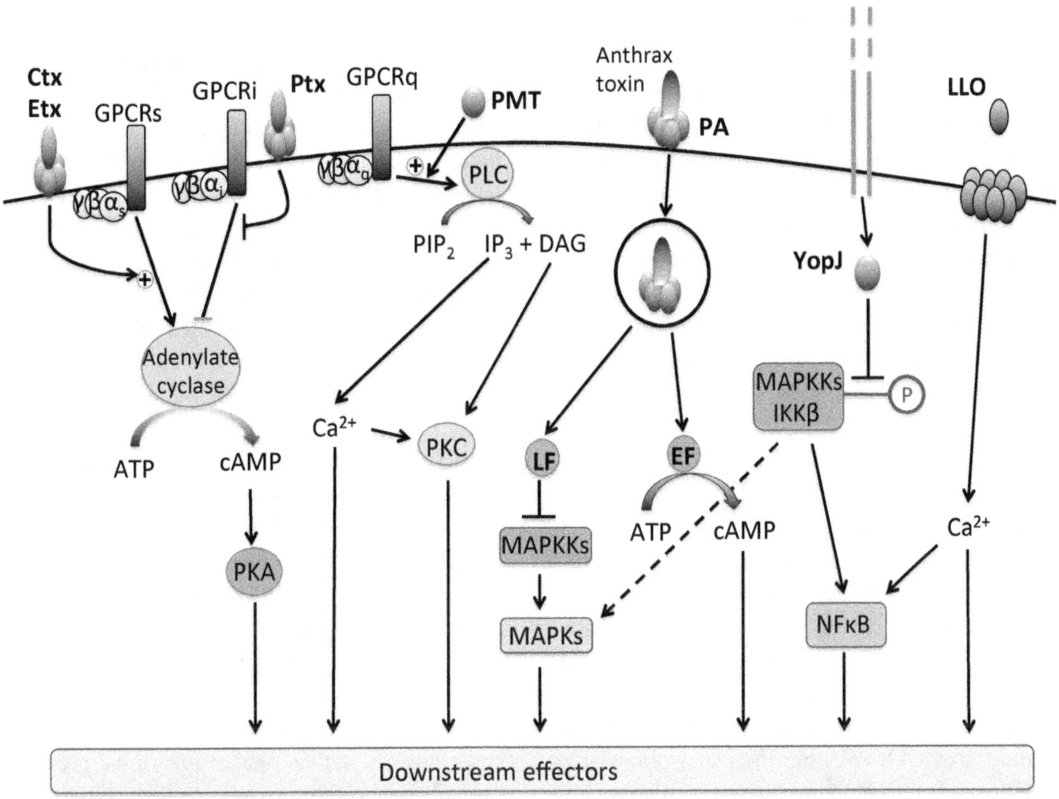

Figure 8.4 Bacterial toxins interfering with cell signalling. Cholera toxin (Ctx) and *Escherichia coli* enterotoxin (Etx) activate Gs proteins, thereby activating adenylate cyclase that will increase cyclic AMP (cAMP) synthesis. cAMP then stimulates protein kinase A (PKA), which will activate a myriad of downstream effectors. Pertussis toxin (Ptx) inhibits Gi proteins, thereby indirectly activating also adenylate cyclase. *Pasteurella multocida* toxin (PMT) activates Gq proteins, and thereby activates phospholipase C (PLC), which results in inositol tri-phosphate (IP_3) and diacylglycerol (DAG) synthesis, increase of intracellular calcium (Ca^{2+}) concentration and protein kinase C (PKC) activation. Anthrax toxin is composed of protective antigen (PA), lethal factor (LF) and oedema factor (EF). PA forms a membrane channel that mediates entry of EF and LF into the cell. LF will cleave mitogen-activated protein kinase kinases (MAPKKs) and subsequently inhibit MAPK pathways, while EF is itself an adenylate cyclase that will increase cellular cAMP levels. The Yops effectors proteins from *Yersinia pestis* are injected into the cell through type III secretion system. YopJ prevents MAPKKs and IKKβ phosphorylation by acetylating critical Ser and Thr residues, which results in inhibition of the following MAPK and NFκB pathways. Listeriolysin O (LLO) forms Ca^{2+}-permeable pores, inducing oscillations in intracellular Ca^{2+} concentration, which results in a multitude of downstream effects on cellular signalling.

specific signalling cascades. While pertussis, cholera and *Pasteurella* toxins have been extensively used to study G proteins and adenylate cyclase-mediated signalling pathways (Hepler and Gilman, 1992; Wilson and Ho, 2004), the Lethal Factor of Anthrax toxin can be used as a powerful tool to inhibit MAPK pathways (Turk, 2007). Some other interesting toxins, such as adenylate cyclase toxins, YopJ from *Yersinia pestis* or listeriolysin from *Listeria monocytogenes* may also provide interesting tools in cell signalling biology.

Studying G proteins signalling

G proteins constitute a large family of GTPases that plays an essential role in transducing external signals by linking many cell surface receptors to effector proteins. They are heterotrimers, composed of three distinct subunits: α, β and γ. While β and γ are tightly associated and function as a unit, α has a high-affinity binding site for guanine nucleotides and possess intrinsic GTPase activity. When being bound to GDP, the α subunit binds to βγ and is inactive, whereas the GTP-bound form of α dissociates from βγ and regulates effector proteins. G proteins are classified into four main subfamilies: G_s, G_i, G_{12-13} and G_q (Gilman, 1987).

G proteins can be modified by bacterial toxins, which results in their activation or inhibition. While cholera toxin and *E. coli* enterotoxin ADP-ribosylate α-subunits of G_s proteins, constitutively activating these proteins by blocking their GTPase activity, pertussis toxin ADP-ribosylates α-subunits of other G proteins, including G_i and G_o, uncoupling them from their receptors and preventing their activation (Casey and Gilman, 1988).

Effects on cAMP pathway

G-protein-coupled receptors (GPCRs), such as hormone or neurotransmitter receptors interact with members of the G_s or G_i family (among others) to respectively stimulate or inhibit adenylate cyclase. Activation of this enzyme induces the synthesis of cyclic AMP (cAMP), which then acts as a second messenger to activate protein kinase A (PKA). PKA can then phosphorylate a plethora of downstream targets (Walsh and Van Patten, 1994).

Stimulation of many hormone receptors results in inhibition of adenylate cyclase activity. The fact that this response is blocked by treatment with Ptx showed that G_i proteins were involved in this inhibitory pathway. Therefore, as Ptx prevents the inhibition of this signalling cascade by inactivating G_i proteins, treatment with this toxin induces its indirect activation (Fig. 8.4).

Treatment with Ctx leads to the constitutive activation of adenylate cyclase, and to the following rapid elevation of cAMP levels in the cell. As Ctx was shown to have an activatory effect on G_s protein, this helped understanding that the GPCRs activating adenylate cyclase were coupled to G_s proteins. Thus, Ctx induces and enhances the adenylate cyclase pathway (Fig. 8.4) (Hepler and Gilman, 1992).

Effects on PLC pathway

Receptors for many hormones, neurotransmitters and growth factors are coupled to G_q proteins. Stimulation of G_q-coupled receptors results in activation of phospholipase C (PLC) that cleaves phosphatidylinositol 4,5-biphosphate (PIP_2) into diacyl glycerol (DAG) and inositol 1,4,5-triphosphate (IP_3). DAG remains bound to the membrane, and IP_3 is released into the cytosol. IP_3 then acts as a second messenger to increase calcium concentration in the cytosol, which induces downstream signalling events. In addition, Ca^{2+} and DAG together activate protein kinase C (PKC), which will in turn phosphorylate other molecules (Berridge, 1993).

Some G proteins were shown to be refractory to Ptx and Ctx treatment. Of these 'Ptx-insensitive G proteins', as they were called, a subgroup has been involved in receptor-mediated activation of PLC, and was later identified as the G_q family. Ptx was then used to distinguish this group of G proteins (Fields and Casey, 1997). Another bacterial toxin, the dermonecrotic toxin produced by *Pasteurella multocida* (PMT), was later shown to interfere with G_q proteins (Wilson and Ho, 2004). PMT deamidates a specific glutamine residue in the α-subunit of G proteins, resulting in an inhibition of their GTPase activity and leading to a constitutively active phenotype of G proteins. The substrates of this toxin include the G_q, the G_{13} and the G_i proteins (Orth and Aktories, 2010). PMT has then been used to discriminate between signalling events involving G_q proteins ('Ptx-insensitive') and G_i proteins ('Ptx-sensitive'), and to study these G_q signalling events.

Among others, PMT was a powerful tool to selectively activate G_q coupled PLC signalling, verify the functionality of this pathway in some cells, identify the involvement of PLC in given phenomena, explore the interaction between G_q and effectors of other signalling pathways such as Rho GTPases, and to study all processes in general that require the activation of the G_q-PLC pathway such as dendritic cells maturation, circadian rhythms, mitogenic signalling and apoptosis (Wilson and Ho, 2004). A prolonged treatment with PMT was shown to desensitize G_q dependent PLC signalling. PMT can thus be used as both an activator and a down-regulator of G_q-PLC signalling, according to the duration of toxin treatment. In addition to its action on PLC signalling, the activation of G_q by PMT may also modulate some MAPK pathways. For instance, PMT transactivates the EGF receptor, which leads to ERK activation (Sabri *et al.*, 2002).

G proteins are also involved in the regulation of ion channels. Because of their action on G proteins, Ptx and Ctx can modulate the activation or inhibition of ion channels. For example, Ptx inhibits the activation of K^+ channels by muscarinic cholinergic receptors as well as Ca^{2+} channels, whereas Ctx activates Ca^{2+} channels and probably others ion channels dependent on G_s protein signal transduction, such as cardiac Na^+ channels (Brown and Birnbaumer, 1988; Rosenthal *et al.*, 1988; Yatani *et al.*, 1988; Schubert *et al.*, 1989; Suga *et al.*, 2004).

Given their activation or inhibition properties on different subclasses of G proteins, Ptx, Ctx and PMT have been extensively used as tools to modulate G protein-dependent signalling pathways and to discriminate between the involvement of one type of G protein or another.

Modulation of signalling pathways by anthrax toxins

Anthrax toxin is composed of three proteins: protective antigen (PA), oedema factor (EF) and lethal factor (LF). Upon binding to cell surface receptors (see section 'Microbial toxins as tools to target cell surface receptors'), PA forms a membrane channel that mediates the entry of EF and LF into the cell. PA can combine either with LF to form the lethal toxin (LeTx), or with EF to form the oedema toxin (EdTx). Both LF and EF have been shown to interfere with cell signalling (Turk, 2007).

Lethal factor of anthrax toxin as a tool to block MAPK pathway

Simultaneously, two research groups have discovered that LF inactivates MAP-Kinase-Kinases (MAPKKs) by proteolytic cleavage, thereby inhibiting the downstream MAPK signalling pathway (Duesbery *et al.*, 1998; Vitale *et al.*, 1998).

The mitogen-activated protein kinase (MAPK) pathway is a cellular signalling cascade that can be activated in response to a wide variety of extracellular stimuli, such as mitogens, cytokines, growth-factors and stress, and which modulates cellular activities such as proliferation, differentiation, cell survival and apoptosis (Pearson *et al.*, 2001). Following stimulation, a MAP-Kinase-Kinase-Kinase (MAPKKK) is activated and phosphorylates a MAPKK, which phosphorylates in turn a MAPK. MAPKs then phosphorylate and activate a large number of downstream effectors and cellular targets. There are several groups of MAPKs, the best known being ERK1/2 (extracellular-regulated kinase), p38, and JNK (c-Jun N-terminal Kinase). The family of MAPKKs contains seven members, activated by at least 20 MAPKKKs (Pearson *et al.*, 2001).

As LF contains the characteristic zinc binding motif of zinc endopeptidases, Vitale *et al.* (1998) wondered about the potential substrates of LF and identified MAPKK1 and MAPKK2 by two-hybrid screen as proteins interacting with LF. Having noticed, while screening a database of antineoplastic drugs, that LF has an activity profile similar to that of compound inhibiting selectively the MAPK pathway, Duesbery *et al.* (1998) decided to investigate the effect of LF on the MAPK pathway. Both groups found that LF is a metalloendopeptidase that cleaves MAPKKs in their N-terminal region.

All MAPKKs family members can be cleaved by LF, with the exception of MAPKK5 (Vitale *et al.*, 2000). A consensus motif for the cleavage site, consisting of four positively charged residues and two hydrophobic residues at defined positions has been identified by alignments and was confirmed by artificial substrates (Tonello and Montecucco, 2009). While MAPKK 1, 2 and 6 are cleaved at single sites, MAPKK4 and 7 are cleaved at two close peptide bonds (Vitale *et al.*, 2000). An LF-interacting region (LFIR) has been identified in the C-terminal region of the catalytic domain of MAPKKs, which is essential for LF-mediated proteolysis (Chopra *et al.*, 2003). N-terminal proteolysis of MAPKKs removes a MAPK docking motif. As a consequence, MAPKKs have impaired binding to their MAPKs and cannot phosphorylate them (Bardwell *et al.*, 2004; Bardwell and Shah, 2006). Thus, by cleaving MAPKKs, LF inhibits MAPK phosphorylation.

Inhibition of MAPK pathway by LF results in several cellular and physiological consequences, most of them aiming at impairing host defence and immunity as a way to establish infection. It was described, among others, that LF treatment blocks neutrophil chemotaxis and superoxide production (Wright and Mandell, 1986; Crawford *et al.*, 2006), inhibits secretion of TNF-α as well as several cytokines such as IL-1β, IL-6, IL-8 and IL-10 (Popov *et al.*, 2002), induces cell death in macrophages (Collier and Young, 2003), impairs dendritic cells functions such as up-regulation of co-stimulatory molecules, production of inflammatory cytokines and T cell priming (Agrawal *et al.*, 2003), blocks proliferation of B and T cells (Comer *et al.*, 2005; Fang *et al.*, 2005, 2006), and affects gene production mediated by nuclear hormone receptors, such as glucocorticoid receptor (Webster and Sternberg, 2005).

Because of its capacity to inhibit MAPK pathways, LF has shown interesting anti-tumoral properties, notably by inhibiting tumour angiogenesis (Depeille *et al.*, 2007; Alfano *et al.*, 2008). Given the known roles of MAPKs in cell proliferation, it was not surprising that LF inhibits tumour growth *in vivo*. However, the facts that only a modest inhibition was observed *in vitro* on LF-treated cells and that tumours treated with LF showed an important reduced vascular content, led to the identification of the important role played by MAPKs signalling pathways in tumour vascularization. Several MAPKs inhibitors targeting different MAPK pathways have also inhibitory effects on vascularization, as it was shown for LF, confirming the critical role played by MAPK signalling in angiogenesis cascades. Treatment of cells with LF reduces the expression of key cytokines promoting angiogenesis, such as VEGF and IL-8 (Depeille *et al.*, 2007), and also induces apoptosis in endothelial cells forming vessels (Kirby, 2004), as well as in tumour cells. These observations led to the engineering of a modified lethal toxin (LeTx), which is cleaved and activated specifically by matrix metalloproteinases enriched in the region of tumour angiogenesis instead of ubiquitous protease furin, and thus show reduced toxicity. This engineered toxin has been used successfully in pre-clinical studies (Liu *et al.*, 2008).

In conclusion, LF of anthrax toxin can be used as a powerful tool to inhibit MAPK pathways, and so to study the implication and involvement of these signalling cascades in various cellular and physiological situations.

Oedema factor of anthrax toxin as a tool to modulate cAMP-dependent pathways

Many bacterial toxins induce increase of cAMP in the host cells (Ahuja *et al.*, 2004). If pertussis toxin and cholera toxin are doing so indirectly by acting on G proteins, others are themselves adenylate cyclase enzymes.

The oedema factor of anthrax toxin (EF) is a calcium/calmodulin-dependent adenylate cyclase that causes an increase in the intracellular cAMP levels of the host cells (Leppla, 1982). EF requires the presence of calmodulin and calcium to be catalytically active. Binding of calmodulin induces conformational changes in EF, which results in the formation of a high-affinity catalytic site. ATP can then bind and be transformed into cAMP that will activate PKA and downstream effectors (Fig. 8.4).

The increase of intracellular cAMP triggered by EF leads to several cellular responses, most of them aiming at affecting the host immune response as a way to establish infection. EF modulates the production of cytokines induced by lipopolysaccharides (LPS) in monocytes, it increases for instance the production of IL-6 while it reduces the production of TNF-α (Hoover *et al.*, 1994). This regulation may be the result of an inhibition of some MAPK pathways, since it has been shown that rising of intracellular cAMP levels in monocytes blocks LPS-induced activation of ERK and JNK (Willis and Nisen, 1996). Treatment with EF inhibits phagocytosis in neutrophils and macrophages, and inhibits also neutrophil chemotaxis and superoxide production (Wright and Mandell, 1986; Crawford *et al.*, 2006). In addition to innate immunity, EF can also exert an effect on adaptive immune responses. Indeed, treatment of lymphocytes by EF results in alteration of critical immunomodulatory genes, apoptosis and decrease in T cell proliferation (Skalhegg *et al.*, 1992; Bodor *et al.*, 2001; Colic *et al.*, 2001).

cAMP dynamics and the subsequent activation of PKA are precisely controlled spatiotemporally *in vivo*, which is crucial for the regulation of the signalling cascades induced (DiPilato *et al.*, 2004). For instance, most of the cAMP responses occur at the cell membrane, and an asymmetrical PKA activity is required for actin cytoskeletal polarization and migration during neutrophil chemotaxis (Jones and Sharief, 2005).

EF generates intracellular cAMP concentration gradients from perinuclear foci, probably corresponding to late endosomes, to the cell periphery (Dal Molin *et al.*, 2006). Thus, EF may generate high levels of mislocalized cAMP, disturbing cAMP signalling and inhibiting some cAMP-dependent processes (Turk, 2007).

Thus, as the other adenylate cyclase toxins (e.g. YopJ from *Yersinia pestis* or ExoY from *Pseudomonas aeruginosa*; see below), EF could be used as a tool to study cAMP signalling events.

Others microbial toxins influencing signalling

In addition to toxins acting on G protein signalling and anthrax toxins, other bacterial toxins have been described to exert modulatory effects on signalization and could thus be used also as tools to study the implication of some signalling pathways in biological processes.

Interference of *Yersinia* Yop effector proteins with signalling

Yersinia pestis is the bacterial pathogen responsible for plague. Upon host cell contact, these bacteria inject effector proteins, referred to as Yops (*Yersinia* outer proteins), into the cell through a type III secretion system (Viboud and Bliska, 2005).

Several Yops have been shown to inactivate host proteins involved in signalling, thereby disturbing the dynamics of cytoskeleton and blocking the production of cytokines, which protect the bacteria from the immune system and allow their survival (Fallman and Gustavsson, 2005). YopH for instance is a protein tyrosine phosphatase that has been shown to interact with and dephosphorylate some host cell proteins, such as Crk-associated substrates (p130Cas), Fyn binding protein (FYB), paxillin, FAK and SKAP-HOM. These proteins have been implicated in β1-integrin signalling. Furthermore, YopH has been shown to block early Ca^{2+} signalling and phagocytosis, events that are also controlled by β1-integrin signalling. One might then postulate that YopH inhibits specifically the β1-integrin signalling pathway. Although the tyrosine phosphatase activity of YopH on the above cited proteins is rapid, it was also observed that after several hours of infection, YopH leads to a complete dephosphorylation of almost all tyrosine phosphorylated proteins in the cell (Fallman and Gustavsson, 2005). Whether the tyrosine phosphatase activity of YopH may be interesting to use in signalling studies remains to be investigated.

YopJ is an ubiquitin-like protein protease that has been shown to exert inhibitory effects on MAPK and NFkB signalling pathways (Viboud and Bliska, 2005). Interestingly, it was discovered that this effector protein is an effective acetyltransferase that catalyses the transfer of acetyl moieties on Ser and Thr residues that are normally phosphorylated on critical kinases, thereby preventing their activation and blocking the downstream pathway (Mukherjee *et al.*, 2006). YopJ acetylates on this way Ser and Thr residues in the activation loops of MAPKKs and IKKβ, kinases that activate the MAPK and NFkB pathways, respectively. This discovery has allowed to understand that not only lysines can be acetylated as it was previously thought, but also that acetylation of serine and threonine residues may be a commonly used eukaryotic mechanism to regulate phosphorylation. In any case, the acetyltransferase activity of YopJ could be used as a very efficient tool to inhibit cell signalling, and in particular the MAPK and NFκB pathways.

Listeriolysin as a tool to modulate Ca^{2+} increase

Listeriolysin O (LLO) is a major virulence factor of *Listeria monocytogenes*. This is a pore-forming toxin belonging to the family of cholesterol-dependent cytolysins (see 'Microbial toxins as tools to permeabilize membranes; β-sheet structures) (Portnoy *et al.*, 1988). It has been reported that LLO has an action on several signalling pathways, such as activation of MAPKs, production of IP3 and DAG through PI-PLC activation, or activation of NFκB, thereby activating cellular responses such as modulation of cytokine expression, expression of cell adhesion molecules on endothelial cells, delivery of heterologous antigens to the MHC class I processing and presentation pathway, or apoptosis (Darji *et al.*, 1995; Guzman *et al.*, 1996; Sibelius *et al.*, 1996; Tang *et al.*, 1996; Krull *et al.*, 1997; Kayal *et al.*, 1999; Kohda *et al.*, 2002). The signalling cascades that are activated by LLO are all dependent on increase of intracellular Ca^{2+} levels (Berridge *et al.*, 1998).

A study of Repp *et al.* (2002) showed that LLO are forming Ca^{2+}-permeable pores, thereby leading to intracellular Ca^{2+} oscillations that are caused by an influx of extracellular Ca^{2+}. Pore formation and pore size depends on the concentration of LLO and is transient. While high toxin concentration leads to cytolytic effects, sublytic toxin concentration modulates cellular signalling. The effects on signalling are generated by LLO-induced intracellular Ca^{2+} oscillations; moreover, it is also known that oscillation frequency contributes to the specificity of gene expression (Dolmetsch *et al.*, 1997, 1998). The increase in cellular Ca^{2+} concentration induced by LLO has been reported to induce a variety of cellular responses, such as degranulation of mast cells, cytokine synthesis, inhibition of the differentiation of Th2 cells and NFκB activation (Yamamoto *et al.*, 2005; Kayal and Charbit, 2006; Gekara *et al.*, 2007). LLO also induces the release of Ca^{2+} from intracellular stores, notably by activating intracellular Ca^{2+} channels through G proteins, tyrosine kinases and PLC activation (Gekara *et al.*, 2007).

Through its NFκB activation property, LLO has been shown to induce dendritic cell maturation, resulting in effective stimulation of T cells (Peng *et al.*, 2004). LLO has thus been tested as an adjuvant for the elaboration of anti-tumoral vaccines (Xu *et al.*, 2009; Xu and Dai, 2010). Thus, LLO could be used as an interesting tool to study Ca^{2+}-dependent signalling events.

The adenylate cyclase toxins

Increasing cAMP levels in the cells is a common mechanism used by bacterial toxins in order to perturb signalling events involved in host defence and to facilitate establishment of infection. As described for EF of anthrax toxin (see above), some bacteria are producing toxins that possess themselves an adenylate cyclase activity (Ahuja *et al.*, 2004). In addition to EF, three additional toxins have been described: the adenylate cyclase of *Bordetella pertussis*, the ExoY of *Pseudomonas aeruginosa*, and the adenylate cyclase of *Yersinia pestis*. All these adenylate cyclase toxins require eukaryotic factors such as calmodulin to become activated and to synthesize cAMP; in the case of ExoY it is a yet unidentified cytosolic protein distinct from calmodulin. Thus, these adenylate cyclase toxins provide interesting tools to study the regulation of cellular processes by cAMP and to modulate them.

In order to escape the host defence and corrupt the immune response, bacterial pathogens have evolved toxins that have specific enzymatic activities targeting crucial elements of signalling cascades. These bacterial toxins are thus for life scientists powerful tools to study cell-signalling events. If some of them have already been used to this purpose since quite a long time now, others that exert an interesting interfering action on signalling would deserve to be considered as tools, too. Pertussis toxin and cholera toxin first helped to decipher G proteins signalling, and are still used today to study these signalling cascades and discriminate between them. The lethal factor of anthrax toxin can be used very efficiently to inhibit MAPK pathways. The Yops proteins of *Yersinia pestis* can interfere with phosphorylation, which is a crucial phenomenon is cell signalling, and in particular YopJ leads this way to the inhibition of MAPK and NFκB pathways. Listeriolysin from *Listeria monocytogenes* could be used to play with Ca^{2+} influx, which is a central phenomenon in a lot of signalling pathways. Finally, the adenylate cyclase toxins could be used to study cAMP-dependent signalling events. These toxins constitute a palette of tools that allow to target more or less specifically given signalling pathways, and there are probably a lot more interesting microbial toxins that could enlarge the bacterial toolkit.

Conclusion

In this chapter we have described specific functions of toxins that make them valuable as tools in cell biology. However, being highly optimized virulence factors, most toxins cause multiple effects on cells. Some of these effects are protection mechanisms from the host cell, but the toxin in itself may have several different functions. As an example, Shiga toxin does not only modify ribosomes, but also triggers numerous signalling pathways, starting already at the plasma membrane (Johannes and Romer, 2010).

Acknowledgements

This work was supported by the Swedish Research Council grant 2010-878 (G.R.), by a starting grant of the European Research Council to W.R. (Programme 'Ideas' – call identifier: ERC-2011-StG 282105-lec&lip2invade) and by the German Excellence Initiative of the Deutsche Forschungsgemeinschaft (EXC 294). Molecular graphics and analyses were performed with the UCSF Chimera package. Chimera is developed by the Resource for Biocomputing, Visualization, and Informatics at the University of California, San Francisco, with support from the National Institutes of Health (National Center for Research Resources grant 2P41RR001081, National Institute of General Medical Sciences grant 9P41GM103311).

References

Abrami, L., Liu, S., Cosson, P., Leppla, S.H., and van der Goot, F.G. (2003). Anthrax toxin triggers endocytosis of its receptor via a lipid raft-mediated clathrin-dependent process. J. Cell Biol. *160*, 321–328.

Abrami, L., Lindsay, M., Parton, R.G., Leppla, S.H., and van der Goot, F.G. (2004). Membrane insertion of anthrax protective antigen and cytoplasmic delivery of lethal factor occur at different stages of the endocytic pathway. J. Cell Biol. *166*, 645–651.

Abrami, L., Reig, N., and van der Goot, F.G. (2005). Anthrax toxin: the long and winding road that leads to the kill. Trends Microbiol. *13*, 72–78.

Agrawal, A., Lingappa, J., Leppla, S.H., Agrawal, S., Jabbar, A., Quinn, C., and Pulendran, B. (2003). Impairment of dendritic cells and adaptive immunity by anthrax lethal toxin. Nature *424*, 329–334.

Ahuja, N., Kumar, P., and Bhatnagar, R. (2004). The adenylate cyclase toxins. Crit. Rev. Microbiol. *30*, 187–196.

Alfano, R.W., Leppla, S.H., Liu, S., Bugge, T.H., Duesbery, N.S., and Frankel, A.E. (2008). Potent inhibition of tumor angiogenesis by the matrix metalloproteinase-activated anthrax lethal toxin: implications for broad anti-tumor efficacy. Cell Cycle *7*, 745–749.

Alouf, J.E. (2000a). Bacterial protein toxins. An overview. Methods Mol. Biol. *145*, 1–26.

Alouf, J.E. (2000b). Cholesterol-binding cytolytic protein toxins. Int. J. Med. Microbiol. *290*, 351–356.

Atlas, R.M. (1998). The medical threat of biological weapons. Crit. Rev. Microbiol. *24*, 157–168.

Bardwell, A.J., Abdollahi, M., and Bardwell, L. (2004). Anthrax lethal factor-cleavage products of MAPK (mitogen-activated protein kinase) kinases exhibit reduced binding to their cognate MAPKs. Biochem. J. *378*, 569–577.

Bardwell, L., and Shah, K. (2006). Analysis of mitogen-activated protein kinase activation and interactions with regulators and substrates. Methods *40*, 213–223.

Beddoe, T., Paton, A.W., Le Nours, J., Rossjohn, J., and Paton, J.C. (2010). Structure, biological functions and applications of the AB5 toxins. Trends Biochem. Sci. *35*, 411–418.

Bennett, M.J., and Eisenberg, D. (1994). Refined structure of monomeric diphtheria toxin at 2.3 A resolution. Protein Sci. *3*, 1464–1475.

Berridge, M.J. (1993). Inositol trisphosphate and calcium signalling. Nature *361*, 315–325.

Berridge, M.J., Bootman, M.D., and Lipp, P. (1998). Calcium – a life and death signal. Nature *395*, 645–648.

Bhakdi, S., Bayley, H., Valeva, A., Walev, I., Walker, B., Kehoe, M., and Palmer, M. (1996). Staphylococcal alpha-toxin, streptolysin-O, and *Escherichia coli* hemolysin: prototypes of pore-forming bacterial cytolysins. Arch. Microbiol. *165*, 73–79.

Bhakdi, S., and Tranum-Jensen, J. (1991). Alpha-toxin of *Staphylococcus aureus*. Microbiol. Rev. *55*, 733–751.

Birkeli, K.A., Llorente, A., Torgersen, M.L., Keryer, G., Tasken, K., and Sandvig, K. (2003). Endosome-to-Golgi transport is regulated by protein kinase A type II alpha. J. Biol. Chem. *278*, 1991–1997.

Blewitt, M.G., Chung, L.A., and London, E. (1985). Effect of pH on the conformation of diphtheria toxin and its implications for membrane penetration. Biochemistry *24*, 5458–5464.

Bodor, J., Feigenbaum, L., Bodorova, J., Bare, C., Reitz, Jr, M.S.., and Gress, R.E. (2001). Suppression of T-cell responsiveness by inducible cAMP early repressor (ICER). J. Leukoc. Biol. *69*, 1053–1059.

Boll, W., Ehrlich, M., Collier, R.J., and Kirchhausen, T. (2004). Effects of dynamin inactivation on pathways of anthrax toxin uptake. Eur J. Cell Biol. *83*, 281–288.

Bonifacino, J.S., and Hurley, J.H. (2008). Retromer. Curr. Opin. Cell Biol. *20*, 427–436.

Bradley, K.A., Mogridge, J., Mourez, M., Collier, R.J., and Young, J.A. (2001). Identification of the cellular receptor for anthrax toxin. Nature *414*, 225–229.

Brown, A.M., and Birnbaumer, L. (1988). Direct G protein gating of ion channels. Am. J. Physiol. *254*, H401–410.

Brunger, A.T., and Rummel, A. (2009). Receptor and substrate interactions of clostridial neurotoxins. Toxicon *54*, 550–560.

Byres, E., Paton, A.W., Paton, J.C., Lofling, J.C., Smith, D.F., Wilce, M.C., Talbot, U.M., Chong, D.C., Yu, H., Huang, S., *et al.* (2008). Incorporation of a non-human glycan mediates human susceptibility to a bacterial toxin. Nature *456*, 648–652.

Casey, P.J., and Gilman, A.G. (1988). G protein involvement in receptor-effector coupling. J. Biol. Chem. *263*, 2577–2580.

Chopra, A.P., Boone, S.A., Liang, X., and Duesbery, N.S. (2003). Anthrax lethal factor proteolysis and inactivation of MAPK kinase. J. Biol. Chem. *278*, 9402–9406.

Colic, M., Vucevic, D., Jandric, D., Medic-Mijacevic, L., and Rakic, L. (2001). 8-Chloro-cAMP modulates apoptosis of thymocytes and thymocyte hybridoma. Transplant Proc. 33, 2347–2349.

Collier, R.J., and Young, J.A. (2003). Anthrax toxin. Annu. Rev. Cell Dev. Biol. *19*, 45–70.

Comer, J.E., Chopra, A.K., Peterson, J.W., and Konig, R. (2005). Direct inhibition of T-lymphocyte activation by anthrax toxins *in vivo*. Infect. Immun. *73*, 8275–8281.

Conner, S.D., and Schmid, S.L. (2003). Regulated portals of entry into the cell. Nature *422*, 37–44.

Crawford, M.A., Aylott, C.V., Bourdeau, R.W., and Bokoch, G.M. (2006). *Bacillus anthracis* toxins inhibit human neutrophil NADPH oxidase activity. J. Immunol. *176*, 7557–7565.

Dal Molin, F., Tonello, F., Ladant, D., Zornetta, I., Zamparo, I., Di Benedetto, G., Zaccolo, M., and Montecucco, C. (2006). Cell entry and cAMP imaging of anthrax edema toxin. EMBO J. *25*, 5405–5413.

Darji, A., Chakraborty, T., Wehland, J., and Weiss, S. (1995). Listeriolysin generates a route for the presentation of exogenous antigens by major histocompatibility complex class I. Eur. J. Immunol. *25*, 2967–2971.

Del Nery, E., Miserey-Lenkei, S., Falguieres, T., Nizak, C., Johannes, L., Perez, F., and Goud, B. (2006). Rab6A and Rab6A' GTPases play non-overlapping roles in membrane trafficking. Traffic *7*, 394–407.

Depeille, P., Young, J.J., Boguslawski, E.A., Berghuis, B.D., Kort, E.J., Resau, J.H., Frankel, A.E., and Duesbery, N.S. (2007). Anthrax lethal toxin inhibits growth of and vascular endothelial growth factor release from endothelial cells expressing the human herpes virus 8 viral G protein coupled receptor. Clin. Cancer Res. *13*, 5926–5934.

DiPilato, L.M., Cheng, X., and Zhang, J. (2004). Fluorescent indicators of cAMP and Epac activation reveal differential dynamics of cAMP signaling within discrete subcellular compartments. Proc. Natl. Acad. Sci. U.S.A. *101*, 16513–16518.

Doherty, G.J., and McMahon, H.T. (2009). Mechanisms of endocytosis. Annu. Rev. Biochem. *78*, 857–902.

Dolmetsch, R.E., Lewis, R.S., Goodnow, C.C., and Healy, J.I. (1997). Differential activation of transcription factors induced by Ca^{2+} response amplitude and duration. Nature *386*, 855–858.

Dolmetsch, R.E., Xu, K., and Lewis, R.S. (1998). Calcium oscillations increase the efficiency and specificity of gene expression. Nature *392*, 933–936.

Donta, S.T., Tomicic, T.K., and Donohue-Rolfe, A. (1995). Inhibition of Shiga-like toxins by brefeldin A. J. Infect. Dis. *171*, 721–724.

Dragneva, Y., Anuradha, C.D., Valeva, A., Hoffmann, A., Bhakdi, S., and Husmann, M. (2001). Subcytocidal attack by staphylococcal alpha-toxin activates NF-kappaB and induces interleukin-8 production. Infect. Immun. *69*, 2630–2635.

Duesbery, N.S., Webb, C.P., Leppla, S.H., Gordon, V.M., Klimpel, K.R., Copeland, T.D., Ahn, N.G., Oskarsson, M.K., Fukasawa, K., Paull, K.D., *et al.* (1998). Proteolytic inactivation of MAP-kinase-kinase by anthrax lethal factor. Science *280*, 734–737.

Duran, J.M., Valderrama, F., Castel, S., Magdalena, J., Tomas, M., Hosoya, H., Renau-Piqueras, J., Malhotra, V., and Egea, G. (2003). Myosin motors and not actin comets are mediators of the actin-based Golgi-to-endoplasmic reticulum protein transport. Mol. Biol. Cell *14*, 445–459.

Eierhoff, T., Stechmann, B., and Römer, W. (2012). Pathogen and toxin entry – how pathogens and toxins induce and harness endocytotic mechanisms. In Molecular Regulation of Endocytosis, Ceresa, B., ed. (InTech, Rijeka), pp. 249–276.

Ewers, H., Romer, W., Smith, A.E., Bacia, K., Dmitrieff, S., Chai, W., Mancini, R., Kartenbeck, J., Chambon, V., Berland, L., *et al.* (2010). GM1 structure determines SV40-induced membrane invagination and infection. Nat. Cell Biol. *12*, 11–18; sup pp 11–12.

Fallman, M., and Gustavsson, A. (2005). Cellular mechanisms of bacterial internalization counteracted by Yersinia. Int. Rev. Cytol. *246*, 135–188.

Fang, H., Cordoba-Rodriguez, R., Lankford, C.S., and Frucht, D.M. (2005). Anthrax lethal toxin blocks MAPK kinase-dependent IL-2 production in CD4+ T cells. J. Immunol. *174*, 4966–4971.

Fang, H., Xu, L., Chen, T.Y., Cyr, J.M., and Frucht, D.M. (2006). Anthrax lethal toxin has direct and potent inhibitory effects on B cell proliferation and immunoglobulin production. J. Immunol. *176*, 6155–6161.

Fujinaga, Y., Wolf, A.A., Rodighiero, C., Wheeler, H., Tsai, B., Allen, L., Jobling, M.G., Rapoport, T., Holmes, R.K., and Lencer, W.I. (2003). Gangliosides that associate with lipid rafts mediate transport of cholera and related toxins from the plasma membrane to endoplasmic reticulm. Mol. Biol. Cell *14*, 4783–4793.

Gekara, N.O., Westphal, K., Ma, B., Rohde, M., Groebe, L., and Weiss, S. (2007). The multiple mechanisms of Ca^{2+} signalling by listeriolysin O, the cholesterol-dependent cytolysin of *Listeria monocytogenes*. Cell. Microbiol. *9*, 2008–2021.

Gilman, A.G. (1987). G proteins: transducers of receptor-generated signals. Annu. Rev. Biochem. *56*, 615–649.

Girod, A., Storrie, B., Simpson, J.C., Johannes, L., Goud, B., Roberts, L.M., Lord, J.M., Nilsson, T., and Pepperkok, R. (1999). Evidence for a COP-I-independent transport route from the Golgi complex to the endoplasmic reticulum. Nat. Cell Biol. *1*, 423–430.

Gonnord, P., Blouin, C.M., and Lamaze, C. (2012). Membrane trafficking and signaling: two sides of the same coin. Semin. Cell Dev. Biol. *23*, 154–164.

Gordon, V.M., and Leppla, S.H. (1994). Proteolytic activation of bacterial toxins: role of bacterial and host cell proteases. Infect. Immun. *62*, 333–340.

Guerra, L., Teter, K., Lilley, B.N., Stenerlow, B., Holmes, R.K., Ploegh, H.L., Sandvig, K., Thelestam, M., and Frisan, T. (2005). Cellular internalization of cytolethal distending toxin: a new end to a known pathway. Cell. Microbiol. *7*, 921–934.

Guerra, L., Cortes-Bratti, X., Guidi, R., and Frisan, T. (2011). The biology of the cytolethal distending toxins. Toxins (Basel) *3*, 172–190.

Guzman, C.A., Domann, E., Rohde, M., Bruder, D., Darji, A., Weiss, S., Wehland, J., Chakraborty, T., and Timmis, K.N. (1996). Apoptosis of mouse dendritic cells is triggered by listeriolysin, the major virulence determinant of *Listeria monocytogenes*. Mol. Microbiol. *20*, 119–126.

Hazes, B., and Read, R.J. (1997). Accumulating evidence suggests that several AB-toxins subvert the endoplasmic reticulum-associated protein degradation pathway to enter target cells. Biochemistry *36*, 11051–11054.

Hepler, J.R., and Gilman, A.G. (1992). G proteins. Trends Biochem. Sci. *17*, 383–387.

Herz, J., and Strickland, D.K. (2001). LRP: a multifunctional scavenger and signaling receptor. J. Clin. Invest. *108*, 779–784.

Heuck, A.P., Tweten, R.K., and Johnson, A.E. (2001). Beta-barrel pore-forming toxins: intriguing dimorphic proteins. Biochemistry *40*, 9065–9073.

Himmelweit, F. (1956). The Collected Papers of Paul Ehrlich, Vol 1 (New York, Pergamon Press).

Holmgren, J., Lonnroth, I., and Svennerholm, L. (1973). Tissue receptor for cholera exotoxin: postulated structure from studies with GM1 ganglioside and related glycolipids. Infect. Immun. *8*, 208–214.

Hoover, D.L., Friedlander, A.M., Rogers, L.C., Yoon, I.K., Warren, R.L., and Cross, A.S. (1994). Anthrax edema toxin differentially regulates lipopolysaccharide-induced monocyte production of tumor necrosis factor alpha and interleukin-6 by increasing intracellular cyclic AMP. Infect. Immun. *62*, 4432–4439.

Inoue, T., Moore, P., and Tsai, B. (2011). How viruses and toxins disassemble to enter host cells. Annu. Rev. Microbiol. *65*, 287–305.

Jackson, M.E., Simpson, J.C., Girod, A., Pepperkok, R., Roberts, L.M., and Lord, J.M. (1999). The KDEL retrieval system is exploited by Pseudomonas exotoxin A, but not by Shiga-like toxin-1, during retrograde transport from the Golgi complex to the endoplasmic reticulum. Cell. J. Sci. *112 (Pt 4)*, 467–475.

Johannes, L., and Popoff, V. (2008). Tracing the retrograde route in protein trafficking. Cell *135*, 1175–1187.

Johannes, L., and Romer, W. (2010). Shiga toxins – from cell biology to biomedical applications. Nat. Rev. Microbiol. *8*, 105–116.

Johannes, L., and Wunder, C. (2011). Retrograde transport: two (or more) roads diverged in an endosomal tree? Traffic *12*, 956–962.

Johannes, L., Tenza, D., Antony, C., and Goud, B. (1997). Retrograde transport of KDEL-bearing B-fragment of Shiga toxin. J. Biol. Chem. *272*, 19554–19561.

Jones, S.L., and Sharief, Y. (2005). Asymmetrical protein kinase A activity establishes neutrophil cytoskeletal polarity and enables chemotaxis. J. Leukoc. Biol. *78*, 248–258.

Kayal, S., and Charbit, A. (2006). Listeriolysin O: a key protein of *Listeria monocytogenes* with multiple functions. FEMS Microbiol. Rev. *30*, 514–529.

Kayal, S., Lilienbaum, A., Poyart, C., Memet, S., Israel, A., and Berche, P. (1999). Listeriolysin O-dependent activation of endothelial cells during infection with *Listeria monocytogenes*: activation of NF-kappa B and up-regulation of adhesion molecules and chemokines. Mol. Microbiol. *31*, 1709–1722.

Kirby, J.E. (2004). Anthrax lethal toxin induces human endothelial cell apoptosis. Infect. Immun. 72, 430–439.

Kirkham, M., Fujita, A., Chadda, R., Nixon, S.J., Kurzchalia, T.V., Sharma, D.K., Pagano, R.E., Hancock, J.F., Mayor, S., and Parton, R.G. (2005). Ultrastructural identification of uncoated caveolin-independent early endocytic vehicles. J. Cell Biol. *168*, 465–476.

Kohda, C., Kawamura, I., Baba, H., Nomura, T., Ito, Y., Kimoto, T., Watanabe, I., and Mitsuyama, M. (2002). Dissociated linkage of cytokine-inducing activity and cytotoxicity to different domains of listeriolysin O from *Listeria monocytogenes*. Infect. Immun. *70*, 1334–1341.

Kounnas, M.Z., Morris, R.E., Thompson, M.R., FitzGerald, D.J., Strickland, D.K., and Saelinger, C.B. (1992). The alpha 2-macroglobulin receptor/low density lipoprotein receptor-related protein binds and internalizes Pseudomonas exotoxin A. J. Biol. Chem. *267*, 12420–12423.

Krull, M., Nost, R., Hippenstiel, S., Domann, E., Chakraborty, T., and Suttorp, N. (1997). *Listeria monocytogenes* potently induces up-regulation of endothelial adhesion molecules and neutrophil adhesion to cultured human endothelial cells. J. Immunol. *159*, 1970–1976.

Lamaze, C., and Johannes, L. (2006). Intracellular trafficking of bacterial and plant protein toxins. In Comprehensive Sourcebook of Bacterial Protein Toxins, Alouf, J.E., and Popoff, M.R., eds. (London, Academic Press), pp. 135–153.

Lauvrak, S.U., Torgersen, M.L., and Sandvig, K. (2004). Efficient endosome-to-Golgi transport of Shiga toxin is dependent on dynamin and clathrin. J. Cell Sci. *117*, 2321–2331.

Lencer, W.I., Constable, C., Moe, S., Jobling, M.G., Webb, H.M., Ruston, S., Madara, J.L., Hirst, T.R., and Holmes, R.K. (1995). Targeting of cholera toxin and *Escherichia coli* heat labile toxin in polarized epithelia: role of COOH-terminal KDEL. J. Cell Biol. *131*, 951–962.

Leppla, S.H. (1982). Anthrax toxin edema factor: a bacterial adenylate cyclase that increases cyclic AMP concentrations of eukaryotic cells. Proc. Natl. Acad. Sci. U.S.A. *79*, 3162–3166.

Lesieur, C., Vecsey-Semjen, B., Abrami, L., Fivaz, M., and Gisou van der Goot, F. (1997). Membrane insertion: The strategies of toxins (review). Mol. Membr. Biol. *14*, 45–64.

Lindberg, A.A., Brown, J.E., Stromberg, N., Westling-Ryd, M., Schultz, J.E., and Karlsson, K.A. (1987). Identification of the carbohydrate receptor for Shiga toxin produced by *Shigella dysenteriae* type 1. J. Biol. Chem. *262*, 1779–1785.

Liu, S., Wang, H., Currie, B.M., Molinolo, A., Leung, H.J., Moayeri, M., Basile, J.R., Alfano, R.W., Gutkind, J.S., Frankel, A.E., *et al.* (2008). Matrix metalloproteinase-activated anthrax lethal toxin demonstrates high potency in targeting tumor vasculature. J. Biol. Chem. *283*, 529–540.

Lu, L., Tai, G., and Hong, W. (2004). Autoantigen Golgin-97, an effector of Arl1 GTPase, participates in traffic from the endosome to the trans–golgi network. Mol. Biol. Cell *15*, 4426–4443.

Luna, A., Matas, O.B., Martinez-Menarguez, J.A., Mato, E., Duran, J.M., Ballesta, J., Way, M., and Egea, G. (2002). Regulation of protein transport from the Golgi complex to the endoplasmic reticulum by CDC42 and N-WASP. Mol. Biol. Cell *13*, 866–879.

Madshus, I.H., Stenmark, H., Sandvig, K., and Olsnes, S. (1991). Entry of diphtheria toxin-protein A chimeras into cells. J. Biol. Chem. *266*, 17446–17453.

Majoul, I., Schmidt, T., Pomasanova, M., Boutkevich, E., Kozlov, Y., and Soling, H.D. (2002). Differential expression of receptors for Shiga and Cholera toxin is regulated by the cell cycle. J. Cell Sci. *115*, 817–826.

Mallard, F., Antony, C., Tenza, D., Salamero, J., Goud, B., and Johannes, L. (1998). Direct pathway from early/recycling endosomes to the Golgi apparatus revealed through the study of shiga toxin B-fragment transport. J. Cell Biol. *143*, 973–990.

Mallard, F., and Johannes, L. (2003). Shiga toxin B-subunit as a tool to study retrograde transport. Methods Mol. Med. *73*, 209–220.

Mallard, F., Tang, B.L., Galli, T., Tenza, D., Saint-Pol, A., Yue, X., Antony, C., Hong, W., Goud, B., and Johannes, L. (2002). Early/recycling endosomes-to-TGN transport involves two SNARE complexes and a Rab6 isoform. J. Cell Biol. *156*, 653–664.

Mammen, M., Choi, S.K., and Whitesides, G.M. (1998). Polyvalent interactions in biological systems: implications for design and use of multivalent ligands and inhibitors. Angew Chem. Int. Ed. *37*, 2754–2794.

McKenzie, J., Johannes, L., Taguchi, T., and Sheff, D. (2009). Passage through the Golgi is necessary for Shiga toxin B subunit to reach the endoplasmic reticulum. FEBS J. *276*, 1581–1595.

McMahon, H.T., and Boucrot, E. (2011). Molecular mechanism and physiological functions of clathrin-mediated endocytosis. Nat. Rev. Mol. Cell Biol. *12*, 517–533.

Merritt, E.A., Sarfaty, S., Jobling, M.G., Chang, T., Holmes, R.K., Hirst, T.R., and Hol, W.G. (1997). Structural studies of receptor binding by cholera toxin mutants. Protein Sci. *6*, 1516–1528.

Millen, S.H., Lewallen, D.M., Herr, A.B., Iyer, S.S., and Weiss, A.A. (2010). Identification and characterization of the carbohydrate ligands recognized by pertussis toxin via a glycan microarray and surface plasmon resonance. Biochemistry *49*, 5954–5967.

Mobley, J.A. (1995). Biological warfare in the twentieth century: lessons from the past, challenges for the future. Mil. Med. *160*, 547–553.

Montecucco, C., Schiavo, G., and Tomasi, M. (1985). pH-dependence of the phospholipid interaction of diphtheria-toxin fragments. Biochem. J. *231*, 123–128.

Montesano, R., Roth, J., Robert, A., and Orci, L. (1982). Non-coated membrane invaginations are involved in binding and internalization of cholera and tetanus toxins. Nature *296*, 651–653.

Mukherjee, S., Keitany, G., Li, Y., Wang, Y., Ball, H.L., Goldsmith, E.J., and Orth, K. (2006). Yersinia YopJ acetylates and inhibits kinase activation by blocking phosphorylation. Science *312*, 1211–1214.

Mukhopadhyay, S., and Linstedt, A.D. (2012). Manganese blocks intracellular trafficking of Shiga toxin and protects against Shiga toxicosis. Science *335*, 332–335.

Munro, S. (2003). Lipid rafts: elusive or illusive? Cell *115*, 377–388.

Nambiar, M.P., Oda, T., Chen, C., Kuwazuru, Y., and Wu, H.C. (1993). Involvement of the Golgi region in the intracellular trafficking of cholera toxin. J. Cell. Physiol. *154*, 222–228.

Natarajan, R., and Linstedt, A.D. (2004). A cycling cis-Golgi protein mediates endosome-to-Golgi traffic. Mol. Biol. Cell *15*, 4798–4806.

Nichols, B.J., Kenworthy, A.K., Polishchuk, R.S., Lodge, R., Roberts, T.H., Hirschberg, K., Phair, R.D., and Lippincott-Schwartz, J. (2001). Rapid cycling of lipid raft markers between the cell surface and Golgi complex. J. Cell Biol. *153*, 529–541.

Nishimoto-Morita, K., Shin, H.W., Mitsuhashi, H., Kitamura, M., Zhang, Q., Johannes, L., and Nakayama, K. (2009). Differential effects of depletion of ARL1 and ARFRP1 on membrane trafficking between the trans-Golgi network and endosomes. J. Biol. Chem. *284*, 10583–10592.

Oh, K.J., Senzel, L., Collier, R.J., and Finkelstein, A. (1999). Translocation of the catalytic domain of diphtheria toxin across planar phospholipid bilayers by its own T domain. Proc. Natl. Acad. Sci. U.S.A. *96*, 8467–8470.

Orlandi, P.A., and Fishman, P.H. (1998). Filipin-dependent inhibition of cholera toxin: evidence for toxin internalization and activation through caveolae-like domains. J. Cell Biol. *141*, 905–915.

Orth, J.H., and Aktories, K. (2010). Pasteurella multocida toxin activates various heterotrimeric G proteins by deamidation. Toxins (Basel) *2*, 205–214.

Parker, M.W. (1997). More than one way to make a hole. Nat. Struct. Mol. Biol. *4*, 250–253.

Parton, R.G., Molero, J.C., Floetenmeyer, M., Green, K.M., and James, D.E. (2002). Characterization of a distinct plasma membrane macrodomain in differentiated adipocytes. J. Biol. Chem. *277*, 46769–46778.

Pearson, G., Robinson, F., Beers Gibson, T., Xu, B.E., Karandikar, M., Berman, K., and Cobb, M.H. (2001). Mitogen-activated protein (MAP) kinase pathways: regulation and physiological functions. Endocr. Rev. *22*, 153–183.

Peng, X., Hussain, S.F., and Paterson, Y. (2004). The ability of two *Listeria monocytogenes* vaccines targeting human papillomavirus-16 E7 to induce an antitumor response correlates with myeloid dendritic cell function. J. Immunol. *172*, 6030–6038.

Perez-Victoria, F.J., Mardones, G.A., and Bonifacino, J.S. (2008). Requirement of the human GARP complex for mannose 6-phosphate-receptor-dependent sorting of cathepsin D to lysosomes. Mol. Biol. Cell *19*, 2350–2362.

Pettersen, E.F., Goddard, T.D., Huang, C.C., Couch, G.S., Greenblatt, D.M., Meng, E.C., and Ferrin, T.E. (2004). UCSF Chimera – a visualization system for exploratory research and analysis. J. Comput. Chem. *25*, 1605–1612.

Popoff, V., Mardones, G.A., Tenza, D., Rojas, R., Lamaze, C., Bonifacino, J.S., Raposo, G., and Johannes, L. (2007). The retromer complex and clathrin define an early endosomal retrograde exit site. J. Cell Sci. *120*, 2022–2031.

Popoff, V., Mardones, G.A., Bai, S.K., Chambon, V., Tenza, D., Burgos, P.V., Shi, A., Benaroch, P., Urbe, S., Lamaze, C., *et al.* (2009). Analysis of articulation between clathrin and retromer in retrograde sorting on early endosomes. Traffic *10*, 1868–1880.

Popov, S.G., Villasmil, R., Bernardi, J., Grene, E., Cardwell, J., Popova, T., Wu, A., Alibek, D., Bailey, C., and Alibek, K. (2002). Effect of *Bacillus anthracis* lethal toxin on human peripheral blood mononuclear cells. FEBS Lett. *527*, 211–215.

Portnoy, D.A., Jacks, P.S., and Hinrichs, D.J. (1988). Role of hemolysin for the intracellular growth of *Listeria monocytogenes*. J. Exp. Med. *167*, 1459–1471.

Proux-Gillardeaux, V., and Galli, T. (2008). Targeting the epithelial SNARE machinery by bacterial neurotoxins. Methods Mol. Biol. *440*, 187–201.

Proux-Gillardeaux, V., Gavard, J., Irinopoulou, T., Mege, R.M., and Galli, T. (2005). Tetanus neurotoxin-mediated cleavage of cellubrevin impairs epithelial cell migration and integrin-dependent cell adhesion. Proc. Natl. Acad. Sci. U.S.A. *102*, 6362–6367.

Ren, J., Kachel, K., Kim, H., Malenbaum, S.E., Collier, R.J., and London, E. (1999). Interaction of diphtheria toxin T domain with molten globule-like proteins and its implications for translocation. Science *284*, 955–957.

Repp, H., Pamukci, Z., Koschinski, A., Domann, E., Darji, A., Birringer, J., Brockmeier, D., Chakraborty, T., and Dreyer, F. (2002). Listeriolysin of *Listeria monocytogenes* forms Ca^{2+}-permeable pores leading to intracellular Ca^{2+} oscillations. Cell. Microbiol. *4*, 483–491.

Römer, W., Berland, L., Chambon, V., Gaus, K., Windschiegl, B., Tenza, D., Aly, M.R., Fraisier, V., Florent, J.C., Perrais, D., *et al.* (2007). Shiga toxin induces tubular membrane invaginations for its uptake into cells. Nature *450*, 670–675.

Römer, W., Pontani, L.L., Sorre, B., Rentero, C., Berland, L., Chambon, V., Lamaze, C., Bassereau, P., Sykes, C., Gaus, K., *et al.* (2010). Actin dynamics drive membrane reorganization and scission in clathrin-independent endocytosis. Cell *140*, 540–553.

Rosenthal, W., Hescheler, J., Trautwein, W., and Schultz, G. (1988). Control of voltage-dependent Ca^{2+} channels by G protein-coupled receptors. FASEB J. *2*, 2784–2790.

Roux, E., and Yersin, A. (1888). Contribution à l'étude de la diphthérie. Ann. Inst. Pasteur *2*, 629–661.

Sabharanjak, S., Sharma, P., Parton, R.G., and Mayor, S. (2002). GPI-anchored proteins are delivered to recycling endosomes via a distinct cdc42-regulated, clathrin-independent pinocytic pathway. Dev. Cell *2*, 411–423.

Sabri, A., Wilson, B.A., and Steinberg, S.F. (2002). Dual actions of the Galpha(q) agonist Pasteurella multocida toxin to promote cardiomyocyte hypertrophy and enhance apoptosis susceptibility. Circ. Res. *90*, 850–857.

Saint-Pol, A., Yelamos, B., Amessou, M., Mills, I.G., Dugast, M., Tenza, D., Schu, P., Antony, C., McMahon, H.T., Lamaze, C., *et al.* (2004). Clathrin adaptor epsinR is required for retrograde sorting on early endosomal membranes. Dev. Cell *6*, 525–538.

Sandvig, K., and Olsnes, S. (1980). Diphtheria toxin entry into cells is facilitated by low pH. J. Cell Biol. *87*, 828–832.

Sandvig, K., and van Deurs, B. (2002). Membrane traffic exploited by protein toxins. Annu. Rev. Cell Dev. Biol. *18*, 1–24.

Sandvig, K., and van Deurs, B. (2005). Delivery into cells: lessons learned from plant and bacterial toxins. Gene Ther. *12*, 865–872.

Sandvig, K., Olsnes, S., Brown, J.E., Petersen, O.W., and van Deurs, B. (1989). Endocytosis from coated pits of Shiga toxin: a glycolipid-binding protein from F 1. J. Cell Biol. *108*, 1331–1343.

Sandvig, K., Garred, O., Prydz, K., Kozlov, J.V., Hansen, S.H., and van Deurs, B. (1992). Retrograde transport of endocytosed Shiga toxin to the endoplasmic reticulum. Nature *358*, 510–512.

Sandvig, K., Torgersen, M.L., Raa, H.A., and van Deurs, B. (2008). Clathrin-independent endocytosis: from nonexisting to an extreme degree of complexity. Histochem. Cell Biol. *129*, 267–276.

Schapiro, F.B., Lingwood, C., Furuya, W., and Grinstein, S. (1998). pH-independent retrograde targeting of glycolipids to the Golgi complex. Am. J. Physiol. *274*, C319–332.

Schiavo, G., and van der Goot, F.G. (2001). The bacterial toxin toolkit. Nat. Rev. Mol. Cell Biol. *2*, 530–537.

Schmitt, C.K., Meysick, K.C., and O'Brien, A.D. (1999). Bacterial toxins: friends or foes? Emerg. Infect. Dis. *5*, 224–234.

Schubert, B., VanDongen, A.M., Kirsch, G.E., and Brown, A.M. (1989). Beta-adrenergic inhibition of cardiac sodium channels by dual G-protein pathways. Science *245*, 516–519.

Scobie, H.M., Rainey, G.J., Bradley, K.A., and Young, J.A. (2003). Human capillary morphogenesis protein 2 functions as an anthrax toxin receptor. Proc. Natl. Acad. Sci. U.S.A. *100*, 5170–5174.

Sekiya, K., Satoh, R., Danbara, H., and Futaesaku, Y. (1993). A ring-shaped structure with a crown formed by streptolysin O on the erythrocyte membrane. J. Bacteriol. *175*, 5953–5961.

Sens, P., Johannes, L., and Bassereau, P. (2008). Biophysical approaches to protein-induced membrane deformations in trafficking. Curr. Opin. Cell Biol. *20*, 476–482.

Sibelius, U., Rose, F., Chakraborty, T., Darji, A., Wehland, J., Weiss, S., Seeger, W., and Grimminger, F. (1996). Listeriolysin is a potent inducer of the phosphatidylinositol response and lipid mediator generation in human endothelial cells. Infect. Immun. *64*, 674–676.

Simons, K., and Gerl, M.J. (2010). Revitalizing membrane rafts: new tools and insights. Nat. Rev. Mol. Cell Biol. *11*, 688–699.

Simpson, J.C., Smith, D.C., Roberts, L.M., and Lord, J.M. (1998). Expression of mutant dynamin protects cells against diphtheria toxin but not against ricin. Exp. Cell Res. *239*, 293–300.

Skalhegg, B.S., Landmark, B.F., Doskeland, S.O., Hansson, V., Lea, T., and Jahnsen, T. (1992). Cyclic AMP-dependent protein kinase type I mediates the inhibitory effects of 3′,5′-cyclic adenosine monophosphate on cell replication in human T lymphocytes. J. Biol. Chem. *267*, 15707–15714.

Song, L., Hobaugh, M.R., Shustak, C., Cheley, S., Bayley, H., and Gouaux, J.E. (1996). Structure of staphylococcal alpha-hemolysin, a heptameric transmembrane pore. Science *274*, 1859–1866.

Stenmark, H. (2009). Rab GTPases as coordinators of vesicle traffic. Nat. Rev. Mol. Cell Biol. *10*, 513–525.

Suga, S., Takeo, T., Nakano, K., Sato, T., Igarashi, T., Yamana, D., and Wakui, M. (2004). Pertussis toxin-sensitive pathway inhibits glucose-stimulated Ca^{2+} signals of rat islet beta-cells by affecting L-type Ca^{2+} channels and voltage-dependent K+ channels. Cell Calcium *36*, 469–477.

Tang, P., Rosenshine, I., Cossart, P., and Finlay, B.B. (1996). Listeriolysin O activates mitogen-activated protein kinase in eucaryotic cells. Infect. Immun. *64*, 2359–2361.

Tangvoranuntakul, P., Gagneux, P., Diaz, S., Bardor, M., Varki, N., Varki, A., and Muchmore, E. (2003). Human uptake and incorporation of an immunogenic nonhuman dietary sialic acid. Proc. Natl. Acad. Sci. U.S.A. *100*, 12045–12050.

Teter, K., Allyn, R.L., Jobling, M.G., and Holmes, R.K. (2002). Transfer of the cholera toxin A1 polypeptide from the endoplasmic reticulum to the cytosol is a rapid process facilitated by the endoplasmic reticulum-associated degradation pathway. Infect. Immun. *70*, 6166–6171.

Thomsen, P., Roepstorff, K., Stahlhut, M., and van Deurs, B. (2002). Caveolae are highly immobile plasma membrane microdomains, which are not involved in constitutive endocytic trafficking. Mol. Biol. Cell *13*, 238–250.

Tonello, F., and Montecucco, C. (2009). The anthrax lethal factor and its MAPK kinase-specific metalloprotease activity. Mol. Aspects Med. *30*, 431–438.

Torgersen, M.L., Skretting, G., van Deurs, B., and Sandvig, K. (2001). Internalization of cholera toxin by different endocytic mechanisms. J. Cell Sci. *114*, 3737–3747.

Turk, B.E. (2007). Manipulation of host signalling pathways by anthrax toxins. Biochem. J. *402*, 405–417.

Valderrama, F., Duran, J.M., Babia, T., Barth, H., Renau-Piqueras, J., and Egea, G. (2001). Actin microfilaments facilitate the retrograde transport from the Golgi complex to the endoplasmic reticulum in mammalian cells. Traffic *2*, 717–726.

Viboud, G.I., and Bliska, J.B. (2005). Yersinia outer proteins: role in modulation of host cell signaling responses and pathogenesis. Annu. Rev. Microbiol. *59*, 69–89.

Vitale, G., Pellizzari, R., Recchi, C., Napolitani, G., Mock, M., and Montecucco, C. (1998). Anthrax lethal factor cleaves the N-terminus of MAPKKs and induces tyrosine/threonine phosphorylation of MAPKs in cultured macrophages. Biochem. Biophys. Res. Commun. *248*, 706–711.

Vitale, G., Bernardi, L., Napolitani, G., Mock, M., and Montecucco, C. (2000). Susceptibility of mitogen-activated protein kinase kinase family members to proteolysis by anthrax lethal factor. Biochem. J. *352 Pt 3*, 739–745.

Walev, I., Bhakdi, S.C., Hofmann, F., Djonder, N., Valeva, A., Aktories, K., and Bhakdi, S. (2001). Delivery of proteins into living cells by reversible membrane permeabilization with streptolysin-O. Proc. Natl. Acad. Sci. U.S.A. *98*, 3185–3190.

Walsh, D.A., and Van Patten, S.M. (1994). Multiple pathway signal transduction by the cAMP-dependent protein kinase. FASEB J. *8*, 1227–1236.

Watkins, E.B., Miller, C.E., Majewski, J., and Kuhl, T.L. (2011). Membrane texture induced by specific protein binding and receptor clustering: active roles for lipids in cellular function. Proc. Natl. Acad. Sci. U.S.A. *108*, 6975–6980.

Webster, J.I., and Sternberg, E.M. (2005). Anthrax lethal toxin represses glucocorticoid receptor (GR) transactivation by inhibiting GR-DNA binding *in vivo*. Mol. Cell Endocrinol. *241*, 21–31.

Willis, S.A., and Nisen, P.D. (1996). Differential induction of the mitogen-activated protein kinase pathway by bacterial lipopolysaccharide in cultured monocytes and astrocytes. Biochem. J. *313 (Pt 2)*, 519–524.

Wilson, B.A., and Ho, M. (2004). Pasteurella multocida toxin as a tool for studying Gq signal transduction. Rev. Physiol. Biochem. Pharmacol. *152*, 93–109.

Wolf, A.A., Jobling, M.G., Wimer-Mackin, S., Ferguson-Maltzman, M., Madara, J.L., Holmes, R.K., and Lencer, W.I. (1998). Ganglioside structure dictates signal transduction by cholera toxin and association with caveolae-like membrane domains in polarized epithelia. J. Cell Biol. *141*, 917–927.

Wright, G.G., and Mandell, G.L. (1986). Anthrax toxin blocks priming of neutrophils by lipopolysaccharide and by muramyl dipeptide. J. Exp. Med. *164*, 1700–1709.

Xu, M., and Dai, M.S. (2010). Recombinant *E. coli* LLO/OVA induces murine BMDCs maturation via TLR4 and NOD1 receptor and promotes specific cytotoxic T cell immunity. Biomed. Environ. Sci. *23*, 350–356.

Xu, M., Dai, M.S., and Mi, C. (2009). Expression of cytokines in murine BMDCs induced by recombinant E. coli LLO/OVA via TLR4 and NOD1 receptors. Xi Bao Yu Fen Zi Mian Yi Xue Za Zhi *25*, 9–12.

Yamamoto, K., Kawamura, I., Tominaga, T., Nomura, T., Kohda, C., Ito, J., and Mitsuyama, M. (2005). Listeriolysin O, a cytolysin derived from *Listeria monocytogenes*, inhibits generation of ovalbumin-specific Th2 immune response by skewing maturation of antigen-specific T cells into Th1 cells. Clin. Exp. Immunol. *142*, 268–274.

Yamane, J., Kubo, A., Nakayama, K., Yuba-Kubo, A., Katsuno, T., and Tsukita, S. (2007). Functional involvement of TMF/ARA160 in Rab6-dependent retrograde membrane traffic. Exp. Cell Res. *313*, 3472–3485.

Yang, Z., Bergstrom, J., and Karlsson, K.A. (1994). Glycoproteins with Gal alpha 4Gal are absent from human erythrocyte membranes, indicating that glycolipids are the sole carriers of blood group P activities. J. Biol. Chem. *269*, 14620–14624.

Yatani, A., Mattera, R., Codina, J., Graf, R., Okabe, K., Padrell, E., Iyengar, R., Brown, A.M., and Birnbaumer, L. (1988). The G protein-gated atrial K+ channel is stimulated by three distinct Gi alpha-subunits. Nature *336*, 680–682.

Yoshida, T., Chen, C.C., Zhang, M.S., and Wu, H.C. (1991). Disruption of the Golgi apparatus by brefeldin A inhibits the cytotoxicity of ricin, modeccin, and Pseudomonas toxin. Exp. Cell Res. *192*, 389–395.

Yoshino, A., Setty, S.R., Poynton, C., Whiteman, E.L., Saint-Pol, A., Burd, C.G., Johannes, L., Holzbaur, E.L., Koval, M., McCaffery, J.M., *et al.* (2005). tGolgin-1 (p230, golgin-245) modulates Shiga-toxin transport to the Golgi and Golgi motility towards the microtubule-organizing centre. J. Cell Sci. *118*, 2279–2293.

Young, J.A., and Collier, R.J. (2007). Anthrax toxin: receptor binding, internalization, pore formation, and translocation. Annu. Rev. Biochem. *76*, 243–265.

Yu, M., and Haslam, D.B. (2005). Shiga toxin is transported from the endoplasmic reticulum following interaction with the luminal chaperone HEDJ/ERdj3. Infect. Immun. *73*, 2524–2532.

Zhang, R.G., Scott, D.L., Westbrook, M.L., Nance, S., Spangler, B.D., Shipley, G.G., and Westbrook, E.M. (1995). The three-dimensional crystal structure of cholera toxin. J. Mol. Biol. *251*, 563–573.

Zolov, S.N., and Lupashin, V.V. (2005). Cog3p depletion blocks vesicle-mediated Golgi retrograde trafficking in HeLa cells. J. Cell Biol. *168*, 747–759.

The Toxins of *Clostridium difficile*

9

Glen P. Carter, Milena M. Awad, Julian I. Rood and Dena Lyras

Abstract

Clostridium difficile infections represent a range of antibiotic-associated diarrhoea syndromes that are caused by the Gram-positive, spore-forming anaerobe *C. difficile*. This bacterium is the most significant cause of hospital-acquired diarrhoea in many countries and the emergence of variant strains with enhanced virulence capacity over the last decade has intensified this problem, resulting in major worldwide epidemics. Upon colonization of a susceptible host *C. difficile* produces two large clostridial cytotoxins, toxin A (TcdA) and toxin B (TcdB), which are monoglucosyltransferases that irreversibly modify members of the Rho family of host regulatory proteins, leading to disruption of downstream signalling pathways and cell death, with disease manifesting as diarrhoea. Recent studies have indicated that toxin B plays a much more important role in disease than early studies had suggested. Comparison of the more recent epidemic isolates with historical strains has identified numerous differences that may contribute to increased virulence, including the production of binary toxin (CDT), which is an actin-specific ADP-ribosyltransferase. However, the role of CDT in disease pathogenesis remains unclear. This review will discuss our current understanding of the role of the *C. difficile* toxins in disease, with a particular emphasis on results obtained from studying these toxins using animal disease models.

Introduction

Hospital-acquired infection with the bacterium *Clostridium difficile* is a major global public health concern. *C. difficile* is a Gram-positive, spore-forming bacterium which causes severe gastrointestinal illness and death in humans. It is the most significant cause of hospital-acquired diarrhoea in many countries, leading to substantial morbidity and mortality and placing a considerable economic burden on healthcare systems. In the USA, for example, infections with *C. difficile* cause as many as 500,000 illnesses per year (Kuehn, 2011) at an estimated cost of $3.2 billion per annum to treat these patients (O'Brien *et al.*, 2007).

Host susceptibility to *C. difficile* is usually the consequence of antibiotic treatment of a patient for another, unrelated condition (Borriello, 1998). Disruption of the natural gastrointestinal microbiota, which provides 'colonization resistance' and protection against *C. difficile* infection, allows this bacterium to colonize and flourish. Unlike other enteric pathogens, disease is almost always associated with antimicrobial therapy. Importantly, the increase in antibiotic resistant bacteria in recent years has led to much higher usage of broad spectrum antibiotics. Paradoxically, treating these bacteria has left patients vulnerable to infection by bacteria such as *C. difficile*. *C. difficile* causes a spectrum of gastrointestinal diseases, collectively known as *C. difficile* infections (CDI), which can range from mild diarrhoea, through moderately serious disease, to severe pseudomembranous colitis (Borriello, 1998). It is of particular concern that morbidity and mortality resulting from CDI have increased significantly over the last decade, most likely due to changes in the virulence of the causative strains, and the resulting spread of these epidemic strains, coupled with changes in antibiotic usage patterns (McDonald *et al.*, 2005; Pépin *et al.*, 2005; Warny *et al.*, 2005; Redelings *et al.*, 2007).

The mechanism by which *C. difficile* causes disease is complex and reflects the intricate life cycle of this bacterium. *C. difficile* is an anaerobic bacterium that adapts to changing environmental conditions by producing highly resistant and persistent spores, which enable *C. difficile* to survive in aerobic environments outside the host. These spores are very resistant to commonly used disinfectants and alcohol-based hand washes (Gerding *et al.*, 2008; Dawson *et al.*, 2011) and act as a reservoir for further infections and disease transmission, allowing *C. difficile* to become endemic in many hospitals and making outbreaks very difficult to contain (Lawley *et al.*, 2009). Once spores are ingested by a susceptible patient, germination into vegetative cells occurs, leading to colonization of the gut and toxin production.

CDI is primarily a toxin-mediated disease, with most disease symptoms observed being directly linked to the production of toxins (Carter *et al.*, 2011b). Virulent *C. difficile* strains produce one or both of two major toxins that are members of the large clostridial cytotoxin family, designated toxin A (TcdA) and toxin B (TcdB), encoded by the *tcdA* and *tcdB* genes, respectively (Voth and Ballard, 2005). These toxins are potent monoglucosyltransferases that irreversibly modify members of the Rho family of host regulatory proteins, leading to disruption of downstream signalling pathways and cell death (Just *et al.*, 1995). Infection with toxigenic *C. difficile* strains therefore causes extensive colonic inflammation and epithelial tissue damage. The net effect is rapid fluid loss into the intestinal lumen, which manifests as diarrhoea (Voth and Ballard, 2005). A number of strains, including recent epidemic isolates and many animal isolates, also produce a third toxin known as CDT binary toxin, which is an ADP-ribosyltransferase that causes actin depolymerization, resulting in disruption of the host cell cytoskeleton and cell death (Perelle *et al.*, 1997; Schwan *et al.*, 2009). The exact role of this toxin in disease is, however, currently unknown. The following chapter will discuss our current understanding of the toxins of *C. difficile*, with a particular emphasis on the role that the toxins play in disease pathogenesis.

Epidemic or hypervirulent strains of *C. difficile*

The number of CDI cases and severity of disease caused by *C. difficile* has dramatically increased over the past 10 years, and is in part associated with the emergence of strains with enhanced virulence traits and extended antibiotic resistance characteristics. The indiscriminate use of broad spectrum antibiotics, such as fluoroquinolones, has most likely played an important role in the emergence of these strains through the provision of an ideal niche within which these strains can evolve (Pépin *et al.*, 2005; Carter *et al.*, 2012). The most important of these strains are classified as the BI/NAP1/027 strains which are often described as being 'hypervirulent'. These strains were first reported in Canada in 2003 and are now the most prevalent epidemic strain isolated from healthcare facilities within the USA and Canada, despite being rarely isolated prior to 2001 (McDonald *et al.*, 2005; Janezic *et al.*, 2012). Similarly, ribotype 078 isolates (toxinotype V), traditionally associated with CDI in animals, are increasingly emerging as a cause of CDI in humans, particularly in the Netherlands and England (Wilcox and Planche, 2009; Bakker *et al.*, 2010). Ribotype 078 isolates are also considered to be 'hypervirulent' and appear to cause disease as severe as that associated with ribotype 027 isolates (Goorhuis *et al.*, 2008; Stabler *et al.*, 2012). Hypervirulence in both ribotype 027 and 078 strains is thought to be due to the increased levels of toxin A and toxin B produced by these strains (McDonald *et al.*, 2005; Dingle *et al.*, 2011), but may also be due to the production of a third toxin called binary toxin, or CDT (McDonald *et al.*, 2005; Paltansing *et al.*, 2007). As discussed later, CDT is an ADP ribosyltransferase that might cause adjunctive damage to the host in addition to that caused by the two major toxins, and may also lead to enhanced host colonization (Schwan *et al.*, 2009, 2011).

The pathogenicity locus (PaLoc)

Toxigenic strains of *C. difficile* are characterized by the presence of a chromosomal region known as the pathogenicity locus or PaLoc (Cohen, 2000). PaLoc, originally identified in 1995 (Hammond and Johnson, 1995), is 19.6 kb in size and contains the genes *tcdA* and *tcdB* which encode the two major *C. difficile* toxins, toxins A and B, respectively. PaLoc also contains three accessory genes, *tcdR* and *tcdC*, which encode two proteins involved in the regulation of toxin production, and *tcdE*, which encodes a protein that shows similarity to bacteriophage holin-like proteins (Fig. 9.1). The expression of the PaLoc genes has been shown to be growth dependent, with *tcdA, tcdB, tcdR* and *tcdE* all transcribed during stationary phase (Hundsberger *et al.*, 1997; Dupuy and Sonenshein, 1998), whilst *tcdC* is expressed during the exponential phase of growth (Hundsberger *et al.*, 1997). The integration site for PaLoc was delineated in 1996 (Braun *et al.*, 1996) and was shown to be located between the *cdu1* and *cdd1* genes, which appear to be of bacteriophage origin (Himeno *et al.*, 1986; Rowland and Dyke, 1990; Tesch *et al.*, 1990). In non-toxigenic isolates, PaLoc is replaced by a 115 bp fragment that shows no homology with PaLoc (Fig. 9.1) (Braun *et al.*, 1996).

The *tcdA, tcdB, tcdR* and *tcdE* PaLoc genes are positively regulated at the transcriptional level by the TcdR protein, which has been shown to be an RNA polymerase alternative sigma factor (Moncrief *et al.*, 1997; Mani and Dupuy, 2001). TcdR is a member of Group 5 of the σ^{70}- family, together with TetR from *Clostridium tetani*, BotR from *Clostridium botulinum* and UviA from *Clostridium perfringens* (Dupuy *et al.*, 2006), with each being structurally and functionally distinct from other members of the σ^{70}- family of alternative sigma factors (Raffestin *et al.*, 2005). The regulatory role of TcdR on toxin production was first shown in 1997 (Moncrief *et al.*, 1997) and has since been supported by studies performed in *C. perfringens* and *C. difficile* (Mani and Dupuy, 2001; Mani *et al.*, 2002).

Whilst the regulatory role of TcdR is well established, the exact role of TcdC as a putative negative regulator of toxin production remains controversial. A recent *in vivo* study showed that a naturally occurring mutation in *tcdC* could potentiate virulence in an epidemic PCR ribotype 027 isolate of *C. difficile*, presumably as a result of increased levels of toxin production in this strain (Carter *et al.*, 2011c). The emergence of 'hypervirulent' strains that harbour mutations in *tcdC* and produce elevated levels of toxins A and B, coupled with the original observation that *tcdC* transcription occurs during early growth phases and is repressed during stationary phase, which is in contrast to the other PaLoc genes (Hundsberger *et al.*, 1997), gave rise to the hypothesis that TcdC might be an anti-sigma factor that acts as a negative regulator of toxin production. Further support for this hypothesis came from a study which showed that TcdC was able to significantly repress the activity of TcdR when both proteins were co-expressed in the surrogate host *C. perfringens* (Matamouros *et al.*, 2007). However, several studies have noted that a non-functional *tcdC* gene is not always correlated with increased toxin

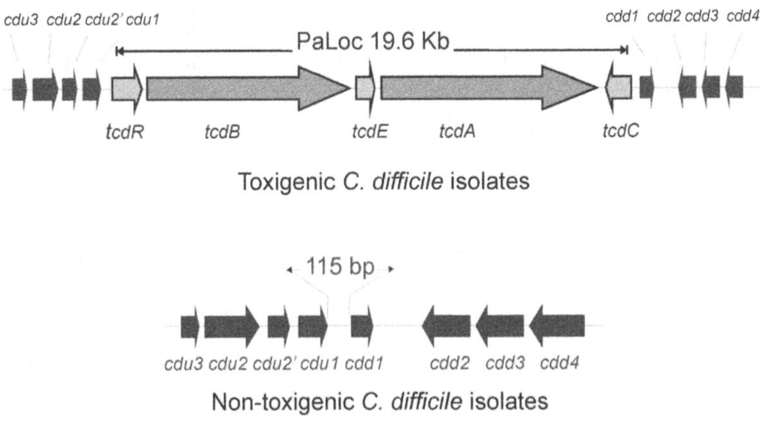

Figure 9.1 Genetic organization of the 19.6kb pathogenicity locus (PaLoc) and flanking regions in *C. difficile* (top panel). Boxes indicate open reading frames with arrows showing the direction of transcription. Toxin genes (*tcdB, tcdA*), regulatory genes (*tcdR, tcdC*) and the putative holin gene, *tcdE*, are indicated. Genes located outside the PaLoc are in black. The integration site for PaLoc is located between the *cdu1* and *cdd1* genes and PaLoc is replaced by a 115bp fragment that shows no homology with the pathogenicity locus in non-toxigenic isolates (bottom panel) (Hundsberger *et al.*, 1997).

production (Verdoorn *et al.*, 2010; Goldenberg and French, 2011; Sirard *et al.*, 2011) which is at odds with the suggestion that TcdC negatively regulates toxin production. Furthermore, a recent study showed that neither the restoration to wild-type of the naturally occurring *tcdC* mutation in *C. difficile* strain R20291, a PCR ribotype 027 isolate, nor the introduction of a frameshift mutation into the intact *tcdC* gene of *C. difficile* strain 630 had any observable effect on toxin production (Cartman *et al.*, 2012), suggesting that TcdC may play a modulatory role in the control of toxin production (Merrigan *et al.*, 2010; Vohra and Poxton, 2011).

The *tcdE* gene is not as well studied as the other gene PaLoc genes. This gene encodes a protein with similarity to the class 1 family of bacteriophage holins, such as the λ S protein (Grundling *et al.*, 2000). Holins are small cytolytic proteins that are involved in the release of bacteriophage progeny from host cells, however, some studies have suggested that holin-like proteins may also be involved in bacterial protein secretion (Desvaux *et al.*, 2005, 2009; Desvaux and Hebraud, 2006). The structural similarity of TcdE to holins, and the localization of *tcdE* in close proximity to the toxin genes within PaLoc, has led to the prediction that TcdE may be involved in the release of toxins A and B from the cell, particularly since these toxins do not have a recognizable export sequence or signal peptide to direct export or secretion (von Eichel-Streiber *et al.*, 1990). In keeping with this hypothesis, an isogenic *tcdE* mutant in *C. difficile* strain 630 was recently shown to exhibit a marked reduction in the accumulation of toxin A and toxin B in culture supernatants compared with the wild-type strain, a defect that was not attributable to altered levels of toxin gene expression in the mutant. Importantly, this phenotype was restored upon complementation with the intact *tcdE* gene (Govind and Dupuy, 2012). A second study, however, suggested that TcdE is not involved in toxin release from the cell since the amount of toxin present in the culture supernatant of an independently derived *tcdE* mutant was indistinguishable from that of the wild-type strain (Olling *et al.*, 2012). The reasons for the disparity between these two studies remain unclear.

The structure and function of toxin A and toxin B

Toxins A and B from *C. difficile* are both members of the large clostridial toxin (LCT) family, which also includes alpha toxin (TcnA) from *Clostridium novyi* (Busch *et al.*, 2000), the lethal and haemorrhagic toxins (TcsL and TcsH, respectively) of *C. sordellii* (Voth *et al.*, 2006), and TpeL from particular strains of *Clostridium perfringens* (Amimoto *et al.*, 2007). The LCTs are very large, ranging in size from 250 to 308 kDa, and share a high degree of structural similarity and sequence identity. These toxins are important bacterial virulence factors, being monoglycosyltransferases that are proficient at inactivating Rho family GTPases such as Rho, Rac, Ras, Ral and Cdc42, which are found in eukaryotic host cells. Inactivation of these proteins occurs through the covalent transfer of a glucose moiety, or in the case of TcnA and TpeL an N-acetylglucosamine moiety, at position Thr 35 or Thr 37 of the Rho GTPase proteins. This effectively locks Rho GTPases into an inactive conformation, therefore disrupting all downstream cellular activities involving these proteins, which includes organization of the cellular actin cytoskeleton, maintenance of epithelial barrier function and immune cell signalling (Pothoulakis *et al.*, 1998).

Each of the LCTs is a potent bacterial toxin which result in the rapid death of mice when they are injected with purified toxin (Carter *et al.*, 2010). The importance of the LCTs in bacterial virulence in an infection context is also now beginning to be understood. For example, work has shown that TcsL is essential for virulence in the *C. sordellii* type strain ATCC9714, since isogenic *tcsL* mutants were avirulent in the mouse infection model whereas the wild-type strain resulted in the death of all infected animals (Carter *et al.*, 2011a). Similarly, when *C. novyi* was cured of the bacteriophage encoding TcnA it was rendered avirulent in a mouse model of infection (Dang *et al.*, 2001). As will be discussed later, toxin B appears to be the major virulence factor of *C. difficile*, however, the individual role of this toxin as well as that of toxin A remains controversial and further work is needed to conclusively determine the role of these toxins in disease. Similarly, the role of TpeL in the virulence of *C. perfringens* remains to be determined.

As stated previously, the large clostridial toxins are all structurally very similar and have considerable amino acid sequence identity (Just and Gerhard, 2004). They each possess multiple conserved domains including an N-terminal catalytic or glycosyltransferase domain, an autocatalytic cysteine protease domain, which is required for delivery of the catalytically active region of the toxin into the cytosol of the cell, a centrally located hydrophobic translocation domain that is thought to mediate insertion of the toxin into the endosomal membrane and may be involved in host cell receptor binding, and finally a C-terminal receptor binding domain (Jank and Aktories, 2008) (Fig. 9.2). Although, the original publication identifying TpeL suggested that this toxin does not contain a C-terminal receptor binding domain (Amimoto *et al.*, 2007), a more recent paper has identified a variant TpeL toxin which possesses a C-terminal extension (Guttenberg *et al.*, 2012). It is not clear at this time whether both of these proteins are functional and if they are both biologically relevant.

The process by which a host cell becomes intoxicated by one of the LCTs is best characterized for *C. difficile* toxins A and B and is thought to involve a number of important steps. The process begins with the binding of the C-terminal region of toxin A or B to one or more receptors present on the surface of a host cell. Once toxin binding has taken place, the toxin is endocytosed and trafficked to early endosomes within the cell (Jank and Aktories, 2008). The acidic endosomal environment is then thought to bring about a conformational change within the toxin which facilitates the insertion of the translocation domain into the endosomal membrane, resulting in the formation of a pore within the membrane (Genisyuerek *et al.*, 2011). Binding of host cell

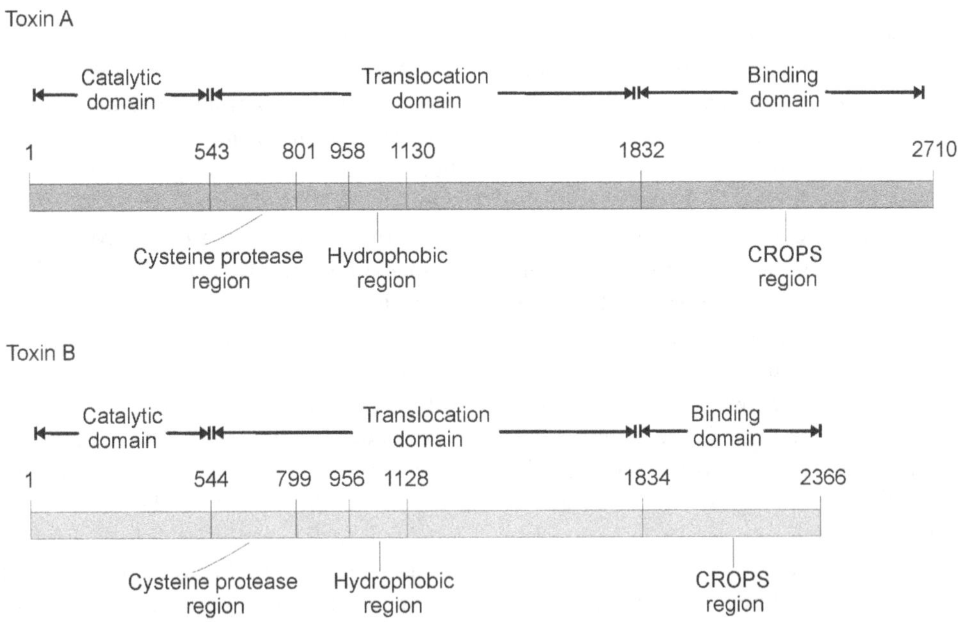

Figure 9.2 The domain structure of toxin A (top panel) and toxin B (bottom panel) is shown. Numbers indicate amino acids that mark domain boundaries. The catalytic glucosyltransferase (GT) domain is located at the N-terminus. The translocation domain is in the central portion of the protein and includes the cysteine protease (CP) region, required for toxin processing and cleavage, and a hydrophobic domain, which may be involved in pore formation, delivery of the catalytic domain into the host cell cytosol, and receptor binding. The C-terminal portion of the protein consists of polypeptide repeats (CROPS region) and is putatively involved in binding to host cell receptors (Jank and Aktories, 2008; Genisyuerek *et al.*, 2011).

inositol hexakisphosphate by the toxin then leads to a further conformational change, specifically within the cysteine protease domain, leading to an autocatalytic cleavage event occurring between the catalytic and the cysteine protease domains of the toxin. This finally results in the release of the active glycosyltransferase domain into the host cell cytosol, which is then able to inactivate Rho family GTPases (Guttenberg *et al.*, 2011).

Rho GTPases are important in many processes that occur within the target cell, including maintenance of an ordered actin cytoskeleton; they also play an important role in controlling epithelial barrier function and in the signalling and motility of immune cells (Jank and Aktories, 2008). Inactivation of Rho GTPases by toxins A or B therefore has a profound effect on the host cell, leading to the complete loss of cellular structural integrity, primarily as a consequence of toxin-mediated dysregulation of actin depolymerization (Voth and Ballard, 2005). This loss of structural integrity is characterized by cell rounding and normally precedes apoptosis and, finally, death of the intoxicated cell which is thought to occur via a caspase-3 and caspase-9 dependent pathway (Voth and Ballard, 2005). Colonocyte death caused by toxins A and B leads to a loss of intestinal epithelial barrier function through impaired tight junction integrity. The toxins also cause cytokine production by intestinal epithelial cells resulting in the activation of enteric nerves and sensory neurons within the lamina propria and the recruitment of neutrophils to the site of infection (Pothoulakis *et al.*, 1998). This in turn results in the release of neuropeptides and pro-inflammatory cytokines, which leads to an acute inflammatory response that further compromises the integrity of the epithelial barrier (Pothoulakis *et al.*, 1998). Ultimately, toxin-mediated loss of epithelial barrier function results in an increase in intestinal permeability, which leads to fluid accumulation followed by the onset of diarrhoea which is the most characteristic symptom of CDI (Pothoulakis, 2000).

Toxin gene variants of *C. difficile*

As a species, *C. difficile* exhibits extensive chromosomal variation between strains, including within the genes of PaLoc, resulting in the development of a *C. difficile* typing scheme known as 'toxinotyping' (Rupnik *et al.*, 1998). Toxinotyping differentiates strains based on genetic variation within PaLoc, particularly in the *tcdA* and *tcdB* genes (Rupnik *et al.*, 1998). Currently, in excess of 30 toxinotypes exist in addition to toxinotype 0, to which the reference strain VPI 10463 belongs (Carter *et al.*, 2011c). Not surprisingly, the number of toxinotypes continues to increase, with new toxinotypes regularly being identified around the world (Dingle *et al.*, 2011; Elliott *et al.*, 2011).

Mutations located within PaLoc can have a profound effect on toxin production by *C. difficile*, in some cases leading to a loss of toxin production. Isolates which carry these mutations, and which no longer produce toxin A or toxin B, are generally classified as toxinotype class XIa and XIb and are considered avirulent since they do not cause disease in animal infection models (Geric *et al.*, 2006). Toxin A-negative, toxin B-positive strains are another important class of toxin variant. These isolates are usually classed as toxinotypes VIII, X, XVI and XVII (Rupnik, 2008) but more recently have also been found to include toxinotypes XXX and XXXI (Elliott *et al.*, 2011). These strain variants appear to be clinically relevant since they are associated with *C. difficile* infections (CDI) in humans (Alfa *et al.*, 2000; Kuijper *et al.*, 2001; Sato *et al.*, 2004; Drudy *et al.*, 2007). Interestingly, these strains often cause severe disease in humans and their prevalence appears to be rising (Goorhuis *et al.*, 2009). In addition to the loss of one or both toxins, genetic variation can also result in the production of toxins with altered substrate specificity and antigenicity (Rupnik, 2007, 2008). For example, the variant toxin B protein produced by toxinotype VIII, X and XIV isolates modifies Rap, Ral and R-Ras but does not modify Rho. Interestingly, this toxin was found to be approximately 8-fold more toxic when injected into mice than the equivalent toxin produced by reference strain VPI 10463 (Rupnik, 2008; Carter *et al.*, 2012). The clinical impact of this increased toxicity is, however, not known.

Perhaps the best-characterized toxin variant is the toxin B protein produced by BI/NAP1/027 isolates of *C. difficile*. This toxin, known as $TcdB_{HV}$ (Lanis *et al.*, 2010), exhibits greater tissue tropism and cytotoxicity in a zebrafish model when compared to TcdB from the historical isolate strain 630 ($TcdB_{HIST}$) (Lanis *et al.*, 2010). Furthermore, $TcdB_{HV}$ is able to enter cells more rapidly and at an earlier stage in the endocytic pathway when compared with $TcdB_{HIST}$. This rapid cell entry rate may provide the toxin with refuge from lysosomal proteases (Lanis *et al.*, 2010) and enable it to more rapidly cause cell damage. In addition, earlier studies showed this toxin to be more cytotoxic to a wider range of cell types than historical toxins, most likely as a result of variation within the receptor binding domain of the toxin (Stabler *et al.*, 2006, 2008). Although not experimentally proven, it seems plausible that these phenotypic and functional differences may at least partly be responsible for the increased virulence associated with the BI/NAP1/027 *C. difficile* isolates.

The role of toxin A and toxin B in disease

A historical perspective

For many years toxin A was thought to be the major virulence factor of *C. difficile* despite the observation that most clinical isolates produced both toxin A and toxin B. This hypothesis was based on experiments that were designed to assess the effects of purified toxins on hamsters following intragastric administration. The critical observation in these experiments was that when hamsters received purified toxin A the animals succumbed to disease and displayed symptoms typical of *C. difficile* infection, such as fluid accumulation and intestinal tissue necrosis. By contrast, intragastric administration of purified toxin B resulted in no observable disease unless prior intestinal damage was present or if sublethal concentrations of toxin A were co-administered with toxin B. These experiments therefore also indicated that the toxins work synergistically, with toxin B-mediated effects occurring only after toxin A had caused prior tissue damage (Lyerly *et al.*, 1985). Experiments using rabbit ileal loops provided further evidence that toxin A might be the major virulence factor of *C. difficile* since purified toxin A alone was able to cause haemorrhage, severe epithelial necrosis and increased intestinal permeability, whilst toxin B alone showed little enterotoxic activity (Mitchell *et al.*, 1986; Lima *et al.*, 1988).

Finally, immunological findings and animal vaccination experiments also provided evidence for the importance of toxin A in *C. difficile* infections since a strong humoral response against toxin A was found to be an important predictor of reduced disease severity in patients with CDI (Bacon and Fekety, 1994; Kyne *et al.*, 2000; Drudy *et al.*, 2007). Furthermore, hamsters vaccinated with toxin A-derived toxoids were fully protected against *C. difficile* infection (Ketley *et al.*, 1987; Kim *et al.*, 1987; Corthier *et al.*, 1991).

The last 30 years have, however, seen the emergence of clinical strains of *C. difficile* that do not produce toxin A but still cause symptomatic, and severe, disease (Drudy *et al.*, 2007). The observation that these strains can elicit the full spectrum of symptoms associated with CDI and are clinically indistinguishable from toxin A- and B-positive *C. difficile* strains is clearly at odds with the hypothesis that toxin A is the major virulence factor of *C. difficile*. On the one hand, laboratory based experiments with the purified toxins showed that toxin A was the major mediator of disease yet the occurrence of toxin A-negative, toxin B-positive isolates suggested that toxin A was not essential for the onset of disease and, additionally, that this toxin was not needed for the *in vivo* activity of toxin B. The isolation of these strains therefore suggested that toxin B might play a much more important role in CDI than had been previously thought.

Subsequent experiments provided the first laboratory-based evidence that toxin B might play a more important role in CDI than was previously thought. In these experiments, the effects of intragastric administration of purified toxin A and toxin B on mice carrying human intestinal explants was determined. In contrast to the previous experiments performed with hamsters, these experiments showed that toxin B was the more potent enterotoxin, resulting in severe damage to the intestinal epithelium even in the absence of toxin A (Savidge *et al.*, 2003). As a consequence of this work, the authors questioned the generally accepted view of the time that toxin A was the major virulence factor.

More recent experiments using the zebrafish

embryo model of intoxication provided further verification of the important role of toxin B in disease. In these experiments, toxin B was shown to be cardiotoxic, with cardiovascular damage seen in fish exposed to Toxin B, resulting in a significant reduction in both systemic blood flow and heart rate (Hamm *et al.*, 2006). These findings suggest that toxin B can cause systemic damage to the host in addition to the localized gut damage that is most often associated with CDI (Lanis *et al.*, 2010), which may go some way towards explaining why *C. difficile* infection can result in multiple organ dysfunction syndrome in humans (Dobson *et al.*, 2003; Hariri *et al.*, 2011).

Functional analysis of toxins using isogenic mutants

Despite the insight gained from previous studies using purified toxins, conclusively assigning individual roles for toxin A and toxin B in disease was not possible from this work primarily because the laboratory findings did not accurately reflect the clinical observations of CDI. Most importantly, until this time it had not been possible to generate isogenic strain toxin mutants in *C. difficile*. This problem was overcome in 2009 when we generated the first isogenic *C. difficile* toxin mutants (Lyras *et al.*, 2009). The construction of these mutants was a major breakthrough since for the first time they allowed the individual role of each toxin to be determined using the natural infection process, and in the context of the whole bacterium rather than as isolated toxins. In these studies, Syrian golden hamsters were infected with the wild-type *C. difficile* strain or one of the isogenic toxin mutants. As expected, when infected with the wild-type strain, the majority of animals (90%) succumbed to fatal disease. Similarly, 94% of hamsters infected with the toxin A mutants died, providing clear evidence that toxin A was not needed for disease. By contrast, only 22% of animals displayed signs of disease when infected with the mutants that no longer produced toxin B. The subsequent analysis of faecal pellets collected from the hamsters that did succumb to infection revealed the presence of toxin B, suggesting that reversion of the mutants to wild-type had occurred in the strains infecting these animals. Of particular relevance is that these observations are in keeping with clinical observations of CDI and with the occurrence of virulent toxin A-negative, toxin B-positive isolates, all of which strongly suggest that toxin B is the major virulence factor of *C. difficile* (Lyras *et al.*, 2009).

A more recent study using independently derived isogenic toxin mutants in *C. difficile* has, however, brought into question the hypothesis that toxin A is not a major virulence factor in *C. difficile* (Kuehne *et al.*, 2010). In this study, both a toxin A and a toxin B mutant caused fulminant disease in the hamster model of infection, leading to the suggestion that these experiments re-established the importance of both toxin A and toxin B in *C. difficile* disease.

It is currently not clear why the toxin B mutants used in the first study were essentially avirulent whilst the mutant used in the second study was able to cause disease. However, the toxin B mutant in the second study did appear to be less virulent since the mean time to death of hamsters infected with this strain was 4 days compared with 1 and 1.3 days, respectively, for the wild-type strain and the toxin A mutant (Kuehne *et al.*, 2010). This result seems to be in keeping with the hypothesis that toxin B is the major virulence factor of *C. difficile* but further experimentation is required to gain a better understanding of the role of the major toxins during CDI.

C. difficile binary toxin (CDT)

Structure and mode of action

The clostridial ADP-ribosyltransferase family of binary toxins comprise a closely related family of toxins that have the ability to transfer ADP-ribose to actin. Members of this family include ι-toxin from *C. perfringens* type E, *C. spiroforme* toxin (CST), *C. botulinum* C2 toxin and *C. difficile* binary toxin (CDT). These toxins all consist of two separate polypeptides: an A, or active component, that has ADP-ribosyltransferase activity and a B, or binding component, that is responsible for binding to specific host cell surface receptors (Barth *et al.*, 2004; Davies *et al.*, 2011). The B component forms a heptameric pore in the host cell membrane and stimulates the formation of an endocytic vacuole. The A component is then translocated through the heptameric pore into the cytoplasm, where it catalyses the transfer of ADP-ribose residues from NAD onto G-actin, thereby inhibiting actin polymerization and disrupting the host cytoskeleton.

All of these toxins are chromosomally encoded, with the exception of the *C. perfringens* ι-toxin structural genes, which are located on large plasmids (Li *et al.*, 2007). CDT is encoded by an operon that contains two separate genes, *cdtA* and *cdtB* (Fig. 9.3A), which encode the CDTa active component and the CDTb binding component, respectively. CDTa and CDTb have high levels of amino acid sequence identity to the respective Ia and Ib components of *C. perfringens* ι-toxin. These proteins are produced as precursor proteins, each containing signal peptides that are cleaved upon transit through the bacterial cell membrane (Davies *et al.*, 2011).

CDTb is activated by further proteolytic cleavage, it then forms a heptameric pore complex and binds to a receptor on the host cell surface (Davies *et al.*, 2011). Recent studies that involved mutagenesis of a near-haploid human cell line and selection for resistance to exposure to CDT led to the identification of the host cell surface receptor for CDTb, the LSR (lipolysis-stimulated lipoprotein receptor) protein (Papatheodorou *et al.*, 2011). LSR has been proposed to be involved in the uptake of triglyceride-rich lipoproteins, chylomicrons, and low-density lipoproteins (Yen *et al.*, 1999). *C. perfringens* ι-toxin and CST from *C. spiroforme*, but not *C. botulinum* C2 toxin, also bind to this receptor (Papatheodorou *et al.*, 2011, 2012).

CDTa binds to the CDTb complex at the cell surface, which then is internalized by endocytosis. Acidification of the resultant endosome leads to a conformational change in the heptameric complex, the formation of a pore in the endosomal membrane and translocation of the active component, CDTa, into the cytoplasm of the host cell (Davies *et al.*, 2011; Papatheodorou *et al.*, 2011). The translocation of both CDTa, and Ia, is

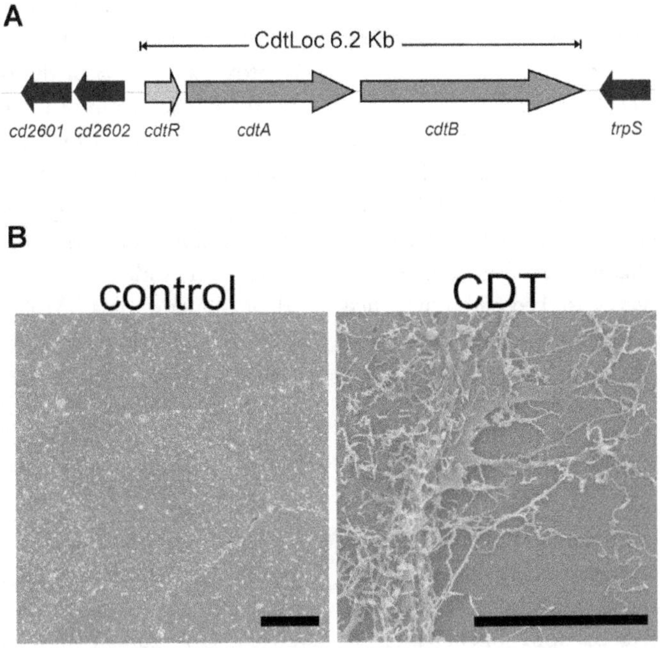

Figure 9.3 (A) Structure of the 6.2kb CdtLoc (binary toxin locus) and flanking regions. Boxes indicate open reading frames with arrows showing the direction of transcription. Toxin genes (*cdtA*, *cdtB*) and the regulatory genes (*cdtR*) are indicated. Genes located outside the CdtLoc are in black (Carter *et al.*, 2007). (B) Scanning electron microscopy of Caco-2 cells treated with CDT (20 ng/ml CDTa and 40 ng/ml CDTb) for 2h. Control is untreated Caco-2 cells. CDT treated cells show strong formation of protrusions (reproduced from Schwan *et al.*, 2009).

dependent upon the heat-shock protein HSP90 and cytophilin (Kaiser *et al.*, 2011).

CDTa comprises two major domains, with the N-terminal domain being responsible for binding to CDTb and the C-terminal domain carrying the ADP-ribosyltransferase active site (Gülke *et al.*, 2001). Based on the similarity of CDTa to the Ia component of ι-toxin (Tsuge *et al.*, 2008) it was predicted that CDTa ADP-ribosylates Arg-177 of G-actin, thereby irreversibly inactivating its polymerization and disrupting the host cell's cytoskeleton (Davies *et al.*, 2011).

The crystal structure of CDTa has been determined, both in its native form and complexed with its substrates NAD and NADPH (Sundriyal *et al.*, 2009). Both the N-terminal and C-terminal domains consist of five α-helices and eight β-sheets. Comparison with the structure of Ia (Tsuge *et al.*, 2008) revealed that these proteins adopt very similar folds, although some differences were observed when the residues involved in the active site were compared (Sundriyal *et al.*, 2009). For example, although Glu-378 and Glu-380 of Ia appear to interact with the substrate, the equivalent residues of CDTA, Glu-385 and Glu-387 do not interact with either NAD or NADPH

Regulation of CDT production

In strains that produce CDT the *cdtA* and *cdtB* toxin structural genes are located in a 6.2 kb region known as the CdtLoc (Carter *et al.*, 2007) (Fig. 9.3A). *C. difficile* strains that do not produce CDT either have a locus that has non-functional *cdtAB* pseudogenes or have a conserved 68-bp sequence in place of the CdtLoc. In a recent survey of 80 *C. difficile* isolates obtained from faecal samples, 50 isolates carried the complete CdtLoc, 88% of which produced CDT in broth culture. The remaining 30 isolates did not produce CDT; 25 of these strains had a CdtLoc carrying *cdt* pseudogenes, the remaining five isolates did not have the CdtLoc (Carman *et al.*, 2011). This study also showed that using an enzyme immunoassay CdtB could be detected in faeces from human patients.

CDT production is not regulated by the genes that regulate the production of toxin A and toxin B, instead it is controlled at the level of transcription by the product of the third gene located in the CdtLoc, *cdtR* (Carter *et al.*, 2007). The CdtR protein is a member of the LytTR family of response regulators and is related to AgrA from *Staphylococcus aureus* and VirR from *C. perfringens*, both of which regulate toxin production. It contains an N-terminal response regulator receiver domain and a C-terminal DNA binding domain. LytTR-like response regulators usually are located in an operon with a gene that encodes a cognate sensor histidine kinase and are often integral components of a quorum sensing system that enables the host bacterium to respond to changes in population density. However, the CdtLoc does not contain a gene that encodes a sensor kinase; the *cdtR* gene is an orphan response regulator gene. Therefore the mechanism by which CdtR is activated is unknown, as is the growth phase or environmental signal(s) that through CdtR leads to the activation of CDT production (Carter *et al.*, 2007).

Our initial analysis of *C. difficile* genomic microarrays performed by other workers (Stabler *et al.*, 2006) revealed that approximately 96% of *cdtR*-positive isolates also carried the *cdtAB* genes or pseudogenes, providing evidence that *cdtR* and the *cdtAB* genes were widely disseminated and genetically linked (Carter *et al.*, 2007). In addition, approximately 82% of the *cdtR*-negative isolates were clustered within the toxin A^- toxin B^+ clade of *C. difficile* isolates, for reasons that are not clear. Subsequent sequence analysis of the regulatory genes present in 57 *C. difficile* isolates showed that these strains all had the *cdtR* gene, except the toxin A^- toxin B^+ variants (Bouvet and Popoff, 2008). Their sequence data revealed that the *cdtR* genes were very closely related; they detected only five separate clades, with 3–22 nucleotide sequence differences. Comparison with the sequence data obtained from the corresponding *tcdC* and *tcdR* genes indicated that the *cdtR* alleles had evolved independently.

Other workers analysed the CdtLoc from 10 isolates, nine from humans and one from a pig, all from Ontario, Canada (Metcalf and Weese, 2011). One human ribotype 078 toxinotype V isolate had a mutated *cdtR* gene with a stop codon located after codon 107 of the normal 248 codon gene. Therefore, the resultant truncated CdtR protein would not contain the normal C-terminal LytTR DNA binding domain. It was reported that qRT-PCR analysis showed that the *cdtA* gene was still transcribed in this strain (Metcalf and Weese, 2011), but the actual quantitative data were not presented and no Western blots or functional assays for ADP-ribosyltransferase activity were carried out. It was suggested that either the truncated CdtR protein was functional, which seems highly unlikely, or that *cdtA* expression was controlled by another regulator in this strain. Sequence analysis of the *cdtR* genes from 86 *C. difficile* isolates, including 57 toxinotype V strains, 51 of which were ribotype 078, revealed that the same *cdtR* truncation allele was present in 44 of these toxinotype V isolates. It was not found in any of the other isolates that were sequenced in this study (Metcalf and Weese, 2011), but previously was detected by other workers (Bouvet and Popoff, 2008). In the latter study ADP-ribosyltransferase activity was detected in strains carrying this allele, in agreement with the suggestion that in this subset of strains CDT production may be regulated by a different signal transduction system.

The role of CDT in virulence

Although many clinical isolates of *C. difficile* do not produce CDT (Geric *et al.*, 2003, 2004), implying that this toxin is not essential for disease, it has been suggested that CDT may increase the severity of *C. difficile* infections (Geric *et al.*, 2003; Barth *et al.*, 2004; Goncalves *et al.*, 2004; McDonald *et al.*, 2005; Carman *et al.*, 2011). There is considerable variation in the incidence of CDT production in *C. difficile* isolates. Analysis of 158 isolates from a single hospital in the USA revealed that only nine isolates (5.8%), from different toxinotypes, produced CDT (Geric *et al.*, 2004). Other studies that examined strains from several hospitals, obtained CDT positive incidence rates of 13.2% of 51 Italian isolates (Spigaglia and Mastrantonio, 2002), 6% of 369 isolates from France (Goncalves *et al.*, 2004) and 5.5% of 2246 strains from the United Kingdom (Stubbs *et al.*, 2000). By contrast, epidemic toxinotype III/ribotype 027/NAP1/BI isolates of *C. difficile* all appear to produce CDT (Martin *et al.*, 2008; Tenover *et al.*, 2011). It remains to be determined whether CDT makes a contribution to the virulence of these epidemic isolates.

Unexpectedly, it appears that CDT may contribute to the adherence of *C. difficile* to its target epithelial cells (Schwan *et al.*, 2009). Schwan and co-workers used time-lapse and immunofluorescence microscopy to show that treatment of human colon carcinoma (Caco-2) epithelial cells with low concentrations of purified CDTa and CDTb led to restructuring of the actin cytoskeleton, the redistribution of α-tubulin molecules and the formation of dynamic microtubule-based protrusions at the cell surface (Fig. 9.3B). Subsequent experiments showed that the formation of these protrusions was dependent on the presence of cholesterol- and sphingolipid-rich lipid microdomains in the cell membrane (Schwan *et al.*, 2011). In separate experiments it was shown that formation of these microtubular protrusions led to a 5-fold increase in the adherence of *C. difficile* cells to the epithelial cell surface, presumably by the formation of a dense lattice of fibres that trap the bacterial cells at the host cell surface (Schwan *et al.*, 2009). Similar results were obtained with other members of the clostridial binary toxin family. Although the significance of this phenomenon for disease pathogenesis remains to be determined, it is reasonable to postulate that it may increase the adhesion of *C. difficile* cells to the colonic epithelium and lead to increased colonization, providing a plausible mechanism by which CDT may contribute to the severity of disease.

Animal models for studying virulence in *C. difficile*

For many years, the most widely used animal model for studying *C. difficile* colonization and disease was the hamster infection model. However, there are a number of significant limitations associated with the use of this model. Perhaps of foremost importance is the fact that hamsters are exquisitely sensitive to both *C. difficile* toxins A and B (Lyerly *et al.*, 1985). Following infection with *C. difficile*, hamsters invariably succumb to a fulminant form of disease which results in the rapid death of infected animals. By contrast, severe cases of CDI resulting in death in humans are considered rare. In addition, the site of infection in hamsters is the caecum and ascending colon whilst in humans CDI almost exclusively affects the descending colon (Songer and Anderson, 2006). For these reasons, the hamster disease model may not accurately reflect human disease and the results obtained from experiments performed using this model must be extrapolated to human disease with caution.

A mouse model of *C. difficile* infection has recently been developed (Chen *et al.*, 2008). Mice are less sensitive to toxins A and B than hamsters and display the full spectrum of CDI, ranging from mild through to fulminant and life threatening disease, which more accurately reflects clinical human disease. Furthermore, unlike hamsters, mice are susceptible to recurrent infection (Sun *et al.*, 2011; Deakin *et al.*, 2012) which is an important aspect of CDI in humans. The mouse model of infection may therefore more closely reflect human disease and many research groups are now using this model in preference to the hamster disease model to study CDI. However, the earlier observation that purified toxin B elicits severe enterotoxic effects only in a humanized mouse model (Savidge *et al.*, 2003) reinforces the case that important differences exist between humans and all of the available animal models.

Of note, a piglet model of *C. difficile* infection has also been developed (Steele *et al.*, 2010). Although this model has been used in a limited capacity to date, it appears to closely resemble human disease. In a similar way to the mouse infection model, piglets succumb to the full spectrum of clinical disease that is seen in humans (Steele *et al.*, 2010). The piglet model has also enabled studies on systemic toxin effects to be performed (Steele *et al.*, 2012). Importantly, *C. difficile* often infects piglets in the natural environment and is emerging as a significant problem in piggeries, causing disease symptoms that are very similar to those seen in humans (Songer and Anderson, 2006). For this reason, it may be that the piglet model represents a more accurate animal model for the study of human CDI and will be valuable in future studies on *C. difficile* disease pathogenesis.

Conclusions and future perspectives

Genetic and functional studies on the *C. difficile* toxins, coupled with the development of numerous animal disease models, have begun to unravel the role that each of the toxins play in CDI. Understanding the functional role of the *C. difficile* toxins is critical for strategically designing vaccines or other therapeutic treatments that can effectively be used to combat these often devastating infections. The recent finding that toxin B is the major virulence factor suggests that targeting this toxin would be a more attractive option for developing treatments for CDI, particularly for infections involving toxin A-negative, toxin B-positive variant strains, which are becoming increasingly common. Undoubtedly, further experiments are required to accurately determine the relative roles of each of the toxins in CDI, especially in clinically relevant isolates and toxin variant strains. In this context, the availability of numerous animal infection models of CDI will be invaluable in understanding disease pathogenesis and for designing specific strategies to combat infections caused by *C. difficile*.

References

Alfa, M.J., Kabani, A., Lyerly, D., Moncrief, S., Neville, L.M., Al-Barrak, A., Harding, G.K., Dyck, B., Olekson, K., and Embil, J.M. (2000). Characterization of a toxin A-negative, toxin B-positive strain of *Clostridium difficile* responsible for a nosocomial outbreak of *Clostridium difficile*-associated diarrhea. J. Clin. Microbiol. *38*, 2706–2714.

Amimoto, K., Noro, T., Oishi, E., and Shimizu, M. (2007). A novel toxin homologous to large clostridial cytotoxins found in culture supernatant of *Clostridium perfringens* type C. Microbiology *153*, 1198–1206.

Bacon, A.E. 3rd, and Fekety, R. (1994). Immunoglobulin G directed against toxins A and B of *Clostridium difficile* in the general population and patients with antibiotic-associated diarrhea. Diagn. Microbiol. Infect. Dis. *18*, 205–209.

Bakker, D., Corver, J., Harmanus, C., Goorhuis, A., Keessen, E.C., Fawley, W.N., Wilcox, M.H., and Kuijper, E.J. (2010). Relatedness of human and animal *Clostridium difficile* PCR ribotype 078 isolates determined on the basis of multilocus variable-number tandem-repeat analysis and tetracycline resistance. J. Clin. Microbiol. *48*, 3744–3749.

Barth, H., Aktories, K., Popoff, M.R., and Stiles, B.G. (2004). Binary bacterial toxins: biochemistry, biology, and applications of common *Clostridium* and *Bacillus* proteins. Microbiol. Mol. Biol. Rev. *68*, 373–402.

Borriello, S.P. (1998). Pathogenesis of *Clostridium difficile* infection. J. Antimicrob. Chemother. *41*, 13–19.

Bouvet, P.J., and Popoff, M.R. (2008). Genetic relatedness of *Clostridium difficile* isolates from various origins determined by triple-locus sequence analysis based on toxin regulatory genes *tcdC*, *tcdR*, and *cdtR*. J. Clin. Microbiol. *46*, 3703–3713.

Braun, V., Hundsberger, T., Leukel, P., Sauerborn, M., and Eichel-Streiber, C.v. (1996). Definition of the single integration site of the pathogenicity locus in *Clostridium difficile*. Gene *181*, 29–38.

Busch, C., Hofmann, F., Gerhard, R., and Aktories, K. (2000). Involvement of a conserved tryptophan residue in the UDP-glucose binding of large clostridial cytotoxin glycosyltransferases. J. Biol. Chem. *275*, 13228–13234.

Carman, R.J., Stevens, A.L., Lyerly, M.W., Hiltonsmith, M.F., Stiles, B.G., and Wilkins, T.D. (2011). *Clostridium difficile* binary toxin (CDT) and diarrhea. Anaerobe *17*, 161–165.

Carter, G.P., Lyras, D., Allen, D.L., Mackin, K.E., Howarth, P.M., O'Connor, J.R., and Rood, J.I. (2007). Binary toxin production in *Clostridium difficile* is regulated by CdtR, a LytTR family response regulator. J. Bacteriol. *189*, 7290–7301.

Carter, G.P., Rood, J.I., and Lyras, D. (2010). The role of toxin A and toxin B in *Clostridium difficile*-associated disease: Past and present perspectives. Gut Microbes *1*, 58–64.

Carter, G.P., Awad, M.M., Hao, Y., Thelen, T., Bergin, I.L., Howarth, P.M., Seemann, T., Rood, J.I., Aronoff, D.M., and Lyras, D. (2011a). TcsL is an essential virulence factor in *Clostridium sordellii* ATCC 9714. Infect. Immun. *79*, 1025–1032.

Carter, G.P., Awad, M.M., Kelly, M.L., Rood, J.I., and Lyras, D. (2011b). TcdB or not TcdB: a tale of two *Clostridium difficile* toxins. Future Microbiol. *6*, 121–123.

Carter, G.P., Douce, G.R., Govind, R., Howarth, P.M., Mackin, K.E., Spencer, J., Buckley, A.M., Antunes, A., Kotsanas, D., Jenkin, G.A., *et al.* (2011c). The anti-sigma factor TcdC modulates hypervirulence in an epidemic BI/NAP1/027 clinical isolate of *Clostridium difficile*. PLoS Pathog. *7*, e1002317.

Carter, G.P., Rood, J.I., and Lyras, D. (2012). The role of toxin A and toxin B in the virulence of *Clostridium difficile*. Trends Microbiol. *20*, 21–29.

Cartman, S.T., Kelly, M.L., Heeg, D., Heap, J.T., and Minton, N.P. (2012). Precise manipulation of the *Clostridium difficile* chromosome reveals a lack of association between the *tcdC* genotype and toxin production. Appl. Environ. Microbiol. *78*, 4683–4690.

Chen, X., Katchar, K., Goldsmith, J.D., Nanthakumar, N., Cheknis, A., Gerding, D.N., and Kelly, C.P. (2008). A mouse model of *Clostridium difficile*-associated disease. Gastroenterology *135*, 1984–1992.

Cohen, M.L. (2000). Changing patterns of infectious disease. Nature *406*, 762–767.

Corthier, G., Muller, M.C., Wilkins, T.D., Lyerly, D., and L'Haridon, R. (1991). Protection against experimental pseudomembranous colitis in gnotobiotic mice by use of monoclonal antibodies against *Clostridium difficile* toxin A. Infect. Immun. *59*, 1192–1195.

Dang, L.H., Bettegowda, C., Huso, D.L., Kinzler, K.W., and Vogelstein, B. (2001). Combination bacteriolytic therapy for the treatment of experimental tumors. Proc. Natl. Acad. Sci. U.S.A. *98*, 15155–15160.

Davies, A.H., Roberts, A.K., Shone, C.C., and Acharya, K.R. (2011). Super toxins from a super bug: structure and function of *Clostridium difficile* toxins. Biochem. J. *436*, 517–526.

Dawson, L.F., Valiente, E., Donahue, E.H., Birchenough, G., and Wren, B.W. (2011). Hypervirulent *Clostridium difficile* PCR-ribotypes exhibit resistance to widely used disinfectants. PLoS One *6*, e25754.

Deakin, L.J., Clare, S., Fagan, R.P., Dawson, L.F., Pickard, D.J., West, M.R., Wren, B.W., Fairweather, N.F., Dougan, G., and Lawley, T.D. (2012). The *Clostridium difficile spo0A* Gene Is a Persistence and Transmission Factor. Infect. Immun. *80*, 2704–2711.

Desvaux, M., and Hebraud, M. (2006). The protein secretion systems in *Listeria*: inside out bacterial virulence. FEMS Microbiol. Rev. *30*, 774–805.

Desvaux, M., Khan, A., Scott-Tucker, A., Chaudhuri, R.R., Pallen, M.J., and Henderson, I.R. (2005). Genomic analysis of the protein secretion systems in *Clostridium acetobutylicum* ATCC 824. Biochim. Biophys. Acta *1745*, 223–253.

Desvaux, M., Hebraud, M., Talon, R., and Henderson, I.R. (2009). Secretion and subcellular localizations of bacterial proteins: a semantic awareness issue. Trends Microbiol. *17*, 139–145.

Dingle, K.E., Griffiths, D., Didelot, X., Evans, J., Vaughan, A., Kachrimanidou, M., Stoesser, N., Jolley, K.A., Golubchik, T., Harding, R.M., *et al.* (2011). Clinical *Clostridium difficile*: clonality and pathogenicity locus diversity. PLoS One *6*, e19993.

Dobson, G., Hickey, C., and Trinder, J. (2003). *Clostridium difficilE. coli*tis causing toxic megacolon, severe sepsis and multiple organ dysfunction syndrome. Intensive Care Med. *29*, 1030.

Drudy, D., Fanning, S., and Kyne, L. (2007). Toxin A-negative, toxin B-positive *Clostridium difficile*. Int. J. Infect. Dis. *11*, 5–10.

Dupuy, B., and Sonenshein, A. (1998). Regulated transcription of *Clostridium difficile* toxin genes. Mol. Microbiol. *27*, 107–120.

Dupuy, B., Raffestin, S., Matamouros, S., Mani, N., Popoff, M.R., and Sonenshein, A.L. (2006). Regulation of toxin and bacteriocin gene expression in *Clostridium* by interchangeable RNA polymerase sigma factors. Mol. Microbiol. *60*, 1044–1057.

Elliott, B., Squire, M.M., Thean, S., Chang, B.J., Brazier, J.S., Rupnik, M., and Riley, T. (2011). New types of toxin A-negative, toxin B-positive strains among clinical isolates of *Clostridium difficile* in Australia. J Med. Microbiol. *60*, 1108–1111.

Genisyuerek, S., Papatheodorou, P., Guttenberg, G., Schubert, R., Benz, R., and Aktories, K. (2011). Structural determinants for membrane insertion, pore formation and translocation of *Clostridium difficile* toxin B. Mol. Microbiol. *79*, 1643–1654.

Gerding, D.N., Muto, C.A., and Owens, Jr, R.C.. (2008). Measures to control and prevent *Clostridium difficile* infection. Clin. Infect. Dis. *46 Suppl 1*, S43–49.

Geric, B., Johnson, S., Gerding, D.N., Grabnar, M., and Rupnik, M. (2003). Frequency of binary toxin genes among *Clostridium difficile* strains that do not produce large clostridial toxins. J. Clin. Microbiol. *41*, 5227–5232.

Geric, B., Rupnik, M., Gerding, D.N., Grabnar, M., and Johnson, S. (2004). Distribution of *Clostridium difficile* variant toxinotypes and strains with binary toxin genes among clinical isolates in an American hospital. J Med. Microbiol. *53*, 887–894.

Geric, B., Carman, R.J., Rupnik, M., Genheimer, C.W., Sambol, S.P., Lyerly, D.M., Gerding, D.N., and Johnson, S. (2006). Binary toxin-producing, large clostridial toxin-negative *Clostridium difficile* strains are enterotoxic but do not cause disease in hamsters. J. Infect. Dis. *193*, 1143–1150.

Goldenberg, S.D., and French, G.L. (2011). Lack of association of *tcdC* type and binary toxin status with disease severity and outcome in toxigenic *Clostridium difficile*. J. Infect. *62*, 355–362.

Goncalves, C., Decre, D., Barbut, F., Burghoffer, B., and Petit, J.C. (2004). Prevalence and characterization of a binary toxin (actin-specific ADP-ribosyltransferase) from *Clostridium difficile*. J. Clin. Microbiol. *42*, 1933–1939.

Goorhuis, A., Debast, S.B., van Leengoed, L.A., Harmanus, C., Notermans, D.W., Bergwerff, A.A., and Kuijper, E.J. (2008). *Clostridium difficile* PCR ribotype 078: an emerging strain in humans and in pigs? J. Clin. Microbiol. *46*, 1157; author reply 1158.

Goorhuis, A., Legaria, M.C., van den Berg, R.J., Harmanus, C., Klaassen, C.H., Brazier, J.S., Lumelsky, G., and Kuijper, E.J. (2009). Application of multiple-locus variable-number tandem-repeat analysis to determine clonal spread of toxin A-negative *Clostridium difficile* in a general hospital in Buenos Aires, Argentina. Clin. Microbiol. Infect. *15*, 1080–1086.

Govind, R., and Dupuy, B. (2012). Secretion of *Clostridium difficile* toxins A and B requires the holin-like protein TcdE. PLoS Pathog. *8*, e1002727.

Grundling, A., Blasi, U., and Young, R. (2000). Genetic and biochemical analysis of dimer and oligomer interactions of the lambda S holin. J. Bacteriol. *182*, 6082–6090.

Gülke, I., Pfeifer, G., Liese, J., Fritz, M., Hofmann, F., Aktories, K., and Barth, H. (2001). Characterization of the enzymatic component of the ADP-ribosyltransferase toxin CDTa from *Clostridium difficile*. Infect. Immun. *69*, 6004–6011.

Guttenberg, G., Papatheodorou, P., Genisyuerek, S., Lu, W., Jank, T., Einsle, O., and Aktories, K. (2011). Inositol hexakisphosphate-dependent processing of *Clostridium sordellii* lethal toxin and *Clostridium novyi* alpha-toxin. J. Biol. Chem. *286*, 14779–14786.

Guttenberg, G., Hornei, S., Jank, T., Schwan, C., Lu, W., Einsle, O., Papatheodorou, P., and Aktories, K. (2012). Molecular characteristics of *Clostridium perfringens* TpeL toxin and consequences of mono-O-GlcNAcylation of Ras in living cells. J. Biol. Chem. *287*, 24929–24940.

Hamm, E.E., Voth, D.E., and Ballard, J.D. (2006). Identification of *Clostridium difficile* toxin B cardiotoxicity using a zebrafish embryo model of intoxication. Proc. Natl. Acad. Sci. U.S.A. *103*, 14176–14181.

Hammond, G.A., and Johnson, J.L. (1995). The toxigenic element of *Clostridium difficile* strain VPI 10463. Microb. Pathog. *19*, 203–213.

Hariri, S., Gouin, P., Tuech, J.J., Veber, B., and Dureuil, B. (2011). *Clostridium difficile* infection causing multiple organ failure and small-bowel enteritis. Clin. Res. Hepatol. Gastroenterol. *35*, 142–144.

Himeno, T., Imanaka, T., and Aiba, S. (1986). Nucleotide sequence of the penicillinase repressor gene *penI* of *Bacillus licheniformis* and regulation of *penP* and *penI* by the repressor. J. Bacteriol. *168*, 1128–1132.

Hundsberger, T., Braun, V., Weidmann, M., Leukel, P., Sauerborn, M., and Eichel-Streiber, C.v. (1997). Transcription analysis of the genes *tcdA-E* of the pathogenicity locus of *Clostridium difficile*. Eur. J. Biochem. *244*, 735–742.

Janezic, S., Ocepek, M., Zidaric, V., and Rupnik, M. (2012). *Clostridium difficile* genotypes other than ribotype 078 that are prevalent among human, animal and environmental isolates. BMC Microbiol. *12*, 48.

Jank, T., and Aktories, K. (2008). Structure and mode of action of clostridial glucosylating toxins: the ABCD model. Trends Microbiol. *16*, 222–229.

Just, I., and Gerhard, R. (2004). Large clostridial cytotoxins. Rev. Physiol. Biochem. Pharmacol. *152*, 23–47.

Just, I., Selzer, J., Wilm, M., von Eichel-Streiber, C., Mann, M., and Aktories, K. (1995). Glucosylation of Rho proteins by *Clostridium difficile* toxin B. Nature *375*, 500–503.

Kaiser, E., Kroll, C., Ernst, K., Schwan, C., Popoff, M., Fischer, G., Buchner, J., Aktories, K., and Barth, H. (2011). Membrane translocation of binary actin-ADP-ribosylating toxins from *Clostridium difficile* and *Clostridium perfringens* is facilitated by cyclophilin A and Hsp90. Infect. Immun. *79*, 3913–3921.

Ketley, J.M., Mitchell, T.J., Candy, D.C., Burdon, D.W., and Stephen, J. (1987). The effects of *Clostridium difficile* crude toxins and toxin A on ileal and colonic loops in immune and non-immune rabbits. J Med. Microbiol. *24*, 41–52.

Kim, P.H., Iaconis, J.P., and Rolfe, R.D. (1987). Immunization of adult hamsters against *Clostridium difficile*-associated ileocecitis and transfer of protection to infant hamsters. Infect. Immun. *55*, 2984–2992.

Kuehn, B.M. (2011). Scientists seek strategies to prevent *Clostridium difficile* infections. JAMA *306*, 1849–1850.

Kuehne, S.A., Cartman, S.T., Heap, J.T., Kelly, M.L., Cockayne, A., and Minton, N.P. (2010). The role of toxin A and toxin B in *Clostridium difficile* infection. Nature *467*, 711–713.

Kuijper, E.J., de Weerdt, J., Kato, H., Kato, N., van Dam, A.P., van der Vorm, E.R., Weel, J., van Rheenen, C., and Dankert, J. (2001). Nosocomial outbreak of *Clostridium difficile*-associated diarrhoea due to a clindamycin-resistant enterotoxin A-negative strain. Eur. J. Clin. Microbiol. Infect. Dis. *20*, 528–534.

Kyne, L., Warny, M., Qamar, A., and Kelly, C.P. (2000). Asymptomatic carriage of *Clostridium difficile* and serum levels of IgG antibody against toxin A. N. Engl. J. Med. *342*, 390–397.

Lanis, J.M., Barua, S., and Ballard, J.D. (2010). Variations in TcdB activity and the hypervirulence of emerging strains of *Clostridium difficile*. PLoS Pathog. *6*.

Lawley, T.D., Clare, S., Walker, A.W., Goulding, D., Stabler, R.A., Croucher, N., Mastroeni, P., Scott, P., Raisen, C., Mottram, L., *et al.* (2009). Antibiotic treatment of *Clostridium difficile* carrier mice triggers a supershedder state, spore-mediated transmission, and severe disease in immunocompromised hosts. Infect. Immun. *77*, 3661–3669.

Li, J., Miyamoto, K., and McClane, B.A. (2007). Comparison of virulence plasmids among *Clostridium perfringens* type E isolates. Infect. Immun. *75*, 1811–1819.

Lima, A.A.M., Lyerly, D.M., Wilkins, T.D., Innes, D.J., and Geurrant, R.L. (1988). Effects of *Clostridium difficile* toxins A and B in rabbit small and large intestine *in vivo* and on cultured cells *in vitro*. Infect. Immun. *56*, 582–588.

Lyerly D.M., Saum, K.E., MacDonald D.K., and D., W.T. (1985). Effects of *Clostridium difficile* toxins given intragastrically to animals. Infect. Immun. *47*, 349–352.

Lyras, D., O'Connor, J.R., Howarth, P.K., Sambol, S.P., Carter, G.P., Phumoonna, T., Poon, R., Adams, V., Vedantam, G., Johnson, S., *et al.* (2009). Toxin B is essential for virulence of *Clostridium difficile*. Nature *458*, 1176–1179.

Mani, N., and Dupuy, B. (2001). Regulation of toxin synthesis in *Clostridium difficile* by an alternative RNA polymerase sigma factor. Proc. Nat. Acad. Sci. U.S.A. *98*, 5844–5849.

Mani, N., Lyras, D., Barroso, L., Howarth, P., Wilkins, T., Rood, J.I., Sonenshein, A.L., and Dupuy, B. (2002). Environmental response and autoregulation of *Clostridium difficile* TxeR, a sigma factor for toxin gene expression. J. Bacteriol. *184*, 5971–5978.

Martin, H., Willey, B., Low, D.E., Staempfli, H.R., McGeer, A., Boerlin, P., Mulvey, M., and Weese, J.S. (2008). Characterization of *Clostridium difficile* strains isolated from patients in Ontario, Canada, from 2004 to 2006. J. Clin. Microbiol. *46*, 2999–3004.

Matamouros, S., England, P., and Dupuy, B. (2007). *Clostridium difficile* toxin expression is inhibited by the novel regulator TcdC. Mol. Microbiol. *64*, 1274–1288.

McDonald, L.C., Killgore, G.E., Thompson, A., Owens, R.C., Kazakova, S.V., Sambol, S.P., Johnson, S., and Gerding, D.N. (2005). An epidemic, toxin gene–variant strain of *Clostridium difficile*. New Engl. J. Med. *353*, 2433–2441.

Merrigan, M., Venugopal, A., Mallozzi, M., Roxas, B., Viswanathan, V.K., Johnson, S., Gerding, D.N., and Vedantam, G. (2010). Human hypervirulent *Clostridium difficile* strains exhibit increased sporulation as well as robust toxin production. J. Bacteriol. *192*, 4904–4911.

Metcalf, D.S., and Weese, J.S. (2011). Binary toxin locus analysis in *Clostridium difficile*. J Med. Microbiol. *60*, 1137–1145.

Mitchell, T.J., Ketley, J.M., Haslam, S.C., Stephen, J., Burdon, D.W., Candy, D.C., and Daniel, R. (1986). Effect of toxin A and B of *Clostridium difficile* on rabbit ileum and colon. Gut *27*, 78–85.

Moncrief, J., Barroso, L., and Wilkins, T. (1997). Positive regulation of *Clostridium difficile* toxins. Infect. Immun. *65*, 1105–1108.

O'Brien, J.A., Lahue, B.J., Caro, J.J., and Davidson, D.M. (2007). The emerging infectious challenge of *Clostridium difficile*-associated disease in Massachusetts hospitals: clinical and economic consequences. Infect. Control Hosp. Epidemiol. *28*, 1219–1227.

Olling, A., Seehase, S., Minton, N.P., Tatge, H., Schroter, S., Kohlscheen, S., Pich, A., Just, I., and Gerhard, R. (2012). Release of TcdA and TcdB from *Clostridium difficile* cdi 630 is not affected by functional inactivation of the *tcdE* gene. Microb. Pathog. *52*, 92–100.

Paltansing, S., van den Berg, R.J., Guseinova, R.A., Visser, C.E., van der Vorm, E.R., and Kuijper, E.J. (2007). Characteristics and incidence of *Clostridium difficile*-associated disease in The Netherlands, 2005. Clin. Microbiol. Infect. *13*, 1058–1064.

Papatheodorou, P., Carette, J.E., Bell, G.W., Schwan, C., Guttenberg, G., Brummelkamp, T.R., and Aktories, K. (2011). Lipolysis-stimulated lipoprotein receptor (LSR) is the host receptor for the binary toxin *Clostridium difficile* transferase (CDT). Proc. Natl. Acad. Sci. U.S.A. *108*, 16422–16427.

Papatheodorou, P., Wilczek, C., Nolke, T., Guttenberg, G., Hornuss, D., Schwan, C., and Aktories, K. (2012). Identification of the cellular receptor of *Clostridium spiroforme* toxin. Infect. Immun. *80*, 1418–1423.

Pépin, J., Saheb, N., Coulombe, M.-A., Alary, M.-E., Corriveau, M.-P., Authier, S., Leblanc, M., Rivard, G., Bettez, M., Primeau, V., *et al.* (2005). Emergence of fluoroquinolones as the predominant risk factor for *Clostridium difficile*-associated diarrhea: a cohort study during an epidemic in Quebec. Clin. Infect. Dis. *41*, 1254–1260.

Perelle, S., Gibert, M., Bourlioux, P., Corthier, G., and Popoff, M. (1997). Production of a complete binary toxin (actin-specific ADP-ribosyltransferase) by *Clostridium difficile* CD196. Infect. Immun. *65*, 1402–1407.

Pothoulakis, C. (2000). Effects of *Clostridium difficile* toxins on epithelial cell barrier. Ann. N. Y. Acad. Sci. *915*, 347–356.

Pothoulakis, C., Castagliuolo, I., and LaMont, J.T. (1998). Nerves and intestinal mast cells modulate responses to enterotoxins. News Physiol. Sci. *13*, 58–63.

Raffestin, S., Dupuy, B., Marvaud, J.C., and Popoff, M.R. (2005). BotR/A and TetR are alternative RNA polymerase sigma factors controlling the expression of the neurotoxin and associated protein genes in *Clostridium botulinum* type A and *Clostridium tetani*. Mol. Microbiol. *55*, 235–249.

Redelings, M.D., Sorvillo, F., and Mascola, L. (2007). Increase in *Clostridium difficile*-related mortality rates, United States, 1999–2004. Emerg. Infect. Dis. *13*, 1417–1419.

Rowland, S.J., and Dyke, K.G. (1990). Tn552, a novel transposable element from *Staphylococcus aureus*. Mol. Microbiol. *4*, 961–975.

Rupnik, M. (2007). Is *Clostridium difficile*-associated infection a potentially zoonotic and foodborne disease? Eur. Soc. Clin. Microbiol. Infect. Dis. *13*, 457–459.

Rupnik, M. (2008). Heterogeneity of large clostridial toxins: importance of *Clostridium difficile* toxinotypes. FEMS Microbiol. Rev. *32*, 541–555.

Rupnik, M., Avesani, V., Janc, M., von Eichel-Streiber, C., and Delmee, M. (1998). A novel toxinotyping scheme and correlation of toxinotypes with serogroups of *Clostridium difficile* isolates. J. Clin. Microbiol. *36*, 2240–2247.

Sato, H., Kato, H., Koiwai, K., and Sakai, C. (2004). A nosocomial outbreak of diarrhea caused by toxin A-negative, toxin B-positive *Clostridium difficile* in a cancer center hospital. Kansenshogaku Zasshi *78*, 312–319.

Savidge, T.C., Pan, W.H., Newman, P., O'Brien, M., Anton, P.M., and Pothoulakis, C. (2003). *Clostridium difficile* toxin B is an inflammatory enterotoxin in human intestine. Gastroenterology *125*, 413–420.

Schwan, C., Stecher, B., Tzivelekidis, T., van Ham, M., Rohde, M., Hardt, W.D., Wehland, J., and Aktories, K. (2009). *Clostridium difficile* toxin CDT induces formation of microtubule-based protrusions and increases adherence of bacteria. PLoS Pathog. *5*, e1000626.

Schwan, C., Noelke, T., Kruppke, A.S., Schubert, D.M., Lang, A.E., and Aktories, K. (2011). Cholesterol- and sphingolipid-rich microdomains are essential for microtubule-based membrane protrusions induced by *Clostridium difficile* transferase CDT. J. Biol. Chem. *286*, 29356–29365.

Sirard, S., Valiquette, L., and Fortier, L.C. (2011). Lack of association between clinical outcome of *Clostridium difficile* infections, strain type, and virulence-associated phenotypes. J. Clin. Microbiol. *49*, 4040–4046.

Songer, J.G., and Anderson, M.A. (2006). *Clostridium difficile*: an important pathogen of food animals. Anaerobe *12*, 1–4.

Spigaglia, P., and Mastrantonio, P. (2002). Molecular analysis of the pathogenicity locus and polymorphism in the putative negative regulator of toxin production (TcdC) among *Clostridium difficile* clinical isolates. J. Clin. Microbiol. *40*, 3470–3475.

Stabler, R.A., Gerding, D.N., Songer, J.G., Drudy, D., Brazier, J.S., Trinh, H.T., Witney, A.A., Hinds, J., and Wren, B.W. (2006). Comparative phylogenomics of *Clostridium difficile* reveals clade specificity and microevolution of hypervirulent strains. J. Bacteriol. *188*, 7297–7305.

Stabler, R.A., Dawson, L.F., Phua, L.T., and Wren, B.W. (2008). Comparative analysis of BI/NAP1/027 hypervirulent strains reveals novel toxin B-encoding gene (*tcdB*) sequences. J Med. Microbiol. *57*, 771–775.

Stabler, R.A., Dawson, L.F., Valiente, E., Cairns, M.D., Martin, M.J., Donahue, E.H., Riley, T.V., Songer, J.G., Kuijper, E.J., Dingle, K.E., *et al.* (2012). Macro and micro diversity of *Clostridium difficile* isolates from diverse sources and geographical locations. PLoS One *7*, e31559.

Steele, J., Feng, H., Parry, N., and Tzipori, S. (2010). Piglet models of acute or chronic *Clostridium difficile* illness. J. Infect. Dis. *201*, 428–434.

Steele, J., Chen, K., Sun, X., Zhang, Y., Wang, H., Tzipori, S., and Feng, H. (2012). Systemic dissemination of *Clostridium difficile* toxins A and B is associated with severe, fatal disease in animal models. J. Infect. Dis. *205*, 384–391.

Stubbs, S., Rupnik, M., Gibert, M., Brazier, J., Duerden, B., and Popoff, M. (2000). Production of actin-specific ADP-ribosyltransferase (binary toxin) by strains of *Clostridium difficile*. FEMS Microbiol. Lett. *186*, 307–312.

Sun, X., Wang, H., Zhang, Y., Chen, K., Davis, B., and Feng, H. (2011). Mouse relapse model of *Clostridium difficile* infection. Infect. Immun. *79*, 2856–2864.

Sundriyal, A., Roberts, A.K., Shone, C.C., and Acharya, K.R. (2009). Structural basis for substrate recognition in the enzymatic component of ADP-ribosyltransferase toxin CDTa from *Clostridium difficile*. J. Biol. Chem. *284*, 28713–28719.

Tenover, F.C., Akerlund, T., Gerding, D.N., Goering, R.V., Bostrom, T., Jonsson, A.M., Wong, E., Wortman, A.T., and Persing, D.H. (2011). Comparison of strain typing results for *Clostridium difficile* isolates from North America. J. Clin. Microbiol. *49*, 1831–1837.

Tesch, W., Ryffel, C., Strassle, A., Kayser, F.H., and Berger-Bachi, B. (1990). Evidence of a novel staphylococcal *mec*-encoded element (*mecR*) controlling expression of penicillin-binding protein 2'. Antimicrob. Agents Chemother. *34*, 1703–1706.

Tsuge, H., Nagahama, M., Oda, M., Iwamoto, S., Utsunomiya, H., Marquez, V.E., Katunuma, N., Nishizawa, M., and Sakurai, J. (2008). Structural basis of actin recognition and arginine ADP-ribosylation by *Clostridium perfringens* iota-toxin. Proc. Natl. Acad. Sci. U.S.A. *105*, 7399–7404.

Verdoorn, B.P., Orenstein, R., Rosenblatt, J.E., Sloan, L.M., Schleck, C.D., Harmsen, W.S., Nyre, L.M., and Patel, R. (2010). High prevalence of *tcdC* deletion-carrying *Clostridium difficile* and lack of association with disease severity. Diagn. Microbiol. Infect. Dis. *66*, 24–28.

Vohra, P., and Poxton, I.R. (2011). Comparison of toxin and spore production in clinically relevant strains of *Clostridium difficile*. Microbiology *157*, 1343–1353.

von Eichel-Streiber, C., Laufenberg-Feldmann, R., Sartingen, S., Schulze, J., and Sauerborn, M. (1990). Cloning of *Clostridium difficile* toxin B gene and demonstration of high N-terminal homology between toxin A and B. Med. Microbiol. Immunol. *179*, 271–279.

Voth, D.E., and Ballard, J.D. (2005). *Clostridium difficile* toxins: mechanism of action and role in disease. Clin. Microbiol. Rev. *18*, 247–263.

Voth, D.E., Martinez, O.V., and Ballard, J.D. (2006). Variations in lethal toxin and cholesterol-dependent cytolysin production correspond to differences in cytotoxicity among strains of *Clostridium sordellii*. FEMS Microbiol. Lett. *259*, 295–302.

Warny, M., Pepin, J., Fang, A., Killgore, G., Thompson, A., Brazier, J., Frost, E., and MacDonald, L.C. (2005). Toxin production by an emerging strain of *Clostridium difficile* associated with outbreaks of severe disease in North America and Europe. Lancet *366*, 1079–1084.

Wilcox, M.H., and Planche, T. (2009). *Clostridium difficile* infection. BMJ *338*, b2528.

Yen, F.T., Masson, M., Clossais-Besnard, N., Andre, P., Grosset, J.M., Bougueleret, L., Dumas, J.B., Guerassimenko, O., and Bihain, B.E. (1999). Molecular cloning of a lipolysis-stimulated remnant receptor expressed in the liver. J. Biol. Chem. *274*, 13390–13398.

Index

N

P

S

www.ingramcontent.com/pod-product-compliance
Lightning Source LLC
LaVergne TN
LVHW062123200926
848484LV00007B/1745
* 9 7 8 1 9 0 8 2 3 0 2 8 7 *